CONTROL OF NONLINEAR MECHANICAL SYSTEMS

APPLIED INFORMATION TECHNOLOGY

Series Editor:

M. G. SINGH
UMIST, Manchester, England

Editorial Board:

K. ASTROM
Lund Institute of Technology, Lund, Sweden
S. J. GOLDSACK
Imperial College of Science and Technology, London, England
M. MANSOUR
ETH-Zentrum, Zurich, Switzerland
G. SCHMIDT
Technical University of Munich, Munich, Federal Republic of Germany
S. SETHI
University of Toronto, Toronto, Canada
J. STREETER
GEC Research Laboratories, Great Baddow, England
A. TITLI
LAAS, CNRS, Toulouse, France

CONTROL OF NONLINEAR MECHANICAL SYSTEMS

Janislaw M. Skowronski
University of Southern California
Los Angeles, California

SPRINGER SCIENCE+BUSINESS MEDIA, LLC

Library of Congress Cataloging-in-Publication Data

Skowroński, Janisław M.
    Control of nonlinear mechanical systems / Janisław M. Skowronski.
        p.   cm. -- (Applied information technology)
    Includes bibliographical references and index.
    ISBN 978-1-4613-6656-0   ISBN 978-1-4615-3722-9 (eBook)
    DOI 10.1007/978-1-4615-3722-9
    1. Automatic control. 2. Nonlinear mechanics.   I. Title.
II. Series.
TJ213.S47475   1991
629.8--dc20   90-25787
                                                                    CIP

ISBN 978-1-4613-6656-0

© 1991 Springer Science+Business Media New York
Originally published by Plenum Press in 1991
Softcover reprint of the hardcover 1st edition 1991

PREFACE

A modern mechanical structure must work at high speed and with high
precision in space and time, in cooperation with other machines and systems.
All this requires accurate dynamic modelling, for instance, recognizing
Coriolis and centrifugal forces, strong coupling effects, flexibility of
links, large angles articulation. This leads to a motion equation which
must be highly nonlinear to describe the reality. Moreover, work on the
manufacturing floor requires coordination between machines, between each
machine and a conveyor, and demands robustness of the controllers against
uncertainty in payload, gravity, external perturbations etc. This requires
adaptive controllers and system coordination, and perhaps a self organizing
structure. The machines become complex, strongly nonlinear and strongly
coupled mechanical systems with many degrees of freedom, controlled by
sophisticated mathematical programs. The design of such systems needs basic
research in Control and System Dynamics, as well as in Decision Making
Theory (Dynamic Games), not only in the use of these disciplines, but in
their adjustment to the present demand. This in turn generates the need to
prepare engineering students for the job by the teaching of more sophisticated techniques in Control and Mechanics than those contained in previous
curricula.

On the other hand, all that was mentioned above regarding the design of
machines applies equally well to other presently designed and used mechanical structures or systems. We have the same fundamental problems in active
control of flexible large space structures (LSS) and high rise building
structures, as well as in the flight control of air or spacecraft, including
air traffic control and air combat games.

Working on basic methodology in all these directions makes one realize how valuable the interface between the applications may be to each of them separately. Techniques used to design flexible links in manipulators and LSS are presently developed jointly. It is perhaps still not sufficiently realized that coordination control of robotic manipulators may be obtained by methods used in air combat games, and that such games may be used in robot decision making. The investigation of such an interface may result in the means to overcome many problems in present design practice. The book attempts to give the fundamental background for such investigation.

As mentioned, the dynamical models in all of the above applications, in order to be realistic, must recognize untruncated nonlinearity of the acting forces and be robust against uncertainties hidden in modelling and/or external perturbations. The study of the interface makes it obvious that in order to handle such models, one must seek for methods entirely different from those used in classical Control Theory, which approximates reality with linearized models. Although it might not be immediately visible from behind the Laplace transformation, Control Theory had been born out of Mechanics, particularly Nonlinear Mechanics. The latter has been developing quite rapidly for the last twenty or thirty years, but this was somehow unnoticed by the control theorists. Now, with the applications mentioned, Nonlinear Mechanics may no longer be ignored in Control Dynamics, and the demand for it is growing rapidly. There is no applied text available which would deal with control of fully nonlinear, uncertain mechanical systems. This book has been written to fill the gap.

The thirty years of research work which this author has devoted to the subject gives him the advantage of knowing what is needed, but also the disadvantage of habitually favoring some of the topics. This bias has proved useful, considering the space limitations which must be imposed on any text. I hope, however, that the book is a healthy compromise between the needs and the bias.

The first chapter outlines the models of mechanical systems used, the second and third introduce the reader to the energy relations and the Liapunov design technique applied later. Chapter Four specifies the objectives of control and the types of controllers used in the three basic directions of study: robotics, spacecraft structures and air games. Sufficient conditions, control algorithms and case studies in these three directions are covered by Chapters 5, 6 and 7, in terms of control (collision, avoidance and tracking, respectively), while Chapter 8 deals with the same problems but subject to conflict.

The text has grown up from lecture notes for junior graduate and senior undergraduate courses taught at Mechanical Engineering, University of Southern California, Los Angeles, in Advanced Mechanics, Analytic Methods of Robotics, Control of Robotic Systems, and at University of Queensland, Australia, in Control Theory, Systems Dynamics and Robot Theory. Apart from natural use in such courses, the book may serve as reference for the design of control algorithms for nonlinear systems.

The author is indebted to Professors M.D. Ardema, A. Blaquière, M.J. Corless, H. Flashner, E.A. Galperin, W.J. Grantham, R.S. Guttalu, G. Leitmann, W.E. Schmitendorf, R.J. Stonier and T.L. Vincent for cooperation leading to results included in this book, as well as to some of the mentioned colleagues for comments improving the text. Thanks are also due to my wife Elzbieta Skowronski, and to the graduate students Harvinder Singh and Nigel Greenwood for solving some problems and proof-reading, as well as to Mrs Marie Stonier for careful and patient typing.

LOS ANGELES, JANUARY 1989

J.M. SKOWRONSKI

CONTENTS

Chapter 1
MECHANICAL SYSTEMS

1.1 SIMPLE MECHANICAL SYSTEMS

It seems both convenient and illustrative to introduce some of our later defined notions on simple but typical examples of mechanical systems. Perhaps the simplest and, at the same time, most typical mechanical system is a single link mathematical (idealized) pendulum discussed in the following example.

EXAMPLE 1.1.1. Consider the simple pendulum shown in Fig. 1.1, swinging about the base point 0 in the Cartesian plane Oxy, by the angle $\theta(t)$, for all $t \geq t_0$, where $t_0 \in \mathbb{R}$ is an initial time instant. The plane is a part of the *Cartesian* physical *coordinates space* $Oxyz$ where the position of the point-mass m is specified by the current values of $x(t)$, $y(t)$, $z(t)$ subject to obvious *constraints*: $z = \text{const}$, $x^2 + y^2 = \ell^2$. Under such constraints only one variable can be independent, and thus we say that the system has a single *degree of freedom* (DOF). It is more convenient to choose $\theta(t)$ as the *generalized* (lagrangian) *coordinate* describing such DOF, rather than $x(t)$ or $y(t)$, although the choice of either of these two is obviously possible. We thus define $q(t) \triangleq \theta(t)$, $t \geq t_0$. The point-mass m is considered an *object* in a *point-mass model* of a mechanical system. In our simple case the model consists of the object concerned with the single DOF specified by $q(t)$.

The generalized variable $q(t)$ is free of the Cartesian constraints mentioned but it has its own work limitations in some interval Δ of its

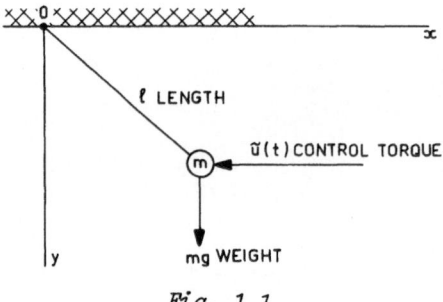

Fig. 1.1

values. For instance, when the pendulum is suspended from a ceiling which is non-penetrable, we must impose $\Delta : -\pi \leq q(t) \leq \pi$.

In a symbolic way, we represent the system as a single mass cube railed to move in one direction only, generally subject to gravity $\tilde{G}$, spring or elastic (in link) forces $\tilde{K}$ and damping force $\tilde{D}$, as well as to an external input-control force (torque) $\tilde{u}$, see Fig. 1.2. This representation is well known as the *schematic diagram* of the system structure. The arrows crossing the symbols of elastic and damping connections indicate that the corresponding forces may be represented by nonlinear functions. The damping symbol is shown as a damper open from above if the damping is positive, and from below it if is negative.

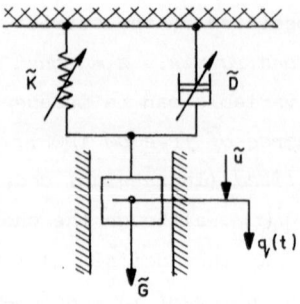

Fig. 1.2

The point mass subject to weight $\tilde{G}$ is restricted to the vertical motion only as shown in Fig. 1.2, modelling the single degree of freedom.

For simplicity of exposition, we ignore the elastic force in the link. The damping force $\widetilde{D}$ is made dependent upon the velocity $\dot{q}(t) \triangleq dq(t)/dt$ and is specified by the function $\widetilde{d}|\dot{q}|\dot{q}$, $\widetilde{d} > 0$ to be positive damping. Since we ignored elastic forces, the potential force acting upon the mass m reduces to gravity, specified by its component $mg \sin \theta$, where g is the earth acceleration. Then the Lagrange equation of motion gives

$$m\ell^2\ddot{q} + \widetilde{d}|\dot{q}|\dot{q} + mg\ell \sin q = \widetilde{u} . \tag{1.1.1}$$

With m, ℓ constant and measured, it is convenient to rewrite (1.1.1) in terms of the forces per coefficients of inertia which we call *characteristics* of the forces involved. We obtain

$$\ddot{q} + D(\dot{q}) + \Pi(q) = u \tag{1.1.2}$$

where $D(\dot{q}) \triangleq \widetilde{d}|\dot{q}|\dot{q} / m\ell^2$, $\Pi(q) \triangleq (g/\ell) \sin q$ and $u \triangleq \widetilde{u}/m\ell^2$. Introducing the force characteristics frees the acceleration term in (1.1.1) from the inertia coefficient. For multidimensional systems, such a procedure is connected with decoupling the equations inertially (dividing by the matrix of inertia), which then makes it possible to apply the results of control theory, usually formalized in terms of the normal form of differential equations, see later examples.

The potential energy of the pendulum is then expressed by

$$V(q) = V^0 + \int_{q(t)} \Pi(q)\,dq \tag{1.1.3}$$

with the initial storage of energy $V^0 = V(q^0)$, where $q^0 = q(t_0)$, $t_0 = 0$ being the initial instant of time. The equilibria of the pendulum occur at rest positions $\dot{q} = 0$ coinciding with the extrema of the function $V(\cdot)$, i.e. $\sin q = 0$ or $q^e = n\pi$, $n = 0, \pm 1, \pm 2, \ldots$. As is well known from elementary mechanics, the minima correspond to the Dirichlet stable equilibria occurring at the downward positions of the pendulum, after each full rotation by 2π . The maxima correspond to Dirichlet unstable equilibria occurring at the upwards positions obtained on every half-turn by π from the preceding stable equilibrium. The maximal values of $V(\cdot)$ at these positions form the energy thresholds which have to be passed before another stable equilibrium is attained, i.e. before the pendulum realizes a rotation.

Obviously the gravity characteristic $\Pi(q) = (g/\ell) \sin q$ is a highly nonlinear function. It can, however, be expanded as a Taylor or power series

$$\Pi(q) = (g/\ell) \left[q - \frac{q^3}{3!} + \frac{q^5}{5!} - \cdots \right] .$$
(1.1.4)

Since q is bounded for physical reasons, the equilibria which are zeros of (1.1.4) become zeros of some polynomial with a number of terms related to the number of rotations performed, see Fig. 1.3.

As there are many cases in engineering design where truncation of the series is necessary, if only for computational reasons, it is of interest to see when and to which extent such an operation is physically justified.

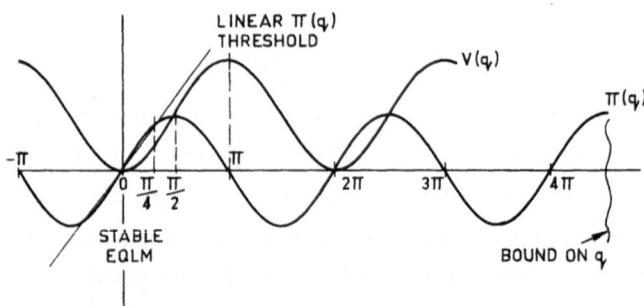

Fig. 1.3

For $q \in [-(\pi/4),(\pi/4)]$, the linear approximation $\Pi(q) = (g/\ell)q$ may be satisfactory, see Fig. 1.3. However, for larger swing angles, we need to include the nonlinear terms in (1.1.4) (the more of them the larger the swing angle). For the pendulum turning upwards, but not falling down again: $q = \pm\pi$, we need at least

$$\Pi(q) = (g/\ell)(q - \frac{1}{6} q^3) ,$$
(1.1.5)

for the pendulum falling down again: $q = \pm 2\pi$, we need

$$\Pi(q) = (g/\ell)(q - \frac{1}{6} q^3 + \frac{1}{120} q^5) , \quad \cdots \text{ etc.}$$

Consequently the equation of motion (1.1.2) becomes nonlinear and must be treated as such, if we do not want our model to disagree principally with the physical reality of the rotating pendulum. The nonlinear terms of the force characteristics cannot be truncated. This also means that we must recognize the existence of several stable equilibria separated by thresholds. With positive damping, these equilibria will attract motion trajectories in the *phase-space* (state space) $0q\dot{q}$ from specific regions of attraction, see Fig. 1.4, and will in fact be in competition as to their attracting role. Each attracting equilibrium (attractor) will have its own (winning) region of attraction. Given the same controller, depending upon where the trajectory starts (initial conditions) it will land in a corresponding attractor. By truncating the nonlinearities, we ignore all the equilibria except the single basic equilibrium at q = 0 and we may be led into a false sense of security, assuming that trajectories from everywhere will land in that equilibrium. Such a conclusion may be true only in some neighbourhood of the basic equilibrium (below the thresholds) but a slight change of initial conditions beyond this neighbourhood, i.e. beyond a threshold, may produce unstable trajectories tending somewhere else than intended. Then we may need a power expensive controller to rectify the situation. Moreover, the further from the threshold we are, the more costly such a controller becomes, possibly beyond its saturation value.

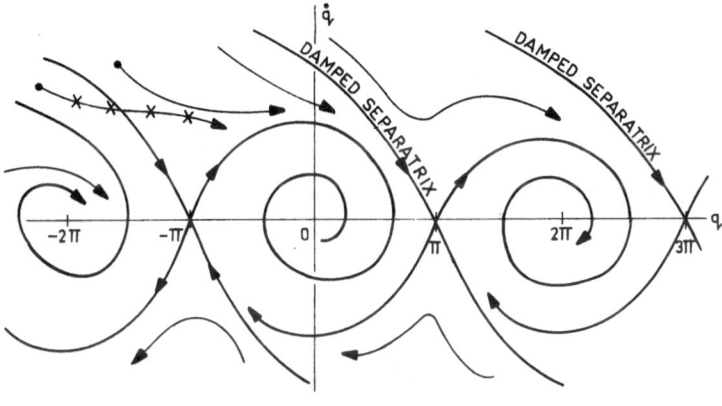

Fig. 1.4

Let us have a closer look at the trajectories. In our case it will be possible to do it directly, as the equation (1.1.2) with u ≡ 0 is integrable in closed form. We choose the state variables x_1, x_2 by substituting $x_1 \triangleq q$, $x_2 \triangleq \dot{q}$ and rewrite (1.1.2) with $d = \tilde{d}/m\ell^2$ as

$$\left. \begin{aligned} \dot{x}_1 &= x_2 \\ \dot{x}_2 &= -(g/\ell) \sin x_1 - d|x_2|x_2 + u \ , \end{aligned} \right\} \tag{1.1.6}$$

or in terms of the directional field in the phase-plane (state-plane) $0x_1x_2$:

$$\frac{dx_2}{dx_1} = \frac{-(g/\ell) \sin x_1 - d|x_2|x_2 + u}{x_2} \ . \tag{1.1.7}$$

For the moment let us free the system from control, i.e. assume $u(t)$ to be a given function of time, in particular $u(t) \equiv 0$. Since $x_2 = |x_2| \, \text{sign}\, x_2$, we can rewrite (1.1.7) as

$$\frac{dx_2^2}{dx_1} \pm 2dx_2^2 = -2(g/\ell) \sin x_1 \tag{1.1.8}$$

where the plus sign is used for $x_2 > 0$, and the minus whenever $x_2 < 0$. Except for these changes in sign, (1.1.8) is a linear equation in x_2^2 with x_1 as an independent variable. Hence the solutions of (1.1.8) are elementary:

$$x_2^2 = Ce^{\pm 2dx_1} + \frac{2g \cos x_1}{\ell(1 + 4d^2)} \pm \frac{4gd \sin x_1}{\ell(1 + 4d^2)} \tag{1.1.9}$$

where C is a constant of integration and where the signs $\pm$ are interpreted as in (1.1.8).

The above first integral of (1.1.8) is a curve $x_2(x_1)$ in the phase-plane $0x_1x_2$ of the pendulum, already called the *trajectory*. A suitable choice of the constant C makes the pieces of trajectories (1.1.9) due to the $\pm$ sign fit together at the points of intersection with the x_1-axis. The trajectories are shown in Fig. 1.4. The unstable equilibria correspond to *saddle points* at odd multiples of π, the stable equilibria correspond to *foci* at even multiples of π. The trajectories that cross the unstable equilibria (energy thresholds) become *damped separatrices*, i.e. lines separating the families of trajectories attracted to a particular stable equilibrium.

The number of full rotations exhibited by the pendulum depends upon the initial magnitude of the velocity x_2. The greater this initial speed, the greater the number of full rotations, provided the system is free. The region between the separatrices enclosing the corresponding stable equilibrium is the region of attraction to this equilibrium considered the attractor. As mentioned, for the free system, the trajectories from outside this region will be attracted to some other attractor, and we may never be able to attain the target of a trajectory unless we start from a suitable region of attraction. The trajectories from beyond the threshold can also become entirely unstable and unbounded.

Let us now consider the controlled system, i.e. when $u(t)$, generally non-zero, has been determined by a specified control program which is a function of the state x_1, x_2:

$$u(t) = P(x_1(t), x_2(t)), \quad t \geq t_0. \tag{1.1.10}$$

Between the separatrices, there is little need for a controller to produce any inputs in order to lead the pendulum to the corresponding stable equilibrium. However, if the initial values x_1^0, x_2^0 lie between separatrices other than these bounding the desired stable equilibrium (see trajectory denoted by crosses in Fig. 1.4), i.e. outside the region of attraction to the desired equilibrium, an additional force is needed in order for a trajectory to pass over the energy threshold, which corresponds to the separatrix to be crossed. In terms of the equation (1.1.2) it means to produce the control program for $u(t)$ which cancels some (one or more) nonlinear terms of the polynomial by which the gravity force is represented, say for instance,

$$u(t) = - (g/\ell) (\tfrac{1}{6} \theta^3 - \tfrac{1}{120} \theta^5 + \cdots) . \tag{1.1.11}$$

It cuts off the thresholds or, in physical terms, forces the rotation back to the basic region of $\theta \in [-\pi, \pi]$. It may prove expensive in terms of power supply. In fact, it is the more expensive, the more thresholds must be cut between the given initial conditions and the region of attraction attempted.

When linearizing the system by the choice of a suitable control, we have exactly the above case when forcing the trajectories to the region of attraction of the basic equilibrium. Then obviously the cost is the greater, the more nonlinear terms (equilibria) we have to cancel.

The same effect may be obtained in a less costly manner by a *gravity* or *spring compensation*, for example, such as shown in Fig. 1.5(a) below. A counterweight Mg on the radius L adds to $M(\theta) = (g/\ell) \sin \theta$ an additional $M'(\theta) = (ML/m\ell) \sin (-\pi)$ which, with a suitable choice of M and L, holds the link swinging about its upward position instead of downward. Moreover, the above is done not by control torque but by structural means. This reduces the control torque needed to that generating a small swing about the new equilibrium. A very similar effect could have been obtained by inserting a spring about the suspension point of the equilibrium, the spring supporting the link in the desired position, see Fig. 1.5(b). For the linearisation effect, the spring characteristic may be specified, for instance, by $K(\theta) = -(kg/\ell) \cdot \theta - (g/6\ell) \theta^3 + \dots$, $k < 1$, again depending upon the number of nonlinear terms we wish to cancel.

The control or structural cancellation of the nonlinear terms is however not possible when the *nonlinearity is a target of our design* – as for instance if we want to produce a system that will aim at achieving a specified equilibrium, say second or third from the basic, i.e. will aim at work after a specific number of rotations. The latter case often applies in engineering. □

The gravity or spring compensation mentioned in the above example apply generally, even for a system of much higher dimensions. In some cases the designer has also an option of nonpotential compensation. He may use extra damping - positive by inserting fluid dampers or negative by designing self-oscillatory devices.

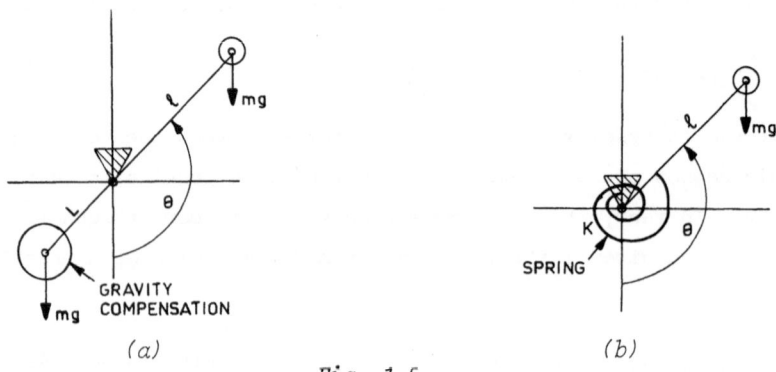

GRAVITY COMPENSATION

SPRING

(a)   (b)

Fig. 1.5

As mentioned, the model of Example 1.1.1 was a special simple version of the class of mechanical models called *point-mass representation*, where the mass or more generally the inertia of the system is reduced to a finite number of material points, each with at most three degrees of freedom (DOF) - translations in the three dimensional Cartesian space $Oxyz$ of physical coordinates. Such masses are considered reduced objects of the system. In Example 1.1.1 we had a single object moving with a single DOF. Let us consider now some cases with two DOF.

EXAMPLE 1.1.2. Two simple pendula of Example 1.1.1 are coupled by a spring at the given distance from their joint suspension base, see Fig. 1.6. Their basic equilibria are attained in the vertical downwards positions of the pendula. The system now has two objects, point-masses m_1, m_2 , with positions specified by coordinates x^i, y^i, z^i , $i = 1, 2$ measured from their corresponding basic equilibria in $Oxyz$. As the pendula move in the vertical plane, we have constraints $z^i = $ const , $(x^i)^2 + (y^i)^2 = \ell^2$, and moreover the coupling generated constraint $y^i > 0$, $i = 1, 2$, as full rotation is not possible. Again generally, the two point-mass modelled object will have six DOF $(3 + 3)$, but the constraint relations reduce the DOF of the system to two. Similarly as in Example 1.1.1, it is more convenient to choose the

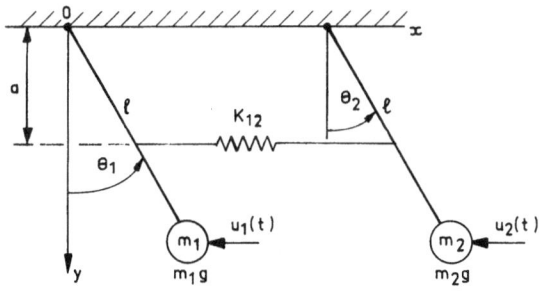

Fig. 1.6

angles of articulation θ_1, θ_2 to be the generalized (lagrangian) coordinates q_1, q_2 for these two DOF rather than any of the Cartesian (physical) coordinates chosen from $x_1, \ldots, z_2$. The latter option is obviously possible, but leads to more complicated motion equations following the insertion of constraints. The Lagrange equations of motion give the following motion equations of the system:

$$\left. \begin{array}{l} m_1 \ell^2 \ddot{q}_1 + d_1 |\dot{q}_1| \dot{q}_1 + m_1 g \ell \sin q_1 - a^2 k (q_2 - q_1) = \tilde{u}_1 \\ m_2 \ell^2 \ddot{q}_2 + d_2 |\dot{q}_2| \dot{q}_2 + m_2 g \ell \sin q_2 + a^2 k (q_2 - q_1) = \tilde{u}_2 \end{array} \right\} \qquad (1.1.12)$$

or '

$$\left. \begin{array}{l} \ddot{q}_1 + D_1 (\dot{q}_1) + G_1 (q_1) + K_{12} (q_1 - q_2) = u_1 \\ \ddot{q}_2 + D_2 (\dot{q}_2) + G_2 (q_2) + K_{21} (q_2 - q_1) = u_2 \end{array} \right\} \qquad (1.1.13)$$

in terms of the force characteristics: $D_i (\dot{q}_i) = d_i |\dot{q}_i| \dot{q}_i / m_i \ell^2$ for damping, $G_i (q_i) = (g/\ell) \sin q_i$ for gravity, $K_{ij} (q_i - q_j) = a^2 k (q_i - q_j)$, $k > 0$, $i, j = 1, 2$, $i \neq j$ for the spring coupling, with $K_{12} = -K_{21}$ assumed. The spring coupling characteristic is a linear function, with the spring coupling coefficient $k_{ij} / m_i \ell^2 = k$ being the same for both interactions: first object on second and conversely. The input characteristics are obviously represented by $u_i = \tilde{u}_i / m_i \ell^2$, $i = 1, 2$.

The characteristics obtained under the passage from (1.1.12) to (1.1.13) are relatively simple functions, since the equations (1.1.12) are already dynamically (inertially) decoupled. Normally, such passage would have been obtained by multiplying the motion equations, resulting from the Lagrange format, by an inverse of an inertia matrix, which in general will not be diagonal as in (1.1.12), see Section 1.5.

The schematic diagram corresponding to the system (1.1.13), i.e. to the inertially decoupled form, is shown in its general version in Fig. 1.7.

9

Although it is inertially decoupled, it maintains the static couplings of
both damping and spring forces. Of these only the spring coupling is
relevant to our case. Again the nonlinearity of the characteristics is
indicated by the crossing arrows. The point masses move in the vertical
direction only, as marked. Thus each mass exhibits a single DOF. Apart
from the coupling connection, each of the masses is connected to the
external frame of reference (suspended). The latter represents the
"eigen" characteristics.

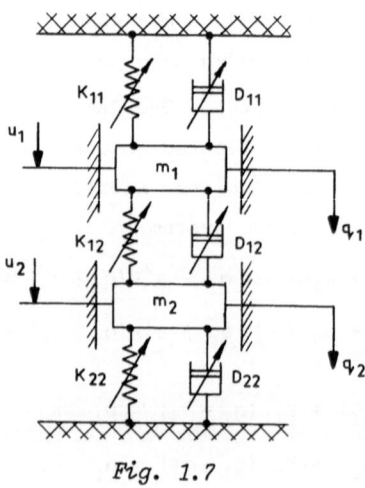

Fig. 1.7

Generally, the characteristics of a two DOF system may be expressed
by the following functions of velocities and displacements:

$$D_i(q_1,q_2,\dot{q}_1,\dot{q}_2) = D_{ii}(q_i,\dot{q}_i) + D_{ij}(q_i - q_j , \dot{q}_i - \dot{q}_j) \qquad (1.1.14)$$

$$K_i(q_1,q_2) = K_{ii}(q_i) + K_{ij}(q_i - q_j) , \quad i,j = 1,2 , \quad i \neq j , \qquad (1.1.15)$$

the first terms with the subscript ii corresponding to the object m_i
itself and also referring to the interaction between the object and the
environment (reference frame). The terms with the mixed subscripts ij
correspond to the coupling between the objects, referring to m_i acting
upon m_j, $i \neq j$. In our particular case, as seen above, $D_{ij} = D_{ji} = 0$
so there is no coupling in damping, and $K_{11} = K_{22} = 0$, so there is no
elastic suspension of objects, with the spring forces reduced to coupling

only. As we shall see later in this chapter, the general case of the mechanical model includes the kinetic (inertial) coupling as well, i.e. the inertial terms in the motion equation will depend, generally, upon all acceleration components $\ddot{q}_i$, $i = 1,2$. However, in most cases, these terms are linear in $\ddot{q}_i$ and there is a straightforward method of decoupling these equations, see Example 1.1.3.

The relationship of the force characteristics concerned to the configuration variables and (formally) the velocities is indeed static. In order to investigate it, we have to put the system at rest, i.e. into its equilibrium. The functions represent the capacity of the forces to act, and the collection of such functions determines the static *organization* or *structure of the system*. Given the characteristics,.we know what it is that moves and how it may move when it starts, but until solutions to the motion equations are found, we will not know how it actually moves. The spring K_{ii} extends under load. A certain static extension $q_i^s = const$, measured along the configuration coordinate q_i , corresponds to the weight $m_i g$. Let k_i be the load necessary to produce a unit extension. Then $K_{ii} = m_i g = k_i q_i^s$ and $q_i^s = m_i g / k_i$. The static extension is also the equilibrium position of the point mass or cube m_i .

Suppose the load changes producing a deflection $\pm \delta q_i$ from the equilibrium position defined by $q_i^e = q_i^s$. It makes the total extension of the spring equal to some $q_i^e \pm \delta q_i = q_i \in \mathbb{R}$. Then the corresponding restoring force in the spring is $-k_i q_i$. Without narrowing generality, we may conveniently place the origin of $\mathbb{R}$ at the equilibrium position (as in fact is done above) $q_i^s = q_i^e = 0$, yielding $q_i = \pm \delta q_i$. Obviously the values $k_i q_i$ represent the capacity of the spring to bear various loads $m_i g$, i.e. the capacity to attain the goals which may be expected from the system element K_{ii} . The function $K_{ii}(q_i) = -k_i q_i$ is the *static characteristic of the element* concerned. Obviously it may be arbitrarily nonlinear, this depending upon our design. The spring coupling elements K_{ij} , specified by $K_{ij}(q_i - q_j)$, in our case $K_{12}(q_1 - q_2) = -a^2 k(q_2 - q_1)$, are obtained in exactly the same way as K_{ii} but referring to the relative position of the two objects $q_i - q_j$. They are *static characteristics of the spring coupling*. Again $K_{ij}(\cdot)$ may be arbitrarily nonlinear and give the capacity of the system for the action of the spring coupling forces. Note that considering the force per coefficient of inertia versus the displacement is conventional. For some purposes, just the opposite may be more suitable, both K_{ii} and q_i being what we shall call *system variables*. The inverse function giving the displacement versus the force/inertia will be later called the *co-characteristic*. All the above comments may be, *ex-equo*, made with respect to the damping characteristics $D_{ii}(\cdot)$, $D_{ij}(\cdot)$. The vectors of the characteristic forces $\bar{K} = (K_1, K_2)^T$, $\bar{D} = (D_1, D_2)^T$ together, form the *system characteristic*, which determines the *organization* or *static structure* of the system.

Similarly as in (1.1.3), the potential energy of the system will now be expressed by the integral of the sum of the potential force, i.e. spring and gravity:

$$V(q_1, q_2) = V^0 + \sum_{i=1}^{2} \int_{q_i(t)} [G_i(q_1, q_2) + K_i(q_1, q_2)] dq_i \qquad (1.1.16)$$

with the equilibria specified by $\dot{q}_1 = 0$, $\dot{q}_2 = 0$ and $G_i(q_1, q_2) = 0$, $K_i(q_1, q_2) = 0$, $i = 1,2$. Obviously the underlying axis of Fig. 1.3 will now become an underlying plane $0q_1q_2$ and the potential surface will be located in three dimensions, see reference to it later in Chapter 2.

The generalized coordinates of the system q_1, q_2 which represent its deviation from the equilibrium move in time $q_1(t)$, $q_2(t)$, $t \geq t_0$, while the vector $\bar{q}(t) = (q_1(t), q_2(t))^T$ describes the instantaneous configuration of the system. Thus it is often called a *configuration vector* and $q_1(t)$, $q_2(t)$ the configuration variables. The configuration variables are independent, but obviously bounded, see Fig. 1.6, thus we shall have a bounded set $\Delta_q \subset \mathbb{R}^2$ such that $\bar{q}(t) \in \Delta_q$, $\forall t \geq t_0$. The set Δ_q forms the *configuration envelope* while $\mathbb{R}^2$ is the *Configuration Space* of the system.

The motion equations corresponding to the full system represented in Fig. 1.7 can be written as

$$\left.\begin{aligned} \ddot{q}_i + D_{ii}(\bar{q}_i, \dot{q}_i) + D_{ij}(q_i - q_j, \dot{q}_i - \dot{q}_j) + G_i(q_1, q_2) + K_{ii}(q_i) \\ + K_j(q_i - q_j) = u_i, \qquad i,j = 1,2, \quad i \neq j \end{aligned}\right\} \qquad (1.1.17)$$

or in more general notation,

$$\ddot{q}_i + D_i(\bar{q}, \dot{\bar{q}}) + G_i(\bar{q}) + K_i(\bar{q}) = u_i, \qquad i = 1,2, \qquad (1.1.18)$$

which will later be called the *symplectic* or *Newton form* of the motion equations. We immediately observe that (1.1.13) is a special case of (1.1.17) and thus also (1.1.18). Note that this form is inertially decoupled, the property which will enable it to become transformed in the system state equation format, which is a set of first order differential equations in normal form.

According to Mechanics, the motion is fully described by registering its velocities as well as positions. Hence we shall need

$$\dot{\bar{q}}(t) = (\dot{q}_1, \dot{q}_2)^T \in \Delta_{\dot{q}} \ ,$$

$\Delta_{\dot{q}}$ the bounded velocity envelope in the tangent space $\mathbb{R}^2$ at (q_1, q_2), to form the state description of the system, which will thus be given by the *state vector* $\bar{x}(t) = (x_1, \ldots, x_4)^T \triangleq (q_1, q_2, \dot{q}_1, \dot{q}_2)^T$, and will vary during motion in the state work region $\Delta \triangleq \Delta_q \times \Delta_{\dot{q}} \in \mathbb{R}^4$, which is the *state space* of the system, while $x_1, \ldots, x_4$ are called the *state variables*. Then the state equations corresponding to (1.1.3) become

$$\left.\begin{aligned}
\dot{x}_1 &= x_3 \\
\dot{x}_2 &= x_4 \\
\dot{x}_3 &= -D_1(x_3) - G_1(x_1) - K_{12}(x_1 - x_2) = u_1 \\
\dot{x}_4 &= -D_2(x_4) - G_2(x_2) - K_{21}(x_2 - x_1) = u_2
\end{aligned}\right\} \ , \qquad (1.1.19)$$

that is, four equations of the first order defined (hopefully) on Δ and with unique solutions on this set. $\qquad\qquad\qquad\qquad\qquad\qquad$ $\square$

EXAMPLE 1.1.3. Let us now double the pendulum of Example 1.1.1, not in parallel as in Example 1.1.2, but in series, making the system a double pendulum with two DOF, see Fig. 1.8, coupled not only statically by potential forces (this time gravity, not spring) but also dynamically (inertial coupling). The pendulum moves in the Cartesian plane Oxy, with constraints $x_1^2 + y_1^2 = \ell_1^2$, $(x_2 - x_1)^2 + (y_2 - y_1)^2 = \ell_2^2$, $z_1 = z_2 = 0$. Ignoring the spring forces and considering the system subject to viscous damping $-d|\dot{\theta}_i|\dot{\theta}_i$, $i = 1, 2$, and external input $F_1 = u_1 \ell_1$, the Lagrange equations of motion produce the following system of equations:

$$\left.\begin{aligned}
(m_1 + m_2)\ell_1^2 \ddot{\theta}_1 &+ m_2 \ell_1 \ell_2 \ddot{\theta}_2 \cos(\theta_2 - \theta_1) - m_2 \ell_1 \ell_2 \dot{\theta}_2^2 \sin(\theta_2 - \theta_1) \\
&= -(m_1 + m_2)\ell_1 g \sin\theta_1 - d|\dot{\theta}_1|\dot{\theta}_1 + u_1\ell_1 \ , \\[2mm]
m_2 \ell_2^2 \ddot{\theta}_2 &+ m_2 \ell_1 \ell_2 \ddot{\theta}_1 \cos(\theta_2 - \theta_1) + m_2 \ell_1 \ell_2 \dot{\theta}_1^2 \sin(\theta_2 - \theta_1) \\
&= -m_2 \ell_2 g \sin\theta_2 - d|\dot{\theta}_2|\dot{\theta}_2 \ .
\end{aligned}\right\} \quad (1.1.20)$$

It is immediately seen that these motion equations are much more complicated than those of the system with two parallel pendula of Example 1.1.2, where the system had been coupled statically only (gravity, spring).

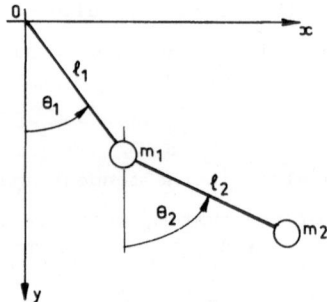

Fig. 1.8

Here the system is dynamically (inertially) coupled, but since the inertia forces are linear in $\ddot{\theta}_1, \ddot{\theta}_2$ we may decouple it by solving the equations (1.1.20) for $\ddot{\theta}_1, \ddot{\theta}_2$ to obtain:

$$
\begin{aligned}
\ddot{\theta}_1 = M_1^{-1}\{ & m_2 \ell_1 \sin(\theta_2 - \theta_1)[\ell_1 \dot{\theta}_1^2 \cos(\theta_2 - \theta_1) + \ell_2 \dot{\theta}_2^2] - d|\dot{\theta}_1|\dot{\theta}_1 \\
& + d|\dot{\theta}_2|\dot{\theta}_2(\ell_1/\ell_2)\cos(\theta_2 - \theta_1) - (m_1 + m_2)g\ell_1 \sin\theta_1 \\
& + m_2 g\ell_1 \sin\theta_1 \cos(\theta_2 - \theta_1) + u_1\ell_1\} \\[2ex]
\ddot{\theta}_2 = M_2^{-1}\Big\{ & -m_2 \ell_2 \sin(\theta_2 - \theta_1)\left[\frac{m_2 \ell_2}{m_1 + m_2}\dot{\theta}_2 \cos(\theta_2 - \theta_1) + \ell_1 \dot{\theta}_1^2\right] \\
& - d|\dot{\theta}_2|\dot{\theta}_2 + \frac{m_2 \ell_2}{(m_1 + m_2)\ell_1}d|\dot{\theta}_1|\dot{\theta}_1 \cos(\theta_2 - \theta_1) \\
& - m_2 \ell_2 g[\sin\theta_2 - \sin\theta_1 \cos(\theta_2 - \theta_1)] + \frac{u_1 m_2 \ell_2}{m_1 + m_2}\cos(\theta_2 - \theta_1)\Big\}
\end{aligned}
$$

$$(1.1.21)$$

where

$$
M_1 = \ell_1^2[m_1 + m_2 \sin^2(\theta_2 - \theta_1)]
$$

$$
M_2 = m_2 \ell_2^2 \left[1 - \frac{m_2}{m_1 + m_2}\cos^2(\theta_2 - \theta_1)\right] .
$$

There are simpler methods for inertial decoupling, for instance as shown in our next example, the purpose of the above exercise being only to indicate that solving for the linearly appearing acceleration is possible though complicated. Obviously for any number of DOF higher than two, we would have to use matrix solutions.

The format (1.1.21) enables us to represent the system in terms of the schematic diagram in Fig. 1.7, with $q_1 = \theta_1$, $q_2 = \theta_2$ as in Example 1.1.2. □

14

So far we have used the class of Cartesian (physical) coordinate models called the point-mass representation. There is an alternative class which may be called a *multi-body model*, in which the interconnected objects are interpreted as solid rigid bodies, each with generally six DOF, that is, three translations and three rotations. It is a matter of convenience which of the class of models we use in the controlling process of a mechanical system. In fact, as will be indicated later, the multi-body representation may be turned into an equivalent point-mass model by a suitable selection of the point-masses, usually by increasing the number of objects of the model in *Oxyz* with rotations replaced with translations. This, however, should not increase the overall dimensionality of the phase or state spaces, which matters for computational purposes.

EXAMPLE 1.1.4. The solid, rigid rod shown in Fig. 1.9 with the mass m is suspended on two springs separated by a constant distance $r_{12} = \ell_1 + \ell_2$, where ℓ_1, ℓ_2 locate the center of gravity as indicated in Fig. 1.9. The rod moves in the vertical plane *Oxy* with two DOF (translation and rotation) when either an initial displacement or a control torque is applied to the body. When the actuator and sensor of motions are applied arbitrarily, a natural way to choose the independent lagrangian coordinates for the DOF is to take $q_1 = y$ which is the vertical translation at the point of action of the actuator u, a distance ℓ from the center of gravity, and take $q_2 = \theta$ which is the rotation angle about that point, both marked in Fig. 1.9.

As we shall see below, this is only one of at least two possible ways of choosing the lagrangian coordinates in this case.

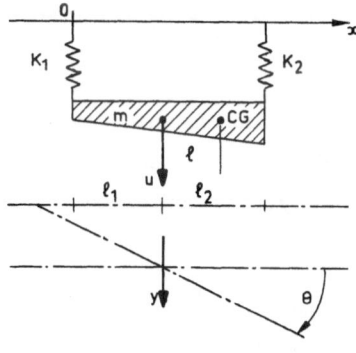

Fig. 1.9

15

Ignoring the damping and assuming the springs linear $K_{ii} = k_i q_i$, $i = 1,2$, for the sake of simplicity, the motion equations become

$$\left. \begin{array}{l} m\ddot{q}_1 + m\ell\ddot{q}_2 + (k_1 + k_2)q_1 + (k_2\ell_2 - k_1\ell_1)q_2 = u \ , \\[2mm] J\ddot{q}_2 + m\ell\ddot{q}_1 + (k_1\ell_1^2 + k_2\ell_2^2)q_2 + (k_2\ell_2 - k_1\ell_1)q_1 = u\ell \ , \end{array} \right\} \qquad (1.1.22)$$

where J is the moment of inertia concerned. The system obviously possesses both the static and the kinetic (inertial) coupling terms. The inertial decoupling can be obtained, as mentioned before, by multiplying the equations by the inverse of the inertia matrix M immediately visible from the equations (1.1.22). In our particular simple case, the procedure may be realized by calculating $\ddot{q}_2$ from the second equation and substituting the result into the first, thus obtaining the equation for q_1 in an inertially decoupled format. A similar calculation yields such an equation for q_2 . Even in this simple two DOF case, the computation is, however, complicated. It can be totally avoided, provided the kinetic energy can be written as a sum of squares of the lagrangian velocities, which means that the inertia matrix becomes diagonal. This may be achieved in a simple way by actuating and sensing the motion at CG, thus obtaining $\ell = 0$. Then (1.1.22) become

$$\left. \begin{array}{l} m\ddot{q}_1 + (k_1 + k_2)q_1 + (k_2\ell_2 - k_1\ell_1)q_2 = u \ , \\[2mm] J\ddot{q}_2 + (k_1\ell_1^2 + k_2\ell_2^2)q_2 + (k_2\ell_2 - k_1\ell_1)q_1 = 0 \ , \end{array} \right\} \qquad (1.1.23)$$

which expressed in terms of characteristics, i.e. subdivided by m,J respectively, can be immediately written in the symplectic form (1.1.18). Then, choosing the states as in Example 1.1.2, we obtain the state format (1.1.19).

The static - spring decoupling is attained, when the potential energy function $V(q_1, q_2)$ can be made a sum of squares of the lagrangian coordinates (which is typical for the linear systems) and then if the restoring forces (in our example, the spring) may be weighted by their distances to C.G., in particular $k_1\ell_1 = k_2\ell_2$.

The choice of lagrangian variables which satisfy both inertial and static spring decoupling produces the so called *normal* or *principal coordinates* essential for the systems concerned. They totally decouple the linear equations of motion.

An equivalent model of the rod is obtained in terms of the two point-masses m_1 and m_2 shown in Fig. 1.10. They are interconnected by a massless rigid frame at a constant distance r_{12} with the actuators of motion acting upon the two masses by controls u_1, u_2 and the configuration variables chosen as vertical translations of each of the masses from their equilibrium $q_i = y_i$, $i = 1,2$. To compare with the previous type actuation u, the controls u_1, u_2 may be weighted by the size and location of m_1, m_2. This immediately gives the motion equations in the inertially decoupled form corresponding to (1.1.23), which makes it easier to rewrite these equations in the symplectic format of (1.1.17). □

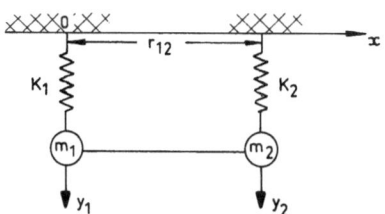

Fig. 1.10

We hope that the examples of this introductory section give an intuitive feeling of the kind of structures discussed in this book. We proceed now to a more general and systematic description of the systems concerned.

1.2 WHAT MOVES AND HOW?

Klir [1] lists 25 substantially different definitions of the system met in the contemporary technical literature, thus the choice is ours but rather difficult. Barnett [1] seems to be close enough to the natural definition when, quoted loosely, he says that a system is a *collection of formally organized objects related by interactions which produce various outputs in response to different inputs, the latter generated by specified external and/or internal programs.* Still, at least until some of the notions of this definition are given a definite meaning for a specified purpose, the word "system" as well as the other terms involved in the quotation are best left to be understood colloquially.

In mechanical interpretation, the objects will mean material elements (substructures and structures) acted upon and interacting via forces which generate motion. The external, or external and internal, forces that are programmed produce control inputs. The interacting forces may come with the type of substructures involved or may also be designed to suit a specific purpose. In the latter case, they will be a subject of *design synthesis*. The design synthesis ought not to be confused with the optimal control synthesis, which is simply a feedback control. Control and design synthesis can supplement each other, as illustrated in Example 1.1.1, see Fig. 1.5. In fact, for mechanical systems, there cannot be any sharp division between these two procedures.

Investigating the control problems of mechanical systems, it is logical and convenient to separate three sets of questions:

(i)   *What is moving,*
(ii) *how it may move under given forces, and*
(iii) *what design or control to use in order to achieve a desired motion?*

In terms of Mechanics, the questions of (ii) refer to the Second Problem of Mechanics: *given forces, find motion*, while the questions of (iii) refer to the First Problem: *given motion, find the acting forces.*

Although our final task is to obtain answers to (iii), it obviously cannot be done without answering (i) and (ii). To investigate what is moving, it is best to look at the mechanical structures and their inter-acting forces at rest, i.e. at an equilibrium state. Thus the natural set up for investigating the question (i) is the *static study* of objects and force characteristics, which interrelate the objects. The collection of both objects and characteristics gives the system its *organization* or *static structure* and this is what will, in this text, form the system. The system will then be put into motion by inputs and controls. Since the objects may, under specific interpretation, be considered a type of force characteristics ("eigen" interactions, see Skowronski [32]), like those defined by kinetic or "fictitious" forces: inertia, gyro, Coriolis, ... , etc., we may think of the organization as of a collection of force characteristics alone. Then the characteristics have the system-theoretic-meaning of a class of reflex-ive relations that define the system, see again Skowronski [32]. For our purpose, however, separating the objects from other static characteristics is convenient and more instructive.

A description of the organization is considered a *structural study.* It must be supported by investigating the properties of materials concerned, most often in terms of specific rheological models, see Section 1.5. For instance, flexibility of links, and viscous or dry friction damping in connecting joints as well as slip, other boundary shear effects on matting surfaces and, in general, negative damping all give structural characteristics, while internal damping, internal crystal sliding effect, etc. are specifically material in character. They must be either included in the structural interactions or made into a separate characteristic. Whatever is the case, both the structural and material features must be represented in the physical models of mechanical systems which we use for the structural study.

A physical model is a mathematical description of a real structure in the Cartesian three dimensional physical Coordinates Space $Oxyz$ and thus it is also called a *Cartesian Model*. The space $Oxyz$ is considered inertial, unless separate notation $OXYZ$ is introduced for the latter. The Cartesian model forms the bridge between the real structure and the corresponding motion equations. The format of the model depends upon the problem and the aim of intended investigations, both structural (static) and kinetic.

With the present technological demands on structures to work at high speed, under heavy payloads, in difficult, uncertain conditions, and yet with high precision, the designer must look at the mechanical structure as on a complicated machine. Consequently the investigator and thus also the modeller must make the model robust against uncertainties in system parameters as well as against errors, both in modelling and design. At the same time, the model must be complex enough to accommodate *all the options* of the designer within each class of problems and design objectives to make the study useful for solving such problems. The latter may also include changing the design choice, when the system is scheduled for different operations, for instance in flexible manufacturing. Consequently the size of the class of problems concerned depends in turn on the variety of applications we like to embrace by the model. This specifies the needed degree of abstraction of the model. The more general the model is, the wider its applications and the better its robustness against modelling errors, but the less concrete information for the designer may be obtained by using it.

Having accepted some degree of the trade-off, we must admit that every model will be somehow idealized, although idealization does not mean simplification. Rather it means the selection of features essential to the aim of study. For instance, a continuous medium model may be redundant for modelling an aircraft or a rocket when investigating their trajectories but not for the aeroelasticity study. In the first case, the reduction of DOF in the point-mass representation may mean already simplifying. Indeed, simplification is usually done after a type of the model has been selected and mathematical difficulties arise when solving its motion equations, cf. Erdman-Sandor [1]. Linearization of models is here a typical example.

Needless to say that giving undue preference to mathematical aspects against physical requirements is dangerous in modelling and later in control, see Kosut et al [1], Belytschko-Hughes [1].

As the reader might have observed, in spite of the fact that the aim of our investigation is usually very precisely posed, the modelling procedure is still very much an art of the designer. Nevertheless, he has a considerable help from the abundant literature on the topic. Modelling is not our subject here and thus in this book we could give only a very short account of it. The remaining sections of this chapter describe the models needed in this book. Moreover, below there is a selected list of review works which successively bring the description and study of the modelling methods almost up-to-date: Myklestad [1], Skowronski [10,12,20], Koenig-Tokad-Kesavan [1], Lowen-Jandrasits [1], Erdman-Sandor [1], Likins [1,2], Hooker-Margulies [1], Hooker [1,2], Ho [1], Wittenberg [1], Hughes [1], Jerkowsky [1], Huston [1], and relatively recent books edited by Magnus [1] and Atluri-Amos [1], Brogan [1], Frederick [1], Robertson-Schwertassek [1].

Once the model is selected and the number of DOF established, we transform the model from $Oxyz$ to the Configuration Space of lagrangian coordinates and into the format of the schematic diagram illustrated in Fig. 1.7 for the case study concerned. Such a transfer may be direct, as in this text, or indirect through some intermediate variables, see Simo - Vu Quoc [1].

We then call the new format the Generalized or *Lagrangian Model*. The equations of motion transformed from $Oxyz$ to the Configuration Space will have the simplectic form illustrated by (1.1.17) in Example 1.1.2.

From this point on, our study is kinetic. With given forces as well as controls it answers the type of questions (ii): How the system may move under given forces. It is the discipline of *Kinetic Analysis* which reveals the options for the motions, thus giving the background for the choice of designing the system and its control programs. Quite naturally, it must rely heavily upon the results obtained in the theory of differential equations, both ordinary and partial, usually in the normal state format, see Examples 1.1.3 and 1.1.4, with the characteristics becoming right hand sides of such equations and with solutions describing the motion. As we do not allow truncation of the nonlinearity in the functions of force characteristics, the equations must be nonlinear, in fact accommodating an arbitrary nonlinearity.

The theory of nonlinear differential equations, together with its more applied part, Nonlinear Mechanics, has a long and well established tradition

dating from before the turn of this century, beginning with the works by
Poincaré [1,2], Hadamard [1], Bendixon [1], Liapunov [1], Van der Pol [1],
Lienard [1], Duffing [1], through the mathematical theory of dynamical
systems formalized by Birkhoff [1,2,3] and through the further development
of Nonlinear Mechanics, right after the Second World War, in at least
several main research centers in Russia, United States, Italy, Poland and
the United Kingdom.

The list of leading names is long and we can only quote a few repre-
senting schools of thought, like Andronov, Chetaiev, Stepanov, Nemitzky,
Markov, Alexandrov, Biebutov, Barbashin, Krasnoselsky, Krassovski, Krylov,
Bogolubov, Mitropolsky in Russia; Timoshenko, Den Hartog, Klotter, Minorsky,
Smith with the support of the Lefschetz school in Topological Dynamics:
Antosiewicz, La Salle, Bellman, Cesari, Seibert, all in the United States;
then Levi-Civita, Sansone, Graffi, Caciopolli, Conti, Szegö in Italy;
Wazewski, Pliss, Bielecki, Albrecht, Opial in Poland; and Cartwright, Little-
wood, Whittacker, Cherry in England. The basic works of the above centers
are listed in our References under the cited names. A considerable number of
monographs or reviews discussing these works appeared at the time or slightly
later, see Nemitzky-Stepanov [1], Nemitzky [1], Krassovski [1], Stocker [1], Den
Hartog [1], Synge [1], Minorsky [2], Kauderer [1], Sansone-Conti [1], Antosiewicz
[1], Zubov [1,2], Bulgakov [1], Andronov-Vitt-Chaikin [1], and later Bhatia-
Szegö [1], Blaquière [1], Barbashin [1], Lefschetz [1], La Salle - Lefschetz
[1], Cesari [1], Reissig-Sansone-Conti [1], Pliss [1], Hajek [1], Struble
[1], Hayashi [1], Yoshizawa [1], Lanczos [1], Ziemba [1], Skowronski [12,14,
20], Starzhinskii [1], Rouche-Habets-Laloy [1], and a few others.

It may seem that the results reviewed and described in these works should
now have only a historic value. However, at closer inspection, it is not the
case. Much of this material remains valid to date, and in some directions
not much more research has been done since these books were written. We use
a considerable amount of these results in this text. A decade or two later,
in the 70's, two popular sidelines appeared on the horizon of Nonlinear
Mechanics: Dynamical Systems on Differentiable Manifolds, centered in the
United States and pioneered by Abraham [1,2], Markus [1], Nitecki [1] and Smale
[1,2]; and the Catastrophe Theory originated by Thom [1] and the Warwick
Center, see Zeemann [1], and for review Gilmore [1] and Zeemann [2]. Then the
last decade brought in a marked intensification of the study on bifurcation
of parameters which began a long time ago with works by Abraham [1],
Andronov-Leontovich [1], Andronov-Vitt-Chaikin [1] and Minorsky [1], see
the recent review by Chow-Hale [1], and which branched, starting with

Lorenz [1], into the present explosion of works on the problems with chaos. On the latter topic, we can quote only some basic monographs, like Irvin [1], Sparrow [1], Thompson-Stewart [1], Moon [1], Guckenheimer-Holmes [1], Devaney [1], Lichtenberg-Lieberman [1], and recently Seydel [1], many of them based on the recent numerical and experimental work of Ueda-Akamatsu-Hayashi [1] and Ueda [1,2].

The above obviously give a very rough picture of the main research avenues in the subject. We shall return to this topic later, in a selective way, when needed.

The structural studies and the kinetic analysis give the preparatory background for answering our questions in (iii), i.e. *how to design the system forces*, both interacting (internal) and control (input), in order to obtain in response a desired motion of the system. The above problem is the subject of this book, thus we leave the discussion on it, including the review of the past and present results, to the next chapters. The only comment which we make in closing this section is that these problems refer to both *control*, for which we must design programs, and to *kinetic synthesis* which requires design of force characteristics, and that these two ways of generating the desired motion are complementary, as shown in Example 1.1.1.

1.3 THE POINT-MASS CARTESIAN MODEL

In the last section we stated that the main features of the physical model are selected due to the aim of study and that, in particular for some cases, a continuous medium model may be redundant. Indeed, the first question asked in modelling is whether the system substructures are *rigid* or *flexible* bodies, or perhaps a mixture of both (a *hybrid* system), cf. Leipholz[1], Meirovitch-Quinn [1]. In the first case, the model is immediately reducible either to the *point-mass* model or to a *multi-body* model. In the second case we face another question, whether the model should be *continuous* or *discrete*. In fact, there is a problem whether there is such a thing at all as a system with an infinite number of DOF, see Hughes [2]. Indeed, between the continuous dislocation of the point-masses (distributed parameters) and their discrete distribution, there is no qualitative difference as far as the modelled physical reality is concerned. Both involve a similar phenomenological approach, as has been known since Livesley [1]. The difference is rather quantitative, referring to the question of how many discretely distributed masses we may technically consider in order to secure the objective

of our investigation and/or design. Studying the hybrid system, may we
perhaps ignore the dynamics which lends itself to continuous medium modelling
(infinite DOF), if the frequency of motion of such a flexible part of the
structure is significantly higher than that of the lumped part? We deal
then with the so called *fast dynamics* in the flexible substructures. Incidentally, it is not an easy task to estimate how "fast" the dynamics should
be to become ignorable, see Corless-Leitmann-Ryan [2], Corless [2]. We
shall return to this problem in Chapter 7.

The above question may also lead to a specific degree of concentration
of the discrete point-masses into larger lumped masses or bodies forming
substructures, see Sadler-Sandor [1] and Sadler [1]. On the other hand,
the usefulness of the model, at least for the initial stages of design,
tends to increase with its simplicity. Moreover, the model that attempts
to simulate too many details quite often cannot answer many questions related to the principal objective of design. Fig. 1.11 illustrates the point
by presenting a set of vertical milling machine models, in the form of the
so called "augmented bodies", see Liegeois-Khalil-Dumas-Renaud [1] and
Hooker-Margulies [1], which gets more detailed modelling along successive
steps of the design procedure. Selecting such augmentation belongs to
practical case-design techniques. Control Theory considers already well
specified models, but it is useful to know what arguments have been used
for specifying a particular case, if only to accommodate them in the
control investigation.

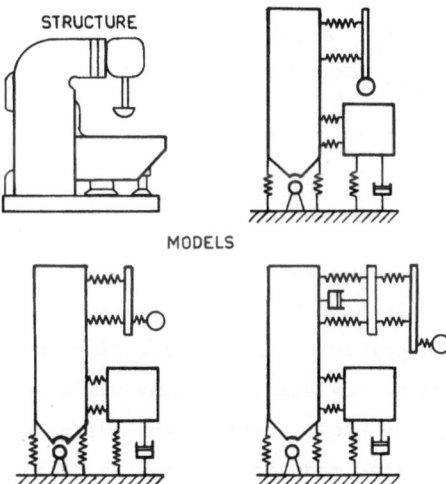

Fig. 1.11

Investigating the vibrations of an automobile, see Fig. 1.12, we may
for instance disregard the masses of axles and tires, the chassis can be
regarded as a rigid beam supported by a flexible foundation. In some cases
even this simple model can be further simplified. Reduction in coupling may
illustrate the case. If the stiffness ratio for both front and rear axle is
approximately inverse proportional to the distance of the gravity center of
the two axles, then the vertical vibrations of the chassis is independent
of its longitudinal angular vibration, see Example 1.1.2. In view of the
symmetry of the structure relative to the longitudinal axis, perpendicular
angular vibrations of the chassis seem to be independent of the vibrations
in the longitudinal plane. If the mass of the axles is small as compared
to the mass of the chassis or the rigidity of the foundation (absorbers)
less than that of the tires, then the vibration of the axle can be regarded
as independent of the motion of the chassis. We can notice this, for
instance, when driving a car at high speed on roads with an unpleasant surface. The small amplitude and higher frequency of the motion of the axles
do not disturb the fluent motion of the chassis. Here again, we may
apply the mentioned augmented body modelling, introducing successive versions of partitioning of the whole body into a hierarchy of subsystems,
as seen from Fig. 1.12.

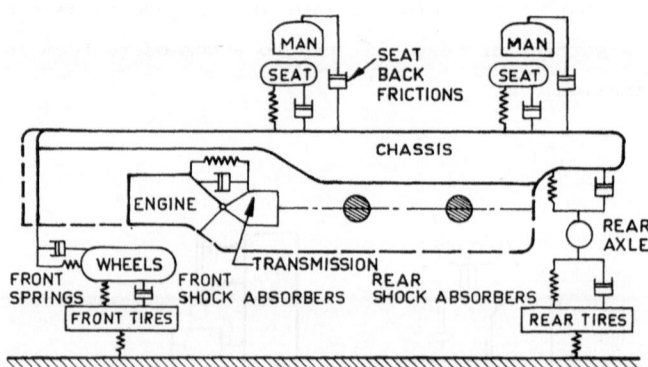

Fig. 1.12

Quite often the problem dictates a selection of some substructure only.
Dealing, for instance, with vibrations of a transmission, we may choose
shafts with gears as the complete system, excluding connections with the
chassis etc. If we want to investigate shimmy vibrations in the wheels,
for independent wheels, the wheel assembly including suspension (absorbers
etc.) may suffice. A similar case applies to the machining process with
the self-excited vibration of a tool investigated. We may and usually do

isolate the tool assembly blending the rest into an unspecified environment.

The literature on lumped mass models is as old as Mechanics, and thus classical and too vast to be even quoted. The interested reader may look for instance into any texts on modelling methods quoted in Section 1.2. From our viewpoint it may be useful to browse through Koenig-Tokad-Kesavan [1], Wittenberg [1], Magnus [2,3], Meirovitch [1], Hagedorn [1] and Skowronski [32]. There obviously is a large and significant literature on discretisation of either the flexible or hybrid structures, we shall briefly refer to it at the opening of Section 1.8.

Granted that we agree on the discrete model, here again are the two options mentioned already: either the point-mass model with objects possessing three DOF each (translations) or the multi-body model, with objects being rigid bodies, in general of six DOF each (translations and rotations). In this section we introduce the point-mass model, leaving its multi-body counterpart to Section 1.4. As mentioned, modelling is not our subject (see Section 1.2), thus the system introduced will be heavily biased towards our further control goals, without pretending to general excellence.

The standard point-mass model is based on the traditional set of particles of classical Mechanics located in $Oxyz$. In fact it is a natural system-theoretic interpretation of such a set, and as such has been used for many years in practically all case studies where reduction to point-masses is feasible. The description produced below is a modification of the model introduced in the 1960's by Skowronski-Ziemba [3,6], Skowronski [10,12,14,20], Muszynska [1] and Jewusiak-Bigley [1]. We see the model as a three dimensional mechanical network shown in Fig. 1.13, with M vertices m_i , i = 1,...,M , interconnected by branches, each vertex connected with all the others, and with an additional vertex m_∞ considered the reference. The M vertices represent the objects of the system, in this case point-masses, allowed to translate but not rotate (3 DOF each), and the branches represent interactions within the system, specified by the force characteristics. The reference vertex m_∞ models the non-structural environment and the branches connecting it with each of the M vertices $\equiv$ objects, model external forces determined by input programs. The vertex m_∞ is not covered by the system physical interpretation and its connections with the system are of a different, often non-structural, nature, hence they are indicated in Fig. 1.13 by dashed lines. What belongs to m_∞ is relative to our aim of study in particular cases of investigation for which the model has been designed.

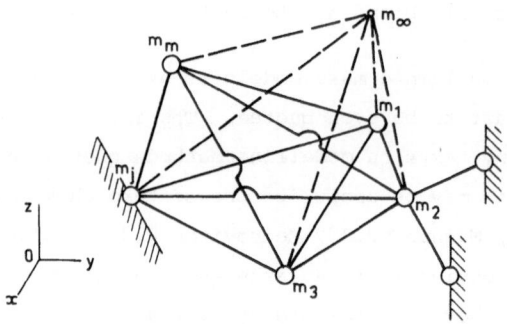

Fig. 1.13

The number of masses m_i may be very small or infinite, depending upon the aim of our study and acceptability of a given approximation. The model may be deformed and/or moving with the point-masses moving relatively to each other and *Oxyz*, see Skowronski-Ziemba [3], Frackiewicz [1]. However, no mechanical system moves in the air, unsuspended. To model such a suspension, we may have two options. Assume some, in general all, of the vertices to be *suspension points* somehow attached to the *structural environment* of the system, see mass m_j in Fig. 1.13, or alternatively introduce for each of the objects, that is, point-masses, a number of suspension points attached to the structural environment and connected to the point-mass concerned by special suspension characteristics, see mass m_2 in Fig. 1.13. The choice between these two versions depends on the convenience in design. The second version is particularly useful when we need the inertial representation of the suspension to be distinctively separate from the rest of the system, like for instance in the case of wheels in a vehicle. In general, however, the second version is easily reduced to the first by assuming the structural environment a part of the structure, that is, the suspension point-masses included in the number M with some of the interconnections missing.

Whichever version is used, the suspension point-masses are restricted by constraints, with some of them resting and/or some moving along prescribed surfaces or lines in *Oxyz*. The constraints are defined by constraint equations imposed upon the coordinates of these masses in *Oxyz*. Then obviously the masses lose some of their DOF. In general, the constraints may also move, forcing a change in the motion of the system.

Replacing the constraints by reaction forces, as it is classical in Mechanics, we are making them part of m_∞ with suspension characteristics serving as an input, provided the motion of the constraints is programmed.

The branches between the vertices of the network will generally be multiple, owing to various types of *interactions*, each specified by a function. Some will be the "eigen" interactions defining the object itself (inertia, or mass, or weight), see Fig. 1.14(a). Basically we divide the interactions into two groups: structural and material. A rather typical representation is illustrated in Fig. 1.14(b) generalizing the known Voigt-type phenomenological model by parallel connections of the massless non-linear spring A representing the potential restitutive forces, two nonlinear viscous dampers: positive B and negative C, the symbolic mark D of the interior slidings in the material depending on displacement, and E the dry friction, damping. Using the symbol "-" for series and "/" for parallel ordering, the connection A/B-D is material, the connection C/E is struc-tural. The material connection is that used for an elastoplastic solid body rheological model, see Reiner [1]. Both the dry friction and the negative viscous damping in the structural connection may appear when self-sustained oscillations are observed in the system (cf. flutter, shimmy). The above example ignores many features like the memory of material, retardation of the acting forces, temperature, change of mass or inertia, etc. We shall return to the discussion on the interactions later, calling them character-istics. The characteristics have physical meaning of force per inertia rather than directly that of force. However, as long as the mass remains constant, the difference is only with regard to scale and may be ignored in qualitative investigation.

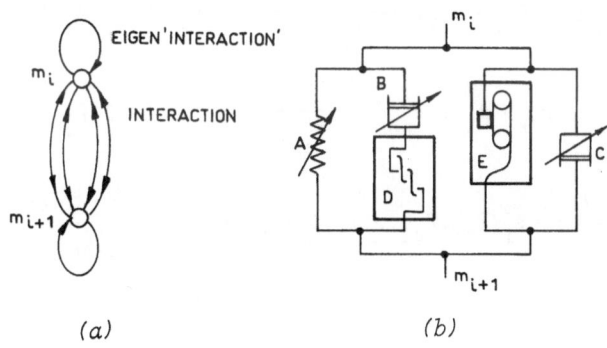

(a)   (b)

Fig. 1.14

Let $\widetilde{x}_i$, $\widetilde{y}_i$, $\widetilde{z}_i$ be coordinates of m_i. We introduce the standard simplifying transformation $x_i = \sqrt{m_i}\,\widetilde{x}_i$, $y_i = \sqrt{m_i}\,\widetilde{y}_i$, $z_i = \sqrt{m_i}\,\widetilde{z}_i$ and consider x_i, y_i, z_i the components in $Oxyz$ of the radius vector $\bar{r}_{0i}$ of the ith vertex of our network. For the corresponding vectors of the suspension points of the jth object, if introduced, we shall have the notation of $\bar{r}_{i\nu}$, $\nu = 1,\ldots,\ell < \infty$, with components $x_{i\nu}, y_{i\nu}, z_{i\nu}$. Then we make the system move in time t, $t \geq t_0$, with $t_0 \in \mathbb{R}$ the initial instant, whence $\bar{r}_{0j}$, $\bar{r}_{j\nu}$ become functions of time $\bar{r}_{0i}(t)$, $\bar{r}_{i\nu}(t)$, and so do their components. Moreover, let d_{ik} = const be the distance between two vertices i and k in the undeformed state, at rest. The displacement of these vertices during motion will then be determined by $\delta d_{ik} \triangleq r_{ik}(t) - d_{ik}$, where $r_{ik}^2 = (x_i - x_k)^2 + (y_i - y_k)^2 + (z_i - z_k)^2$ is the distance between these points at the instant t.

Consider now the forces applied upon each of the point-masses m_i. Without narrowing the applicability of our model, we can split them into the following classes:

Potential forces: $P_{ik}(\bar{r}_{ik})$, $P_{i\nu}(\bar{r}_{i\nu})$, energy conservative, representing gravity and spring forces;

Damping forces: $D_{ik}(\bar{r}_{ik},\dot{\bar{r}}_{ik})$, $D_{i\nu}(\bar{r}_{i\nu},\dot{\bar{r}}_{i\nu})$, energy dissipative (positive damping) or accumulative (negative damping);

Input forces: $F_i(\bar{r}_{0i},\dot{\bar{r}}_{0i},u_i)$, with $u_i(t)$ a control variable produced at the vertex i by an actuator attached there;

External perturbation: $R_i(\bar{r}_{0i},\dot{\bar{r}}_{0i},t)$ acting upon m_i.

The reader is invited to look for closer discussion on the shapes of the force functions in Benedict-Tesar [1], Skowronski-Ziemba [3] and Skowronski [10,12,15]. Then the Newtonian format of the motion equations of the model becomes:

$$
\begin{aligned}
\ddot{x}_i = &-\sum_{k=1}^{M} P_{jk}\,(\bar{r}_{ik} - d_{ik})\,\frac{(x_i - x_k)}{|\bar{r}_{ik}|} - \sum_{\nu=1}^{\ell} P_{i\nu}(\bar{r}_{i\nu} - d_{i\nu})\,\frac{(x_i - x_\nu)}{|\bar{r}_{j\nu}|} \\[2mm]
&-\sum_{k=1}^{M} D_{ik}\,[\,(\bar{r}_{ik} - d_{ik}),\dot{\bar{r}}_{ik}]\,\frac{(x_i - x_k)}{|\bar{r}_{ik}|} - \sum_{\nu=1}^{\ell} D_{j\nu}\,[\,(\bar{r}_{i\nu} - d_{i\nu}),\dot{\bar{r}}_{i\nu}] \\[2mm]
&\cdot \frac{(x_i - x_\nu)}{|\bar{r}_{i\nu}|} + [F_i(\bar{r}_{0i},\dot{\bar{r}}_{0i},u_i) + R_i(\bar{r}_{0i},\dot{\bar{r}}_{0i},t)]\,\frac{(x_i - x_0)}{|\bar{r}_{0i}|},
\end{aligned}
$$

$$i = 1,\ldots,M, \qquad (1.3.1)$$

and identically for the coordinates y_i, z_i. The motion is then determined by the equations (1.3.1) applied to x_i, y_i, z_i and the equations of the suspension constraints.

Examples 1.1.1 - 1.1.4 give a very good illustration of the point-mass models. Derivation of the motion equations (1.3.1) is left to the reader. We may also illustrate our case with the following example.

EXAMPLE 1.3.1. The typical flight dynamics model of an aircraft is a point-mass type with the lumped mass centered at CG . With the geometry shown in Fig. 1.15 the Cartesian motion equations (1.3.1) become

$$
\left.
\begin{aligned}
\dot{x} &= v \cos \gamma \cos \chi \\
\dot{y} &= v \cos \gamma \sin \chi \\
\dot{h} &= v \sin \gamma \\
\dot{v} &= g(T-D)/W - g \sin \gamma \\
\dot{\chi} &= gL \sin \theta / Wv \cos \gamma \\
\dot{\gamma} &= gL \cos \theta / Wv \cos \gamma / v
\end{aligned}
\right\}
\qquad (1.3.2)
$$

where $T = T(\pi, h, v)$ is the thrust, $D = D(h, v, \alpha)$ is the drag and $L = L(h, v, \alpha)$ is the lift force, with the control parameters: angle of attack α, bank angle θ, and throttle coefficient π. The weight W is assumed constant, and the thrust is assumed aligned with the velocity vector $\bar{v}$. The earth is taken to be flat, non-rotating and producing constant gravitational attraction.

The first three equations in (1.3.2) describe the kinematics of the aircraft, while the last three refer to its dynamics.

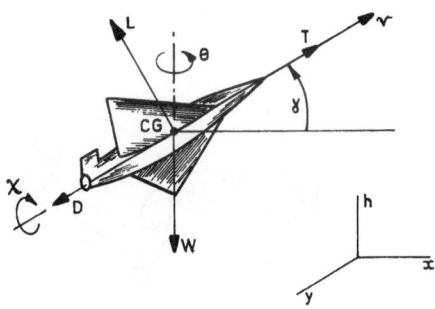

Fig. 1.15

The control variables influence the dynamics only, the kinematics follow from such action. When the aircraft is considered kinematically only (forces ignored), the controls appear in the first three equations.

1.4 THE MULTI-BODY CARTESIAN MODEL

Let us augment the point-masses of the network in Section 1.3 to M' rigid bodies B1,...,BM' interconnected by the branches, still each with all the others, in order to provide the complete set of options for the designer. The branches are now reduced to what will be called *hinges* H1,...,HM' between the bodies, see Fig. 1.16, understood rather generally, i.e. in such a way that with a suitable choice of succession by the designer, each of the interconnected bodies is capable of three translations and three rotations about the hinge that joints it with its predecessor. This suggests that the designer will choose a single tree or several trees (no closed loops). Compare, for instance, the model of a car in Section 1.2 with wheels and suspensions branching out of a chassis. In fact, it can be shown that if a network is not a tree (has closed loops) then it can be equivalently made into one, by cutting any of the hinges of any of the loops, see Jerkowsky [1]. It is thus traditionally assumed that a multi-body model is a collection of trees, cf. Ho-Herbert [1], Vu Quoc-Simo [1], Ibrahim-Modi [1]. When some B1 is directly connected to the rest of the bodies, we have a *cluster tree*; when bodies are connected in series, we have a *chain tree*. These are the two extremes. In Fig. 1.17 they are marked by the dashed and full lines respectively. The shape of any other substructure falls between these two types.

Once the designer made his choice, the model is obviously fixed. It must be both fixed and "well designed" before we may proceed to a further

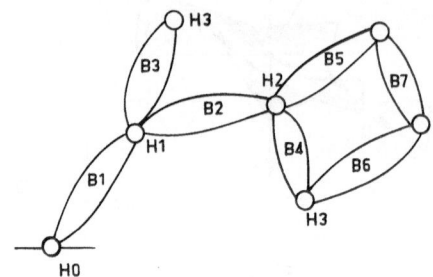

Fig. 1.16

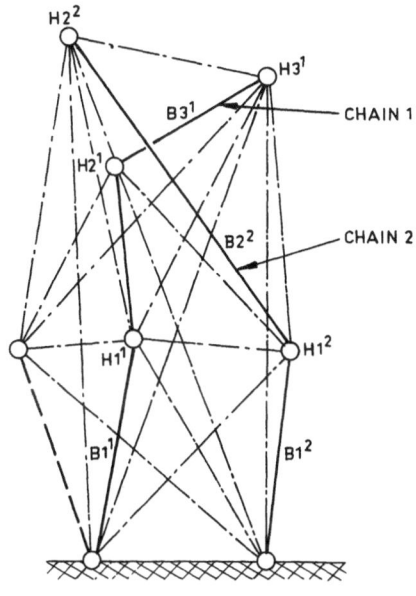

$Fig.$ 1.17

study of selecting suitable motion controllers. By "well designed" we mean not only a definite selection of the tree, but also a specification which of the physical features are expressed by the characteristics and how.

On the other hand, in order to secure the freedom of the all-option choice for the designer, the model must consist of all the possible trees growing up from a given *main body* B1 , as to which in turn the designer must also have a free choice. Thus the model consists of all possible trees that can be selected in the network.

Then the information as to which tree has been selected and how it is interconnected in the particular case investigation is stored in the so called *structural matrix* S which specifies both the geometry and the organization of the network. The matrix is defined by $S_{ij} = 1$ when there is the connection $B_i \rightarrow B_j$, $S_{ij} = -1$ when the connection is reverse; $B_j \rightarrow B_i$, and $S_{ij} = 0$ when there is no connection. There is considerable literature on forming such matrices and incorporating them in the motion equations to follow. As the latter is not of our concern in this text, we restrict ourselves to referring the reader to a good review on the topic in Coiffet [1].

In designing our model, we refer to the Cartesian space $Oxyz$ as an inertial frame, cf. Example 1.1.4. Moreover, we establish the origin 0 as the first hinge H0 joining some B1 either to a suspension body or directly to the environment. The body B1 with the mass m_1 has the body coordinate system $O_1 x_1 y_1 z_1$ fixed at the next hinge H1 to the body B2, with the axis Oz along the line joining hinges H0 and H1. Obviously the location of O_j is conventional and we could pose it anywhere in B1. Positioning it in CG of B1 will decouple the system dynamically, see Example 1.1.4. For our general all-options case, we do not do the latter, and pose O_1 at H1. The same geometry and kinematics refers to B2 with respect to B1, and all subsequent pairs. The body coordinates $O_1 x_1 y_1 z_1$ translate and rotate with respect to $Oxyz$ modelling the motion of B1. For instance, if B1 rotates only, say about the horizontal axis Oy, as shown in Fig. 1.18, we have its current position and velocity specified by

$$
\left.
\begin{aligned}
x(t) &= r_{01} \cos \theta(t) , & \dot{x}(t) &= -r_{01}\dot{\theta}(t) \sin \theta(t) , \\
y(t) &\equiv 0 , & \dot{y}(t) &\equiv 0 , \\
z(t) &= r_{01} \sin \theta(t) , & \dot{z}(t) &= r_{01}\dot{\theta}(t) \cos \theta(t) .
\end{aligned}
\right\}
\qquad (1.4.1)
$$

The distance r_{01} between the hinges is kept constant as there is no translation. The reader may immediately write the corresponding expressions for a rotating B2, having in mind the following feature which refers to all successive bodies in the tree.

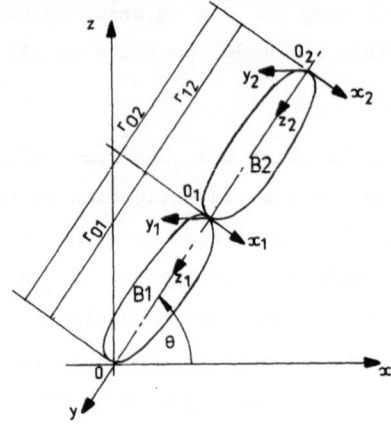

Fig. 1.18

While B1 may obviously be referenced only to $Oxyz$, the position and orientation of each next Bi, $i > 1$, may be referenced either *directly* to $Oxyz$, or *relative* to its predecessor $0_j x_j y_j z_j$, $j = i-1$.

Continuing our example of Fig. 1.18, let us assume that B2 translates only, say along its own axis $0z_1$ which lies in Oxy. Then its direct reference to $Oxyz$ is expressed by

$$\left. \begin{aligned} x(t) &= r_{02}(t) \cos \theta(t) , \\ y(t) &\equiv 0 , \\ z(t) &= r_{02}(t) \sin \theta(t) , \end{aligned} \right\} \tag{1.4.2}$$

and

$$\left. \begin{aligned} \dot{x}(t) &= \dot{r}_{02}(t) \cos \theta(t) - r_{02}(t) \dot{\theta}(t) \sin \theta(t) , \\ \dot{y}(t) &\equiv 0 , \\ \dot{z}(t) &= \dot{r}_{02}(t) \sin \theta(t) + r_{02}(t) \dot{\theta}(t) \cos \theta(t) , \end{aligned} \right\} \tag{1.4.3}$$

while the relative reference to $0_1 x_1 y_1 z_1$ is expressed by

$$\left. \begin{aligned} x_1(t) &\equiv 0 , & y_1(t) &\equiv 0 , & z_1(t) &= r_{02}(t) - r_{01} , \\ \dot{x}_1(t) &\equiv 0 , & \dot{y}_1(t) &\equiv 0 , & \dot{z}_1(t) &= \dot{r}_{02}(t) . \end{aligned} \right\} \tag{1.4.4}$$

In order to produce the all-option equations of motion, we look now at an arbitrary pair of successive bodies Bj, Bi, $i > j$, shown in Fig. 1.19. The body Bi is translated and rotated both relatively with respect to $0_j x_j y_j z_j$ and directly with respect to $Oxyz$. The translation is measured by the vector $\bar{r}_{ji}(t)$ of relative position, with initial value $r_{ji}^0 = \bar{r}_{ji}(t_0)$, $t_0 \in \mathbb{R}$ and the distance between 0_j and 0_i,

$$(\bar{r}_{ji})^2 = (x_j - x_i)^2 + (y_j - y_i)^2 + (z_j - z_i)^2 .$$

The motion of Bi referred to $Oxyz$ is expressed as

$$\left. \begin{aligned} \bar{r}_{0i}(t) &= \bar{r}_{0j}(t) + \bar{r}_{ji}(t) \\ \dot{\bar{r}}_{0i}(t) &= \dot{\bar{r}}_{0j}(t) + \dot{\bar{r}}_{ji}(t) + \bar{\omega}_i(t) \times \bar{r}_{ji}(t) \end{aligned} \right\} \tag{1.4.5}$$

with $\bar{\omega}_i$ the angular velocity of Bi. To formalize the dynamics we calculate

$$\ddot{\bar{r}}_{0i} = \ddot{\bar{r}}_{0j} + \bar{\omega}_i \times (\bar{\omega}_i \times \bar{r}_{ji}) + \dot{\bar{\omega}}_i \times \bar{r}_{ji} + \ddot{\bar{r}}_{ji} + 2\bar{\omega}_i \times \dot{\bar{r}}_{ji} . \tag{1.4.6}$$

The first and fourth term on the right hand side represent the translational acceleration, the second and third centrifugal and tangential, and the last term Coriolis acceleration.

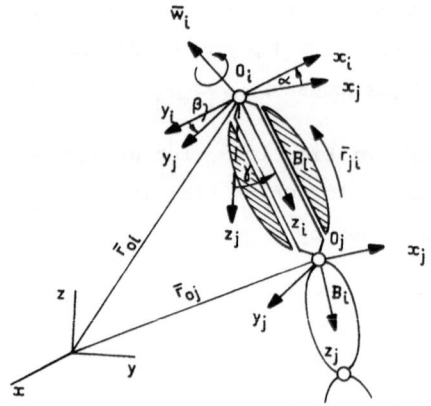

Fig. 1.19

Multiplying the acceleration (1.4.6) by corresponding coefficients of inertia of a body, and equalizing the result to the resultant of forces acting upon this body, we obtain the standard Newton equations of motion.

Referring now to the choice of applied forces we have used in the point-mass model specified in Section 1.3, the Newton equations of motion for each body B_i following all options B_j become

$$\left. \begin{aligned} &\sum_j \left[m_i (\ddot{\bar{r}}_{0j} + \ddot{\bar{r}}_{ji}) + m_i \bar{\omega}_i \times (\bar{\omega}_i \times \bar{r}_{ji}) + m_i \dot{\bar{\omega}}_i \times \bar{r}_{ji} + 2 m_i \bar{\omega}_i \times \dot{\bar{r}}_{ji} \right] \\ &= \sum_j \left[P_{ji} (\bar{r}_{0i}) + D_{ji} (\bar{r}_{0i}, \dot{\bar{r}}_{0i}) \right] + F_i (\bar{r}_{0i}, \dot{\bar{r}}_{0i}, \bar{u}) + R_i (\bar{r}_{0i}, \dot{\bar{r}}_{0i}, t) , \end{aligned} \right\} \quad (1.4.7)$$

$$i, j = 1, \ldots, M'$$

with the first term on the left hand side representing translative inertia force, the second and third terms centrifugal and tangential forces and the fourth term the Coriolis forces, which are all called the *fictitious* or *kinetic* forces as opposed to the *applied* forces $P_i = \sum_j P_{ji}$, $D_i = \sum_j D_{ji}$, F_i and the perturbations R_i. In the applied forces used presently, we obviously amalgamate the interacting forces between the bodies of the network and the *suspension* forces, which may appear between each of the bodies of the network and its environment. The latter may obviously be considered as an extra body in the network, a *suspension body*, entirely analogous to the suspension point of Section 1.3.

Now, the reader may note that a rigid body is a system of particles whose distances are locked constant. In particular, such a body may be represented by two rigidly connected point-masses, as illustrated in

Example 1.1.4, see Figs. 1.9, 1.10 and our double interpretation of the rod, there. Hence each body in the multi-body network may be replaced by two point-masses of the point-mass model, subject to constraints $|\bar{r}_{ij}|$ = const or

$$(x_i - x_j)^2 + (y_i - y_j)^2 + (z_i - z_j)^2 = \text{const} = d_{ij}^2 \qquad (1.4.8)$$

with m_i, m_j denoting the selected pair. Then our M'-body network becomes an M = 2M' point-mass model with the branches corresponding to bodies subject to the constraints (1.4.8). In the above sense, our *multi-body model becomes a special version of the point-mass model.*

On the other hand, it is obvious that when the multi-body model is deprived of the rotational DOF, it formally reduces to the point-mass network. To show this, it suffices to let $\bar{\omega}_i(t) \equiv 0$ in (1.4.7) obtaining $\ddot{\bar{r}}_{0i} = \ddot{\bar{r}}_{0j} + \dot{\bar{r}}_{ji}$ and thus the vectorial form of (1.3.1). Consequently we may use equivalently both the models for structural representation, adjusting only the number of objects.

The work done by the particular classes of applied forces on virtual (without change in time) displacements is

$$\left. \begin{aligned} \delta W_P &= \sum_i P_i(\bar{r}_{0i}) \, \delta\bar{r}_{0i} \\ \delta W_D &= \sum_i D_i(\bar{r}_{0i}, \dot{\bar{r}}_{0i}) \, \delta\bar{r}_{0i} \\ \delta W_F &= \sum_i F_i(\bar{r}_{0i}, \dot{\bar{r}}_{0i}, u_i) \, \delta\bar{r}_{0i} \\ \delta W_R &= \sum_i R_i(\bar{r}_{0i}, \dot{\bar{r}}_{0i}, t) \, \bar{r}_{0i} \end{aligned} \right\} \qquad (1.4.9)$$

where from the total virtual work is

$$\delta W = \sum_i (P_i + D_i + F_i + R_i) \, \delta\bar{r}_{0i} \qquad (1.4.10)$$

with the reactions producing no work, as we assumed them ideal. From (1.4.9) we have the *potential of the conservative forces* $P = \sum_i P_i$

$$U_K = \sum_i \int P_i(\bar{r}_{0i}) \, d\bar{r}_{0i} \qquad (1.4.11)$$

with their components

$$P_x = \frac{\partial U_K}{\partial x}, \qquad P_y = \frac{\partial U_K}{\partial y}, \qquad P_z = \frac{\partial U_K}{\partial z} \qquad (1.4.12)$$

and the *potential energy* $V(\bar{r}_{0i}) = -U_K(\bar{r}_{0i})$. Similarly (1.4.9) produce the generalized *dissipation function*

$$U_D = \sum_i \int D_i(\bar{r}_{0i}, \dot{\bar{r}}_{0i}) \, d\bar{r}_{0i} \qquad (1.4.13)$$

provided

$$\mathcal{D}_x = \frac{\partial U_D}{\partial x}\ , \qquad \mathcal{D}_y = \frac{\partial U_D}{\partial y}\ , \qquad \mathcal{D}_z = \frac{\partial U_D}{\partial z} \tag{1.4.14}$$

where $\mathcal{D}_x, \mathcal{D}_y, \mathcal{D}_z$ are the components of $\mathcal{D} = \sum_i \mathcal{D}_i$. The function U_D becomes the known *Rayleigh damping function* when (1.4.9) are linear. For detailed derivation of the above based on (1.3.1) and (1.4.7), the reader may look in Skowronski-Ziemba [3].

Let us now suppose for the moment that the Cartesian coordinate space *Oxyz* containing our multi-body network moves, not with the inertial motion, but with some stipulated motion. Referring to our example in Fig. 1.18, let us say that the origin 0 of the plane *Oxz* moves with inertial velocity (v_x, v_z) and constant acceleration (a_x, a_z) . Then the motion described by (1.4.1) becomes

$$x(t) = v_x t + \tfrac{1}{2} a_x t^2 + r_{0i} \cos \theta(t)$$
$$z(t) = v_z t + \tfrac{1}{2} a_z t^2 + r_{0i} \sin \theta(t) \ .$$

In general, when reference frames are moving, each coordinate of one frame becomes a function of each and every coordinate of the other, and time:

$$\left.\begin{aligned}
x'(t) &= x'(x,y,z,t)\\
y'(t) &= y'(x,y,z,t)\\
z'(t) &= z'(x,y,z,t)
\end{aligned}\right\} \tag{1.4.15}$$

for *Oxyz* moving in *O'x'y'z'* .

As mentioned, the equations (1.4.7) and then subsequently the shapes of the forces embrace all options for the design. In particular cases, the model will be specified by the structural matrix S , which is successively applied to all the forces, both kinetic and applied, thus zeroing the non-existent interactions in each of these classes of forces.

In particular, the designer may like to differentiate between the substructures, from within a single or several trees available, by selecting a specific chain of bodies along a prescribed *structural path* through the network. In general, these chains may be dependent, i.e. the forces involved may interact across them. Let us say that we wish to split the model into m such chains, each chain j enclosing M^j bodies $B1^j,\ldots,BM^j$, $j = 1,\ldots,m$. In the same way any of the motions discussed before may be referred to the chain j by labelling it with the superscript j , e.g. $0_i^j\ x_i^j\ y_i^j\ z_i^j$ for the body reference system, etc. Then the equations (1.4.7)

36

become a set of equations for each Bi^j in the particular chain j, with the coupling forces referring now to coupling both within the chain j and across the m chains. We obtain

$$\sum_\nu [m_i^j(\ddot{\bar{r}}_{0\nu}^j + \ddot{\bar{r}}_{\nu i}^j) + m_i^j \bar{\omega}_i^j \times (\bar{\omega}_i^j \times \bar{r}_{\nu i}^j) + m_i^j \dot{\bar{\omega}}_i^j \times \bar{r}_{\nu i}^j + 2m_i^j \bar{\omega}_i^j \times \dot{\bar{r}}_{\nu i}^j]$$

$$= \sum_j [\mathbf{P}_{\nu i}^j(\bar{r}_{0i}^j) + \mathbf{D}_{\nu i}^j(\bar{r}_{0i}^j, \dot{\bar{r}}_{0i}^j)] + \mathbf{F}_i^j(\bar{r}_{0i}^j, \dot{\bar{r}}_{0i}^j, \bar{u}^j) + \mathbf{R}_i^j(\bar{r}_{0i}^j, \dot{\bar{r}}_{0i}^j, t), \quad (1.4.7)'$$

$$\nu = 1,\dots,M'\,; \quad i = 1,\dots,M^j\,; \quad j = 1,\dots,m\,; \quad \sum_j M^j = M'\,,$$

for the motion equations referring to chains. These equations will again simplify depending upon the matrix S.

On the other hand, the designer may want to select several *independent chains*. It applies when the substructures concerned are disjoint or at least separated except for a common Bl. The latter case occurs, for instance, when designing appendages of a spacecraft (antennae, platforms, etc.), or when coordinating several robotic arms, or when guiding a rendezvous between two or more spacecraft or ships, or when forming the model for an independent wheel suspension in an automobile.

An independent chain may be obtained within any tree when a structural *direct path* is drawn in the network from some Bl upwards, see the sequence Bl,B2,B4,B6,... in Fig. 1.16. The term "direct path" is frequently used, but may be confused with the state trajectory or "path" in dynamics. Thus we leave the name of "chain", but will talk about the *direct path method*.

The method is already classical in the multi-body modelling, cf. Ho-Herber [1], Ibrahim-Modi [1], Jerkowsky [1]. It has several distinctive advantages over the other means of modelling the multi-body systems. First and foremost, it allows us to use equally well both of the most popular methods of deriving the motion equations: the *Lagrange formalism*, easy to obtain from the kinetic and potential energies, and the *numerical* design of the so called *Newton-Euler model* which uses the successive calculation of system parameters from Bl upwards. Moreover, the direct path method opens the way to introduce the continuous mass distribution in the body integrating the mass elements along the chain, which is a significant feature of modelling flexible systems, see Section 1.8.

It follows from the above discussion that the Newton-Euler formalism is based upon the relative reference within the sequence of a chain, while the Lagrange formalism, as we shall see, may use both relative and direct references. That is, in fact, why the Newton-Euler formalism is considered convenient for a fast numerical simulation of the kinematics, see Hollerbach [1], while the Lagrangian formalism gives much better physical insight into the system which moves and serves dynamics. For a comparison of these two methods, see Snyder [1].

Since the chains are independent, for the general purpose, investigation of an arbitrary chain will do. Similarly as with the structural matrix for the network, the structural case study of our chain is based on two matrices. The first is the so called *incidence matrix*, say S^j, $j = 1, \ldots, m$, m the number of chains. It is defined by $S^j_{ij} = 1$ when a body Bi belongs to the chain $Bl^j \to BM^j$ and $S^j_{ij} = 0$ otherwise. Since we are dealing with a single chain now, the superscript j will be dropped if no confusion may occur. The second matrix is the limb-branch matrix A_i defined by $a_{ij} = 1$ when Bi is the follower of Bj, and $a_{ij} = 0$ otherwise. The limb-branch matrix is also called a *transformation matrix* between the body coordinates of Bi and Bj, $j = i - 1$. The name is particularly often used in the modelling of robotic manipulators, see Paul [1], Snyder [1], Skowronski [32]. The matrix is a basic tool in the Newton-Euler method of modelling.

In order to illustrate the procedure closer, the reader may recall the example of Fig. 1.18 with the kinematics determined by (1.4.2) – (1.4.4), noting the relative reference of the body coordinates: B1 to $Oxyz$, and B2 to $Ox_1y_1z_1$. The equations (1.4.2) – (1.4.4) can be easily rewritten in the matrix format.

Generalizing the above example, we conclude that the use of the transformation matrix A_i may be considered a routine. Each of these matrices describes the relative translation and/or rotation between $Ox_iy_iz_i$ and its predecessor $O_jx_jy_jz_j$. Then the position and orientation of Bi in $Oxyz$ are given by the matrix product $T_i = A_1 \cdot A_2 \cdots \cdot A_i$, with A_i transforming the position vector $\bar{r}_{0j}$ of Bj into $\bar{r}_{0i}$ of Bi: $\bar{r}_{0i} = A_i \bar{r}_{0j}$. The matrix may include rotation matrices (direction cosines) and/or translation matrices (components of translation vector). The matrices will normally be 3×3 dimensional, but the practice of using them in calculating robotic manipulator kinematics suggested an artificial augmentation to 4×4 with $\bar{r}_{0i} = (a_i, b_i, c_i)$ represented by the matrix $[x_i y_i z_i, d_i]$ where $a_i = x_i/d_i$, $b_i = y_i/d_i$, $c_i = z_i/d_i$, $d_i = \text{const}$. For an excellent detailed instruction, see Paul [1]. The augmentation is purely technical and the matrices generated are called *homogeneous*. In our simple example on Fig. 1.18 following (1.4.1), (1.4.4), we have two such matrices

$$A_1 = \begin{bmatrix} \cos\theta & -\sin\theta & 0 & 0 \\ \sin\theta & \cos\theta & 0 & 0 \\ 0 & 0 & 1 & 0 \\ 0 & 0 & 0 & 1 \end{bmatrix}, \quad A_2 = \begin{bmatrix} 1 & 0 & 0 & r_{02} \\ 0 & 1 & 0 & 0 \\ 0 & 0 & 1 & 0 \\ 0 & 0 & 0 & 1 \end{bmatrix}, \quad (1.4.16)$$

the first being rotation matrix only, the second translation only. Then the transformation matrix for the chain is

$$T_2 = A_1 \cdot A_2 = \begin{bmatrix} \cos\theta & -\sin\theta & 0 & r_{02}\cos\theta \\ \sin\theta & \cos\theta & 0 & r_{02}\sin\theta \\ 0 & 0 & 1 & 0 \\ 0 & 0 & 0 & 1 \end{bmatrix}. \quad (1.4.17)$$

It is worth noting that T_2 specifies the case of rotation first and translation second, along the path concerned. The translation gives the fourth column in T_2.

We may have a similar arrangement about velocities, with the transformation matrices A_i replaced by Jacobian matrices:

$$(\dot{\bar{r}}_{ji}, \bar{\omega}_{ji}) = J_i \cdot (\dot{\theta}_1, \ldots, \dot{\theta}_i, \dot{\bar{r}}_{01}, \ldots, \dot{\bar{r}}_{0i}),$$

$$J_i : (\dot{\theta}_1, \ldots, \dot{\theta}_i, \dot{\bar{r}}_{01}, \dot{\bar{r}}_{0i}) \longrightarrow (\dot{x}_i, \dot{y}_i, \dot{z}_i),$$

where $\bar{\omega}_i$ are the vectors of angular velocities in the rotation of Bi.

Both types of transformation specified by A_i and J_i, $i = 1, \ldots, m$, give the so called *forward kinematics* along the chain, allowing us to calculate the Cartesian coordinates x_i, y_i, z_i and velocity components $\dot{x}_i, \dot{y}_i, \dot{z}_i$ from $r_{0i}, \theta_i, \dot{r}_{0i}, \dot{\theta}_i$. Choosing the latter as lagrangian coordinates which specify DOF of the system, we may need inverse transformations, i.e. so called *inverse kinematics*, which is not uniquely defined and much more difficult. The reader may observe this for instance by attempting to invert (1.4.1), thus obtaining multivalued inverse trigonometric functions. See also Section 1.6, Example 1.6.1. The names of forward and inverse kinematics come from the theory of manipulators, where the inverse kinematics is fundamental for modelling and design, see Tarn-Bejczy-Yun [1]. There is no general algorithm for calculating the inverses A_i^{-1}, J_i^{-1}, but some numerical solutions are possible, see Lenarcic [1]. Thus, when the Newton-Euler formalism is to be used, which is numerical anyway, the discussed solution is quite feasible.

Passing now over to the dynamics of our arbitrary chain selected from the network, we must distinguish such chains from within the general network equations (1.4.7), or within the assembly of chains (1.4.7)'.

The equations (1.4.7)' simplify, as each body Bi^j in the particular chain j intersects with its predecessor in this chain only, i.e. pairs Bi^j, Bj^j become $Bi^j, B(i-1)^j$. We then obtain

$$\left. \begin{aligned} m_i^j(\ddot{\bar{r}}_{i-1,i}^j + \ddot{\bar{r}}^j) + m_i^j\bar{\omega}_i^j \times (\bar{\omega}_i^j \times \bar{r}_{i-1,i}^j) + m_i^j\dot{\bar{\omega}}_i^j \times \bar{r}_{i-1,j}^j + 2m_i^j\bar{\omega}_i^j \times \dot{\bar{r}}_{i-1,j}^j \\ = \sum_{\mu=i-1}^{\mu=i+1}[\mathbf{P}_{\mu i}^j(\bar{r}_{0i}) + \mathbf{D}_{\mu i}^j(\bar{r}_{0i}, \dot{\bar{r}}_{0i})] + \mathbf{F}_i^j(\bar{r}_{0i}, \dot{\bar{r}}_{0i}, u_i) + \mathbf{R}_i^j(\bar{r}_{0i}) , \end{aligned} \right\} \quad (1.4.18)$$

$$i = 1,\ldots,M^j , \quad j = 1,\ldots,m$$

where M^j is a number of bodies in the chain j, and m is the number of chains.

1.5 LAGRANGIAN MODEL. INERTIAL DECOUPLING

The motions of any of the two Cartesian models discussed are often subject to equality constraints, which may move (rheonomic), but will be assumed holonomic (algebraic or integrable, if differential):

$$\Phi_i(x_1, y_1, z_1, \ldots, x_M, y_M, z_M, t) = 0 , \quad i = 1,\ldots,k < \infty . \quad (1.5.1)$$

Moreover, as mentioned, the Cartesian space frame of reference $Oxyz$ may also move with given motion but not necessarily inertial. Then the $3M$ Cartesian coordinates become dependent and the motion equations of the model must be solved together with (1.5.1). Theoretically, one may solve (1.5.1) and using the solutions in motion equations, say (1.3.1), obtain $3M - k$ independent coordinates which uniquely describe the motion of the network along the constraints, at the same time specifying the DOF of the system. However, it is possible and usually more convenient to choose $n = 3M - k$ independent parameters of the system, for instance like $\theta_i, \bar{r}_{0i}$ in (1.4.1) - (1.4.4), which may describe uniquely the motion along the constraints, each parameter for every DOF. Such parameters are called *generalized coordinates* $q_1,\ldots,q_n$. The Cartesian coordinates $x_1,\ldots,z_M$ of $Oxyz$ become functions of $q_1,\ldots,q_n$, such that substituting these functions into (1.5.1) we obtain the identity for all q_i, $i = 1,\ldots,n$. The vector $\bar{q} \triangleq (q_1,\ldots,q_n)^T$ is called the *configuration vector* of the system and ranges in the *Configuration Space* $\mathbb{R}^n$. The choice of $q_1,\ldots,q_n$ is not unique, as seen in the examples of Section 1.1, in particular in Example 1.1.4. In the substructure modelled as an independent chain, the generalized coordinates are usually the *hinge-variables* (describing the DOF at a hinge), in robotic manipulators they are as a rule the *joint-variables*, cf. Coiffet [1], Paul [1].

In order to satisfy the rheonomic constraints (1.5.1) and accommodate the possible case of a moving $Oxyz$, the mentioned functions of parameters $q_1, \ldots, q_n$ or as we often call them the *transformations* $q_1, \ldots, q_n \rightarrow x_1, \ldots, z_M$ must, in general, be also explicitly time dependent:

$$\bar{r}_{0i} = \bar{r}_{0i}(\bar{q}, t) \ . \tag{1.5.2}$$

Differentiating, we obtain the velocity

$$\dot{\bar{r}}_{0i} = \frac{\partial \bar{r}_{0i}}{\partial t} + \sum_{j=1}^{n} \frac{\partial \bar{r}_{0i}}{\partial q_j} \dot{q}_j \ . \tag{1.5.3}$$

Thus the kinetic energy becomes

$$T = \tfrac{1}{2} \sum_i m_i (\dot{\bar{r}}_{0i})^2 = \tfrac{1}{2} \sum_i m_i \left(\frac{\partial \bar{r}_{0i}}{\partial t} + \sum_j \frac{\partial \bar{r}_{0i}}{\partial q_j} \dot{q}_j \right)^2 \ . \tag{1.5.4}$$

Carrying on the expansion, we obtain

$$\begin{aligned} T(\bar{q}, \dot{\bar{q}}, t) &= T_0 + T_1 + T_2 \\ &= M_0 + \sum_j M_j \dot{q}_j + \tfrac{1}{2} \sum_j \sum_k M_{jk} \dot{q}_j \dot{q}_k \ , \end{aligned} \tag{1.5.5}$$

where

$$M_0 = \sum_i \tfrac{1}{2} m_i \left(\frac{\partial \bar{r}_{0i}}{\partial t} \right)^2 \ ,$$

$$M_j = \sum_i m_i \frac{\partial \bar{r}_{0i}}{\partial t} \cdot \frac{\partial \bar{r}_{0i}}{\partial q_j} \ ,$$

$$M_{jk} = \sum_i m_i \frac{\partial \bar{r}_{0i}}{\partial q_j} \cdot \frac{\partial \bar{r}_{0i}}{\partial q_k} \ ,$$

with $M_{jk}(\bar{q}) = M_{kj}(\bar{q})$ being coefficients of the corresponding positive definite *generalized inertia matrix* $M(\bar{q})$. When the transformations (1.5.2) are stationary (do not depend explicitly on time), we have $T_0 \equiv 0$, $T_1 \equiv 0$ and thus the kinetic energy becomes the square form

$$T(\bar{q}, \dot{\bar{q}}) = \tfrac{1}{2} \sum_j \sum_k M_{jk} \dot{q}_j \dot{q}_k = \tfrac{1}{2} \dot{\bar{q}}^T M(\bar{q}) \dot{\bar{q}} \ . \tag{1.5.6}$$

Similarly to (1.5.3), the arbitrary virtual displacement $\delta \bar{r}_{0i}$ can be connected with the virtual displacement δq_i (without the variation in time) by

$$\delta \bar{r}_{0i} = \sum_j \frac{\partial \bar{r}_{0i}}{\partial q_j} \delta q_j \ . \tag{1.5.7}$$

To transform the inertial forces in (1.3.1) from Cartesian to generalized coordinates, we calculate

$$\sum_i m_i \ddot{\bar{r}}_{0i} \, \delta\bar{r}_{0i} = \sum_i m_i \ddot{\bar{r}}_{0i} \sum_j \frac{\partial \bar{r}_{0i}}{\partial q_j} \, \delta q_j = \sum_j \sum_i m_i \ddot{\bar{r}}_{0i} \frac{\partial \bar{r}_{0i}}{\partial q_j} \, \delta q_j \; . \quad (1.5.8)$$

We may now write formally,

$$\sum_i m_i \ddot{\bar{r}}_{0i} \frac{\partial \bar{r}_{0i}}{\partial q_j} = \sum_i \left\{ \frac{d}{dt} \, (m_i \dot{\bar{r}}_{0i} \frac{\partial \bar{r}_{0i}}{\partial q_j}) - m_i \dot{\bar{r}}_{0i} \frac{d}{dt} \, (\frac{\partial \bar{r}_{0i}}{\partial q_j}) \right\} \; . \quad (1.5.9)$$

In the last term of (1.5.9) the differentiation with respect to t and q_j may be interchanged, for, in analogy to (1.5.3),

$$\frac{d}{dt} \, (\frac{\partial \bar{r}_{0i}}{\partial q_j}) = \frac{\partial \dot{\bar{r}}_{0i}}{\partial q_j} = \sum_k \frac{\partial^2 \bar{r}_{0i}}{\partial q_j \partial q_k} \, \dot{q}_k + \frac{\partial^2 \bar{r}_{0i}}{\partial q_j \partial t} \quad (1.5.9)$$

by (1.5.3). From the latter we have also

$$\frac{\partial \dot{\bar{r}}_{0i}}{\partial \dot{q}_j} = \frac{\partial \bar{r}_{0i}}{\partial q_j} \; . \quad (1.5.10)$$

Substituting the above in (1.5.9),

$$\sum_i m_i \ddot{\bar{r}}_{0i} \frac{\partial \bar{r}_{0i}}{\partial q_j} = \sum_i \left\{ \frac{d}{dt} \, (m_i \dot{\bar{r}}_{0i} \frac{\partial \dot{\bar{r}}_{0i}}{\partial \dot{q}_j}) - m_i \dot{\bar{r}}_{0i} \frac{\partial \dot{\bar{r}}_{0i}}{\partial q_j} \right\} \; ,$$

wherefrom, by virtue of (1.5.4), we obtain for (1.5.8):

$$\sum_j \sum_i m_i \ddot{\bar{r}}_{0i} \frac{\partial \bar{r}_{0i}}{\partial q_j} \, \delta q_j = \sum_j [\frac{d}{dt} \, (\frac{\partial T}{\partial \dot{q}_j}) - \frac{\partial T}{\partial q_j}] \, \delta q_j \; , \quad (1.5.11)$$

as the value of inertia forces in our Cartesian motion equations. On the other hand, the work done by the applied forces on virtual displacement (1.4.10) equals that in generalized coordinates:

$$\sum_i (P_i + D_i + F_i + R_i) \delta\bar{r}_{0i} = \sum_j Q_j \, \delta q_j \; , \quad i,j = 1,\dots,n \; . \quad (1.5.12)$$

In the above $Q_j(\bar{q},\dot{\bar{q}},\bar{u},t)$ is the *generalized applied force* defined by the equality (1.5.12), and $\bar{u}(t) = (u_1(t),\dots,u_r(t))^T \in \mathbb{R}^r$ is a control vector generated by actuators acting in the system. Then by the motion equations (1.3.1) or (1.4.7), (1.5.11) and (1.5.12) result in

$$\sum_j [\frac{d}{dt} \, (\frac{\partial T}{\partial \dot{q}_j}) - \frac{\partial T}{\partial q_j}] \, \delta q_j = \sum_j Q_j \, \delta q_j \; .$$

Since the above must hold for arbitrary δq_j, $j = 1,\dots,n$, we obtain the motion equation in the Lagrange's second form

$$\frac{d}{dt} \, (\frac{\partial T}{\partial \dot{q}_j}) - \frac{\partial T}{\partial q_j} = Q_j \; , \qquad j = 1,\dots,n \; . \quad (1.5.13)$$

Similarly to (1.5.12) obtained from (1.4.10), we can form equalities corresponding to W_K, W_D and W_F in (1.4.9), whence Q_j may be represented by the sum

$$Q_j(\bar{q},\dot{\bar{q}},u_j,t) = Q_j^P(\bar{q}) + Q_j^D(\bar{q},\dot{\bar{q}}) + Q_j^F(\bar{q},\dot{\bar{q}},u_j) + Q^R(\bar{q},\dot{\bar{q}},t) \qquad (1.5.14)$$

of potential, damping, input, and perturbation forces respectively. Swapping the subscripts to more convenient i, we have the Lagrange motion equations

$$\frac{d}{dt}\left(\frac{\partial T}{\partial \dot{q}_i}\right) - \frac{\partial T}{\partial q_i} = Q_i^P(\bar{q}) + Q_i^D(\bar{q},\dot{\bar{q}}) + Q_i^F(\bar{q},\dot{\bar{q}},\bar{u}) + Q_i^R(\bar{q},\dot{\bar{q}},t) \ . \qquad (1.5.15)$$

Then by the above and (1.4.11),

$$Q_i^P(\bar{q}) = \frac{\partial U_p(\bar{q})}{\partial q_i} \ , \qquad i = 1,\ldots,n \qquad (1.5.16)$$

and, introducing the Lagrangian function

$$L(\bar{q},\dot{\bar{q}},t) = T(\bar{q},\dot{\bar{q}},t) - V(\bar{q}) \ ,$$

where $V(\bar{q}) = -U_p(\bar{q})$ is the potential energy of the system, the Lagrange equations obtain their standard format

$$\frac{d}{dt}\left(\frac{\partial L}{\partial \dot{q}_i}\right) - \frac{\partial L}{\partial q_i} = Q_i^D(\bar{q},\dot{\bar{q}}). + Q_i^F(\bar{q},\dot{\bar{q}},\bar{u}) + Q_i^R(\bar{q},\dot{\bar{q}},t) \ , \qquad (1.5.17)$$

with all the total energy changing forces on the right hand side.

Let us now assume that $T \equiv T_2$, that is the kinetic energy is the quadratic form (1.5.6). Then the equations (1.5.15) may be written as

$$\frac{d}{dt}\left(\sum_{j=1}^{n} M_{j,i}\,\dot{q}_j\right) - \frac{1}{2}\sum_{j=1}^{n}\sum_{k=1}^{n}\frac{\partial M_{jk}}{\partial q_i}\,\dot{q}_j q_k = Q_i(\bar{q},\dot{\bar{q}},u,t) \ ,$$

or

$$\sum_{j=1}^{n} M_{ji}\,\ddot{q}_j + \sum_{j=1}^{n}\sum_{k=1}^{n}[j_{\ i}^{\ k}]\,\dot{q}_j\,\dot{q}_k = Q_i(\bar{q},\dot{\bar{q}},u,t) \ , \qquad (1.5.18)$$

where

$$[j_{\ i}^{\ k}] \triangleq \frac{1}{2}\left(\frac{\partial M_{ji}}{\partial q_k} + \frac{\partial M_{ki}}{\partial q_j} - \frac{\partial M_{jk}}{\partial q_i}\right)$$

is the so called Christoffel's symbol, see Whittaker [1]. The second term on the left hand side specifies the Coriolis and centrifugal forces. Denoting it $Q_i^C(\bar{q},\dot{\bar{q}})$ we obtain the dynamically coupled Newtonian form of the Lagrange equations

$$\sum_{j=1}^{n} M_{ji}\,\ddot{q}_j + Q_i^C(\bar{q},\dot{\bar{q}}) - Q_i^D(\bar{q},\dot{\bar{q}}) - Q_i^P(\bar{q}) = Q_i^F(\bar{q},\dot{\bar{q}},\bar{u}) + Q_i^R(\bar{q},\dot{\bar{q}},t). \qquad (1.5.19)$$

Since $T \equiv T_2$ with positive definite matrix M, it follows from the Sylvester criteria that its diagonal minors of M are positive. Hence M

is nonsingular and there is the inverse M^{-1} with coefficients $M_{ij}^{-1} = \tilde{M}_{ji} / \det M$, $i,j = 1,\ldots,n$, where $\tilde{M}_{ji}$ is a cofactor to the element M_{ji} of $\det M$. Multiplying now each of the equations (1.5.19) by M_{ij}^{-1}, summing them up and remembering that

$$\sum_i M_{ki} M_{ij}^{-1} = \frac{1}{\det M} \sum_i M_{ki} \tilde{M}_{ji} = \delta_{kj} = \begin{cases} 0, & k \neq j \\ 1, & k = j \end{cases},$$

we obtain the inertially decoupled symplectic form of the motion equations (see Example 1.1.3):

$$\ddot{q}_i + \Gamma_i(\bar{q},\dot{\bar{q}}) + D_i(\bar{q},\dot{\bar{q}}) + \Pi_i(\bar{q}) = F_i(\bar{q},\dot{\bar{q}},\bar{u}) + R_i(\bar{q},\dot{\bar{q}},t), \qquad (1.5.20)$$

$$i = 1,\ldots,n,$$

where

$$\Gamma_i(\bar{q},\dot{\bar{q}}) = \sum_j M_{ij}^{-1} Q_j^C(\bar{q},\dot{\bar{q}}), \qquad (1.5.21)$$

are the *centrifugal - Coriolis force characteristics*,

$$D_i(\bar{q},\dot{\bar{q}}) = -\sum_j M_{ij}^{-1} Q_j^D(\bar{q},\dot{\bar{q}}) \qquad (1.5.22)$$

are the *damping* force characteristics,

$$\Pi_i(\bar{q}) = -\sum_j M_{ij}^{-1} Q_j^P(\bar{q}) \qquad (1.5.23)$$

are the *potential* force characteristics,

$$F_i(\bar{q},\dot{\bar{q}},\bar{u}) = \sum_j Q_j^F(\bar{q},\dot{\bar{q}},\bar{u}), \qquad (1.5.24)$$

are the *input* force characteristics, and

$$R_i(\bar{q},\dot{\bar{q}},t) = \sum_j Q_j^R(\bar{q},\dot{\bar{q}},t) \qquad (1.5.25)$$

are the external *perturbation* force characteristics.

Forming now the vectors $\bar{\Gamma} = (\Gamma_1,\ldots,\Gamma_n)^T$, $\bar{D} = (D_1,\ldots,D_n)^T$, $\bar{\Pi} = (\Pi_1,\ldots,\Pi_n)^T$, $\bar{F} = (F_1,\ldots,F_n)^T$ and $\bar{R} = (R_1,\ldots,R_n)^T$, the equations (1.5.20) may be written in the vector form

$$\ddot{\bar{q}} + \bar{\Gamma}(\bar{q},\dot{\bar{q}}) + \bar{D}(\bar{q},\dot{\bar{q}}) + \bar{\Pi}(\bar{q}) = \bar{F}(\bar{q},\dot{\bar{q}},\bar{u}) + \bar{R}(\bar{q},\dot{\bar{q}},t). \qquad (1.5.26)$$

Since the work done by potential forces W_P is additive and thus so is the potential energy, then by (1.5.16) and $V(\bar{q}) = -U_P(\bar{q})$, we may separate the *gravity potential forces* $Q_i^G(\bar{q})$ from the *elastic (spring) potential forces* $Q_i^K(\bar{q})$ and write $Q_i^P(\bar{q}) \triangleq Q_i^K(\bar{q}) + Q_i^G(\bar{q})$, $i = 1,\ldots,n$, which yields

$$\Pi_i(\bar{q}) \triangleq K_i(\bar{q}) + G_i(\bar{q}) \tag{1.5.27}$$

with the obvious meaning of the *spring characteristics* $K_i(\bar{q})$ and *gravity characteristics* $G_i(\bar{q})$. Then vectorially

$$\bar{\Pi}(\bar{q}) = \bar{K}(\bar{q}) + \bar{G}(\bar{q}) \tag{1.5.27}'$$

where $\bar{K} \triangleq (K_1, \dots, K_n)^T$, $\bar{G} \triangleq (G_1, \dots, G_n)^T$.

Consider (1.5.20) and move all the characteristics to the right hand side. Then resolving it with respect to $d\dot{q}_i/dt$ and dq_i/dt and dividing, we obtain the so called *phase-space equations*

$$\frac{d\dot{q}_i}{dq_i} = \frac{-\Gamma_i(\bar{q},\dot{\bar{q}}) - D_i(\bar{q},\dot{\bar{q}}) - \Pi_i(\bar{q}) + F_i(\bar{q},\dot{\bar{q}},\bar{u}) + R_i(\bar{q},\dot{\bar{q}},t)}{\dot{q}_i}, \tag{1.5.28}$$

$$i = 1,\dots,n,$$

describing instantaneous slopes of the first integral lines of (1.5.20) in the *phase space* $\mathbb{R}^{2n}$ of points $(\bar{q}(t),\dot{\bar{q}}(t)) \in \Delta_q \times \Delta_{\dot{q}}$. Observe that (1.5.28) can also be written as the system

$$\left.\begin{array}{l}\dfrac{dq_1}{\dot{q}_1} = \cdots = \dfrac{dq_n}{\dot{q}_n} \\[2ex] = \dfrac{d\dot{q}_1}{-\Gamma_1 - D_1 - \Pi_1 + F_1 + R_1} = \dfrac{d\dot{q}_n}{-\Gamma_n - D_n - \Pi_n + F_n + R_n} = dt\end{array}\right\} \tag{1.5.28}'$$

representing the first integral lines in $\mathbb{R}^{2n}$, in general non-unique, see Section 2.1. It thus follows that by virtue of (1.5.28) such lines may be equivalently investigated on n *projection planes* $0q_i\dot{q}_i$, $i = 1,\dots,n$. We shall return to this topic in Section 2.1.

With the DOF established and the generalized or *lagrangian* coordinates selected accordingly, we may now represent the system in terms of the schematic diagram shown in Fig. 1.20(a), see also Example 1.1.2 and Fig. 1.7. The inertial objects, point masses m_i, each with a single DOF, are interconnected - each with all others - by a combination of force characteristics. An object in equilibrium position $q_i = 0$ is shown in Fig. 1.20(b). Depending upon the organization of the model, the characteristic must be divided into those connecting the inertial objects to the frame, call them "eigen" characteristics for want of a better word, and those connecting two arbitrary objects thus *coupling* the corresponding DOF, called coupling characteristics. By the same argument as in justifying (1.5.14), we may write

$$\Gamma_i(\bar{q},\dot{\bar{q}}) = \Gamma_{ii}(q_i,\dot{q}_i) + \sum_j \Gamma_{ij}(q_i,q_j,\dot{q}_i,\dot{q}_j) \tag{1.5.29}$$

$$D_i(\bar{q},\dot{\bar{q}}) = D_{ii}(q_i,\dot{q}_i) + \sum_j D_{ij}(q_i,q_j,\dot{q}_i,\dot{q}_j) \ , \tag{1.5.30}$$

$$\Pi_i(\bar{q}) = \Pi_{ii}(q_i) + \sum_j \Pi_{ij}(q_i,q_j) \tag{1.5.31}$$

with the subscripts ii denoting the eigen-characteristics and ij denoting the coupling characteristics, see Fig. 1.20(a). The latter will be made dependent of the relative displacements and velocities between the DOF concerned, i.e. $(q_i - q_j),(\dot{q}_i - \dot{q}_j)$, wherever it applies.

Note that the form of (1.3.30), (1.3.31) allows for an arbitrary way of coupling the characteristics, that is, for all possible combinations of the interconnections shown in Fig. 1.14 and briefly discussed in Section 1.3.

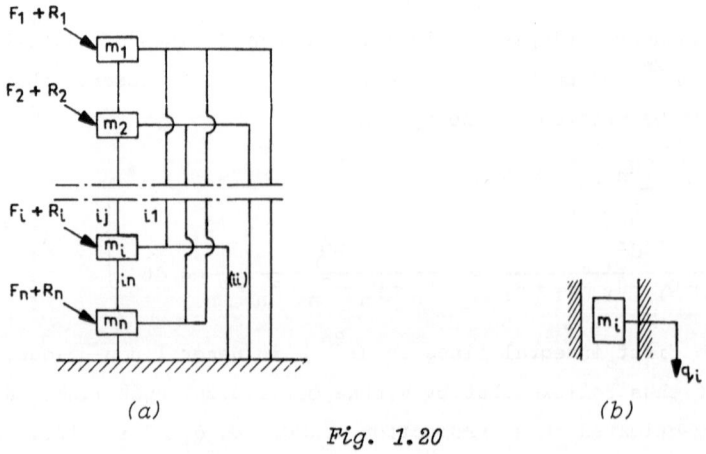

(a)   *(b)*

Fig. 1.20

One of the possible versions of combining the characteristics, both connecting to the frame and coupling, is illustrated in Fig. 1.21. It shows a parallel connection in the Voigt-like manner, summing up the influence of the forces.

The connection is similar to that for the Cartesian forces in Fig. 1.14, with $K_{ij}(q_i - q_j)$ symbolizing the coupling spring forces and with $D_{ij}(q_i - q_j, \dot{q}_i - \dot{q}_j)$ at the present stage symbolizing successively all the types of damping: positive viscous, negative viscous and dry friction.

The spring force characteristics $K_i(\cdot)$ represent usually an *ideally restitutive* spring (without damping) while all the damping effects are modelled by shifting them into $D_{ij}(\cdot)$. A good example here is the wheel suspension shown in Fig. 1.22. The real spring characteristic has obviously a hysteresis loop, since the tire restitutes the force with some lag.

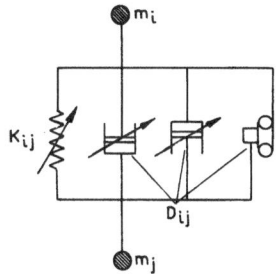

Fig. 1.21

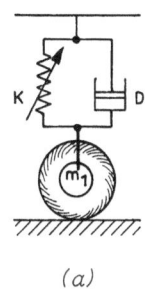

(a)

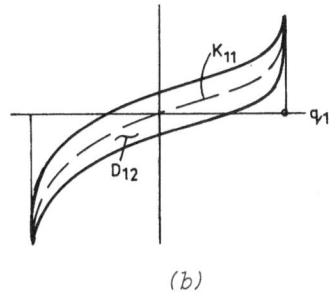

(b)

Fig. 1.22

Thus, in physical reality, damping and spring forces should be combined and presented by a joint function shown in Fig. 1.22(b), with the area of the hysteresis influenced by damping (depending upon the velocity $\dot{q}_i$) which defines the thickness of the loop, while the spring forces generate its length.

Instead, we model the connection by idealized restitution (adiabatic curve, shown as dashed in Fig. 1.22(b)), added to the spring K which results in the characteristic $K_1(\bar{q}) = aq_1 + bq_1^3$, a,b > 0, and by the damping characteristic $D_1(|q_1|, \dot{q}_1)$ measured by the area of the hysteresis loop or by $|q_1|$, recall Example 1.1.1. In almost all cases, the damping added to idealized restitution is measured by the amplitude $|\bar{q}|$ of the motion involved.

The idealized restitution on some region of $\bar{q}$ naturally requires that the direction of the responding force be opposite to the initial deflection, or in terms of characteristics,

$$K_{ii}(q_i) \cdot q_i \geq 0 , \tag{1.5.32}$$

$$K_{ij}(q_i - q_j) \cdot (q_i - q_j) \geq 0 .\qquad(1.5.33)$$

The first inequality refers to the eigen deflections of the objects, the second to the relative deflections in coupling. Obviously, in sum,

$$K_i(\bar{q})\bar{q} \geq 0 ,\qquad(1.5.34)$$

which specifies the region of restitution. We shall return to these properties in Chapter 2. Note here that analogously to Π_i, we have

$$K_i(\bar{q}) = K_{ii}(q_i) + \sum_j K_{ij}(q_i - q_j) , \quad i,j = 1,\ldots,n, \quad j \neq i .\qquad(1.5.35)$$

Similar to the kinetic energy, in general, the potential energy may be nonstationary. If it is not, it can always be reduced to the square form

$$V(\bar{q}) = q^T \tilde{K}(\bar{q})\, \bar{q}\qquad(1.5.36)$$

where $\tilde{K} = [\partial^2 V / \partial q_i \partial q_j]$ is the $n \times n$ *elasticity matrix* representing the organization of spring characteristics in the system. With $\tilde{K} = \text{const}$ the vector $\bar{K} = (K_1,\ldots,K_n)^T$ is measured by the function $\tilde{K}\bar{q}$. It means that spring characteristics become linear and will, as we shall see later, generate only one equilibrium of the Lagrangian model. A very similar description may be introduced for damping

$$\bar{D}(\bar{q},\dot{q}) = \tilde{D}(\bar{q},\dot{q})\dot{\bar{q}}\qquad(1.5.36)'$$

with $\tilde{D}(\bar{q},\dot{q})$ the *matrix damping coefficient*.

When we want to split the Cartesian model into trees or chains of trees which may be interdependent, the couplings of the schematic diagram reduce, but unless we deal with a specific case, see Example 1.1.2, the general diagram stays the same. On the other hand, we will have a need to distinguish the motion equation governing particular chains, cf. (1.4.7)'. To do so, the configuration vector $\bar{q}(t)$ of the system is decomposed into m particular chain vectors: $\bar{q}(t) \triangleq (\bar{q}^1(t),\ldots,\bar{q}^m(t))^T \in \mathbb{R}^n$ with each such *chain-configuration vector* $\bar{q}^j(t) = (q_1^j(t),\ldots,q_{n^j}^j(t))^T \in \mathbb{R}^{n^j}$, $\sum_j n^j = n$, $j = 1,\ldots,m$. Corresponding decomposition is applied to $\dot{\bar{q}}$ and $\ddot{\bar{q}}$, as well as to the force characteristics. Then the motion equations (1.5.20) become

$$\ddot{q}_i^{\,j} + \bar{\Gamma}_i^{\,j}(\bar{q},\dot{q}) + \bar{D}_i^{\,j}(\bar{q},\dot{q}) + \bar{\Pi}_i^{\,j}(\bar{q}) = \bar{F}_i^{\,j}(\bar{q},\dot{q},u^j) + \bar{R}_i^{\,j}(\bar{q},\dot{q},t) ,\qquad(1.5.37)$$

$$i = 1,\ldots,n^j, \quad j = 1,\ldots,m .$$

Here the j-*chain force characteristics* $\Gamma_i^{\,j}(\cdot)$, $\Delta_i^{\,j}(\cdot)$, $\Pi_i^{\,j}(\cdot)$ have the same structure as defined in (1.5.29) - (1.5.31). Due to the possible dependence between chains, each inertial object m_i^j may have several successors in the tree concerned and, in general, each chain coupling force characteristic may depend on an arbitrary number of components of the vectors $\bar{q}(t)$ and $\dot{\bar{q}}(t)$, thus must be related to the latter two full vectors. The system control vector $\bar{u}(t)$ is also decomposed into chain-control vectors $\bar{u}^j(t) = (u_1^j(t),\ldots,u_{r^j}^j)^T \in \mathbb{R}^{r^j}$, $j = 1,\ldots,m$, with $\mathbb{R}^{r^1} \times \ldots \times \mathbb{R}^{r^m} = \mathbb{R}^r$, and we assume each chain to be actuated by a separate $\bar{u}^j(t)$, although these controls may be later generated by coupled programs.

Similarly as in Section 1.4, if we go a little further and assume the Cartesian model in terms of a set of m *independent chains*, the Lagrangian model will have the coupling characteristics still more reduced, to connections between an object i and its immediate predecessor i - 1 and successor i + 1, only, while the eigen-characteristics of the chain all stay the same. The schematic diagram of such an independent chain may be obtained from the general case of Fig. 1.20(a) by deleting all connections ij for each m_i, except those for $j = i-1,i+1$. The motion equations (1.5.20) written in the independent chain related format become:

$$
\left.
\begin{aligned}
&\ddot{q}_i^j + \Gamma_i^j(q_{i-1}^j, q_i^j, q_{i+1}^j, \dot{q}_{i-1}^j, \dot{q}_i^j, \dot{q}_{i+1}^j) \\
&+ D_i^j(q_{i-1}^j, q_i^j, q_{i+1}^j, \dot{q}_{i-1}^j, \dot{q}_i^j, \dot{q}_{i+1}^j) + \Pi_i^j(q_{i-1}^j, q_i^j, q_{i+1}^j) \\
&= F_i^j(q_i^j, \dot{q}_i^j, \bar{u}^j) + R_i^j(q_i^j, \dot{q}_i^j, t), \quad i = 1,\ldots,n^j, \quad j = 1,\ldots,m
\end{aligned}
\right\} \quad (1.5.38)
$$

Here the particular characteristics also will have the format (1.5.29) – (1.5.31), but the summation over j will reduce to two values $j = i-1,i+1$. The reader may in fact observe that (1.5.38) are immediately obtained from (1.5.37) by the above reduced summation and reduction of the vectors $\bar{q},\dot{\bar{q}}$. So there is no need to study (1.5.38) separately. The vectorial forms of (1.5.37) and (1.5.38) are obtainable immediately and it will be assumed that the reader can derive the corresponding notation himself.

1.6 WORK ENVELOPE, TARGETS AND ANTITARGETS

Not all points of *Oxyz* may be reached when either of the two networks in Sections 1.3 and 1.4 moves. The set of all reachable points in this space is usually called a *work envelope* or a Cartesian *work region* W in *Oxyz*. The size of such an envelope must obviously depend on the type of the structure concerned, the suspension points and the constraints imposed, and cannot be specified generally. On the other hand, it certainly is the union of the work envelopes W_i , $i = 1,\ldots,M$, of the particular bodies Bi in the model. If we distinguish particular independent chains j , each with the work envelope W^j , then such an envelope is the union of W_i^j for all bodies in the chain j , and W is the union of all W^j , $j = 1,\ldots,m$.

The sets W^j are at least connected, for otherwise a single chain could not operate on them. The extent of W^j is determined by the static, i.e. equilibrium, configuration in which the top body assumes maximum extension from Bl in the direction of some activity (all forces admitted except gravity and inertia). Since the moment of the force about each hinge axis must be zero in the static configuration, the line of action must intersect all the revolute hinge axes. This conclusion forms the basis of the algorithm for the estimate of the work envelope given in Kunnar-Waldron [1]. Another way of stating the zero-moment condition is that the force and the hinge axes form a reciprocal screw system in the static equilibrium position, see Desa-Roth [1], and the Jacobian of the Cartesian to Lagrangian coordinate space transformation vanishes at the boundary of W^j , see Skowronski [31]. The latter criterion obviously refers to the independent chain structures only.

When the regions W^j are disjoint, each chain operates without the danger of collision with others and there is no obvious need for coordination. In such cases we may select in the work envelope a subregion with each point reachable by the chain top, from an arbitrary direction. We call such a subregion a *dextrous work envelope*, borrowing the phrase from robotics. The remaining part is called *secondary*. The situation becomes different when W^j intersect, we then need *coordination control*, see later text.

Heuristic observation may show that W^j of an independent chain is in fact the work envelope of the top body BM^j , still being the union of all W_i^j , $i = 1,\ldots,M^j$.

The points of W , which are to be attained under a selected controller within a control objective, form the given bounded set $T \subset W$ called the *Cartesian target*. This set does not have to be connected and may take various forms - from a set of points, to a body, a path (curve), or a corridor (several paths), in $Oxyz$. Another part of the control objective may relate to avoidance of some *Cartesian anti-target* T_A , specified in more or less the same way as T . It may be any obstacle stationary or moving in W , and taking various forms in $Oxyz$, from a set of points or a body to a path or a set of paths (corridor). For particular chains j and particular bodies i we may have specified subtargets T_i^j , and sub-anti-targets T_{Ai}^j denoted the same way as the work envelopes W_i^j .

Let us now consider the union Δ_q , in the Configuration Space $\mathbb{R}^n$, of all subsets whose images under transformation (1.5.2) lie in W . Such Δ_q will be bounded by extra restrictions imposed upon the configuration vectors $\bar{q}(t)$ and we shall call it the *configuration envelope* for the system. For instance, the double pendulum of Example 1.1.3 and Fig. 1.8 has its Cartesian coordinates x_1, y_1 and x_2, y_2 calculated from $q_1 = \theta_1$, $q_2 = \theta_2$ through the obvious trigonometric relations that specify (1.5.2) of this case. Due to nonuniqueness of the functions sin and cos , there may be various points $(q_1, q_2) \in \mathbb{R}^2$ corresponding to a particular point in $Oxyz$, but the configuration envelope Δ_q will be assumed to enclose all such points.

When moving over Δ_q , the system will have some velocity restrictions, which in sum will, at each $\bar{q}$, specify the bounded set $\Delta_{\dot{q}}$ in the tangent space to $\mathbb{R}^n$. Again such a maximal set $\Delta_{\dot{q}}$ will be called the *velocity envelope* of the system. Thus in general, the point $(\bar{q}, \dot{\bar{q}})$ representing the motion of the system will move anywhere in $\Delta_q \times \Delta_{\dot{q}}$, in the *phase space* $\mathbb{R}^{2n}$. Very similarly to the Cartesian case, particular bodies $i = 1, \ldots, M$ and chains $j = 1, \ldots, m$ will have their own configuration and velocity envelopes $\Delta_{qi}^j, \Delta_{\dot{q}i}^j$, while $\Delta_q^j = \cap_i \Delta_{qi}^j$, $\Delta_{\dot{q}}^j = \cap_i \Delta_{\dot{q}i}^j$.

The *inverse kinematics* generating the inverse functions (1.5.2) [compare also inverses of (1.4.2), (1.4.3)] produces some sets in $\Delta_q \times \Delta_{\dot{q}}$ that correspond to the Cartesian target T , subject to suitable velocity intervals that are aimed at for the motion over T . It usually suffices to choose one of such sets as the *configuration target* $T_q \subset \Delta_q$ with the corresponding velocity target $T_{\dot{q}} \subset \Delta_{\dot{q}}$ which together will form the phase-space target $T \subset \Delta_q \times \Delta_{\dot{q}} \subset \mathbb{R}^{2n}$ for the system. The target is a *designed, bounded* and *closed* subset of $\mathbb{R}^{2n}$, such that after the transformations (1.5.2), (1.5.3) applied to its points we obtain the desired position of the system in the Cartesian target T subject to the velocities required by the objective. It obviously does not mean that the same T and corresponding Cartesian velocities could not have been obtained by attaining a different T in $\mathbb{R}^{2n}$. It is clearly visible on our double pendulum of Example 1.1.3, where the given position of m_2 may be equally well reached using configuration I and using the alternative configuration II, see Fig. 1.23. The first is attained by reaching the target $T_{qI} : q_1 = q_1'$, $q_2 = q_2'$, the second by reaching the target $T_{qII} : q_2 = q_2''$, $q_2 = q_2''$.

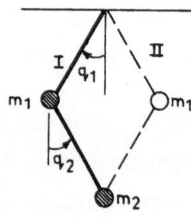

Fig. 1.23

The velocities may in turn form a set in the velocity envelope or, as we may say, the *velocity target* $T_{\dot{q}}$. The technique used to transfer the Cartesian anti-targets into the *configuration and velocity antitargets* T_{Aq}, $T_{A\dot{q}}$ respectively, is very similar except that this time we want to avoid all the images under the transfer concerned, so that $T_{Aq} \times T_{A\dot{q}}$ $\subset \Delta_q \times \Delta_{\dot{q}}$ must cover all the configurations that may lead to the Cartesian antitarget T_A.

EXAMPLE 1.6.1. Consider now the example in Section 1.4 shown in Fig. 1.18, with all DOF locked except for two: $q_1 = \theta$, $q_2 = r_{02}$. The Cartesian target T is then defined as the point (x^f, z^f) in Oxz, while the Cartesian antitarget T_A is specified by the following strip in Oxz: $a \leq x \leq b$, $c \leq z \leq d$. The transformation (1.5.2), (1.5.3) is now defined by (1.4.2), (1.4.3):

$$\left. \begin{array}{ll} x = q_2 \cos q_1 \,, & \dot{x} = \dot{q}_2 \cos q_1 - q_2 \dot{q}_1 \sin q_1 \,, \\ z = q_2 \sin q_1 \,, & \dot{z} = \dot{q}_2 \sin q_1 + q_2 \dot{q}_1 \cos q_1 \,. \end{array} \right\} \tag{1.6.1}$$

The inverse kinematics means to calculate $q_1, q_2, \dot{q}_1, \dot{q}_2$ from (1.6.1):

$$q_1 = \arctan (\frac{z_2}{x_2}) = \arctan (\frac{z_1}{x_1}) \,. \tag{1.6.2}$$

$$q_2 = (x_2^2 + z_2^2)^{\frac{1}{2}} \,. \tag{1.6.3}$$

Substituting the Cartesian target, we obtain

$$q_1 = \arctan (z^f/x^f) \,, \qquad q_2 = [(x^f)^2 + (z^f)^2]^{\frac{1}{2}} \tag{1.6.4}$$

which produces a sequence of values for q_1. The configuration target T_q

must thus be designed based on that sequence. It is virtually our free choice which of the values q_1^f will be taken, while q_2^f is well specified by (1.6.4). For the antitarget T_A we must substitute the band $x \in [a,b]$, $z \in [c,d]$ into (1.6.2), (1.6.3):

$$
\left.
\begin{aligned}
\arctan (c/b) \le q_1 \le \arctan (d/a) \\
(a^2 + c^2)^{\frac{1}{2}} \le q_2 \le (b^2 + d^2)^{\frac{1}{2}}
\end{aligned}
\right\} \tag{1.6.5}
$$

which is to be avoided with all its points, i.e. (1.6.5) defines T_{Aq}.

To invert the velocities in (1.6.1), we differentiate (1.6.2), (1.6.3) with respect to time and obtain

$$
2 q_2 \dot{q}_2 = 2 x_2 \dot{x}_2 + 2 z_2 \dot{z}_2 ,
$$
$$
\dot{q}_1 \sec^2 q_1 = - (z_2/x_2^2) \dot{x}_2 + (1/x_2) \dot{z}_2
$$

which yields

$$
\left.
\begin{aligned}
\dot{q}_1 &= \frac{-z_2}{x_2^2 + z_2^2} \dot{x}_2 + \frac{x_2}{x_2^2 + z_2^2} \dot{z}_2 \\
\dot{q}_2 &= \frac{x_2}{\sqrt{x_2^2 + z_2^2}} \dot{x}_2 + \frac{z_2}{\sqrt{x_2^2 + z_2^2}} \dot{z}_2
\end{aligned}
\right\} . \tag{1.6.6}
$$

To specify the case, let us require that we want the system to be at rest over (x^f, z^f), i.e. that the velocity target $T_{\dot{q}}$ is defined by $\dot{x}_2 = \dot{z}_2 = 0 \Rightarrow \dot{q}_1 = 0$, $\dot{q}_2 = 0$. Then $T = T_q \times T_{\dot{q}}$, with T_q defined by some values q_1 and all q_2 of (1.6.4) and $T_{\dot{q}} : \dot{q}_1 = 0$, $\dot{q}_2 = 0$, is the phase space target of our example. Similarly, specific restrictions upon velocities $\dot{x}_2$, $\dot{z}_2$ inserted into (1.6.6) give the velocity antitarget $T_{A\dot{q}}$ which together with T_{Aq} defined by (1.6.5) produce the phase-space antitarget $T_A = T_{Aq} \times T_{A\dot{q}}$. □

In general, the targets and antitargets in phase-space may not resemble the geometric nature of their Cartesian counterparts. When the transformation (1.5.2) is non-stationary (rheonomic constraints) we must determine the proposed configuration and velocity targets either in a specific time interval in $\mathbb{R}^{2n+1}$, or for all $t \in \mathbb{R}$. The same refers to the anti-targets. Both cases will be illustrated later in particular cases.

1.7 HAMILTONIAN MODEL

The decoupling of the Lagrangian system (1.5.19) that leads to the inertia free coefficients of (1.5.20) is quite often needed - particularly for the applications of control theory - but not so easily obtainable. The latter is mostly due to difficult calculations of the force characteristics (1.5.21) - (1.5.25). We obtain the dynamically decoupled format of the motion equations immediately, if we apply the Hamiltonian model. An extra advantage here lies in the fact that such a model is formed in terms of a scalar function, which is almost always the (calculable) total energy of the system. Thus we may more easily extend the model to the continuous media case. The passage between our Lagrangian equations and the Hamiltonian model is done using the *Legendre transformation* of classical Mechanics. We introduce the new variable

$$p_i = p_i(\bar{q}, \dot{\bar{q}}, t) \triangleq \frac{\partial T(\bar{q}, \dot{\bar{q}}, t)}{\partial \dot{q}_i} \quad , \qquad i = 1, \ldots, n \qquad (1.7.1)$$

called the *generalized momentum*. Then we can prove, under fairly general conditions, see Banach [1], that the relation (1.7.1) can be resolved with respect to $\dot{q}_i$, $i = 1, \ldots, n$:

$$\dot{q}_i = \phi_i(\bar{q}, \bar{p}, t) , \qquad i = 1, \ldots, n \qquad (1.7.2)$$

where $\bar{p} \triangleq (p_1, \ldots, p_n)^T \in \Delta_{\dot{q}}$. Substituting (1.7.2) into $L(\bar{q}, \dot{\bar{q}}, t)$ we obtain $L = \Phi(q_1, \ldots, q_n, \phi_1, \ldots, \phi_n, t)$. Differentiating,

$$\left.\begin{aligned}
\frac{\partial L}{\partial q_i} &= \frac{\partial \Phi}{\partial q_i} + \sum_j \frac{\partial \Phi}{\partial \dot{q}_j} \frac{\partial \dot{q}_j}{\partial q_i} \quad , \\
\frac{\partial \Phi}{\partial p_i} &= \sum_j \frac{\partial \Phi}{\partial \dot{q}_j} \frac{\partial \dot{q}_j}{\partial p_i} \quad .
\end{aligned}\right\} \qquad (1.7.3)$$

Since the potential energy $V(\bar{q})$ does not depend upon $\dot{q}_i$, (1.7.1) gives

$$p_i = \frac{\partial L(\bar{q}, \dot{\bar{q}}, t)}{\partial \dot{q}_i} \quad .$$

Hence by (1.7.1) and $L = \Phi(q_1, \ldots, q_n, \phi_1, \ldots, \phi_n, t)$ we obtain $\partial \Phi / \partial \dot{q}_i = p_i$. From (1.7.3),

$$\sum_j p_j \frac{\partial \dot{q}_j}{\partial q_i} = 0 , \qquad \frac{\partial \Phi}{\partial p_i} = \sum_j p_j \frac{\partial \dot{q}_j}{\partial p_i} \quad . \qquad (1.7.4)$$

Defining now the Hamiltonian function

$$H(\bar{q}, \bar{p}, t) \triangleq \sum_j p_j \dot{q}_j - L(\bar{q}, \dot{\bar{q}}, t) \qquad (1.7.5)$$

and assuming that $\dot{q}_j$, $L(\cdot)$ are functions of $\bar{q},\bar{p},t$, i.e. that they are specified by $\phi_i(\cdot),\Phi(\cdot)$, we have

$$H = \sum_j p_j \dot{q}_j - \Phi(\bar{q},\phi_1,\ldots,\phi_n,t) . \tag{1.7.5}$$

Then

$$\left.\begin{aligned}
\frac{\partial H}{\partial q_i} &= \sum_j p_j \frac{\partial \dot{q}_j}{\partial q_i} - \frac{\partial \Phi}{\partial q_i} , \\[2em]
\frac{\partial H}{\partial p_i} &= \dot{q}_i + \sum_j p_j \frac{\partial \dot{q}_j}{\partial p_i} - \frac{\partial \Phi}{\partial p_i}
\end{aligned}\right\} \tag{1.7.6}$$

and by (1.7.4)

$$\frac{\partial H}{\partial q_i} = -\frac{\partial \Phi}{\partial q_i} , \quad \frac{\partial H}{\partial p_i} = \dot{q}_i . \tag{1.7.7}$$

The Lagrange equations (1.5.17) give $\dot{p}_i = \partial L/\partial q_i + Q_i$ wherefrom by $L = \Phi(\bar{q},\phi_1,\ldots,\phi_n,t)$, we have $\partial \Phi/\partial q_i = \dot{p}_i$. Thus by (1.7.7),

$$\left.\begin{aligned}
\dot{q}_i &= \frac{\partial H(\bar{q},\bar{p},t)}{\partial p_i} , \\[2em]
\dot{p}_i &= -\frac{\partial H(\bar{q},\bar{p},t)}{\partial q_i} + Q_i(\bar{q},\bar{p},t) .
\end{aligned}\right\} \tag{1.7.8}$$

In the above, Q_i represent the sum of energy changing forces in (1.5.17) with (1.7.2) substituted. Thus, in general, we obtain

$$\left.\begin{aligned}
\dot{q}_i &= \frac{\partial H(\bar{q},\bar{p},t)}{\partial p_i} , \\[2em]
\dot{p}_i &= -\frac{\partial H(\bar{q},\bar{p},t)}{\partial q_i} + Q_i^D(\bar{q},\bar{p},t) + Q_i^F(\bar{q},\bar{p},\bar{u},t) + Q^R(\bar{q},\bar{p},t)
\end{aligned}\right\} \tag{1.7.9}$$

forming the *canonical motion equations* of our Hamiltonian model.

Note that the equations (1.7.9) are inertially decoupled, first order differential equations, while at the same time their right hand sides represent forces, not characteristics (forces per inertia). Both these features are distinct advantages of the Hamiltonian model. Note also that for scleronomic constraints when the kinetic energy is not an explicit function of time, and so $L(\cdot)$ and $H(\cdot)$ are not such functions, the system simplifies to the case of $p_i = \phi_i(\bar{q},\dot{\bar{q}})$ generating

$$\left.\begin{aligned}
\dot{q}_i &= \frac{\partial H(\bar{q},\bar{p})}{\partial p_i} , \\[2em]
\dot{p}_i &= -\frac{\partial H(\bar{q},\bar{p})}{\partial q_i} + Q_i^D(\bar{q},\bar{p}) + Q_i^F(\bar{q},\bar{p},\bar{u}) + Q^R(\bar{q},\bar{p},t) ,
\end{aligned}\right\} \tag{1.7.10}$$

$$i = 1,\ldots,n ,$$

which is the most frequently used form of the Hamiltonian model.

EXAMPLE 1.7.1. Consider the ceiling suspended two DOF, PR-manipulator (prismatic, rotary) shown in Fig. 1.24. It works on a horizontal bench. The amalgamated mass centers of links and joints, that is, the positions of m_i, are calculated and located at a fixed distance on the link axis. We also recognize the contribution of the actuator inertia separately. The spring suspending the first link to the base is assumed linear, in view of the design which allowed to take it with constant diameter and constant cross section of the coils, as well as performing relatively small amplitudes of deflection $\bar{r}(t)$.

Following Fig. 1.24, we have

$$\bar{r}_{01} = x_1 \bar{i} , \quad \bar{r}_{02} = (x_1 + \ell_1 + \ell_2 \cos \theta) \bar{i} + (\ell_2 \sin \theta) \bar{j} ,$$

and thus

$$\dot{\bar{r}}_{01} = \dot{x}_1 \bar{i} , \quad \dot{\bar{r}}_{02} = (\dot{x}_1 - \ell_2 \dot{\theta} \sin \theta) \bar{i} + (\ell_2 \cos \theta \cdot \dot{\theta}) \bar{j} .$$

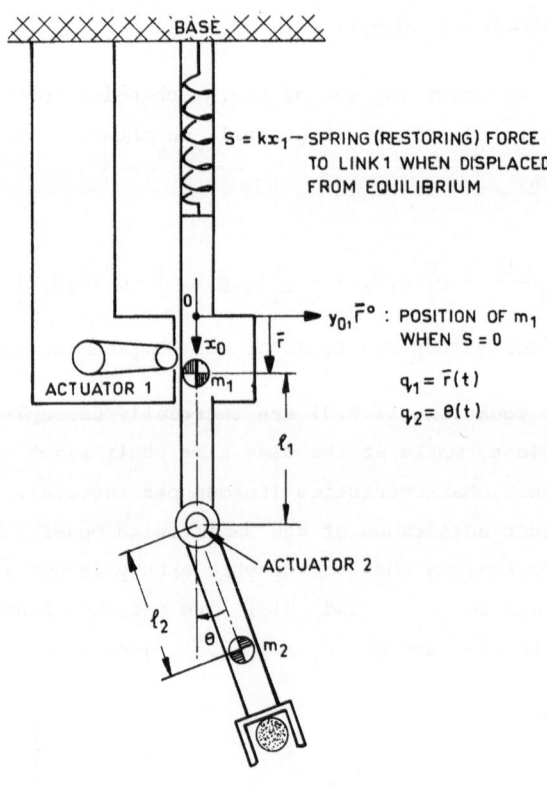

Fig. 1.24

We calculate the kinetic energy for a particular m_i :

$$T_1 = \frac{1}{2} m_1 \dot{\bar{r}}_{01} \cdot \dot{\bar{r}}_{01} = \frac{1}{2} m_1 \dot{q}_1^2 \ ,$$

$$T_2 = \frac{1}{2} m_2 (\dot{q}_1^2 - 2\ell_2 \dot{q}_1 \dot{q}_2 \sin q_2 + \ell_2^2 \dot{q}_2^2) \ ,$$

and then for the actuators

$$T_{1a} = \frac{1}{2} m_{a1} \dot{q}_1^2 \ , \qquad T_{2a} = \frac{1}{2} I_{a2} \dot{q}_2^2 \ ,$$

where I_{a2} is the rotary actuator inertia coefficient. Hence, the total kinetic energy for the manipulator is

$$T = \frac{1}{2}(m_1 + m_{a1} + m_2)\dot{q}_1^2 - m_2 \ell_2 \dot{q}_1 \dot{q}_2 \sin q_2 + \frac{1}{2}(m_2 \ell_2^2 + I_{a2})\dot{q}_2^2 \ , \tag{1.7.11}$$

or in matrix-vector form $T = \frac{1}{2} \dot{\bar{q}}^T M(\bar{q}) \dot{\bar{q}}$, where

$$M(\bar{q}) = \begin{pmatrix} m_1 + m_{a1} + m_2 & -m_2 \ell_2 \sin q_2 \\ -m_2 \ell_2 \sin q_2 & m_2 \ell_2^2 + I_{a2} \end{pmatrix} \tag{1.7.12}$$

is the inertia matrix of our case. Similarly we calculate the potential energy to obtain

$$V = \frac{1}{2} k q_1^2 - 9.81(m_1 + m_2)q_1 - 9.81 m_2 (\ell_2 \cos q_2 + \ell_1) \ . \tag{1.7.13}$$

Hence the Lagrangian $L = T - V$ is

$$\left. \begin{aligned} L = \frac{1}{2} M_{11} \dot{q}_1^2 + M_{12}(q_2) \dot{q}_1 \dot{q}_2 + \frac{1}{2} M_{22} \dot{q}_2^2 + \frac{1}{2} k q_1^2 - 9.81(m_1 + m_2)q_1 \\ - 9.81 m_2 (\ell_2 \cos q_2 + \ell_1) \end{aligned} \right\} \tag{1.7.14}$$

where M_{ij} are the coefficients of the matrix $M(\bar{q})$ of (1.7.12). Introducing the damping forces in each joint as $Q_i^D = \lambda_i |\dot{q}_i| \dot{q}_i$, $\dot{q}_i \neq 0$, $i = 1,2$, and substituting all the above into (1.5.17) we obtain the motion equations

$$\left. \begin{aligned} (m_1 + m_{a1} + m_2)\ddot{q}_1 - m_2 \ell_2 \ddot{q}_2 \sin q_2 - m_2 \ell_2 \dot{q}_2^2 \cos q_2 + k q_1 \\ - 9.81(m_1 + m_2) + \lambda_1 |\dot{q}_1| \dot{q}_1 = Q_1^F \\ (m_2 \ell_2^2 + I_{a2})\ddot{q}_2 - m_2 \ell_2 \ddot{q}_1 \sin q_2 + 9.81 m_2 \ell_2 \sin q_2 + \lambda_2 \dot{q}_2 |\dot{q}_2| \\ - m_2 \ell_2 \dot{q}_1 \dot{q}_2 \cos q_2 = Q_2^F \end{aligned} \right\} \tag{1.7.15}$$

or in vector form

$$\left. \begin{aligned} \begin{pmatrix} m_1 + m_{a1} + m_2 & -m_2 \ell_2 \sin q_2 \\ -m_2 \ell_2 \sin q_2 & m_2 \ell_2^2 + I_{a2} \end{pmatrix} \begin{pmatrix} \ddot{q}_1 \\ \ddot{q}_2 \end{pmatrix} + \begin{pmatrix} \lambda_1 |\dot{q}_1| \dot{q}_1 \\ \lambda_2 |\dot{q}_2| \dot{q}_2 \end{pmatrix} \\ + \begin{pmatrix} k q_1 - 9.81(m_1 + m_2) \\ 9.81 m_2 \ell_2 \sin q_2 \end{pmatrix} \begin{pmatrix} -m_2 \ell_2 \dot{q}_2 \cos q_2 \\ -m_2 \ell_2 \dot{q}_1 \dot{q}_2 \cos q_2 \end{pmatrix} = \begin{pmatrix} Q_1^F \\ Q_2^F \end{pmatrix} \ . \end{aligned} \right\} \tag{1.7.16}$$

This is rather difficult to decouple inertially which, as mentioned, is needed for the control theory study. We may thus use our alternative Hamiltonian model.

From (1.7.1) and $\partial L/\partial \dot{q}_i = \partial T/\partial \dot{q}_i$, $i = 1,2$, we have

$$
\left.
\begin{aligned}
p_1 &= \frac{\partial L}{\partial \dot{q}_1} = M_{11}\dot{q}_1 + M_{12}(q_2)\dot{q}_2 \quad , \\[2mm]
p_2 &= \frac{\partial L}{\partial \dot{q}_2} = M_{12}(q_2)\dot{q}_1 + M_{22}\dot{q}_2 \quad ,
\end{aligned}
\right\}
\tag{1.7.17}
$$

with M_{ij} specified by (1.7.12). Then also, see (1.7.2),

$$
\dot{q}_1 = \frac{M_{22}p_1 - M_{12}(q_2)p_2}{\det M} \quad , \qquad \dot{q}_2 = \frac{-M_{12}(q_2)p_1 + M_{11}p_2}{\det M} \quad .
\tag{1.7.18}
$$

Substituting (1.7.18) into (1.7.11), we have

$$
T = \frac{1}{2(\det M)^2} \left\{ M_{11}\left[\det\begin{pmatrix} p_1 & M_{12} \\ p_2 & M_{22} \end{pmatrix}\right]^2 + 2M_{12}(q_2)\det\begin{pmatrix} p_1 & M_{12} \\ p_2 & M_{22} \end{pmatrix}\det\begin{pmatrix} M_{11} & p_1 \\ M_{12} & p_2 \end{pmatrix} \right.
$$
$$
\left. + M_{22}\left[\det\begin{pmatrix} M_{11} & p_1 \\ M_{12} & p_2 \end{pmatrix}\right]^2 \right\} \quad ,
$$

which gives

$$
T(\bar{q},\bar{p}) = \tfrac{1}{2} M_{11}^{-1} p_1^2 + M_{12}^{-1} p_1 p_2 + \tfrac{1}{2} M_{22}^{-1} p_2^2
\tag{1.7.19}
$$

where

$$
\left.
\begin{aligned}
M_{11}^{-1} &= M_{22} / \det M \\[2mm]
M_{12}^{-1} &= -M_{12} / \det M \\[2mm]
M_{22}^{-1} &= M_{11} / \det M \quad .
\end{aligned}
\right\}
\tag{1.7.20}
$$

Then, recalling that $H(\bar{q},\bar{p}) = T(\bar{q},\bar{p}) + V(\bar{q})$, we have

$$
\frac{\partial H}{\partial p_1} = M_{11}^{-1}p_1 + M_{12}^{-1}p_2 \quad ; \qquad \frac{\partial H}{\partial p_2} = M_{12}^{-1}p_1 + M_{22}^{-1}p_2 \quad ,
\tag{1.7.21}
$$

$$
\frac{\partial H}{\partial q_1} = \frac{\partial V}{\partial q_1} = kq_1 - 9.81(m_1 + m_2) \quad ;
$$

$$
\frac{\partial H}{\partial q_2} = \frac{\partial T}{\partial q_2} + \frac{\partial V}{\partial q_2} = m_2 \ell_2 (9.81 \sin q_2 - \dot{q}_1 \dot{q}_2 \cos q_2) \quad ,
$$

and also for our specified damping forces,

$$
Q_1^D(\bar{q},\bar{p}) = \lambda_1 [M_{22}p_1 - M_{12}(q_2)p_2] \cdot |M_{22}p_1 - M_{12}(q_2)p_2| \quad ,
\tag{1.7.22}
$$

$$
Q_2^D(\bar{q},\bar{p}) = \lambda_2 [M_{12}(q_2)p_1 - M_{11}p_2] \cdot |M_{12}(q_2)p_1 - M_{11}p_2| \quad .
\tag{1.7.23}
$$

Given $Q_i^F(\bar{q},\dot{\bar{q}})$, $i = 1,2$, we can transform them exactly in the same way. All the above gives the Hamilton canonical equations (1.7.10) for our case as the following equations:

$$\left.\begin{aligned}
\dot{q}_1 &= M_{11}^{-1}(q_2)p_1 + M_{12}^{-1}(q_2)p_2 \\[6pt]
\dot{q}_2 &= M_{12}^{-1}(q_2)p_1 + M_{22}^{-1}(q_2)p_2 \\[6pt]
\dot{p}_1 &= kq_1 - 9.81(m_1 + m_2) - Q_1^D(\bar{q},\bar{p}) + Q_1^F \\[6pt]
\dot{p}_2 &= m_2\ell_2(9.81\sin q_2 - \dot{q}_1\dot{q}_2\cos q_2) - Q_2^D(\bar{q},\bar{p}) + Q_2^F
\end{aligned}\right\} \tag{1.7.24}$$

in the inertially decoupled form. Note that multiplication by the scalars M_{ij}^{-1} is a lot easier than by the matrix M^{-1}. □

1.8 FLEXIBLE BODY MODELS. HYBRID SYSTEM

With the increased speed and presently required precision of work done by mechanical system, the bodies must be made stress resistant and thus heavy, in order to withstand the dynamics. This in turn dictates that the actuators must generate more power, be large and heavy as well. All the above contradicts the lightweight requirement for most of the modern structures, particularly those working in space. The obvious solution in the latter cases is better modelling and advanced control algorithms. It is achieved by allowing flexible objects. The advantages of flexible modelling are many, including faster system response, lower energy consumption, smaller actuators and, in general, trimmer design. The tradeoff, however, is the increased difficulty in control, particularly making it robust against the effects of flexure. On the other hand, realistic modelling requires covering both, the objects which are flexible and those which can, and thus should, be considered rigid, as well as the extra lumped masses and control actuators collocated with the hinges. Then the total system comes under the joint format of a *hybrid* (rigid and flexible) model.

The flexible deflections are usually considered a small (linearizable) perturbation of the motion of the rigid substructure which, due to Coriolis, centrifugal forces and large angles of articulation, must be taken highly nonlinear, see Meirovitch-Quinn [1]. This is the same philosophy as used in the so called *shadow beam* method, see Laskin-Likins-Longman [1]. The method is based on introducing a rigid floating frame with large motion, relative to which small strains in the flexible beam are measured.

The flexibility of the objects may be modelled in two basic ways. Either we may consider elastic beams with *continuous mass distribution*, or we may *discretise the object* by various means, for instance taking it as a set of lumped masses spaced in a specific way along the length of the body. It is at least disputable whether the first representation is more accurate, see Hughes-Skelton [1]. Both models are phenomenological and approximate, and their use depends on the aim of our investigation, recall our discussion of Section 1.2. In terms of calculation, only the discrete model, with finite grid of discretisation, is practical.

A very good review on the recent literature dealing with both continuous and discrete systems may be found in Turcic-Midha [1], Ho-Herber [1], and Jerkovsky [1], Geverter [1], Huston [1].

The continuous mass distribution models may in turn be designed in at least two ways, both already classical in Mechanics. The first way refers to *augmented Lagrange and Hamiltonian models* that cover the hybrid system case. Such models do not have to be linearized, although we often do so for calculational purposes. They are particularly convenient as carriers for our type of control study, namely the Liapunov formalism. The work on such models, their stabilization and control began with Pringle [1,2], Wang [1,2] and Meirovitch [2], Meirovitch-Nelson [1], as well as Budynas-Poli [1], with the extension of the Liapunov formalism based on Movchan [1]. The reader may find all the material needed in the book by Leipholz [1].

The second type of continuous mass distribution models is based on extensions and modifications of the classical Timoshenko beam theory, basically linear. There is obviously considerable literature on the topic. Modern results here belong to Book [1,2], Maizza Neto [1], Book-Maizza Neto-Whitney [1], Hughes [1] and later Seraji [2], Book-Majette [1], Yurkovich-Ozgüner-Tzes-Kotnik [1], Simo – Vu Quoc [1], Piedboef-Hurteau [1] and Ahmed-Lim [1], the latter three attacking the problem nonlinearly.

A spatially discrete model of the flexible body makes the hybrid model uniform, in that it brings the flexible part into line with the rigid part, both described in terms of ordinary differential or difference equations. Here also we may have at least several avenues of modelling: reduction to a set of lumped masses, finite element methods, modal methods and the Ritz and/or Galerkin type series representation of the elastic deformations. The literature on the subject is too wide to quote. There are also excellent monographs on discrete modelling. The recent works close to our aims are those by Hughes-Skelton [1], Sunada-Dubovsky [1,2], Usoto-Nadira-Mahil

[1] and Truckenbrodt [1], the latter particularly referring to the series representation. We shall use such a model in this text, mainly discussing the control results obtained by Skowronski [39,41,43].

The rigid part of the model will be represented by nonlinear equations (1.5.19), with the motions considered *nominal*. Abbreviating

$$Q_i^C(\bar{q},\dot{\bar{q}}) - Q_i^D(\bar{q},\dot{\bar{q}}) \triangleq -Q^{CD}(\bar{q},\dot{\bar{q}}) ,$$

we have

$$\sum_{j=1}^n M_{ji}\ddot{q}_j - Q_i^{CD}(\bar{q},\dot{\bar{q}}) - Q_i^P(\bar{q}) = Q_i^F(\bar{q},\dot{\bar{q}},\bar{u}) + Q_i^R(\bar{q},\dot{\bar{q}},t) . \qquad (1.8.1)$$

To accommodate the elasticity of the bodies, we need an extra set of variables considered deviations from the nominal motion. These arise from inertia and flexibility of the links as well as the hinges. Such *compliant variables* are generated, as mentioned, by small (linear) deformations and represent in some sense the vibration modes. The corresponding equations of motion are coupled with (1.8.1) to produce together our hybrid system.

In order to specify the compliant variables, we introduce the deformation coordinates for the ith body as shown in Fig. 1.25. Then partitioning the length of Bi with the grid $1,\dots,M_i$, each element between grids with small deformation, we apply the Ritz-Kantorowitch series expansion

$$r_i(x_i,t) = \sum_{\nu=1}^M r_{i\nu}(x_i) r_{i\nu}(t) = r_i(x_i) r_i(t) , \qquad (1.8.2)$$

and analogously for $v_i(x_i,t)$, $w_i(x_i,t)$. Here x_i is the spatial coordinate along the body, $r_{i\nu}(x_i)$, $v_{i\nu}(x_i)$ and $w_{i\nu}(x_i)$ are the shape functions (spatial dependence), while $r_{i\nu}(t)$, $v_{i\nu}(t)$, $w_{i\nu}(t)$ are the amplitude functions (time dependence), for a particular grid interval ν . The first form the *link shape vector*

$$\bar{\eta}_i(x_i) \triangleq (r_i(x_i),v_i(x_i),w_i(x_i)) ,$$

the second the *compliant vector*

$$\bar{\eta}_i(t) \triangleq (r_i(t),v_i(t),w_i(t))$$

for the body Bi . The grid $\nu = 1,\dots,M_i$ works as a sequence of imaginary joints on the body Bi . Mathematically (1.8.2) means separation of variables and is well known in the theory of linear partial differential equations. The equality holds for $M_i \to \infty$, when the right hand side converges to $r_i(x_i,t)$. To ensure monotonic convergence, the shape functions must meet some conditions:

(a)   be continuous to ensure continuity of the deformations,

(b)   satisfy boundary conditions between the elements, and

(c)   ensure that, when displacements of the element concerned
      correspond to a constant (possibly zero) strain, this
      constant is accommodated by the shape function.

In general, shape functions are difficult to find in any other case
than the standard long slender rod. And a flexible body in a modern
machine is often far removed from such a rod, which makes our assumption
in Fig. 1.25 rather artificial. We leave it as is, since the simplicity
is instructive. For a background in the method, see one of many monographs
on the Finite Elements Method. Getting into more details on this
subject would take us beyond the intended scope of this book. Note,
however, that owing to our approximation all techniques described later
apply to the flexible case as well.

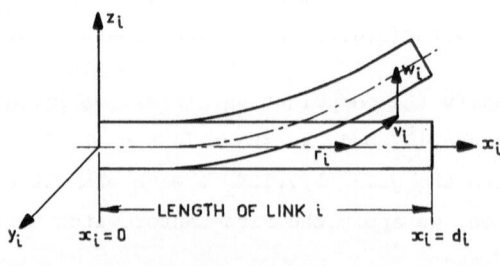

Fig. 1.25

As mentioned, the exact deformation is expected for $M_i \to \infty$. Thus to
justify the linearization physically, one must take M_i sufficiently large.
Technically it means to apply stepwise subdivision of the distances between
grids for as many times as the difference between the obtained successive
approximations becomes acceptably small. Agreeing to this technique, we
shall consider the first step, with links broken up by a single grid-point.
This leads to all (1.8.2) specified jointly by the n - vector of compliant
variables $\bar{\eta}(t) \triangleq (\bar{\eta}_1(t),\ldots,\bar{\eta}_n(t))^T$ where $\bar{\eta}_i(t) \triangleq (\bar{r}_i(t),\bar{v}_i(t),\bar{w}_i(t))$.

Elementary mechanical derivation along the lines shown in Section 1.5
produces the kinetic and potential energies and thus the Lagrangian for the
vectors $(\bar{q},\bar{\eta})^T$, $(\dot{\bar{q}},\dot{\bar{\eta}})^T$ wherefrom we obtain the Lagrange equations of
motion and then the hybrid correspondent of (1.8.1). In the procedure, we
follow Truckenbrodt [1] and obtain the hybrid system in the following
format with untruncated nonlinearities:

$$\begin{bmatrix} M & M_C \\ M_C^T & M_\eta \end{bmatrix}\begin{Bmatrix} \ddot{\bar{q}} \\ \ddot{\bar{\eta}} \end{Bmatrix} + \begin{bmatrix} 0 & D_C \\ 0 & D \end{bmatrix}\begin{Bmatrix} \dot{\bar{q}} \\ \dot{\bar{\eta}} \end{Bmatrix} + \begin{bmatrix} 0 & P_C \\ 0 & P \end{bmatrix}\begin{Bmatrix} \bar{q} \\ \bar{\eta} \end{Bmatrix} + \begin{Bmatrix} \bar{Q}^{CD}(\bar{q},\dot{\bar{q}}) \\ \bar{Q}_\eta^{CD}(\bar{\eta},\dot{\bar{\eta}}) \end{Bmatrix}$$
$$+ \begin{Bmatrix} -\bar{Q}^P(q) \\ -\bar{Q}_\eta^P(\eta) \end{Bmatrix} = \begin{Bmatrix} \bar{Q}^F(\bar{q},\dot{\bar{q}},\bar{u}) \\ 0 \end{Bmatrix} + \begin{Bmatrix} \bar{Q}^R(\bar{q},\dot{\bar{q}},t) \\ 0 \end{Bmatrix} \; . \tag{1.8.3}$$

Here $M_\eta(\bar{\eta})$, $\bar{Q}_\eta^{CD}(\bar{\eta},\dot{\bar{\eta}})$, $\bar{Q}_\eta^P(\bar{\eta})$ are the elastic correspondents of M , $\bar{Q}^{CD}$, $\bar{Q}^P$ of (1.8.1), while $M_C(\bar{q},\bar{\eta})$, $D_C(\bar{q},\dot{\bar{q}},\bar{\eta},\dot{\bar{\eta}})$, $P_C(\bar{q},\bar{\eta})$ and the internal damping coefficient matrix $D(\bar{q},\dot{\bar{q}},\bar{\eta},\dot{\bar{\eta}})$ as well as the hybrid restoring coefficient matrix $P(\bar{q},\bar{\eta})$ represent the coupling between the elastic and joint coordinates. As specified in Truckenbrodt [1], these matrices are formed by integrals over the particular shape functions of each element. Letting

$$M(\bar{q},\bar{\eta}) \triangleq \begin{bmatrix} M & M_C \\ M_C^T & M_\eta \end{bmatrix} \tag{1.8.4}$$

be the *hybrid inertia matrix* of (1.8.3), which is nonsingular and positive definite, we may multiply (1.8.3) by M^{-1}, and substituting the notation

$$\begin{aligned} \bar{D}(\bar{q},\dot{\bar{q}},\bar{\eta},\dot{\bar{\eta}}) &\triangleq M^{-1}[(D_C\dot{\bar{\eta}}+\bar{Q}^{CD})^T, \quad (D\dot{\bar{\eta}}+\bar{Q}_\eta^{CD})^T]^T \\ \bar{P}(\bar{q},\bar{\eta}) &\triangleq M^{-1}[(P_C\bar{\eta}-\bar{Q}^P)^T, \quad (P\bar{\eta}-\bar{Q}_\eta^P)^T]^T \\ \bar{F}(\bar{q},\dot{\bar{q}},\bar{u}) &\triangleq M^{-1}(\bar{Q}^F,0)^T \\ \bar{R}(\bar{q},\dot{\bar{q}},t) &\triangleq M^{-1}(\bar{Q}^R,0)^T \; , \end{aligned} \tag{1.8.5}$$

write (1.8.3) in the inertially decoupled format

$$(\ddot{\bar{q}}^T,\ddot{\bar{\eta}}^T)^T + \bar{D}(\bar{q},\dot{\bar{q}},\bar{\eta},\dot{\bar{\eta}}) + \bar{P}(\bar{q},\bar{\eta}) = \bar{F}(\bar{q},\dot{\bar{q}},\bar{u}) + \bar{R}(\bar{q},\dot{\bar{q}},t) \; . \tag{1.8.6}$$

Obviously $\bar{D}(\cdot)$, $\bar{P}(\cdot)$, $\bar{F}(\cdot)$, $\bar{R}(\cdot)$ do not directly represent forces, but the rates of force per inertia, and as such become the *hybrid system characteristics*. They are 2n - dimensional vectors. With more accurate approximation, the nominal part dimension n will stay, but the dimension of the compliant part will increase to kn , where the integer k shows the density of partitioning the bodies. We have made k the same for all bodies, for convenience. This is by no means necessary, see discussion in Usoto-Nadira-Mahil [1].

So far our partitioning of the bodies is purely technical, based on geometry. Physical arguments may ask for placing the grid between elements i.e. for "flexible hinges" according to the succession of the modes of vibration of the link, say, at modal points. To attain agreement between the latter and our geometry, we may use the method of *assumed modes*,

Meirovitch [1]. We may also justify our truncation, assuming that the amplitudes of higher modes are small compared with the first ones, see Book [1]. The grid $1,\ldots,M_i$, i.e. the modal points, obviously depends upon the *natural modes* of vibration, the shape functions become mode shapes and the amplitude functions become modal displacements. With $\nu = 1,\ldots,M_i$ they build up the variables $\bar{\eta}_i(x_i)$ which may now be called *modal shape vectors* and the variables $\bar{\eta}_i(t)$ called *modal deformation vectors*.

For a more complicated structure of the links, the simple truncation along the natural order, i.e. by increasing frequency as above, may not be adequate. We may be forced to make a selection of modes, see Hughes [1]. There are at least several avenues of the mode selection, of which the most accurate seems to be the mode identification. In classical terms, it means that, granted the output obtained from sensors, one applies either the *mode filtering*, see Meirovitch-Baruch [1], using the orthogonality of eigenfunctions, or the Luenberger-type observers, see Brogan [1], without the orthogonality condition, practically a Luenberger observer for each mode. With the latter, the estimates of the controlled modes are contaminated by the contribution of the unmodelled modes, which is known as the *observation spillover*, see Meirovitch-Baruch [2].

In case studies of the flexible link structures, we may either assume the selection of modes and eigenfunctions as known, or use the modal identification that basically follows the idea of the modal observers, but is implemented in terms of the nonlinear MRAC parameter and state identifiers introduced by Skowronski, [31]. In brief outline, it means the following.

Let $\bar{\eta}(\bar{x}) \triangleq (\bar{\eta}_1(x_1),\ldots,\bar{\eta}_n(x_n))^T$, $\bar{x} \triangleq (x_1,\ldots,x_n)^T$ be the system shape vector. Moreover, let $d_{\nu i}$ be the distance along the body Bi from the joint i to the grid point ν, and let $\bar{d}_i \triangleq (d_{i1},\ldots,d_{iM})$ be the vector of such distances along the body Bi , with the grid location vector in the system represented by the vector $\bar{d} \triangleq (\bar{d}_1,\ldots,\bar{d}_{nm})$. Further we introduce the vector of structural parameters $\bar{k} \triangleq (k_1,\ldots,k_{M_i})$, see Hale-Lisowski [1], each chosen to be proportional to physical quantities such as cross-sectional areas of bar elements, widths of beam elements and thicknesses of membrane or plate elements. Quite naturally the inertia and the restoration matrices in (1.8.3) depend upon the triple $\bar{\eta}(\bar{x}),\bar{d},\bar{k}$. The equations (1.8.6) have been derived with the silent assumption that this triple is known, so that it does not appear explicit under the functions of the hybrid characteristics. In the case of our modal identification, (1.8.6) becomes

$$(\ddot{\bar{q}}^T, \ddot{\bar{\eta}}^T)^T + \bar{\mathcal{V}}(\bar{q}, \dot{\bar{q}}, \bar{\eta}, \dot{\bar{\eta}}, \bar{\eta}(\bar{x}), \bar{a}, \bar{k}) + \bar{P}(\bar{q}, \bar{\eta}, \bar{\eta}(\bar{x}), \bar{a}, \bar{k})$$
$$= \bar{F}(\bar{q}, \dot{\bar{q}}, u) + \bar{R}(\bar{q}, \dot{\bar{q}}, t) \qquad \left.\begin{array}{c} \\ \\ \end{array}\right\} \qquad (1.8.7)$$

with the solutions $[(\bar{q}(t), \bar{\eta}(t))^T, (\dot{\bar{q}}(t), \dot{\bar{\eta}}(t))^T]^T$ dependent upon the quantities $\bar{\eta}(\bar{x}), \bar{a}, \bar{k}$ identified on-line according to the quoted method. It is worth mentioning here that, although the classical Luenberger-type modal observers accrue the observation spillover, such is not the case for our identifiers which, apart from their identifying role, secure a type of stability (Lagrange stability) for the overall system. The reader is referred to Skowronski [31] for details.

No matter which method of hybrid modelling is used, the number of dimensions grows with the presence of flexible bodies. Thus the computing time for solving the equations grows as well, and for a small on-board computer, typical in outdoor working robotic systems and often used elsewhere, may become prohibitively long. This is the case where the observers play a constructive role, if they can be made less dimensional than the plant. The solutions $[(\bar{q}(t), \bar{\eta}(t))^T, (\dot{\bar{q}}(t), \dot{\bar{\eta}}(t))^T]^T$ are simply replaced by those of a suitable identifier for use in the corresponding feedback controller. This role becomes particularly significant when such identifiers can be designed so that they are integrable in closed form. Then the computer works as a calculator with maximal savings of time and effort, see Skowronski [50].

Before closing this section, we must mention two alternatives to the hybrid model (1.8.3). First, let us consider that somebody has assumed the lumped system equations (1.8.1) as independent of any compliant variables, just as it is written in (1.8.1), for whatever such an assumption is worth physically, see Meirovitch-Quinn [1], and Streit-Krousgill-Bajaj [1]. Obtaining the solution of such a lumped system in terms of $\bar{q}(t), \dot{\bar{q}}(t)$, we substitute it into a linear compliant motion equation in the role of time varying parameters, thus obtaining a linear system with parametric excitation (nonautonomous) for the compliant motion, see Sadler [1], Sadler-Sandor [1], Turcic-Midha [1]. It leads to defining $\bar{\eta}(t), \dot{\bar{\eta}}(t)$ needed for the hybrid system, or separately.

The second alternative, of the two mentioned, to writing the full system (1.8.3) is to use the same philosophy as above but inverted. We assume that the compliant equations which are justifiably linear, are solved for $\bar{\eta}(t), \dot{\bar{\eta}}(t)$ with $\bar{q}(t), \dot{\bar{q}}(t)$ as parameters. Then this solution substituted into the lumped equations makes it a parametrically excited

lumped system, which can be discussed instead of the hybrid more dimensional (1.8.6).

Chapter 2
STATE, ENERGY AND POWER

2.1 THE STATE AND PHASE SPACE NOTIONS

If at any instant of time the system dynamics is fully determined by a
finite number of variables $x_1, \ldots, x_N$, we call them *state variables* of the
system. The time instantaneous values of these variables give *information*
on the present state of the system as well as on its past. Thus the state
is represented by a *state vector* $\bar{x}(t) = (x_1(t), \ldots, x_N(t))^T$ corresponding
to a variable point in the *state space* $\mathbf{R}^N$ that describes the motion of the
system. There t is the independent time variable $t \geq t_0$, $\forall\, t_0 \in \mathbf{R}$,
where as before t_0 is the initial instant. We let Δ be a given bounded
set (or its closure) in $\mathbf{R}^N$, representing the formal (mathematical) and
physical constraints imposed upon the state $\bar{x}(t)$. It will often be called
a set of admissible states. The motion may proceed indefinitely in Δ , i.e.
for $t \in \mathbf{R}^+$, $\mathbf{R}^+ = [t_0, \infty)$, or it may terminate at the finite instant
$t_f < \infty$, i.e. $t \in \mathbf{R}^f$, $\mathbf{R}^f = [t_0, t_f]$. The latter is either arbitrary or
stipulated. Correspondingly $\bar{x}^0 \overset{\Delta}{=} \bar{x}(t_0)$, $\bar{x}^f \overset{\Delta}{=} \bar{x}(t^f)$ denote the *initial*
and *terminal states*.

Another set of variables predicting the motion is the· finite set of
control variables $u_1(t), \ldots, u_r(t)$, $t \geq t_0$, representing instantaneously
an input to the system. Such input usually secures some control objective,
usually referring to the motions in $\mathbf{R}^N$. The control variables form a
control vector $\bar{u}(t) = (u_1(t), \ldots, u_r(t))$ ranging in some given bounded and
closed set $U \subset \mathbf{R}^r$ of *control constraints*. Note that the boundedness of
U is natural, as the power of any input is limited, and that without this
boundedness we might have an indefinite behavior of the system in $\mathbf{R}^N$. The
control variables are generated by a *control program* based upon the

(feedback) information represented by $\bar{x}(t)$ and perhaps upon some independent decision at each instant - hence the program will be generally specified by a function of $\bar{x},t$. In order to accommodate the discontinuities of such a function, which appear in many applications, we formalize it generally as the set valued $\bar{P} : \Delta \times \mathbb{R} \to$ set of subsets of U , defined by

$$\bar{u}(t) \in \bar{P}(\bar{x}(t),t) , \quad t \geq t_0 . \tag{2.1.1}$$

One of the cases in which the set valuedness of $P(\cdot)$ appears, is the *relay controller* generating desirable switching in the control action. The controller is usually specified by

$$\bar{u} = \bar{p}(\bar{x}) \operatorname{sgn} \sigma(\bar{x}) \tag{2.1.1}'$$

where $\sigma(\bar{x})$ is a continuous scalar *switching function* with $\sigma(\bar{x}) = 0$ defining the so called *switching surface* in Δ , and $\bar{p}(\bar{x})$ is a single valued branch of the program. The behavior of systems under relay control were described at length by Flüge-Lotz [1], among others, for wider classes of switching functions. It is traditional to define

$$\operatorname{sgn} \sigma = \begin{cases} +1 , & \sigma > 0 \\ \delta , & \sigma = 0 , \quad -1 \leq \delta \leq 1 . \\ -1 , & \sigma < 0 \end{cases}$$

Such definition is equivalent to that used for Coulomb friction characteristics and allows for continuous filling up of the control cone $\bar{P}(\bar{x})$ of values $\bar{u}(t)$. In this Section, however, we narrow our discussion to the single valued particular case $u_i(t) = P_i(\bar{x}(t))$, $t \geq t_0$, $i = 1,\ldots,r$.

The choice of the state variables $x_1,\ldots,x_N$ is by no means unique, and depends very much on our aim of study and perhaps a need to reduce the dimension of the model. We shall have two basic avenues of such choice: *direct* and *relative*. We refer to them successively.

From Mechanics we know that the motion of a system is fully represented by the set of its displacements and velocities, i.e. in lagrangian coordinates by the two vectors $\bar{q}(t),\dot{\bar{q}}(t)$. It is thus natural to choose $x_i \overset{\Delta}{=} q_i$, $x_{n+i} \overset{\Delta}{=} \dot{q}_i$, $i = 1,\ldots,n = N/2$, for the state variables or $\bar{x}(t) = (\bar{q}(t),\dot{\bar{q}}(t))^T \in \Delta_q \times \Delta_{\dot{q}} \overset{\Delta}{=} \Delta \subset \mathbb{R}^N$, $N = 2n$, for the state vector.

Consider now the Lagrangian equations (1.5.26) or (1.8.6) and assume that there are no external perturbations, i.e. $\bar{R}(\bar{q},\dot{\bar{q}},t) \equiv 0$, or $\bar{R}(\bar{q},\dot{\bar{q}},t) \equiv 0$. Then denoting

$$\left.\begin{aligned} f_i(\bar{x},\bar{u}) &= x_{n+i} \\ f_{n+i}(\bar{x},\bar{u}) &= -\Gamma_i(\bar{q},\dot{\bar{q}}) - D_i(\bar{q},\dot{\bar{q}}) - M_i(\bar{q}) + F_i(\bar{q},\dot{\bar{q}},\bar{u}) \end{aligned}\right\} \quad i = 1,\ldots,n \tag{2.1.2}$$

we obtain the so called *normal* form of the *state equations*

$$\dot{x}_i = f_i(\bar{x}, \bar{u}) , \quad i = 1, \ldots, N \tag{2.1.3}$$

or vectorially

$$\dot{\bar{x}} = \bar{f}(\bar{x}, \bar{u}) \tag{2.1.4}$$

with $\bar{f} = (f_1, \ldots, f_N)^T$. This first order system of ordinary differential
equations is fundamental in control theory. For the hybrid system (1.8.6)
we need to choose $x_i \overset{\Delta}{=} q_i$, $x_{n+i} \overset{\Delta}{=} \eta_i$, $x_{2n+i} \overset{\Delta}{=} \dot{q}_i$ and $x_{3n+i} \overset{\Delta}{=} \dot{\eta}_i$ for
the state variables, with the vector

$$\bar{x}(t) = (\bar{q}(t), \bar{\eta}(t), \dot{\bar{q}}(t), \dot{\bar{\eta}}(t))^T \in \Delta_q \times \Delta_\eta \times \Delta_{\dot{q}} \times \Delta_{\dot{\eta}} \overset{\Delta}{=} \Delta$$

and (2.1.2) as

$$\left. \begin{array}{l} f_i(\bar{x}, \bar{u}) = x_{2n+i} \\[2mm] f_{2n+i}(\bar{x}, \bar{u}) = (-\mathcal{D}_i(\bar{q}, \bar{\eta}, \dot{\bar{q}}, \dot{\bar{\eta}}) - P_i(\bar{q}, \bar{\eta}) \cdot + F_i(\bar{q}, \bar{\eta}, \dot{\bar{q}}, \dot{\bar{\eta}}, \bar{u})) \end{array} \right\} \tag{2.1.5}$$

$$i = 1, \ldots, 2n ,$$

in order to obtain (2.1.3) or (2.1.4), now with $N = 4n$. Observe that both
(1.5.26) and (1.8.6) have been obtained by multiplying the Lagrange equations
by the inverse inertia matrix and that (2.1.3) is obtainable directly from
(1.7.10) without any calculations, if we choose $\bar{x}(t) \overset{\Delta}{=} (\bar{q}(t), \dot{\bar{q}}(t))^T$ and
for $f_i(\cdot)$ the right hand sides of (1.7.10). Here we also assume
$\bar{Q}^R(\bar{q}, \bar{p}, t) \equiv 0$ for the use in this Section. There is a direct passage
between (1.7.10) and (2.1.2) via the so called Birkhoff transformation, see
Moser [1]. Finally we wish to comment that in this text we will be using
(1.5.26) and (1.8.6) or their subcases and the *form* (2.1.4) *is introduced*
for abbreviation in writing only.

Let us now discuss the choice of the *relative state variables*. Their
introduction is usually useful when investigating a motion with respect to
an object - either geometric or another system - in Cartesian or state
spaces. The controlled system may be modelled either in terms of the point-
mass or multi-body model. If the motion is considered relative to another
system, we study the mutual behavior of both systems, for instance between
two chains, $j = 1,2$, we may set up $x_i \overset{\Delta}{=} q_i^1 - q_i^2$, $x_{n+i} \overset{\Delta}{=} \dot{q}_i^1 - \dot{q}_i^2$,
$i = 1, \ldots, n$. In the relative motion, the reference object or reference
system is considered a target, by definition located in a neighborhood of
zero-misdistance between the controlled system and the target set, target
curve or target trajectory of a dynamical system in the phase space $\mathbb{R}^{2n}$.
The choice of relative state variables usually reduces the dimension of the
model, but the corresponding right hand sides $f_i(\cdot)$ of (2.1.3) lose their
physical meaning as representatives of a force or force per inertia. We

shall illustrate such a choice on the following special case.

Suppose $\bar{x}_m = \gamma(t)$ is a curve in Δ to which we want to reference the motion of our system and let $\bar{u}_m(t)$ be a control vector selected in such a way that, given suitable functions $\bar{f}(\cdot)$ and $\bar{x}_m^0 = \bar{\gamma}(t_0) = \bar{x}^0$, the curve concerned satisfies (2.1.4): $\dot{\bar{x}}_m = \bar{f}(\bar{x}_m, \bar{u}_m)$. Choosing now a state vector $\bar{z}(t) \overset{\Delta}{=} \bar{x}(t) - \bar{x}_m(t)$ we obtain from (2.1.4)

$$\dot{\bar{z}} = \bar{f}(\bar{z} + \bar{x}_m(t), \bar{u}) - \bar{f}(\bar{x}_m(t), \bar{u}_m) \tag{2.1.6}$$

with the right hand side being of the form (2.1.3) plus a time dependent perturbation $\bar{f}(t) = \bar{f}(x_m(t), u_m(t))$ which is a given function of t, and makes the system nonautonomous - discussed later. We may now rewrite (2.1.3) in terms of the old notation $\bar{x}$, i.e. as $\dot{\bar{x}} = \bar{f}(\bar{x}, \bar{u}, t)$, but with $\bar{f}(0, \bar{u}, t) \equiv 0$, if we want to consider $\bar{x}(t)$ as the state vector measured relative to $\bar{\gamma}(t)$ which presently becomes the trivial solution $\bar{x}(t) \equiv 0$.

Closing our remarks on the choice of state variables, let us comment on the special case when no dynamic study is needed, or for some reason may be ignored. What is then left is the *kinematic investigations*, which are based on the first (nondynamical) relation in (2.1.2), or (2.1.5), either case with both options (direct or relative) of choosing the state variables. With the direct choice $x_i = q_i$, $x_{n+i} = \dot{q}_i$, $i = 1, \ldots, n$, the right hand sides (2.1.2) reduce to

$$f_i(\bar{x}, \bar{u}) = x_{n+i}, \quad i = 1, \ldots, n \tag{2.1.2'}$$

and (2.1.3) become only *kinematic*

$$\dot{\bar{x}}_i = f_i(\bar{x}, \bar{u}), \quad i = 1, \ldots, n. \tag{2.1.3'}$$

Obviously with any other choice (2.1.2)' may be more complicated, but (2.1.3)' preserve the general normal format.

Let us now return to our general discussion and assume the direct choice of the state variables, i.e. either $\bar{x} = (\bar{q}, \dot{\bar{q}})^T$ or $\bar{x} = (\bar{q}, \bar{p})^T$. If, in particular cases, we shall prefer the relative variables, it will be specifically mentioned.

With the external perturbations $\bar{R}(\bar{q}, \dot{\bar{q}}, t)$, $\bar{R}(\bar{q}, \dot{\bar{q}}, t)$ removed, the right hand sides of the state equations (2.1.3) - as indicated in (2.1.2), (2.1.5) respectively - become implicit functions of time t only, namely through $\bar{x}$ and $\bar{u}$. This is the mathematically defining feature of the system that is *autonomous*. The system whose right hand sides of state equations are explicit functions of t is *nonautonomous*. Physically speaking, the autonomous system is governed through its own state $\bar{x}$, as the control vector is generated by a program which is state dependent as well, and only state dependent, $\bar{u} = \bar{P}(\bar{x})$. Traditionally, the system has been called autonomous if it included its own power supply, see Kononenko [1]. However, the control of power supply is possible at the output of the programming device, thus the system is autonomous, if it includes its own control program, see Fig. 2.1.

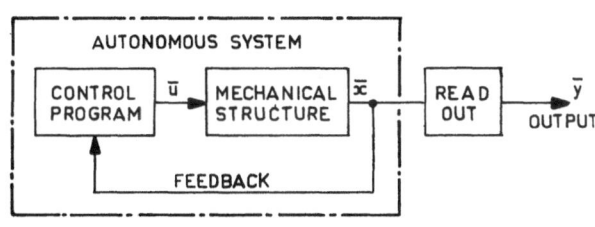

Fig. 2.1

We say that a function $\bar{f}(\cdot)$ satisfies locally the Lipschitz condition on Δ with respect to $\bar{x}$ if and only if for any closed and bounded subset LIP of Δ there is a constant $K(LIP)$ such that $\left| \bar{f}(\bar{x}) - \bar{f}(\bar{y}) \right| \leq K \left| \bar{x} - \bar{y} \right|$ for all $\bar{x}, \bar{y} \in \Delta$. The latter is always satisfied when $\bar{f}(\cdot)$ is continuously differentiable.

Given the function $\bar{f}(\cdot) : \Delta \times U \to \mathbb{R}^N$ and a program $\bar{P}(\bar{x})$ such that $\bar{f}(\bar{x}, P(\bar{x}))$ is locally Lipschitz continuous on Δ , through each $\bar{x}^0 \in \Delta$ there passes a unique solution curve to (2.1.3), $\bar{\phi}(\bar{x}^0, \cdot) : \mathbb{R} \to \Delta$ called a *trajectory* or a *state path* of the system. We will also adapt the notation $\bar{\phi}(\bar{x}^0, \mathbb{R}) = \cup_{t \in \mathbb{R}} \bar{\phi}(\bar{x}^0, t)$ for such a path. Note that $\bar{\phi}(\bar{x}^0, \cdot)$ depends upon $\bar{u}(\cdot)$. The control program $\bar{P}(\cdot)$ generating the trajectory is called *admissible*. Due to the defining property of the autonomous system (not dependent explicitly of t) the trajectory does not depend upon t_0 and thus represents the t_0-family of *motions* of the system, each motion starting at a given $t_0 \in \mathbb{R}$ and continuing along the path. In consequence, to determine a trajectory any t_0 may do, and we will assume the convenient $t_0 = 0$.

In turn, the $\bar{x}^0$-family of trajectories over Δ gives a first integral of (2.1.3). In topological dynamics it is interpreted as a homeomorphic map (one-to-one, onto and continuous) of $\mathbb{R}^N$ into itself and called a *dynamical system*. The basic features of such a map are:-

    (a) *continuity* of $\bar{\phi}(\cdot)$ in $\bar{x}^0, t$;

    (b) *identity* at $\bar{x}^0$: $\bar{\phi}(\bar{x}^0, t_0) = \bar{x}^0$;

    (c) *group property*: $\bar{\phi}[\bar{\phi}(\bar{x}^0, t_1), t_2] = \bar{\phi}(\bar{x}^0, t_1 + t_2)), \forall t_1, t_2 \in \mathbb{R}$.

The latter makes each trajectory retraceable for -t .

In order to describe the state space or phase space pattern of the trajectories, we need some reference sets in these spaces. There are two types of such sets, in fact related: *steady state trajectories* and total *energy levels*. We shall mention the first now, leaving the second to the next Section. A steady state or more formally a *singular trajectory* is a joint name for critical points or equilibria, periodic orbits, almost-periodic, recurrent, and in general *non-wandering trajectories*: for each neighborhood $N(\bar{x}^0)$ there is t such that

$$N(\bar{x}^0) \cap N[\bar{\phi}(\bar{x}^0,t)] \neq \phi .$$

Given $\bar{u}(\cdot)$, the equilibrium is defined by the relation

$$\bar{\phi}(\bar{x}^e,\mathbb{R}) = \bar{x}^e \qquad\qquad (2.1.7)$$

where $\bar{x}^e$ = const is an obvious *rest position* of the system. Moreover, (2.1.7) can equally well be determined by $\bar{\phi}(\bar{x}^e,\mathbb{R}^{\pm}) = \bar{x}^e$, so we have the case of a single point representing the entire trajectory. Thus, given $\bar{u}(\cdot)$, the trajectories are unique (do not cross) and hence any trajectory from $\bar{x}^0 \neq \bar{x}^e$ may approach $\bar{x}^e$ only asymptotically. Also, apart from a few exceptions, which will be mentioned, our equilibria are *isolated* (no other in a neighborhood).

On the other hand, from (2.1.3) we observe that for given $\bar{u}(\cdot)$, $\bar{x}^0$, the vector $\bar{f}$ is tangent to the trajectory concerned at each $\bar{x}(t) = \bar{\phi}(\bar{x}^0,t)$. This vector slides along the trajectory forming a vector field defined everywhere in Δ except at points where

$$\bar{f}(\bar{x},\bar{u}) = 0 \qquad\qquad (2.1.8)$$

which are called *singular points* of the field. If, for the $\bar{u}(\cdot)$ concerned, $\bar{x}^e$ satisfies (2.1.7), then $\bar{\phi}(\bar{x}^e,t) \equiv$ const substituted into (2.1.4) yields (2.1.8). Conversely if, given $\bar{u}(t) \in U$, $\bar{x}^e$ satisfies (2.1.8), then by (2.1.4) it generates (2.1.7). We conclude that (2.1.7) and (2.1.8) are equivalent, i.e. singular points of (2.1.4) are the same as rest positions or equilibria of this system. In particular, $\bar{x}^e = 0$ is called zero-equilibrium or a trivial solution of (2.1.4). In terms of (1.5.19), (1.5.20) (2.1.8) becomes

$$\dot{q}_i = 0 , \quad Q_i^D(\bar{q},\dot{\bar{q}}) + Q_i^P(\bar{q}) + Q_i^F(\bar{q},\dot{\bar{q}},\bar{u}) = 0 , \quad i = 1,\ldots,n \qquad (2.1.9)$$

or

$$\dot{q}_i = 0 , \quad D_i(\bar{q},\dot{\bar{q}}) + \Pi_i(\bar{q}) - F_i(\bar{q},\dot{\bar{q}},\bar{u}) = 0 , \quad i = 1,\ldots,n . \qquad (2.1.10)$$

Similarly in terms of (1.7.10), the equilibria are defined by

$$\frac{\partial H(\bar{q},\bar{p})}{\partial p_i} = 0 \; ; \qquad \frac{\partial H(\bar{q},\bar{p})}{\partial q_i} + Q_i^D(\bar{q},\bar{p}) + Q_i^F(\bar{q},\bar{p},\bar{u}) = 0 \; , \qquad (2.1.11)$$

$$i = 1,\ldots,n \; .$$

The region in Δ which forms a neighborhood of a single isolated equilibrium is called local, whatever its size. In particular, when it covers Δ, it will be called globally local. Any region in Δ which is not local (i.e. includes more than one isolated equilibrium) is called global.

Among the other steady state trajectories mentioned are *periodic orbits* defined by the fact that there is a number $L \neq 0$ such that

$$\bar{\phi}(\bar{x}^0,t) = \bar{\phi}(\bar{x}^0,t+L)$$

which may be shown to mean $\bar{\phi}(\bar{x}^0,L) = \bar{x}^0$, making the equilibrium (2.1.7) trivially periodic. The number L is then a period of $\bar{\phi}(\cdot)$ and may be represented by $n\ell$, $\ell \neq 0$, where n is any positive integer. Then for $n = 1$ or $L = \ell$, the sub-period ℓ is the largest and $\bar{\phi}(\cdot)$ is called harmonic or more precisely *fundamental harmonic*. The reciprocal of the period is the frequency $f = 1/L$ of the L-periodic motion, in particular $f_n = 1/n\ell$, with $f_1 = 1/\ell$ being the *fundamental frequency*. For $n < 1$ we obtain *subharmonic* $\bar{\phi}(\cdot)$ with lower frequencies and larger periods, while $n > 1$, i.e. with $L = n\ell = 2\ell, 3\ell,\ldots$ generates *higher harmonic* $\bar{\phi}(\cdot)$ with sub-periods and *higher frequency*, representing overtones to the fundamental harmonic. As well known, any periodic curve may be expressed in terms of the Fourier trigonometric series of various amplitudes whose frequencies are n-multiples corresponding to the overtones. In particular, $f_n = 1/n\ell = \omega_n/2\pi$, where ω_n is the corresponding angular frequency of traversing the curve. We have

$$x_i(t) = \sum_{n=0}^{\infty} (a_n \cos n\omega_n t + b_n \sin n\omega_n t) \; , \qquad i = 1,\ldots,N$$

where

$$a_0 = (1/n\ell) \int_0^{n\ell} x_i(t)\,dt \; , \qquad a_n = (2/n\ell) \int_0^{n\ell} x_i(t) \cos n\omega_n t\,dt$$

and

$$b_n = (2/n\ell) \int_0^{n\ell} x_i(t) \sin n\omega_n t\,dt$$

are the Fourier coefficients. It also follows from the definition of L-periodic orbit that

$$\bar{\phi}(\bar{x}^0,\mathbb{R}^\pm) = \bar{\phi}(\bar{x}^0,\mathbb{R}) = \bar{\phi}(\bar{x}^0,[0,L])$$

and that these sets are compact.

The equilibria and the periodic orbits are the only steady state trajectories appearing in the state plane $\mathbb{R}^2$. For higher dimensions we have further steady states: almost periodic, recurrent, and Poisson stable trajectories. A trajectory is *almost periodic* if and only if for sufficiently small $\varepsilon \geq 0$ there is a number $L \neq 0$ such that

$$\left| \bar{\phi}(\bar{x}^0, t+L) - \bar{\phi}(\bar{x}^0, t) \right| < \varepsilon , \qquad \forall t \in \mathbb{R} .$$

In non-English literature the name used is *quasi-periodic*. It means that $\bar{x}^0$ returns to itself but only within an approximate distance along the trajectory: for each neighborhood $N(\bar{x}^0)$ there is a compact interval $I \subset \mathbb{R}$ such that $\bar{\phi}(\bar{x}^0, \mathbb{R}) \subset \bar{\phi}(N, I)$. A still weaker property in the same direction is the *recurrence* property of a trajectory defined by the fact that for each $\varepsilon > 0$ there is $L > 0$ such that

$$\bar{\phi}(\bar{x}^0, \mathbb{R}) \subset N_\varepsilon [\bar{\phi}(\bar{x}^0, [t-L, t+L])] , \qquad \forall t \in \mathbb{R} ,$$

where $N_\varepsilon[\cdot]$ denotes ε-neighborhood of the segment of the trajectory between $t-L$ and $t+L$. It may be shown equivalent (Bhatia-Szegö [1]) to the statement that for each two $t_1, t_2 \in \mathbb{R}$ there is $t_3 \in \mathbb{R}$, $t_2 < t_3 < t_2 + L$, such that

$$\rho[\bar{\phi}(\bar{x}^0, t_1) , \bar{\phi}(\bar{x}^0, t_3)] < \varepsilon$$

where $\rho(\bar{x}, \bar{y})$ is the distance between $\bar{x}, \bar{y} \in \mathbb{R}^N$, see Fig. 2.2. Obviously an almost periodic trajectory is recurrent but not vice versa. A still weaker property is the so called *Poisson stability*, possibly the most relaxed within the class of non-wandering, i.e. singular, trajectories. A trajectory is called positively (negatively) Poisson stable, iff for each $t \in \mathbb{R}$ there is some $t_1 > t$ ($t_1 < t$) such that $\bar{\phi}(\bar{x}^0, t_1) \in N(\bar{x}^0)$ with N, some neighborhood of $\bar{x}^0$, called *self-recursive*, see Bhatia-Szegö [1]. The trajectory is *Poisson stable* if and only if it is both positive and negative Poisson stable. It may be shown, see Nemitzky-Stepanov [1], that Poisson stability is equivalent to non-wandering, i.e. any singular trajectory is at least Poisson stable.

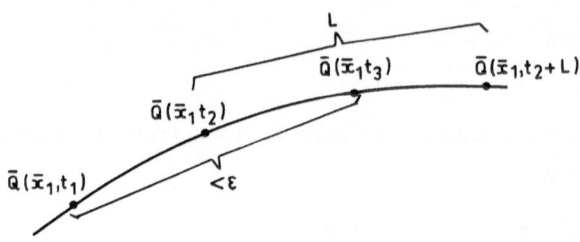

Fig. 2.2

On the other hand, points in Δ which do not belong to a singular trajectory are called *regular*. They are *positively (negatively) wandering*: there is $N(\bar{x}^0)$ and t_1 such that

$$N(\bar{x}^0) \cap N[\bar{\phi}(\bar{x}^0,t_1)] = \phi ,$$

for all $t \geq t_1$ $(t \leq t_1)$. The set of regular points complements the set of steady states.

In terms of the interpretation (1.5.28), the equations (2.1.3) become

$$\frac{d\dot{q}_i}{dq_i} = \frac{-D_i(\bar{q},\dot{\bar{q}}) - \Pi_i(\bar{q},\dot{\bar{q}}) + F_i(\bar{q},\dot{\bar{q}},\bar{u})}{\dot{q}_i} , \qquad i = 1,\ldots,n \qquad (2.1.12)$$

which, as mentioned in Section 1.5, is the phase-space trajectory equation in $\mathbb{R}^{2n}$, for our direct choice of state variables identified with $\mathbb{R}^N$. Again as mentioned, such a trajectory could be investigated on n planes $0 q \dot{q}_i$, $i = 1,\ldots,n$, as long as the calculation is done simultaneously, since the functions $D_i(\cdot)$, $\Pi_i(\cdot)$ and $F_i(\cdot)$ are coupled across the whole vectors $\bar{q},\dot{\bar{q}}$. Each of the equations (2.1.12) presents the ith component of the instantaneous slope of the phase-space trajectory which is identical with $\bar{\phi}(\bar{x}^0,\mathbb{R})$ and constitutes the first integral curve of both (2.1.3) and (2.1.12) through $\bar{x}^0 = (\bar{q}^0,\dot{\bar{q}}^0) \in \Delta$, see Fig. 2.3.

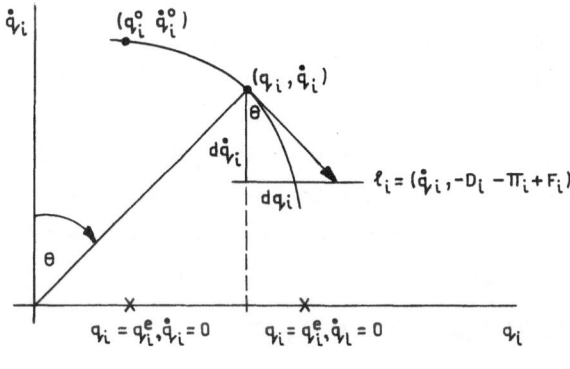

Fig. 2.3

By the uniqueness of the trajectories we have their continuous dependence on $(\bar{q}^0,\dot{\bar{q}}^0)$ and thus (2.1.12) defines the field of slopes $d\dot{q}_i/dq_i$, $i = 1,\ldots,n$ at the regular points of Δ . This field is not defined at the equilibria which, for the (2.1.12) representation, are obviously located in the hypersurface $\dot{\bar{q}} = 0$. Moreover, we observe immediately that all the trajectories must cross $\dot{\ddot{q}} = 0$ vertically, as $d\dot{q}_i/dq_i \to \infty$ when $\dot{q}_i \to 0$, $i = 1,\ldots,n$. The reader may return to Example 1.1.1 for illustration of the above. Finally, (2.1.12) implies in the obvious way, that the motions along our phase-space trajectory proceed clockwise with time, so that the angle θ of the radius vector $(\bar{q},\dot{\bar{q}})^T$ can be taken as a time measure instead of t along the trajectory concerned.

Let us now return to the case of the non-vanishing perturbation $\bar{R}(\bar{q},\dot{\bar{q}},t) \not\equiv 0$, $\bar{R}(\bar{q},\dot{\bar{q}},t) \not\equiv 0$. The right hand sides of the state equations for the perturbed (1.5.26) become

$$\left.\begin{aligned}
f_i(\bar{x},\bar{u},t) &= x_{n+i} \text{ ,}\\
f_{n+i}(\bar{x},\bar{u},t) &= -\Gamma_i(\bar{q},\dot{\bar{q}}) - D_i(\bar{q},\dot{\bar{q}}) - \Pi_i(\bar{q}) + F_i(\bar{q},\dot{\bar{q}},\bar{u}) + R_i(\bar{q},\dot{\bar{q}},t)
\end{aligned}\right\}$$

$$(2.1.13)$$

and the system is nonautonomous. Similar relations may be immediately written for (1.8.6). Consequently the perturbed state equations become

$$\dot{x}_i = f_i(\bar{x},\bar{u},t) \text{ ,} \qquad\qquad (2.1.14)$$

or vectorially

$$\dot{\bar{x}} = \bar{f}(\bar{x},\bar{u},t) \text{ .} \qquad\qquad (2.1.15)$$

Now, the solutions to (2.1.15) will depend upon the initial instant t_0 and *will not be unique in* Δ. Such uniqueness may be however achieved in the so called *space of events* $\mathbb{R}^{N+1} = \mathbb{R}^N \times \mathbb{R}$ enclosing the augmented work envelope $\Delta \times \mathbb{R}$. Given a program $\bar{P}(\cdot)$, stationary or not, and such that the corresponding $\bar{f}(\bar{x},P(\bar{x},t),t)$ is locally Lipschitz continuous in $\bar{x}$ on Δ and measurable in t, through each point $(\bar{x}^0,t_0) \in \Delta \times \mathbb{R}$ there passes a unique solution curve called *motion* $\bar{\phi}(\bar{x}^0,t_0,\mathbb{R}^+) = U_t\phi(\bar{x}^0,t_0,t)$, $t \in \mathbb{R}^+$ of (2.1.15). Mapping the motions into Δ we lose the uniqueness, i.e. the motions may cross each other there and will not be located along the same state path. Again, however, the $\bar{x}^0,t_0$ - family of motions denoted

$$\phi(\Delta,\mathbb{R},\mathbb{R}^+) \overset{\Delta}{=} \{\bar{\phi}(\bar{x}^0,t_0,\mathbb{R}^+) \,|\, x^0 \in \Delta, \, t_0 \in \mathbb{R}\}$$

gives the first integral of (2.1.15), which is also a dynamical system but on the set $\Delta \times \mathbb{R}$ in the time-augmented state space $\mathbb{R}^{N+1}$. Formally, we may in fact introduce the vector of *events* $\bar{e} = (\bar{x},t)^T$ and write (2.1.14) in the autonomous format

$$\left.\begin{aligned}
\frac{de_i}{d\tau} &= f_i(\bar{e},\bar{u})\\
\frac{dt}{d\tau} &= 1
\end{aligned}\right\} \text{ ,} \qquad i = 1,\ldots,N \qquad (2.1.14)'$$

with the "trajectories" $\phi(\bar{e}^0, \cdot) : \Delta \times \mathbb{R} \rightarrow \Delta \times \mathbb{R}$ satisfying the axioms of a dynamical system. However, due to the special shape of the last equation $f_{N+1} \equiv 1$, such a system is "parallelizable" along the time axis, see Nemitzky-Stepanov [1]. Instead of steady state trajectories in Δ we now have to reference the motions to *steady state sets*, augmented into $\Delta \times \mathbb{R}$.

The vector $\bar{f}(\bar{x}, \bar{u}, t)$ of (2.1.15) is still tangent to the motion it generates in $\Delta \times \mathbb{R}$, and forms a vector field along such a motion with singularities

$$\bar{f}(\bar{x}^e, 0, t) = 0 , \qquad \forall t \geq 0 \qquad (2.1.16)$$

at the equilibria, which means $\bar{x} = \bar{x}^e = \text{const}$ is a solution of (2.1.15). It is a point in Δ or a half-line in $\Delta \times \mathbb{R}$. A similar augmentation applies to periodic orbits and any other steady states of (2.1.15). We shall return to this problem in Section 3.2 and later in the text.

The augmentation to (1.5.28) in $\Delta \times \mathbb{R}$ of our comments regarding (2.1.12) in Δ is immediate in view of the above, and it is left to the reader.

Closing this section, let us comment on the case when it is necessary to consider in our general model some chain substructures discussed in Section 1.4, with the Lagrange equations leading to the format (1.5.37) or for independent chains to (1.5.38). Here again we may apply the direct or relative choice of state variables. In the first case $\bar{x}^j \triangleq (\bar{q}^j, \dot{\bar{q}}^j)^T$ is the j - *chain state vector* with the system state represented by the Nm - vector $\bar{x} \equiv (x_1^1, \ldots, x_N^1, \ldots, x_1^m, \ldots, x_N^m)^T$ and $\bar{u}^j = (u_1^j, \ldots, u_r^j)$ is the j - chain control vector. Then letting

$$\bar{f}^j(\bar{x}, \bar{u}^j, t) \triangleq (\dot{\bar{q}}^j, -\bar{\Gamma}^j - \bar{D}^j - \bar{\Pi}^j + \bar{F}^j + \bar{R}^j)$$

we obtain (1.5.37) in the state vector form

$$\dot{\bar{x}}^j = \bar{f}^j(\bar{x}, \bar{u}^j, t) , \qquad j = 1, \ldots, m . \qquad (2.1.17)$$

For the choice of state in relative coordinates the system state vector $\bar{x}$ is selected as an M - vector, $M \leq mN$, with components $x_1, \ldots, x_M$ running over the set of relative configuration variables $q_i^\sigma - q_j^\nu$, $i = 1, \ldots, n^\sigma$, $j = 1, \ldots, n^\nu$, $\sigma, \nu = 1, \ldots, m$ and the set of corresponding relative velocities $\dot{q}_i^\sigma - \dot{q}_j^\nu$. Then the state equation is that of (2.1.15) with the dimension M adjusted, depending on how many relative variables are used. Usually, however, each chain is controlled by a separate controller, hence the control vectors $\bar{u}^j = (u_1^j, \ldots, u_r^j)$, $j = 1, \ldots, m$ will not be amalgamated and we shall write the general format (2.1.15) as

$$\dot{\bar{x}} = \bar{f}(\bar{x}, \bar{u}^1, \ldots, \bar{u}^m, t) \; . \tag{2.1.18}$$

Here, the components $f_1, \ldots, f_M$ of the system vector $\bar{f}$ have a different meaning as in the direct choice of state. They can be considered relative characteristics, not necessarily expressible in terms of the difference between the acting forces per inertia of any pair of the chains. The dynamics simplifies if we set up terms in $q_i^\sigma - q_j^\nu$ or $\dot{q}_i^\sigma - \dot{q}_j^\nu$ to be stipulated, say, as a target.

Obviously, for independent chains, (2.1.17) are state decoupled, i.e. $\bar{f}^j(\cdot)$ become functions of $\bar{x}^j$ only, but (2.1.18) stay coupled by definition.

EXERCISES 2.1

2.1.1 Consider successively the models (1.5.19), (1.5.26), (1.7.9), (1.8.3) and write the equations for equilibria.

2.1.2 Explain why the linear system

$$\dot{\bar{x}} = A(t)\bar{x} + B(t)\bar{u} \; , \quad \bar{x} \in \mathbb{R}^N$$

where A,B matrices of suitable dimensions, is nonautonomous.

2.1.3 State assumptions under which (1.5.26) and (1.7.9) become dynamical systems.

2.1.4 Consider the trajectory equation (2.1.12) representing the instantaneous slope of the trajectory on the $0q_i\dot{q}_i$ - plane. What is this slope at regular points in which the trajectory crosses the axis $\dot{q}_i = 0$? Sketch a few isoclines - lines of equal slope for a family of trajectories.

2.1.5 Show that $\bar{f}(\bar{x}, t)$ is Lipschitz continuous on some Δ , if it is continuous in $(\bar{x}, t)$ and linear in $\bar{x}$ on this set. Give examples of continuous functions which are not Lipschitz continuous.

2.1.6 Can the motions of a nonautonomous system be unique in $\mathbb{R}^N$? Describe the difference between the motions and trajectories.

2.1.7 Consider the system $\ddot{q} + k \sin q = 0$, $k > 0$. Find all the equilibria and sketch the phase-space pattern.

2.1.8 Show that the equation $\dot{x} = |x|^n$, $x(0) = 0$, $n \in (0,1)$, has infinitely many solutions at each x.

2.1.9 Consider the system (*) $\dot{\bar{x}} = \bar{f}(\bar{x})$, $\bar{x} \in \mathbb{R}^2$ with $\bar{f}(\cdot)$ suitable for existence and uniqueness of trajectories on Δ. Set $\bar{x} \stackrel{\Delta}{=} (\phi,\theta)$, $\bar{f} \stackrel{\Delta}{=} (f_1,f_2)$ and assume f_1, f_2 periodic, with period 1, in each of the variables ϕ, θ. Project $\mathbb{R}^2$ on a torus and form the dynamical system corresponding to (*) on such a torus. HINT: convenient representation of the torus:

$$\{(\phi,\theta) \mid 0 \le \phi < 1, \ 0 \le \theta < 1\}$$

in which the pairs of opposite sides $q = 0$, $q = 1$ and $\theta = 0$, $\theta = 1$ are identified, i.e. $(0,0)$, $(0,1)$, $(1,0)$, $(1,1)$ are all identified.

2.1.10 Consider the system

$$\ddot{\bar{q}} + \bar{D}(\bar{q},\dot{\bar{q}}) + \bar{\Pi}(\bar{q}) = 0, \quad \bar{q}(t) \in \Delta \subset \mathbb{R}^n.$$

(i) Show that the trajectories may be defined by their pattern on the planes $0q_i\dot{q}_i$.

(ii) Specify conditions upon $D(\cdot)$, $\Pi(\cdot)$ defining the equilibria. In particular, consider the case $n = 1$, $\Pi(q) = aq - bq^3$, $q \in \mathbb{R}$ and the case $n = 2$ with

$$\Pi_1(q_1,q_2) = a_1 q_1 + b_1 q_1^3 + c_1 q_1^5 + a_{12}(q_1 - q_2)$$
$$\Pi_2(q_1,q_2) = a_2 q_2 + b_2 q_2^3 + c_2 q_2^5 + a_{21}(q_2 - q_1)$$

where $a_i, b_i, c_i > 0$, $i = 1,2$, $a_{12} = -a_{21}$. (Assume same numbers to simplify calculations.)

(iii) For the cases $n = 1$ and $n = 2$ find conditions upon $\bar{D}(\cdot)$, $\bar{\Pi}(\cdot)$ guaranteeing that the trajectories form a dynamical system.

2.2 STATE REPRESENTATION OF UNCERTAIN SYSTEMS

Mechanical systems often work in difficult and unpredictable, or at least partially unpredictable, environments (space, under water, bad weather etc). They are subject to wear and tear during their work time. Moreover, machines are often an expensive investment and must be designed flexible enough to serve a range of operations, and thus robust to varying parameters. Last, but not least, both the designer and the modeller make errors which must be somehow accommodated within the work regime of the machine. All the above factors lead to many system parameters being uncertain, even if they are bounded and their range of values is known. The sources specified above generate uncertainty in external perturbations, payload, and thus also inertia and gravity, but may as well appear in any other type of characteristics. Indeed, if the uncertainty appears in inertia, by definitions (1.5.21)-(1.5.25), it pollutes all of the charac- teristics. A model which is sensitive to uncertainty is thus not worth much in real life applications. We ought to make our control program robust against uncertainty and we can do it since, as mentioned, the range of uncertainty is bounded and often known. Then we do it via the so called *worst-case-design* or a *game-against-nature*, eliminating the uncertainty effects either with the controller alone or more economically with an adaptive controller and adaptation of some parameters of the system, as shown later in this text.

Let us now return to the Lagrange equations (1.5.26) or (1.8.6), initially in autonomous version:

$$\bar{R}(\bar{q},\dot{\bar{q}},t) \equiv 0 \ , \qquad \bar{\bar{R}}(\bar{q},\dot{\bar{q}},\bar{\eta},\dot{\bar{\eta}},t) \equiv 0 \ ,$$

but insert the *uncertainty vector* $\bar{w}(t) \in W \subset \mathbb{R}^s$ ranging in the bounded band W which is known. We obtain

$$\ddot{\bar{q}} + \bar{\Gamma}(\bar{q},\dot{\bar{q}},\bar{w}) + \bar{D}(\bar{q},\dot{\bar{q}},\bar{w}) + \bar{\Pi}(\bar{q},\bar{w}) = \bar{F}(\bar{q},\dot{\bar{q}},\bar{u},\bar{w}) \qquad (2.2.1)$$

and

$$(\ddot{\bar{q}}^T,\ddot{\bar{\eta}}^T)^T + \bar{\mathcal{D}}(\bar{q},\dot{\bar{q}},\bar{\eta},\dot{\bar{\eta}},\bar{w}) + \bar{\mathcal{P}}(\bar{q},\bar{\eta},\bar{w}) = \bar{\mathcal{F}}(\bar{q},\dot{\bar{q}},\bar{\eta},\dot{\bar{\eta}},\bar{w}) \qquad (2.2.2)$$

respectively. One does not expect, in most case studies, the elastic forces to be polluted by uncertainty. However, the gravity forces obviously are, thus so is the potential energy and consequently both the Lagrangian and Hamiltonian. Hence the autonomous version: $Q_i^R(\bar{q},\bar{p},t) \equiv 0$ of (1.7.10) has to be written in the following format

$$\left. \begin{aligned} \dot{q}_i &= \frac{\partial H(\bar{q},\bar{p},\bar{w})}{\partial p_i} \ , \\[2mm] \dot{p}_i &= -\frac{\partial H(\bar{q},\bar{p},\bar{w})}{\partial q_i} + Q_i^D(\bar{q},\bar{p},\bar{w}) + Q_i^F(\bar{q},\bar{p},\bar{u},\bar{w}) \ , \quad i = 1,\ldots,n \end{aligned} \right\} \qquad (2.2.3)$$

Obviously in (2.2.3) neither the damping nor the input forces are polluted by uncertainty owing to the inertial decoupling, as it happened in (2.2.1), (2.2.2). However, they may be, and often are, uncertain on their own account which is then also covered by the influence of the vector $\bar{w}(t)$.

Whatever the choice of state variables, direct or relative, we still finish with the state equation format

$$\dot{x}_i = f_i(\bar{x},\bar{u},\bar{w}) \ , \qquad i = 1,\ldots,N \tag{2.2.4}$$

or vectorially

$$\dot{\bar{x}} = \bar{f}(\bar{x},\bar{u},\bar{w}) \ . \tag{2.2.5}$$

The solutions $\bar{\phi}(\bar{x}^0,\cdot)$ of (2.2.5) will now depend upon $\bar{w}(\cdot)$ and in general may not exist at all, let alone be unique. This, together with the multi-valued control program (2.1.1), see for instance (2.1.1)', which in general will apply for reasons mentioned, leads to the nonuniqueness of the solutions. For instance, the simple system $\dot{x} = u\sqrt{x}$, $x(t) \in \mathbb{R}$, with the program defined by $u = \pm\sqrt{x}$, produces two solutions $x(t) = x^0 e^{\pm t}$ at each $x^0 \in \mathbb{R}$.

In other cases, both the uncertainty and discontinuities in control may produce motion discontinuities which have practical meaning: may produce dangerous vibrations, cf. Red-Truong [2]. To accommodate all the above, we must use the *contingent* format of the state equations:

$$\dot{\bar{x}} \in \{\bar{f}(\bar{x},\bar{u},\bar{w}) \mid \bar{u} \in \bar{P}(\bar{x},t) \ , \ \bar{w} \in W\} \ . \tag{2.2.6}$$

Within the full, nonautonomous versions of the motion equations (1.5.26), (1.8.6) the non-vanishing perturbations $\bar{R}(\cdot)$, $\tilde{R}(\cdot)$ are as much subject to the uncertainty $\bar{w}$ as the characteristics and inputs. Moreover, they themselves may be the source of uncertainty, being the actions of an uncertain environment. The latter role may also be played by the functions $\bar{Q}^R(\cdot)$ in either (1.7.9) or (1.7.10). However, even if uncertain in value, the functions $\bar{R}(\cdot)$, $\tilde{R}(\cdot)$, $\bar{Q}^R(\cdot)$ would have to be assumed of known character as much as $\bar{w}(\cdot)$ was. Hence, there is no narrowing in generality when considering them all known functions of some unknown vector $\bar{w}$ with the dimension s adjusted. This makes (1.5.26) and (1.8.6) become successively

$$\ddot{\bar{q}} + \bar{\Gamma}(\bar{q},\dot{\bar{q}},\bar{w}) + \bar{D}(\bar{q},\dot{\bar{q}},\bar{w}) + \bar{\Pi}(\bar{q},\bar{w}) = \bar{F}(\bar{q},\dot{\bar{q}},\bar{u},\bar{w}) + \bar{R}(\bar{q},\dot{\bar{q}},\bar{w},t) \tag{2.2.1}'$$

and

$$(\ddot{\bar{q}}^T,\ddot{\bar{\eta}}^T)^T + \bar{D}(\bar{q},\dot{\bar{q}},\bar{\eta},\dot{\bar{\eta}},\bar{w}) + \bar{P}(\bar{q},\bar{\eta},\bar{w}) = \bar{F}(\bar{q},\dot{\bar{q}},\bar{\eta},\dot{\bar{\eta}},\bar{w}) + \bar{R}(\bar{q},\dot{\bar{q}},\bar{\eta},\dot{\bar{\eta}},\bar{w},t) \ , \tag{2.2.2}'$$

while (1.7.10) is now written as

$$\dot{q}_i = \frac{\partial H(\bar{q}, \bar{p}, \bar{w})}{\partial p_i}$$

$$\dot{p}_i = - \frac{\partial H(\bar{q}, \bar{p}, \bar{w})}{\partial q_i} + Q_i^D(\bar{q}, \bar{p}, \bar{w}) + Q_i^F(\bar{q}, \bar{p}, \bar{u}, \bar{w}) + Q_i^R(\bar{q}, \bar{p}, \bar{w}, t) \; . \qquad (2.2.3)'$$

Then the state equations become explicit t dependent

$$\dot{\bar{x}} = \bar{f}(\bar{x}, \bar{u}, \bar{w}, t) \qquad\qquad\qquad (2.2.5)'$$

with solutions $\bar{\phi}(\bar{x}^0, t_0, \cdot)$ nonunique in Δ not only because of $\bar{w}(\cdot)$ but also because of the dependence on t_0 . The case is however still covered in Δ by the contingent format

$$\dot{\bar{x}} \in \{\bar{f}(\bar{x}, \bar{u}, t, \bar{w}) \mid \bar{u} \in \bar{P}(\bar{x}, t) , \;\; \bar{w} \in W\} \; . \qquad\qquad (2.2.6)'$$

It is also called the *generalized differential equation* or the *differential inclusion* and has a long mathematical and applicational history. The so called contingent derivative has been introduced by Buligand [1] for geometric purposes, and has been applied to differential equations by Zaremba in the nineteen-thirties. Two decades later, Wazewski [1]-[4] and his Cracow school (Bielecki, Plis, Lasota, Olech) specified sufficient conditions for the existence of solutions to such equations and applied them to Control Theory. A long sequence of improvements in these sufficient conditions followed over the years, until Filippov [1] gave the simplest and most practical result. A very good review can be found in Davy [1].

An elegant geometric interpretation of the application to Control had been brought about by Roxin [1],[2]. The differential game was formalized and developed in contingent terms by Skowronski [22]-[24], Gutman [1] and Stonier [1]. Finally the application of (2.2.6) to the deterministic control of uncertainty (the worst-case-design or the game-against-nature) belongs to Leitmann, see Gutman-Leitmann [1], Gutman [3], Leitmann [3],[4], [6],[7], and Corless-Leitmann [1]. For time-discrete systems the topic has been studied by Bertsekas-Rhodes [1],[2].

We shall now briefly describe the basic features of (2.2.6) and (2.2.6)'. The vector function $\bar{f}(\cdot)$ is called the *selector* of a motion, or t_0- family of motions respectively, from within the *orientor field* described by the right hand side of (2.2.6) for each $t \in \mathbf{R}^+$, see Fig. 2.4(a) for the autonomous and Fig. 2.4(b) for nonautonomous system. When $\bar{P}(\cdot)$ reduces to a single valued function and $\bar{w}$ is given, the equations (1.2.6), (1.2.6)' reduce to (2.2.5), (2.2.5)' respectively, which are called the *selector equations*.

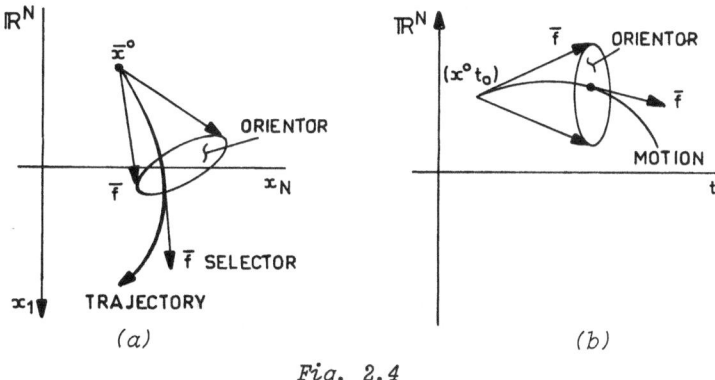

Fig. 2.4

Given any event $(\bar{x}^0, t_0) \in \Delta \times \mathbb{R}$, a solution to (2.2.6)' is an absolutely continuous function $\bar{\phi}(\bar{x}^0, t_0, \cdot) : \mathbb{R}^+ - \Delta$, identified at $\bar{\phi}(\bar{x}^0, t_0, t_0) = \bar{x}^0$, which when substituted satisfies (2.2.6)' almost everywhere on $\mathbb{R}^+$. We designate the family of such solutions by $K(\bar{x}^0, t_0)$. The set of events $(\bar{x}, t) \in \Delta \times \mathbb{R}$ such that $\bar{x} = \bar{\phi}(\bar{x}^0, t_0, t)$, $t \in \mathbb{R}^+$, represents a unique curve in $\Delta \times \mathbb{R}$. We denote it $\bar{\phi}(\bar{x}^0, t_0, \mathbb{R}^+)$ and call it the *motion* from $(\bar{x}^0, t_0)$.

Given t, the set $\bar{\Phi}(\bar{x}^0, t_0, t) = \{\bar{\phi}(\bar{x}^0, t_0, t) | \bar{\phi}(\cdot) \in K(\bar{x}^0, t_0)\}$ in Δ is called the *attainable set* at t from (x^0, t_0). Then the set of events $\bar{\Phi}(\bar{x}^0, t_0, [t_0, t]) = \underset{\tau}{\cup} \bar{\Phi}(\bar{x}^0, t_0, \tau) | \tau \in [t_0, t]$ in $\Delta \times \mathbb{R}$ is the *reachable set* at t from $(\bar{x}^0, t_0)$, while the expression $\bar{\Phi}(\bar{x}^0, t_0, \mathbb{R}^+)$ designates the *reachability cone* from $(\bar{x}^0, t_0)$ in $\Delta \times \mathbb{R}$.

If we choose to leave our model autonomous, we shall have the t_0-family of motions called a *trajectory* in Δ, independent on t_0. The class of such trajectories will be denoted by $K(\bar{x}^0)$.

As mentioned, there are at least several choices of conditions sufficient to the existence of solutions of (2.2.6)' or in particular of (2.2.6) in the above sense, that is, to the fact that $K(\bar{x}^0, t_0)$ is non-void on $\Delta \times \mathbb{R}$. The cited Filippov [1] requires that the set on the right hand side of (2.2.6)' be obtained by a compact set valued function which is continuous and bounded on $\Delta \times \mathbb{R}$. Obviously suitable conditions must be imposed on $\bar{P}(\cdot)$ in order to imply the above. Then, the controlling agent has the convenience of picking up any value of $\bar{u}(t)$ from the corresponding set $\bar{P}(\bar{x}, t)$, and may use any motion from $K(\bar{x}^0, t_0)$ to do the job for him. On the other hand, Filippov also showed that, if $\bar{f}(\cdot)$ is continuous and $\bar{P}(\cdot)$ upper-semi-continuous, there are measurable functions $\bar{u}(\cdot), \bar{w}(\cdot)$ such that $\dot{\phi}(t) = \bar{f}(\phi(t),t,\bar{u}(t),\bar{w}(t))$ for almost every $t \in \mathbb{R}$. Hence the agent can always find control functions that allow us to represent $\bar{\phi}(\bar{x}^0,t_0,\cdot)$ in $\Delta \times \mathbb{R}$ as unique solutions to the selecting equations (2.2.5)', and hence (2.2.6)' as a family of (2.2.5)'. Thus, we have $K(\bar{x}^0,t_0)$ non-void and equipped with a definite differential structure. The programs $\bar{P}(\cdot)$ which allow the above will be called *admissible*. We shall assume all programs admissible, unless otherwise specifically stated.

EXAMPLE 2.2.1 (Stonier [1]). A simple example of (2.2.6) is obtained letting the state equation (2.2.6) be the scalar dynamical system

$$\dot{x} \in \{z = -wu \mid u = P(\bar{x}) \equiv x , w \in [1,2]\} .$$

The attainability set becomes

$$\Phi(x^0,t_0,t) = \begin{cases} (x^0 e^{-2(t-t_0)}, & x^0 e^{-(t-t_0)}) , & x \geq 0 , \\ (x^0 e^{-(t-t_0)}, & x^0 e^{-2(t-t_0)}) , & x < 0 . \end{cases}$$

Considered for all $t \geq t_0$, it carries on all the available motions of the system. □

EXAMPLE 2.2.2 (Hajek [1]). Consider $\dot{\bar{x}} \in \{\bar{z} = A\bar{u} - \bar{w} \mid \bar{w} \in W\}$ with $\bar{x}(t) \in \mathbb{R}^N$, A being some $N \times N$ matrix, and $\bar{u} \equiv \bar{x}$. Then the reachability set at t is

$$\Phi(\bar{x}^0,t_0,[t_0,t]) = \left\{ \int_{t_0}^{t} e^{-A\tau} \bar{w} \, d\tau \mid \bar{w} \in W \right\} .$$

For any $\bar{w}(\cdot)$ the motion is defined by

$$\bar{\phi}(\bar{x}^0,t_0,t) = e^{At} \left[\bar{x}^0 - \int_{t_0}^{t} e^{-A\tau} \bar{w}(\tau) \, d\tau \right]$$

obtainable from the selector equation $\dot{\bar{x}} = A\bar{u} - \bar{w}$ by the method of variation of parameters. □

EXAMPLE 2.2.3. Another example is obtained modifying Hajek [1]. We let the selector equation be scalar,

$$\dot{x} = uw ,$$

with $U : |u| \leq \hat{u}$, $W : |w| \leq \hat{w}$, $x^0 = x(t_0) = 0$. Given $P(\cdot)$ which specifies unique admissible $u = 1$, we ask where the uncertainty w can possibly get the system at the time t? The constraint on w yields

$$|x(t)| \leq \int_{0}^{t} |u(\tau)| \, d\tau \leq \hat{w} t .$$

Conversely, any point x such that $|x| \leq \hat{w}t$ can be obtained using

$$w(\tau) = \begin{cases} \hat{w}\, \dfrac{x}{|x|}\,, & \text{for } 0 \leq \tau \leq |x|\,, \\[2mm] 0\,, & \text{for } |x| < \tau \leq t\,. \end{cases}$$

Thus the available set at t from x^0 is

$$\Phi(x^0, t_0, t) = \{x \mid |x| \leq \hat{w}t\}\quad. \qquad\qquad \square$$

From both, the nature of (2.2.6), (2.2.6)' and the examples, one can conclude that the contingent equations have an alternative representation as differential inequalities:

$$\min_{\bar{u},\bar{w}} \bar{f}(\bar{x},\bar{u},\bar{w},t) \leq \dot{\bar{x}} \leq \max_{\bar{u},\bar{w}} \bar{f}(\bar{x},\bar{u},\bar{w},t) \qquad\qquad (2.2.6)''$$

holding for $\bar{u} \in \bar{P}(\bar{x},t)$, $\bar{w} \in W$ and each $t \in \mathbb{R}^+$.

EXAMPLE 2.2.4. Consider the scalar contingent equation,

$$\dot{x} \in \{z = uwx^2 \mid u \in [-1,1],\ w \in (1,3)\}\quad.$$

Hence also $-3x^2 < \dot{x} < -x^2$ or $x^2 < \dot{x} < 3x^2$. The selector $\dot{x} = x^2$ at $u = 1$, $w = 1$ gives $\dot{x}/x^2 = 1$. Integrating $dx/x^2 = dt$ we obtain $x(t) = x^0[x^0(t_0 - t) + 1]^{-1}$ whence

$$x^0[3x^0(t - t_0) + 1]^{-1} < x < x^0[x^0(t - t_0) + 1]^{-1}\,,$$

or

$$x^0[x^0(t_0 - t) + 1]^{-1} < x < x^0[3x^0(t_0 - t) + 1]^{-1}\,,$$

see Fig. 2.5.

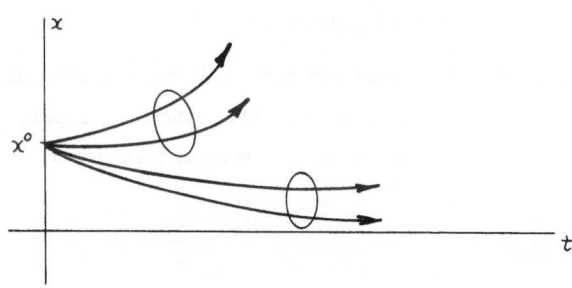

Fig. 2.5

$\square$

If it is necessary to consider some chain substructures with separate dynamics, the contingent format remains the same except for the selector equation replaced by (2.1.17) or (2.1.18) depending whether the choice of state variables is direct or relative.

The alternative dynamics for uncertainty is obtained by making the uncertainty additive and the system linear in control. In general terms, such dynamics is formalized as the *Leitmann's model* of an uncertain dynamical system

$$\dot{\bar{x}} = [A(\bar{x},t) + \delta A(\bar{x},t,F)]\bar{x} + [B(\bar{x},t) + \delta B(\bar{x},t,\bar{s})]\bar{u} + C(\bar{x},t)\bar{w} \qquad (2.2.7)$$

where $\bar{r},\bar{s},\bar{w}$ are uncertain variables assumed to be functions of $\bar{x}$ and t, whose values lie in known closed and bounded sets of suitable spaces:

$$\bar{r}(\bar{x},t) \in R \subset \mathbb{R}^S, \quad \bar{s}(\bar{x},t) \in S \subset \mathbb{R}^k, \quad \bar{w}(\bar{x},t) \in W \subset \mathbb{R}^S.$$

Introduced in 1976, see Gutman-Leitmann [1] and the above cited works, the model had been extensively developed both theoretically and in applications. The latter lead to the wide use of the Leitmann-Gutman controller which we shall discuss in Chapter 5. The basic feature here is the *nominal system* - an unperturbed (2.2.7),

$$\dot{\bar{x}} = A(\bar{x},t)\bar{x} + B(\bar{x},t)\bar{u} \qquad (2.2.8)$$

whose motions have a stipulated performance from which (2.2.7) should not basically differ under the perturbations $\delta A, \delta B$ and $\bar{w}$. The matrix functions $A(\cdot)$, $\delta A(\cdot)$, $B(\cdot)$, $\delta B(\cdot)$ and $C(\cdot)$ are such as to secure the existence of motions of (2.2.7) through each $(\bar{x}^0,t_0) \in \Delta \times \mathbb{R}$, given $\bar{r},\bar{s},\bar{w}$ and $\bar{u}$.

If the so called *matching conditions* are met, that is, there are functions $D(\bar{x},t,\bar{r})$, $E(\bar{x},t,\bar{s})$ and $F(\bar{x},t)$ such that $\delta A = BD$, $\delta B = BE$ and $C = BF$, then (2.2.7) becomes

$$\dot{\bar{x}} = A(\bar{x},t)\bar{x} + B(\bar{x},t)\bar{u} + B(\bar{x},t)c(\bar{x},t) \qquad (2.2.9)$$

with $c(\bar{x},t) = D\bar{x} + E\bar{u} + F\bar{w}$ amalgamating all the uncertainties. Physically it means that the uncertainty is within the range of input. This lumped uncertainty is assumed continuous in $\bar{x}$, measurable in t and bounded by a known function:

$$\|c(\bar{x},t)\| \leq \rho(\bar{x},t) \qquad (2.2.10)$$

with the bound ρ depending also upon the ranges R, S and W of $\bar{r},\bar{s},\bar{w}$ respectively. Moreover all three $\|A(\bar{x},t)\|$, $\|B(\bar{x},t)\|$, $\rho(\bar{x},t)$ are assumed to be majorized by a Lebesgue integrable function of t. One of the advantages of the model lies in the fact that the control program $\bar{P}(\cdot)$ for $\bar{u}$

may include uncertainty as well. The matching conditions can be relaxed, see Barmish-Leitmann [1], but even then there remains a class of mechanical systems which would not satisfy them. The model is, however, wider than a similar model of the "system under persistent perturbations" developed by Barbashin [1]. In the case of our (2.2.1) or (2.2.2) it would require the uncertainty lumped in $\bar{R}$ or $\bar{R}$ and both $\bar{F}(\cdot)$ and $\bar{R}(\cdot)$ to be linear in $\bar{u}$ and $\bar{w}$ respectively, thus excluding the cases of uncertainty in inertia, gravity and damping as well as in the actuator gears.

The adaptation of the contingent format (2.2.6) to the Leitmann model (2.2.7) can be found in Goodall-Ryan [1].

EXERCISES 2.2

2.2.1 Consider the scalar system

$$\dot{x} = x + u\sqrt{x} + wx , \quad w \in [-1,1]$$

governed by the program $P(\cdot)$ defined by $u = \pm\sqrt{x}$. Write the contingent format of the state equation, determine the instantaneous attainability and reachability sets.

2.2.2 Do you know sufficient conditions for the existence of solutions to (2.2.6) other than those by Filippov?

2.2.3 Verify that $x(t) = x^0 \pm t$ are two motions of the contingent equation $\dot{x} \in [-1,1]$. Are there any other motions? What is the attainability set at some t?

2.2.4 Find equilibria of the system

$$\dot{x}_1 = x_2 , \quad \dot{x}_2 = -x_1 - ux_2 + wx_1^3 , \quad w \in [1,2] , \quad u \in [0,1] .$$

2.3 ENERGY SURFACE, CONSERVATIVE FRAME OF REFERENCE

The performance of the motions in state space can be measured and the state space pattern described in terms of a single scalar function, the total energy of our mechanical system, which is the stationary Hamiltonian. Indeed, as such, the energy is a point-function over Δ and mapping its levels into Δ gives a perfect frame of reference for investigating the behavior of state space motions. The topology of energy and power (= time derivative of energy) over Δ is the natural basis not only for the motion's analysis, but their synthesis (system design) and control, see Skowronski [4]-[9], [12]-[18], [20], [30]-[44], Koditchek [1],[2], Takegaki-Arimoto [2], Volpe-Khosla [1]. Effective control programs can be designed using both the energy and the power of the system in various branches of control: feedback, adaptive, coordination, dynamic games, etc. The reader will find many of these controllers later in this text. This Section describes the map of energy levels on Δ obtained directly from the said topology of energy on this set.

Observe that the total energy is the sum of kinetic and potential energies and since the potential forces, particularly gravity, may in general include uncertainty, so does the total energy $E(\bar{q},\dot{\bar{q}},\bar{w})$. We shall later refer such energy and the corresponding power to its *nominal values* (without uncertainty), $E(\bar{q},\dot{\bar{q}})$, that is why the systems without uncertainty (2.1.4), (2.1.15) form the base for describing the reference frame in this Section. Since we want to control our system motion against all the options of the uncertainty, the nominal values of energy for reference are obtained by substituting $\bar{w}^*$ which extremizes $E(\bar{q},\dot{\bar{q}},\bar{w})$. In particular it means minimizing $E(\bar{q},\dot{\bar{q}},\bar{w})$ if we want to keep the motions concerned below E-levels or maximizing $E(\bar{q},\dot{\bar{q}},\bar{w})$ if we want to keep them above E-levels on Δ . The cases will be described in detail in Section 3.2.

Consequently we consider here $E(\bar{q},\dot{\bar{q}}) = T(\bar{q},\dot{\bar{q}}) + V(\bar{q})$, with the obvious alternative notation of $E(\bar{q},\bar{p})$ and $E(\bar{x})$. Observe that in order for the map of the E-levels over Δ to be an effective reference frame in obtaining the full dynamic state information on the system behavior, we need both terms of the sum $T + V$. The potential energy alone gives only static information on the configuration of the system. This is in spite of the fact that we may use the potential energy generated applied forces as feedback controllers, see cited Takegaki-Arimoto [2].

To obtain the hypersurfaces $E(\bar{x}) = const$ we obviously use the energy integral of the conservative subsystem of our system. Let us make a few comments regarding such a subsystem.

Since the potential energy is the negative of potential, then by (1.5.16) the generalized potential forces are defined by the partial derivatives,

$$Q_i^P(\bar{q}) = -\frac{\partial V(\bar{q})}{\partial q_i} \ , \qquad i = 1,\ldots,n \ , \tag{2.3.1}$$

whence the potential energy is specified by

$$V(\bar{q}) = V(\bar{q}^0) - \sum_i \int_{q_i^0}^{q_i} Q_i^P(\bar{q}) dq_i \ . \tag{2.3.2}$$

The function $V(\cdot)$ is assumed defined at least on Δ_q, single valued and smooth to the degree required by (2.3.2) and whatever smoothness is needed for $Q_i^P(\cdot)$. Thus in most cases $V(\cdot)$ is taken as analytic.

By (2.3.1) and (1.5.23) the potential characteristics become

$$\Pi_i(\bar{q}) = \sum_j M_{ij}^{-1}(\bar{q}) \frac{\partial V(\bar{q})}{\partial q_j} \ , \tag{2.3.3}$$

and thus also

$$\Pi_i(\bar{q}) = - \sum_j M_{ij}^{-1}(\bar{q}) \frac{\partial U_P}{\partial p_j} \ . \tag{2.3.4}$$

The potential force Q_i^P is composed of gravity Q_i^G and spring Q_i^K, as indicated in (1.5.27) but may be augmented to include the designed compensation force either in gravity Q_i^{GC} or in spring Q_i^{KC} or both. Such compensation forces added to Q_i^P change the potential energy (2.3.2) and therefore may act as part of a controller. Such a role is particularly effective in changing the equilibria and thus the global state space pattern. The gravity forces Q_i^G, Q_i^{GC} are nonlinear by nature, often trigonometric functions and as such power series developable with some cut-off approximation, see our Example 1.1.1. The spring forces Q_i^K, Q_i^{KC} are often analytic functions as well, with the same possibility of truncating their power series representation. We discuss the problem later.

The subsystem of (1.7.10) or (1.5.20) which contains only kinetic and potential forces is conservative. Indeed, consider (1.7.10) and assume no effect from external forcing or damping forces, i.e. $Q_i^D(\bar{q},\bar{p}) \equiv 0$, $Q_i^F(\bar{q},\bar{p},\bar{u}) \equiv 0$, $Q_i^R(\bar{q},\bar{p},t) \equiv 0$. Substituting (2.3.1) it becomes

$$\dot{q}_i = \frac{\partial H(\bar{q},\bar{p})}{\partial p_i} \ , \qquad \dot{p}_i = - \frac{\partial H(\bar{q},\bar{p})}{\partial q_i} \ , \qquad i = 1,\ldots,n \ . \tag{2.3.5}$$

On the other hand, formally

$$\dot{H}(\bar{q},\bar{p}) = \sum_i \left[\frac{\partial H(\bar{q},\bar{p})}{\partial q_i} \dot{q}_i + \frac{\partial H(\bar{q},\bar{p})}{\partial p_i} \dot{p}_i \right] \ . \tag{2.3.6}$$

Substituting (2.3.5) into (2.3.6), we obtain $\dot{H}(\bar{q},\bar{p}) = 0$ or

$$H(\bar{q},\bar{p}) = \text{const} \ , \tag{2.3.7}$$

and, since the stationary Hamiltonian was equal to the total energy,

$E(\bar{q}, \dot{\bar{q}})$ = const , making (2.3.5) conservative with the total energy representing the first integral.

In terms of the Lagrangian equations (1.5.13) or the Newtonian format (1.5.20) the conservative system (2.3.5) becomes

$$\frac{d}{dt} \frac{\partial T}{\partial \dot{q}_i} - \frac{\partial T}{\partial q_i} = Q_i^P(\bar{q}) , \quad i = 1, \ldots, n , \qquad (2.3.8)$$

or

$$\sum_j M_{ij}(\bar{q}) \ddot{q}_i + \Gamma_i(\bar{q}, \dot{\bar{q}}) - Q^P(\bar{q}) = 0 , \quad i = 1, \ldots, n \qquad (2.3.9)$$

respectively. Consequently to (2.3.7), whatever the amount of $E(\bar{q}, \dot{\bar{q}})$ is used by the kinetic energy in motion, it is balanced by $V(\bar{q})$ and restituted by $\bar{Q}^P$. Consequently the components of $\bar{Q}^P$ are called restitutive (restoring) forces as an alternative name to potential or conservative forces. Note that the centrifugal-Coriolis forces hide in $H(\bar{q}, \bar{p})$ in the passage between (2.3.9) and (2.3.5) and do not affect the change of energy. They are thus called *energy neutral* forces or traditors, see Duinker [1].

The assumed single valuedness of $V(\cdot)$ means that restitution tracks back precisely the graph of usage, see Section 1.5. It also means that Q_i^P, Π_i are single valued which, particularly with spring forces, corresponds to elastic behavior.

In hydraulic systems, the restitutive device of our type means either the adiabatic compressiblity of a fluid or the perfect elasticity of a container. We see both in action when filling a balloon with such fluid. Energy is stored because the fluid is compressed and because the balloon is stretched. It changes back to the kinetic energy when we let the fluid escape. Then the decompression of the fluid and the shrinkage of the tank is supposed to trace back the graphs of compression (adiabatic) and stretching (elastic) respectively.

Such modelling is not entirely realistic. There will be hysteresis in bodies, a non-adiabatic compression in pneumatic or hydraulic suspension etc., so we should apply a multi-valued function $Q_i^K(\cdot)$. Then the system would cease to be conservative. However, one may maintain the restitution property and thus single valuedness of $Q_i^K(\cdot)$ and $V(\cdot)$ by shifting the non-elastic phenomena to the damping characteristics which complement Q_i^K in our overall model, see Skowronski [24]. Without going into details, let us say that this procedure is physically justified and often used, see Christensen [1].

Substituting $Q_i^D \equiv 0$, $Q_i^F \equiv 0$ into (1.7.10), (1.5.20), we have obtained the state representations of the conservative system. Substituting the same into the equilibria equations (2.1.9), (2.1.10), we see that the equilibria of the conservative system are defined by

$$\dot{q}_i = 0 , \quad Q_i^P(\bar{q}) = 0 , \quad i = 1,\ldots,n , \tag{2.3.10}$$

that is, by vanishing of the potential forces. The above implies immediately

$$\dot{q}_i = 0 , \quad \Pi_i(\bar{q}) = 0 , \quad i = 1,\ldots,n . \tag{2.3.11}$$

Conversely, (2.3.11) implies zero acceleration, that is, motion with constant velocity. But *isolated roots* of (2.3.10) have zero-velocity, that is, they *are* rest positions or *equilibria*. Consequently we consider (2.3.10), (2.3.11) equivalent. Obviously the compensation terms in Q_i^P or Π_i may shift the equilibria, as mentioned. In turn, except for the influence of the perturbations $R(q,\dot{q},t)$ and input $\bar{F}(\bar{q},\dot{\bar{q}},\bar{u})$ the equilibria coincide with those of (1.5.26) defined by (2.1.10). Indeed, the viscous or Coulomb type of damping used in most of our case studies implies that there is no damping whenever there is no motion, that is, at rest:

$$D_i(\bar{q},0,\bar{w}) \equiv 0 , \quad \forall \bar{w},\bar{q} , \quad i = 1,\ldots,n . \tag{2.3.12}$$

Moreover, there are also no Coriolis or gyro forces when the system rests,

$$\Gamma_i(\bar{q},0,\bar{w}) \equiv 0 , \quad \forall \bar{w},\bar{q} , \quad i = 1,\ldots,n . \tag{2.3.13}$$

Also, as we deal with the reference system, we want it free from input and perturbations: $\bar{F}(\bar{q},\dot{\bar{q}},\bar{w},\bar{u})$, $\bar{R}(\bar{q},\dot{\bar{q}},\bar{w},t) \equiv 0$. Substituting the latter and (2.3.12), (2.3.13) into (1.5.26) we obtain (2.3.11). A similar argument applies for (2.3.10).

Observe further that (2.3.1) and (2.3.10) together imply that the equilibria lie at the extremal points of the potential energy surface $V : v = V(\bar{q})$ in the space $\mathbb{R} \times \mathbb{R}^n$ over the set Δ_q . At the same time the regular points of Δ are the non-extremal points of V . The surface V is single-sheeted and smooth, since V was assumed single valued and continuously differentiable. There is little chance that any element of the mechanical system may become an energy sink with negative storage. Thus there is no narrowing in assumption that, if V admits negative values at all, they are bounded. With the latter and with the origin placed at an absolute minimum of V over the closure Δ_q , we can always adjust the free constant $V(\bar{q}^0)$ of (2.3.2) such as to obtain V positive semi-definite, if the minimum is not single, or positive definite, if it is.

Since the equilibria coincide with the extrema of V, they are isolated (see our comment in Section 2.1) and there is a finite sequence of them in the bounded Δ. See Fig. 2.6.

As stated by Lagrange, proved by Dirichlet and mentioned before, the isolated minima of V are *Dirichlet stable*. We use this term in the plain meaning of elementary physics in high school. The reader may as well bring back memories of a ball returning to a stable position at the bottom of a convex two-dimensional V.

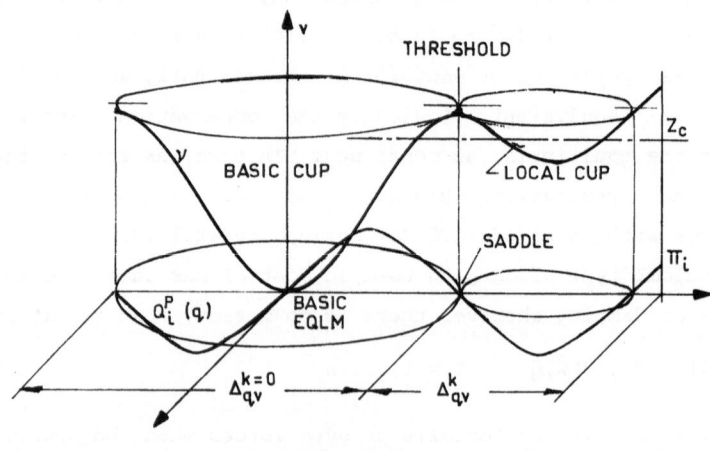

Fig. 2.6

The matrix $(-\partial Q_i^P/\partial q_j)$ is called the *functional coefficient* of *restitution* of the system. The restitution will be said to be positive if the matrix is positive definite, that is, all its principal minors are positive, which suffices for a local minimum. Thus we have what may be called the *Dirichlet Property*: *An equilibrium is Dirichlet stable if the restitution in its neighborhood is positive.*

Note that the matrix $(-\partial Q_i^P/\partial p_j)$ may be immediately replaced by $(\partial \Pi_i/\partial q_j)$, without a change of conditions.

The sections $V(\bar{q}) = \text{constant} = v_c$ of V are the potential energy levels Z_c, with $v_c \in [0, v_\Delta), v_\Delta = \sup V(\bar{q}) |\bar{q} \in \Delta$. We shall assume that each Z_c separates a region of V from its complement. Note that Z_c may be disjoint so that the region separated may be disjoint as well.

The level corresponding to the absolute minimum of V on Δ_q is called *basic* and the corresponding equilibria (one or more) the *basic equilibria*.

We have placed the origin of $\mathbb{R}^n$ at a basic equilibrium and arranged for the constant $V(\bar{q}^0)$ of (2.3.2) to be such that $V(0) = 0$. This means

$$Q_i^P(0) = 0, \quad i = 1,\ldots,n, \tag{2.3.14}$$

and

$$\Pi_i(0) = 0, \quad i = 1,\ldots,n \tag{2.3.15}$$

in agreement with the zero-equilibrium of Section 2.1. Then there is a neighborhood $\Delta_{qv} \subset \Delta_q$ of the origin on which $V(\bar{q})$ increases in $|q_1|,\ldots,|q_n|$. The pattern continues for all q_i up to the lowest level corresponding to the first relative maximum or inflection of V with respect to some q_i. The part of this level neighboring 0 is called the *potential energy threshold*, denoted $z_{cv}, v_c = v_{cv}$. The threshold separates a simply connected region from the rest of V, and is called the *potential energy cup* (basic), denoted Z_v. It is obviously subject to compensation, see Chapter 3.

Consider the continuous v_c- family of levels below z_{cv}, that is, for $v_c \leq v_{cv}$. Such levels have parts enclosed in Z_v, see Fig. 2.6, these parts *form a nest* about 0: they do not cross one another and each of them separates a simply connected region of V. The threshold itself is the upper bound of the family. Indeed, from Morse [1] we know that opening of the level surfaces is possible only at saddles in $\mathbb{R}^n$, which specify the threshold. The threshold projected orthogonally into Δ_q determines Δ_{qv}. The projection of the said family of levels behaves identically in Δ_{qv}, the map is isometric (distance preserving).

By Dirichlet Property we may pick up other stable equilibria. Moving the origin of $\mathbb{R}^n$ by a simple transformation of variables, we may repeat our construction of the local cup about any other Dirichlet stable equilibrium. We then obtain a sequence of z_v^k, $k = 0,1,\ldots,M < \infty$, finite since Δ is bounded, with $V = 0$ referring to the basic cup. Correspondingly there is a sequence of Δ_{qv}^k. It is shown, Shestakov [1], that the stable equilibria and thus Δ_{qv}^k are separated by the Dirichlet unstable equilibria (maxima). Obviously the threshold of $z_v^{k=1}$ will be higher than that of $z_v^{k=0}$ and $\Delta_{qv}^{k=0}$, but on the other hand it may not. It depends upon the case scenario.

Can we estimate or change the size of Δ_{qv}^k? Evidently, there is the possibility of a single equilibrium occuring in Δ_q when the function Q_i^P or Π_i are "hard in the large", that is, with positive partial derivatives yielding a monotone increase on Δ_q. Then no threshold ever appears and

z_v^k stretches indefinitely yielding $\Delta_{qv} = \Delta_q$. If it is not the latter case, one encounters at least two equilibria. So there is at least one that neighbors the basic equilibrium. Thus the threshold must appear, if not sooner, then at this neighboring equilibrium. Since the positive coefficient of restitution suffices to attain the minimum at any equilibrium, thus conversely such a minimum is necessary for positive restitution. Shestakov [1] proved that the neighboring equilibrium cannot be a minimal point again, so the positive restitution is contradicted there. Then, provided the threshold crosses over the said equilibrium which is Dirichlet unstable, the positive coefficient of restitution defines Δ_{qv}. The above proviso must hold for $N = 2$, but quite often holds for $N > 2$. Since $V(q)$ is symmetric by definition, the basic cup is always symmetric and our proviso holds for the basic cup for $N > 2$. We thus have the following.

Property of Restitution: The region Δ_{qv} is defined by the positive coefficient of restitution with accuracy to within the distance between the threshold and the nearest unstable equilibrium.

Note here that the equilibria and the thresholds are subject to compensation control.

We have also an alternative, physically motivated criterion for Δ_{qv}. By the very nature of equilibrium, when a motion goes outwards from a Dirichlet stable equilibrium, the restoring force must tend towards this equilibrium and vice versa: the sense of motion and restoring force vectors are opposite, see (1.5.34),

$$Q_i^P(\bar{q}) \cdot q_i < 0 , \quad q_i \neq 0 , \quad i = 1,\ldots,n \tag{2.3.16}$$

yielding in turn the so called *restitution law*

$$\sum_i Q_i^P(\bar{q}) q_i = \bar{Q}^P(\bar{q})^T \cdot \bar{q} < 0 , \quad \bar{q} \neq 0 . \tag{2.3.17}$$

Correspondingly in terms of characteristics, the above becomes

$$\sum_i \Pi_i(\bar{q}) q_i = \bar{\Pi}(\bar{q})^T \bar{q} > 0 , \quad \bar{q} \neq 0 . \tag{2.3.18}$$

Since (2.3.16) is implied by the stable equilibrium, they are necessary conditions for the minimum. Consider (2.3.16). At the neighboring unstable equilibrium, all Q_i^P cross zero and as they are smooth functions must gain the opposite sign, which contradicts (2.3.16). Contradicting the necessary conditions produces the opposite property, hence the *boundary of the set defined by (2.3.18) determines the threshold:*

$$\bar{\Pi}(\bar{q})^T \bar{q} = 0 . \tag{2.3.19}$$

In multidimensional spaces, however, it may prove difficult to find the lowest unstable equilibrium (saddle) for a cup. Here is the technique which may help. The saddle appears at the opening of the levels Z_c (see Morse [1]) and these levels tend to open at arcs where the growth of $V(\bar{q})$ slows down, that is, where the gradient $\nabla V(\bar{q}) \neq 0$ is minimal. A curve on V that passes through the minimum of z_v^k and minimizes $|\nabla V(\bar{q})|$ intersects Z_{cv} at (2.3.19) and determines the saddle. In order to find such a minimizing curve, we use Lagrange multipliers. The function to be minimized is $\sum_i (\partial V/\partial q_i)^2$, subject to constraints V. Hence the Lagrange's function is $\sum_i (\partial V/\partial q_i)^2 - \lambda V(\bar{q})$. Equalizing partial derivatives to zero, we obtain

$$2\sum_j \frac{\partial^2 V(\bar{q})}{\partial q_i \partial q_j} \frac{\partial V(\bar{q})}{\partial q_j} - \lambda \frac{\partial V(\bar{q})}{\partial q_i} = 0 , \quad i,j = 1,\ldots,n .$$

Eliminating the parameter λ, we have the equation of the curve

$$\left(\sum_{j=1}^n \frac{\partial^2 V}{\partial q_i \partial q_j} \frac{\partial V}{\partial q_j} \right) \frac{\partial V}{\partial q_n} - \left(\sum_{j=1}^n \frac{\partial^2 V}{\partial q_n \partial q_j} \frac{\partial V}{\partial q_j} \right) \frac{\partial V}{\partial q_i} = 0 , \quad i = 1,\ldots,n-1$$

which passes through $(\partial V/\partial q_i) = 0$.

Let us return now to the sequence of cups z_v^k. Since the cups and their corresponding thresholds are defined, so is the highest threshold defined as the maximal over bounded Δ_q. Let z_L be the infimum of the levels Z_c located above the highest threshold. Then z_L separates a region enclosing all the cups z_v^k from the rest of V. This region projected into Δ_q defines the set Δ_{qL} which encloses the union of Δ_{qv}^k, $k = 0,1,\ldots,m$. As there are no thresholds in $\Delta_q - \Delta_{qL}$, this region forms a structure that resembles a local cup, about ∞ and will be thus called the potential energy cup *in the large*.

EXAMPLE 2.3.1. We consider a simple single-DOF system, being an artificially designed perfect subcase of our discussion but matching many applications. Let $\Pi(q) = aq + bq^3 + cq^5$, $q(t), b \in \mathbb{R}$; $a,c > 0$, with certain limitations imposed below on the "softness coefficient" $|b|$. We have immediately

$$V(q) = \tfrac{1}{2} aq^2 + \tfrac{1}{4} bq^4 + \tfrac{1}{6} cq^6 . \tag{2.3.20}$$

The equilibria are obtained from $\partial V/dq = \Pi(q) = q(a + bq^2 + cq^4) = 0$, which yields

$$q^{(0)} = 0 ; \quad q^{(1)}, q^{(3)} = \left(\frac{-b \pm \sqrt{b^2 - 4ac}}{2c} \right)^{\frac{1}{2}} ;$$

$$q^{(2)}, q^{(4)} = -\left(\frac{-b \pm \sqrt{b^2 - 4ac}}{2c} \right)^{\frac{1}{2}} .$$

For $b \geq 0$, the characteristic Π is hard and there should be only one Dirichlet stable equilibrium (basic) $q^{(0)} = 0$. Indeed, by the formulae, the values $q^{(1)}, \ldots, q^{(4)}$ are all imaginary. The same situation occurs for $b < 0$ but sufficiently small in its absolute value, namely $b^2 < 4ac$. On the other hand, for $b < 0$ with $b^2 > 4ac$ we have all five, and with $b^2 = 4ac$, three equilibria $q^{(0)}, q^{(1)} = q^{(3)}, q^{(2)} = q^{(4)}$. The greater is the softening $|b|$ the more equilibria appear. However $|b|$ is limited from above by the physical requirement that (2.3.20) must be positive for all $q \neq 0$.

In order to analyse (2.3.20) in some detail, we let $a = c = 1$, $b = -2.1$, which yields

$$
\left.
\begin{aligned}
V(q) &= \tfrac{1}{2} q^2 - 0.53 q^4 + \tfrac{1}{6} q^6 , \\
\partial V / \partial q &= \Pi(q) = q - 2.1 q^3 + q^5 , \\
\partial^2 V / \partial q^2 &= 1 - 6.3 q^2 + 5 q^4 , \\
\Pi(q) \cdot q &= q^2 - 2.1 q^4 + q^6 .
\end{aligned}
\right\}
\qquad (2.3.21)
$$

The results of calculation are in the following table:

q	-1.17		-0.85		0		0.85		1.17	
$V(q)$	0.12		0.15		0		0.15		0.12	
	↘	min ↗	max	↘	min	↗	max	↘	min	↗
$\dfrac{\partial V}{\partial q}(q) = \Pi(q)$	−	0	+	0	−	0	+	0	−	0 +
$\partial^2 V / \partial q^2$	1.74		-0.94		1		-0.94		1.74	
$\Pi(q) \cdot q$		−				+				−

$\square$

EXAMPLE 2.3.2. We turn now to a "real life" case adapted from Blaquière [1], where also literature on the problem is quoted. The case is that of betatron oscillations of particles in a vacuum chamber of a particle accelerator, the oscillations described in-the-mean by the following model, Hagedorn [1]:

$$
\begin{cases}
\ddot{\xi}_1 + Q_1^2 \xi_1 = \dfrac{\alpha}{2} (\xi_1^2 - \xi_2^2) , \\
\ddot{\xi}_2 + Q_2^2 \xi_2 = -\alpha \xi_1 \xi_2 ,
\end{cases}
\qquad (2.3.22)
$$

where $\xi_1(t), \xi_2(t) \in \mathbb{R}$ are the radial and vertical deviations from a reference orbit (reduced to an equilibrium), α is a coupling coefficient and Q_1, Q_2 are respectively the numbers of betatron oscillations experienced by variables ξ_1, ξ_2 during each revolution of particles in the vacuum chamber. The storage function (potential) used, Hagedorn [1], is such as to lead to

$$V(\xi_1, \xi_2) = \frac{1}{2} \varrho_1^2 \xi_1^2 + \frac{1}{2} \varrho_2^2 \xi_2^2 + \frac{\alpha}{2} (\xi_2^2 \xi_1 - \frac{1}{3} \xi_1^3) \ .$$

To simplify matters, we introduce the following transformation

$$q_1 = -\frac{\alpha}{\varrho_2^2} \xi_1 \ , \qquad q_2 = -\frac{\alpha}{\varrho_2^2} \xi_2 \ ,$$

and let $\mu = \varrho_1^2/\varrho_2^2$. Then

$$\frac{2\alpha^2}{\varrho_2^6} V(\xi_1, \xi_2) = V(q_1, q_2) = \mu q_1^2 + q_2^2 - (q_1 q_2^2 - \frac{1}{3} q_1^3) \ .$$

For the equilibria we require

$$\frac{\partial V}{\partial q_1} = 2\mu q_1 - q_2^2 + q_1^2 = 0 \ , \qquad \frac{\partial V}{\partial q_2} = 2q_2 (1 - q_1) = 0 \ .$$

These two curves cross at the points:

$$q^{(0)} : q_1^{(0)} = 0 \ , \quad q_2^{(0)} = 0 \ ; \quad q^{(1)} : q_1^{(1)} = -2\mu \ , \quad q_2^{(1)} = 0 \ ;$$

$$q^{(2)}, q^{(3)} : q_1^{(2)} = q_1^{(3)} = 1 \ , \quad q_2^{(2)}, q_2^{(3)} = \pm(1 + 2\mu)^{\frac{1}{2}} \ .$$

The equilibria are shown in Fig. 2.7. We may verify easily that $q^{(0)}$ is the absolute minimum and thus the basic potential energy level while the others are saddle points. The extrema are $V(q^{(0)}) = 0$, $V(q^{(1)}) = \frac{4}{3}\mu^3$, $V(q^{(2)}) = V(q^{(3)}) = \mu + \frac{1}{3}$. If we again shall make our life easier by assuming $\mu = 1$, which does no harm to the betatron, Blaquière [1], then we obtain the threshold joining all three saddles at $V_{cv} = 4/3$. The threshold mapped into $\mathbb{R}^2$ produces the boundary

$$\partial \Delta_{qv} : q_1^2 + q_2^2 - q_1 q_2^2 + \frac{1}{3} q_1^3 = \frac{4}{3} \ ,$$

or

$$[3q_2^2 - (q_1 + 2)^2](q_1 - 1) = 0$$

made of three straight lines:

$$(2) - (3) \ q_1 = 1 \ , \quad (1) - (2) \ q_2 = \frac{1}{\sqrt{3}} (q_1 + 2) \ , \quad (1) - (3) \ q_2 = -\frac{1}{\sqrt{3}} (q_1 + 2) \ ;$$

see Fig. 2.7.

Obviously the triangle (1) - (2) - (3) bounds the Δ_{qv} about $q^{(0)} = 0$. Around this basic cup, we find three hills on one threshold level and beyond it, another three cups - open definitely. Observe that the picture is invariant under rotation by $2\pi/3$. $\square$

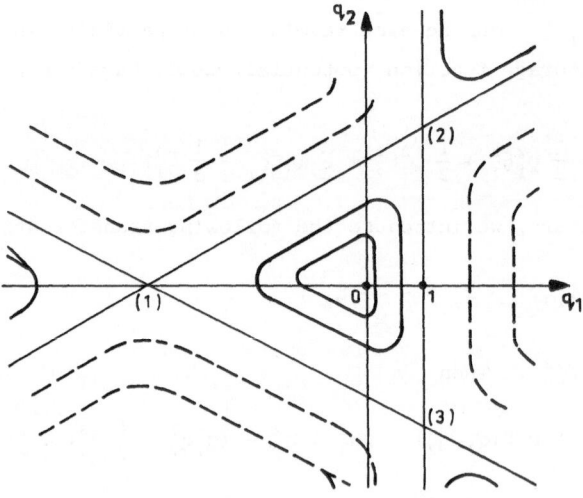

Fig. 2.7

EXAMPLE 2.3.3. Let us now augment the two pendula of Example 1.1.2 to several such pendula with the motion represented by n point-masses m_i rotating along a circle shown in Fig. 2.8. The masses are subject to gravity $\frac{g}{\ell} \sin q_i$ and elastic coupling between each two $K_{ij}(q_i - q_j) = k_{ij} \sin(q_i - q_j)$, with damping ignored. This gives the motion equations

$$\ddot{q}_i + \frac{g}{\ell} \sin q_i + \sum_j a^2 k \sin(q_i - q_j) = 0, \quad i = 1, \ldots, n.$$

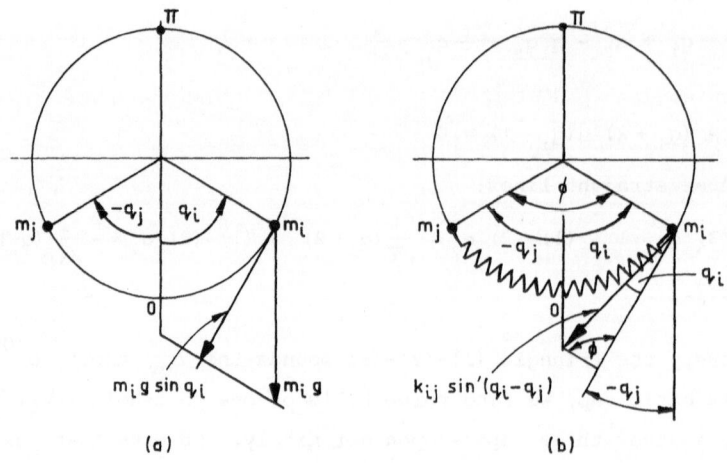

Fig. 2.8

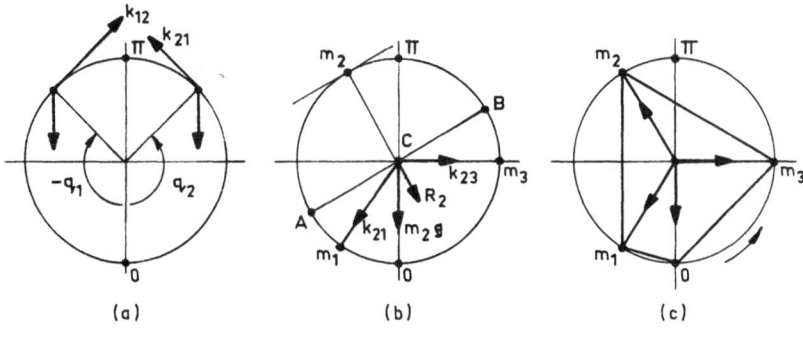

Fig. 2.9

Let us first repeat the case for n = 2. It is seen in Fig. 2.8(b) that the two point masses are pulled towards each other by the elastic coupling, whence by (2.3.18), 0 is a Dirichlet stable equilibrium. There are three more obvious equilibrium positions $(q_1, q_2) = (0, \pi)$, $(\pi, 0)$ and (π, π). If one of the masses is at 0 and the other at π, then any deflection of one of them from its position makes the mass at π pulled down towards 0 by both the spring and the gravity. Hence $(0, \pi), (\pi, 0)$ are unstable. If both masses are at π, any deflection from this equilibrium makes them fall to 0, so (π, π) is unstable as well. Apart from the above, we can have some instantaneous equilibria produced by the balance between gravity and the coupling spring forces. Indeed, it is seen from Fig. 2.8(b) that when both points are located on the upper semicircle, the spring is pulling them towards π while gravity forces them towards 0. It is intuitively obvious, but may also be shown rigorously, that such equilibria are unstable. Hence the only energy cup existing is located about 0, up to the first unstable equilibrium.

Consider now the case n = 3 shown in Fig. 2.9. Here also, 0 is an obvious stable equilibrium and there are no other stable equilibria, so that we still have a single energy cup. To show this we follow the argument of Vaiman [1] and look at the points $m_1, m_2, m_3, 0$ as vertices of some quadrangle inscribed in the circle. We can always find two arcs, smaller than or equal to $\pi/2$ such that the sum of two reciprocal angles equals π.

Consequently, there is at least one point mass, at one of the vertices, such that the corresponding angle is below $\pi/2$. Such an angle opens an arc, smaller than 180°, on which the other three vertices are located. Hence we can always find a semicircle on which there is only one point mass. For instance, consider the semicircle A, m_2, B in Fig. 2.9(b).

The diameter enclosing such a semicircle separates the point mass concerned from the other two and the point O. See the diameter AB separating m_2 in Fig. 2.9(b). Suppose there is an equilibrium as shown in Fig. 2.9(c). We may show that it is unstable. Indeed, consider forces acting upon m_2 and move them to the centre C. They are the gravity $m_2 g$, and the elastic couplings with amplitudes k_{12} and k_{23}. As they successively act along the directions CO, Cm_1, Cm_3, they will be left on one side of AB. Hence their resultant also must be on this side of AB. Since we discuss equilibria the tangential component of the resultant acting upon each of the masses vanishes and the resultant force R_2 on m_2 is directed radially away from C. Consequently the equilibrium defined for m_1, m_2, m_3 is unstable.

$\square$

Let us now introduce the space $h \times \mathbb{R}^N$, $N = 2n$, and consider the surface H in $h \times \mathbb{R}^N$ generated by the function $h = E(\bar{q}, \dot{\bar{q}})$ over $\Delta \subset \mathbb{R}^N$, that is, confined to the set $Z \overset{\Delta}{=} [0, h_\Delta) \times \Delta$, $h_\Delta = \sup E(\bar{x}) | \bar{x} \in \Delta$. For tangibility we shall use $h = E(\bar{q}, \dot{\bar{q}})$, but there is no difference to our further discussion whether we follow $E(\bar{q}, \dot{\bar{q}})$ of (1.5.26) or $H(\bar{q}, \bar{p})$ of (1.7.10). The set Z is bounded since Δ is bounded, see Fig. 2.8.

Next, we introduce a level surface of $E(\cdot)$, $Z_C \overset{\Delta}{=} \{ (h, \bar{q}, \dot{\bar{q}}) \in Z | E(\bar{x}) = $ = constant $= h_C \}$. We say h_C is a *regular value* if $(h, \bar{q}, \dot{\bar{q}}) \in Z_C$ implies $\nabla E(\bar{q}, \dot{\bar{q}}) \neq 0$. It follows from Wilson [1] that almost all values of h are regular. It follows from the derivation of (2.3.7) that the trajectories of the conservative (2.3.5) or (2.3.9) form a h_C-family of curves on the surface H. To every N-tuple $\bar{x}^0 = (x_1^0, \ldots, x_N^0)^T$ of the constants of integration of (2.3.5) in Δ, there corresponds a value $h_C \in [0, h_\Delta)$ which defines some Z_C, called alternatively the integral level. The inverse map is obviously set valued. Nevertheless, exhausting all $\bar{x}^0 \in \Delta$ we obtain the surface H.

Let us project orthogonally the h_C-family of levels onto Δ. For each level we obtain a set E_C in Δ. It follows from the implicit function theorem of Calculus that, given regular h_C, the corresponding E_C is an $(N-1)$-dimensional hypersurface in $\mathbb{R}^N$, with the degree of smoothness determined by the smoothness of H. Such a surface is called *topographic* by analogy to the geographic card images of the height levels in a terrain. According to the above, the first integral of the conservative system (2.3.5) is accommodated by the family of E_C's in Δ, each E_C representing a trajectory picked up by specifying the initial state $\bar{x}^0 = (\bar{q}^0, \dot{\bar{q}}^0)$. The trajectories do not cross, neither do the levels, each defined by $E(\bar{q}, \dot{\bar{q}}) = $ constant $= h_C$.

The shape of H is obviously determined by the fact that its generator $h = E(\bar{q},\dot{\bar{q}})$ is represented by the sum $T(\bar{q},\dot{q}) + V(\bar{q})$. By (1.5.6), $T(\cdot)$ is a single valued, smooth, positive definite function, even and symmetric with respect to the hyper-plane $\dot{\bar{q}} = 0$ in Z . For all practical purposes we can envisage it as an (n-1)-dimensional extension of a parabola. Since the surface H is obtained by superposition of $T(\bar{q},\dot{\bar{q}})$ and $V(\bar{q})$, the latter analysed in terms of the surface V , it is the superposition of the parabolic T over V along a bottom line of the latter, see Fig. 2.10. There are no changes in the notions introduced for V except for expanding them along the additional set of velocity axes $\dot{q}_1,\ldots,\dot{q}_n$ or momentum axes $p_1,\ldots,p_n$, and thus adopting slightly different notation.

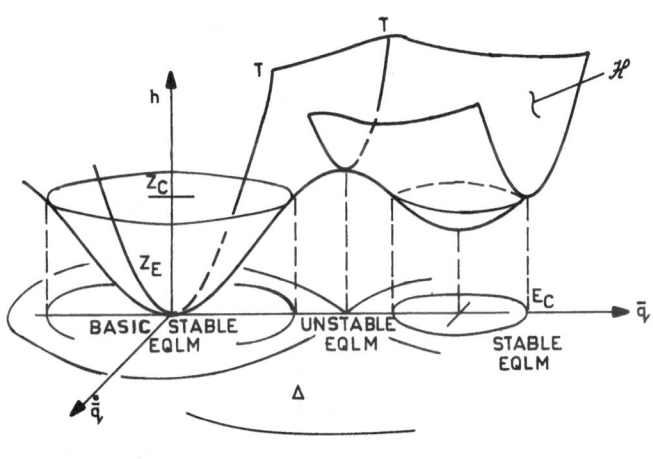

Fig. 2.10

The equilibria have been shown to coincide with the extremal arguments of V . Since $T(\bar{q},0) \equiv 0$, they also coincide with the extremal arguments of $E(\bar{q},\dot{\bar{q}})$ thus underlying the extrema of H . By the shape of $T(\bar{q},\dot{q})$ the Dirichlet stable equilibria of V define minima of H with respect to both V and T , and the Dirichlet unstable equilibria define the saddles of H , that is, maxima of V and minima of T . The iso-energy levels of Z_C are defined by the cuts $T(\bar{q},\dot{q}) + V(\bar{q}) = $ constant $= h_C$ of H and the corresponding topographic surfaces E_C accommodate trajectories of (2.3.5) starting there, and only these trajectories. This is what we meant when saying that the conservative system (2.3.5) serves as the conservative reference system. Since both T and V are positive definite functions, E is positive definite and again we let the origin of Z at the basic equilibrium, corresponding to the absolute minimum of H and to the basic total energy level, briefly *basic energy level* $h = 0$.

As the basic energy level is a double minimum of both V and T, there is a neighborhood $\Delta_E \subset \Delta$ of the basic equilibrium on which E increases in $|q_1|, \ldots, |q_n|, |\dot{q}_1|, \ldots, |\dot{q}_n|$. Since T has no extremal values except 0, it is only the potential energy threshold which may break down the increase. To any potential energy level $V(q) = const$, there corresponds some level z_C in H . Thus the potential energy threshold becomes the *energy threshold* $z_{CE} : E(\bar{q}, \dot{\bar{q}}) = h_{CE}$. Correspondingly, the basic potential energy cup z_v becomes the *basic energy cup* z_E . It corresponds to Δ_E by the fact that the boundary $\partial\Delta_E = E_{CE}$ of Δ_E is an isometric image of z_{CE} . Since the thresholds of V and H coincide, E_{CE} is that E_C which passes through the boundary of Δ_{qv} and since both E_C and Δ_{qv} are defined, the E_{CE} is defined as well, as so is Δ_E . The set E_{CE} is called the *separating set* (generalized separatrix) of Δ_E . Hence, if the potential threshold does not appear: $\Delta_{qv} = \Delta$ (linear or hard Q_i^P), then the energy threshold z_{CE} never appears either, and $\Delta_E = \Delta$.

We may now describe the interior of z_E . Similarly to its potential counterpart z_v, it consists in the continuous h - family, $h \in [0, h_{CE}]$, of components of z_C nested about the basic equilibrium, each separating a simply connected region from the rest of H . The surfaces E_C in Δ_E behave correspondingly. The threshold z_{CE} is the least upper bound of this family of z_C and separates z_E . So does H_{CE} for Δ_E . Since V is symmetric on Δ_E, the threshold z_{CE} coincides with the first maxima with respect to all q_i, $i = 1, \ldots, n$ at the two neighboring Dirichlet unstable equilibria.

Clearly the global shape of H depends on V ($V(\cdot)$ is called the *driving function*). Once $\bar{Q}^P$ and thus V are given (V with accuracy to the constant $V(\bar{q}^0)$), (2.3.15) picks up the equilibria, and (2.3.20) selects the Dirichlet stable ones. Similarly as for V, we may repeat our construction about other Dirichlet stable equilibria transforming the coordinates suitably. We obtain the local cups z_E^k, $k = 0, 1, \ldots, M$ with underlying Δ_E^k .

With all the cups defined, the highest threshold is also defined by $h_{CE}^{max} = max(h_{CE}^1, \ldots, h_{CE}^M)$. Then we again let $z_L | h_L = \inf h_C > h_{CE}^{max}$. The set z_L separates the region $z_{EL} \subset H$ that includes all the local cups z_E^k and defines the corresponding Δ_L, in turn including the union $\underset{k}{\cup} \Delta_E^k$. The shape of V in the large extends to that of h because of the shape of T . Over the complement of Δ_L to Δ, denoted $C\Delta_L$, there are no critical points, and in $Cz_{EL} = Z - z_{EL}$ there are no extrema of H .

Thus the structure of CZ_{EL} resembles that of a local cup with the nest centred about ∞. Therefore CZ_{EL} is called the *energy cup in the large*. The consequences regarding its image in Δ are obvious.

Let us now consider the Z_C all over H, and denote by Δ_C the regions separated by the corresponding $E_C = \partial \Delta_C$ in Δ. It is instructive to envisage H in three dimensions as a well with the bottom line V and walls $T(\bar{q}, \dot{\bar{q}})$ thus consisting of a number of valleys or energy cups separated by saddles. We let the well be exposed to an intensive rain scattered evenly all over the place, thus the well being successively filled up with water. The rising water levels illustrate the varying Z_C, while the area of the free water surface at each level illustrates the corresponding sets Δ_C.

Concluding, in general, the levels Z_C produce the sets Δ_C either simply connected with simply connected complements $C\Delta_C = \Delta - \Delta_C$, or disjoint with the complements multiply connected. For an example, see Fig. 2.6.

As there are no openings between the cups except over the thresholds, the rain fills up the cups quite independently of each other, yielding small disjoint lakes until the free water surface rises above separation sets. The study within cups is local. With rising water, more and more lakes become connected, requiring a global study. Finally at the level Z_L we obtain a large single lake surrounded by the remainder of H without thresholds, that is, the *cup in the large*. The investigation exercised on this remaining part of H, that is, over $C\Delta_L$, will be called the *study in large*.

EXAMPLE 2.3.4. We continue the Example 2.3.1, extending it now to the surface H, the latter derived by

$$E(x) = \tfrac{1}{2}\dot{q}^2 + \int \Pi(q)dq = \tfrac{1}{2}\dot{q}^2 + \tfrac{1}{2}aq^2 + \tfrac{1}{4}bq^4 + \tfrac{1}{6}cq^6 . \qquad (2.3.23)$$

The extremal points are defined by $\partial E/\partial q = aq + bq^3 + cq^5 = 0$, $\partial E/\partial \dot{q} = 0$ thus coinciding with the equilibria $q^{(0)}, \ldots, q^{(4)}$. Using the same data, namely $a = c = 1$, $b = -2.1$, we obtain the threshold $Z_{CE} : E(\bar{x}) = h_{CE}$ where

$$h_{CE} = E(q^{(1)}, 0) = E(q^{(2)}, 0) = 0.15 .$$

The E_{CE} passing through $q^{(1)}, q^{(2)}$ is

$$0.5\dot{q}^2 + 0.5q^2 - 0.53q^4 + 0.17q^6 = 0.15$$

and it encloses not only Δ_E but three disjoint Δ_E^s, the basic Δ_E inclusive. From the table of Example 2.3.1, the threshold $h_{CE} = 0.15$ corresponding to it is the highest of the system on Δ so $\tilde{E}_{CE} = \partial \Delta_L$, and the complement

of sets enclosed by this curve is the region in-the-large $C_{\Delta}\Delta_L$ that under-
lies the cup in-the-large. The reader may himself check, as an exercise,
that (2.3.23) is an integral of

$$\ddot{q} + q - 2.1\,q^3 + q^5 = 0 \ . \tag{2.3.24}$$

We shall turn now to a further extension, namely for the case of
$N = 2n = 4$. We see the reference system now as

$$\ddot{q}_i + \Pi_i(q_1,q_2) = 0 \ , \qquad i = 1,2 \tag{2.3.25}$$

with

$$\Pi_1(q_1,q_2) = \Pi_{11}(q_1) + \Pi_{12}(q_1 - q_2) \ ,$$
$$\Pi_2(q_1,q_2) = \Pi_{22}(q_2) + \Pi_{21}(q_1 - q_2) \ .$$

A large class of physical models allows symmetry in coupling, namely

$$\Pi_{12}(q_1 - q_2) = -\Pi_{21}(q_1 - q_2)$$

which leads to significant consequences.

First, the restitution law becomes simplified and takes the form

$$\Pi(\bar{q})\bar{q} = \Pi_{11}(q_1)q_1 + \Pi_{22}(q_2)q_2 + \Pi_{12}(q_1 - q_2)\cdot(q_1 - q_2) \geq 0 \tag{2.3.26}$$

indicating that the restitutive forces have been directed opposite to the
corresponding displacements, which is a simple physical fact, and must
include the coupling force with respect to the relative displacement
$(q_1 - q_2)$. Obviously (2.3.26) does not imply this fact. It is only the
necessary condition for the minimum, thus also for the stability of equilibrium at the basic level.

Secondly, we obtain for $i = 1,2$,

$$\begin{cases} \Pi_i(-q_1,q_2) = -\Pi_i(q_1,-q_2) \ , \\ \Pi_i(q_1,-q_2) = -\Pi_i(-q_1,q_2) \ . \end{cases}$$

Let us specify now

$$\Pi_1(q_1,q_2) = a_1 q_1 + b_1 q_1^3 + c_1 q_1^5 + a_{12}(q_1 - q_2)$$
$$\Pi_2(q_1,q_2) = a_2 q_2 + b_2 q_2^3 + c_2 q_2^5 - a_{12}(q_1 - q_2) \ ,$$

that is, leaving the restitutive coupling linear.

We let $a_1,a_2,a_{12} > 0$, $c_1,c_2 > 0$ and $b_1,b_2 < 0$ with the same type
of restrictions upon $|b_1|,|b_2|$ as made previously upon $|b|$. The total
energy is

$$E(\bar{q},\dot{\bar{q}}) = \tfrac{1}{2}(\dot{q}_1^2 + \dot{q}_2^2) + \tfrac{1}{2}(a_1 q_1^2 + a_2 q_2^2) + \tfrac{1}{4}(b_1 q_1^4 + b_2 q_2^4) + \tfrac{1}{6}(c_1 q_1^6 + c_2 q_2^6) .$$

The equilibria are defined by the equations

$$\begin{cases} a_1 q_1 + b_1 q_1^3 + c_1 q_1^5 + a_{12} q_1 - a_{12} q_2 = 0 \\ a_2 q_2 + b_2 q_2^3 + c_2 q_2^5 - a_{12} q_2 + a_{12} q_1 = 0 . \end{cases}$$

The first obvious root is the basic equilibrium $q_1 = q_2 = 0$. Further, we add the equations to obtain

$$a_1 q_1 + b_1 q_1^3 + c_1 q_1^5 + a_2 q_2 + b_2 q_2^3 + c_2 q_2^5 = 0 ,$$

which for $q_1 \neq 0$, $q_2 \neq 0$ is true if simultaneously

$$a_i + b_i q_i^2 + c_i q_i^4 = 0 , \qquad i = 1,2 ,$$

that is, if

$$q_i = \pm \left(\frac{-b_i \pm \sqrt{b_i^2 - 4a_i c_i}}{2c_i} \right)^{\tfrac{1}{2}} .$$

Assuming $b_i = -2(a_i c_i)^{\tfrac{1}{2}}$ we shall indicate how to attempt the problem algebraically. The roots obtained are $q_1 = \pm(a_1/c_1)^{\tfrac{1}{4}}$, $q_2 = \pm(a_2/c_2)^{\tfrac{1}{4}}$. On substitution to the equations for equilibria, one obtains

$$q_2^{(1)}, q_2^{(2)} = \pm(a_1/c_1)^{\tfrac{1}{4}}(1 + a_1 + \frac{a_1}{a_{12}} + \sqrt{a_1 b_1})$$

$$q_1^{(1)}, q_1^{(2)} = \pm(a_2/c_2)^{\tfrac{1}{4}}(1 + a_2 + \frac{a_2}{a_{12}} + \sqrt{a_2 b_2})$$

as the coordinates of the two other than stable equilibria. The threshold is obtainable on substituting the values for $q_1^{(1)}, q_2^{(1)}$ into the energy. We have

$$h_{CH} = E(q_1^{(1)}, q_2^{(1)}, 0, 0)$$

$$= \frac{a_1}{2}(a_2/c_2)^{\tfrac{1}{4}}(1 + a_2 + \frac{a_2}{a_{12}} + \sqrt{a_2 c_2})^2 + \frac{a_2}{2}(a_1/c_1)^{\tfrac{1}{4}}(1 + a_1 + \frac{a_2}{a_{12}} + \sqrt{a_1 c_1})^2$$

$$- \frac{a_2\sqrt{a_1 c_1}}{2c_2}(1 + a_2 + \frac{a_2}{a_{12}} + \sqrt{a_2 c_2})^4 - \frac{a_1\sqrt{a_2 c_2}}{2c_1}(1 + a_1 + \frac{a_1}{a_{12}} + \sqrt{a_1 c_1})^4$$

$$+ \frac{c_1}{6}(a_2/c_2)^3(1 + a_2 + \frac{a_2}{a_{12}} + \sqrt{a_2 c_2})^6 + \frac{c_2}{6}(a_1/c_1)^3(1 + a_1 + \frac{a_1}{a_{12}} + \sqrt{a_1 c_1})^6.$$

Then E_{CE} is the 3-dimensional closed surface in R^N

$$\dot{q}_1^2 + \dot{q}_2^2 + a_1 q_1^2 + a_2 q_2^2 - (\sqrt{a_1 c_1}\, q_1^4 + \sqrt{a_2 c_2}\, q_2^4) + \tfrac{1}{3}(c_1 q_1^6 + c_2 q_2^6) = h_{CE}$$

enclosing the region Δ_E . $\qquad\qquad\qquad\qquad$ □

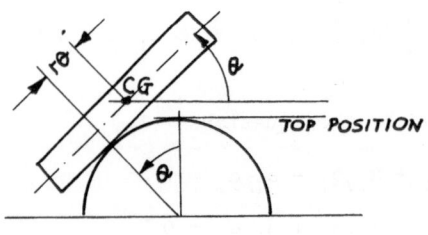

TOP POSITION

Fig. 2.11

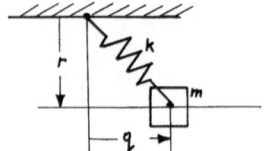

Fig. 2.12

EXERCISES 2.3

2.3.1   A homogeneous wooden plank of length ℓ, thickness s and weight mg
rests on a semicircular support of radius r shown below. If the
plank is tipped slightly, what is the condition for Dirichlet stable
equilibrium at its (balancing) top position? Note that the potential
energy is calculable as $V = mg[(r + \tfrac{1}{2}s) \cos\theta + r\theta \sin\theta]$ and its
minimum is obtained at points θ where $dV/d\theta = 0$, $d^2V/d^2\theta > 0$.
We conclude that the sought condition is $\theta = 0$, $r > \tfrac{1}{2}s$. Granted
the above, show that the motion equation of the plank is

$$\ddot{\theta}[(r\theta)^2 + (s/2)^2 + (s^2 + \ell^2)/12] + [(r\dot{\theta})^2 + rg\cos\theta]\theta - (3g\sin\theta)/2 = 0 .$$

2.3.2   The point mass m may only move along the horizontal axis marked in
the figure below.   It is suspended on a linear spring with characteristic $-kq$ and static length $\ell > r$. With the change of q the
spring is alternatively compressed and extended. Show that the
motion equation can be written as $\ddot{q} - (k/m)q + bq^3 = 0$, b being
a positive constant.

2.4 ENERGY FLOW AND POWER

Let us now investigate how the motions of the nonconservative systems
(1.5.26), (1.7.10) or (1.8.6), or for that matter, (2.2.1), (2.2.2), (2.2.3)
behave with respect to the conservative reference frame on H. The measure
of such behavior will be given in terms of the energy change along such
motions, that is, the instantaneous power. Substituting (1.7.10) into
(2.3.6), we obtain

$$\dot{H}(\bar{q},\bar{p}) = \sum_i \frac{\partial H}{\partial p_i} (Q_i^D + Q_i^F) \ . \tag{2.4.1}$$

Alternatively for the Lagrange equations (1.5.26), in view of (2.3.1) we
obtain

$$\frac{d}{dt} \frac{\partial T}{\partial \dot{q}_i} - \frac{\partial (T - V)}{\partial q_i} = Q_i^F(\bar{q},\dot{\bar{q}},u_i) + Q_i^D(\bar{q},\dot{\bar{q}}) \ , \quad i = 1,\ldots,n \ . \tag{2.4.2}$$

Multiplying it by $\dot{q}_i$ and summing up over i, we have

$$\frac{d}{dt} [\sum_i \frac{\partial T}{\partial \dot{q}_i} - (T - V)] = \sum_i (Q_i^F + Q_i^D)\dot{q}_i \ . \tag{2.4.3}$$

Since

$$\sum_i q_i \frac{\partial T}{\partial \dot{q}_i} = \sum_i q_i \sum_i M_{ij}(\bar{q}) \dot{q}_j = \sum_i \sum_j M_{ij}(\bar{q})\dot{q}_i \dot{q}_j = 2T \ ,$$

and

$$\frac{d}{dt} [T(\bar{q},\dot{\bar{q}}) + V(\bar{q})] = \dot{E}(\bar{q},\dot{\bar{q}}) \ ,$$

(2.4.3) becomes

$$\dot{E}(\bar{q},\dot{\bar{q}}) = \sum_i [Q_i^F(\bar{q},\dot{\bar{q}},u) + Q_i^D(\bar{q},\dot{\bar{q}})]\dot{q}_i \ , \tag{2.4.4}$$

representing the instantaneous rate of change of total energy, obviously
coinciding with (2.4.1). The rate is evaluated by the sum of inner
products

$$\bar{Q}^F(\bar{q},\dot{\bar{q}},\bar{u})^T \dot{\bar{q}} = \sum_i Q_i^F(\bar{q},\dot{\bar{q}},\bar{u})\dot{q}_i \ , \quad \bar{Q}^D(\bar{q},\dot{\bar{q}})^T \dot{\bar{q}} = \sum_i Q_i^D(\bar{q},\dot{\bar{q}})\dot{q}_i \ ,$$

recognized as the *input* and *damping powers*, respectively. When a power is
positive for $\dot{\bar{q}} \neq 0$, it accumulates the energy, and the corresponding
force is called *accumulative*. Non-positive power dissipates or preserves
the energy and the corresponding force is called *dissipative*. In particular, if the force is not potential but still produces a zero-power over
some time interval, it is called neutral.

In the majority of cases, the damping forces are dissipative: dry
friction or viscous damping, caused by slips and other boundary sheer
effects at mating surfaces; oil, water, air resistance in the environment

of the system and/or its hydraulic or pneumatic power supply; structural damping in bodies caused by microscopic interface effects like friction or sliding. We shall specify such forces as *positive damping*, denoted $Q_i^{DD}(\bar{q},\dot{\bar{q}})$ with corresponding characteristics $D_i^D(\bar{q},\dot{\bar{q}})$. Using the same transformation (1.5.7) as for the potential forces from the Cartesian to generalized damping, we may form the *dissipation function*

$$U_D = \sum_i \int Q_i^{DD}(\bar{q},\dot{\bar{q}}) dq_i \qquad (2.4.5)$$

which produces negative work. The power of such positive damping forces is negative:

$$\sum_i Q_i^{DD}(\bar{q},\dot{\bar{q}})\dot{q}_i = \bar{Q}^{DD}(\bar{q},\dot{\bar{q}})^T \dot{\bar{q}} < 0 , \qquad \dot{\bar{q}} \neq 0 , \qquad (2.4.6)$$

or alternatively,

$$\sum_{i=1}^n D_i^D(\bar{q},\dot{\bar{q}})\dot{q}_i = \bar{D}^D(\bar{q},\dot{\bar{q}})^T \dot{\bar{q}} > 0 , \qquad \dot{\bar{q}} \neq 0 , \qquad (2.4.7)$$

complemented by (2.3.12) applicable here as well.

The above assumptions make the positive damping an odd function with respect to the velocities $\dot{q}_i$. By the character of the applied damping forces, either viscous or Coulomb - dry friction, we may also assert that they are at least non-increasing in velocity,

$$-\frac{\partial Q_i^{DD}(\bar{q},\dot{\bar{q}})}{\partial \dot{q}_j} \geq 0 , \qquad \bar{q}_i \neq 0 . \qquad (2.4.8)$$

A similar argument leads to assuming that Q_i^{DD} depends upon $|\bar{q}|$ rather than on an arbitrary $\bar{q}$ and that it monotone decreases with q_i :

$$-\frac{\partial Q_i^{DD}(|\bar{q}|,\dot{\bar{q}})}{\partial q_i} > 0 , \qquad \text{for all } q_i \neq q_i^e, \quad i = 1,\dots,n . \qquad (2.4.9)$$

Whether or not the mechanical system working in certain conditions exhibits auto-oscillations (or more generally any self-sustained motion) remains a matter of discussion, but the possibility is there. In this case the implemented power is positive, at least for some values of the velocities $\dot{q}_i$, and we have the so-called *negative damping* forces. Such forces are obviously energy accumulative and will be denoted $Q_i^{DA}(\bar{q},\dot{\bar{q}})$ with the corresponding characteristics $D_i^A(\bar{q},\dot{\bar{q}})$. Hence

$$\sum_i Q_i^{DA}(\bar{q},\dot{\bar{q}})\dot{q}_i = Q^{DA}(\bar{q},\dot{\bar{q}})^T \dot{\bar{q}} > 0 , \qquad \dot{\bar{q}} \neq 0 , \qquad (2.4.10)$$

and

$$\sum_i D_i^A(\bar{q},\dot{\bar{q}})\dot{q}_i = \sum_\sigma \bar{D}^{A\sigma}(\bar{q},\dot{\bar{q}})^T \dot{\bar{q}}^\sigma = D^A(\bar{q},\dot{\bar{q}})^T \dot{\bar{q}} < 0 , \quad \dot{\bar{q}} \neq 0 . \quad (2.4.11)$$

Self-excitation is known as a so-called "velocity motor", and there is no self-excitation at rest, whence

$$Q_i^{DA}(\bar{q}^e,0) , \quad D_i^A(\bar{q}^e,0) = 0 , \quad i = 1,\ldots,n . \quad (2.4.12)$$

Once a self-sustained oscillation appears, it is reasonable to expect that there will be some neighborhood of $\dot{\bar{q}} = 0$ with the negative damping, and obviously such damping may as well appear on other intervals of velocity.

With all the above, the damping forces result in both types of damping and we assume

$$Q_i^D(\bar{q},\dot{\bar{q}}) = Q_i^{DA}(\bar{q},\dot{\bar{q}}) + Q_i^{DD}(|\bar{q}|,\dot{\bar{q}}) , \quad (2.4.13)$$

and

$$D_i(\bar{q},\dot{\bar{q}}) = D_i^A(\bar{q},\dot{\bar{q}}) + D_i^D(|\bar{q}|,\dot{\bar{q}}) , \quad i = 1,\ldots,n . \quad (2.4.14)$$

There cannot be any blanket assumptions upon the sign of the input power $(\bar{Q}^F)^T \dot{\bar{q}}$ as the controller may be used both ways, either to supply or withdraw the energy from the system. It will also work at equilibria, thus implying possibly non-zero values of Q_i^F there, see Benedict-Tesar [1] for the case in robotics.

On the other hand, since the power from all energy sources (even nuclear) is always eventually limited, there may be no objection to our *axiom of bounded accumulation*, by which we mean the following. For any point $(\bar{q}^0,\dot{\bar{q}}^0) \in \Delta$ there is a number $N > 0$ large enough to secure

$$|\bar{Q}^{DA}(\bar{q},\dot{\bar{q}})^T \dot{\bar{q}} + \bar{Q}^F(\bar{q},\dot{\bar{q}},\bar{u})^T \dot{\bar{q}}| \leq N < \infty , \quad (2.4.15)$$

for all $\bar{q}(t),\dot{\bar{q}}(t) \in \Delta$ along a trajectory from $(\bar{q}^0,\dot{\bar{q}}^0)$. This means that the external input, whatever its source, decreases whenever the velocity of motion increases and vice versa, which gives (2.4.15) a power balancing role.

In a much wider but also more vague sense, the axiom is justified by the so called *le-Chatelier Principle*: every external action produces in a system (body) changes which tend to neutralize that action, building up a resistance to it, see Kononenko [2].

With all the above discussion in mind, we may see that (2.4.4) which may now be written as

$$\dot{E}(\bar{q},\dot{\bar{q}}) = \bar{Q}^{DA}(\bar{q},\dot{\bar{q}})^T \dot{\bar{q}} + Q^F(\bar{q},\dot{\bar{q}},\bar{u})^T \dot{\bar{q}} + \bar{Q}^{DD}(\bar{q},\dot{\bar{q}})^T \dot{\bar{q}} \quad (2.4.16)$$

or, in view of the zero-power of the gyro forces:

$$\dot{E}(\bar{q},\dot{\bar{q}}) = \bar{F}(\bar{q},\dot{\bar{q}},\bar{u})^T \dot{\bar{q}} - \bar{D}^A(\bar{q},\dot{\bar{q}})^T \dot{\bar{q}} - \bar{D}^D(|\dot{\bar{q}}|,\dot{\bar{q}})^T \dot{\bar{q}} \ , \qquad (2.4.17)$$

gives the *power balance* along the trajectories of (1.5.26). Obviously the same discussion refers to the forces of (1.7.10) with $H(\bar{q},\bar{p})$ and with $\bar{p}$ replacing $\dot{\bar{q}}$ in all the assumptions.

We return now to the surface H : $h = E(\bar{x})$ in Z of the previous Section 2.3. Lifting the trajectories $\bar{\phi}(x^0,\mathbb{R})$ of (2.1.4) from Δ into Z , we obtain what may be called *energy-state-trajectories* $h(\bar{u},\bar{x}^0,h^0,\mathbb{R})$, where $h^0 = E(\bar{x}^0)$. Introduce now the scalar function $f_0(\bar{x},\bar{u}) =$ $= \nabla E(\bar{x})^T \cdot \bar{f}(\bar{x},\bar{u}) = \dot{E}(\bar{x})$ called the *power characteristic*, determined by (2.4.4). The energy-state trajectories are defined by the equations

$$\left.\begin{array}{l} \dot{h} = f_0(\bar{x},\bar{u}) \ , \\[2mm] \dot{\bar{x}} = \bar{f}(\bar{x},\bar{u}) \ . \end{array}\right\} \qquad (2.4.18)$$

The first equation is the scalar *energy equation* describing the balance of power or change of E subject to constraints specified by the second equation.

An energy-state trajectory $h(x^0,h^0,\mathbb{R})$ projected into the h-axis of Z produces the one-dimensional set of points $h(h^0,\mathbb{R})$ named the *autonomous energy flow* related continuously to the parameter $h^0 \in [0,h_\Delta)$, describing the initial amount of the total energy contained in the system before motion. The values $h(h^0,t) = h_C \in [0,h^f) \subset [0,h_\Delta)$ define the amount of the energy at t , obtained as the result of the change or *flux* during $[t_0,t)$, either *influx* (increase) or *outflux* (decrease). The latter two are also called respectively the *energy input* and the *work done* by the energy changing actions. The zero flux might be incorporated in either the influx or the outflux. Integrating the energy equation along time up to a given $t < \infty$, we obtain the flow value

$$h(h^0,t) = h^0 + \int_{t_0}^{t} f_0(\bar{x},\bar{u})ds \qquad (2.4.19)$$

with the integral representing the flux with respect to the initial flow value h^0 , contained in the system at t_0 . Note that the flow is additive on $\mathbb{R}$: $h(h^0,t_1) + h(h(h^0,t_1),t_2) = h(h^0,t_2)$, for any $t_1,t_2 \in \mathbb{R}$, $t_2 > t_1$, and that $f_0(\bar{x},\bar{u})$ is bounded, see (2.4.15). We shall also justify below that for intervals with negative values of f_0 , *the flux may not exceed* h^0 *so that there is no chance of* $h(h^0,t)$ *becoming negative*. The surface H is locally diffeomorphic to $\mathbb{R}^N$, that is, a neighborhood of a point in H is the image of $\mathbb{R}^N$ under a one-to-one and onto map which is differentiable together with its inverse. Then it is possible to define a field of directions and thus the integral curves on every such neighborhood and hence on the whole H. Further, since $h = E(\bar{x})$ is single valued on Δ, to every point $\bar{x} \in \Delta$ there corresponds a unique point $(h,\bar{x}) \in H \cap Z_\Delta$ and the energy-state trajectories are unique.

This is not the case with the energy flow. Although to each point $\bar{x}$ of Δ there corresponds a point $h(x^0,h^0,t)$ on the h-axis: $h_C = h(h^0,t)$ of the energy flow, the inverse map is clearly set valued. In other words, the same h_C yields a continuum of points in H, namely the level Z_C and thus the topographic surface E_C in Δ. Thus a trajectory and/or an energy-state trajectory generates the corresponding energy flow, but the latter implements only the change of energy levels along the trajectory, and *we need to refer to known levels to be able to obtain even a qualitative behavior of the trajectory.*

The equations (2.4.18) imply that f_0 is the h-axis component of the vector $\bar{f}^E = (f_0,f_1,\ldots,f_N)^T$, so the rise or fall of the energy values along the trajectory depends upon the sign of f_0. Consider an arbitrary trajectory $\bar{\phi}(\bar{x}^0,\mathbb{R})$, $\bar{x}^0 \in \Delta$, which is in general non-conservative, and given t, consider a regular point $\bar{x}(t)$ on this trajectory. Let E_C be the corresponding topographical (iso-energy) surface through that point, with the projection into Δ shown in Fig. 2.13. It subdivides Δ into two regions. We call the region into which the positive gradient $\nabla E(\bar{x})$ is

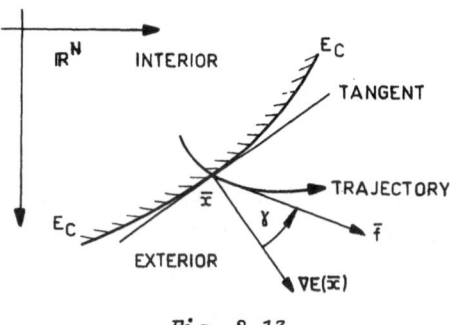

Fig. 2.13

directed the *exterior*, and the other the *interior* of the surface E_C. If given $\bar{u}$, we have $\bar{x}(t)$ such that $f_0(\bar{x},\bar{u}) > 0$, then there is an energy influx at $\bar{x}(t)$ and the corresponding energy flow $h(h^0,\mathbb{R})$ is directed *upwards*, away from $h = 0$ on the h-axis. This means that the other component of $\bar{f}^E$, namely the tangent vector $\bar{f}(\bar{x},\bar{u})$ to the trajectory is directed towards the exterior of E_C, forming a sharp angle γ with the gradient. See Fig. 2.13. If $f_0(\bar{x},\bar{u}) \geq 0$, the thrust of the energy flow is the same, but the trajectory may have *contact* with E_C for longer than the instant t , that is, it may slide on E_C with $\bar{f}(\bar{x},\bar{u})$ parallel to the tangent for some short interval of time. The latter happens at any point where $f_0(\bar{x},\bar{u}) = 0$. On the other hand, when $f_0(\bar{x},\bar{u}) < 0$ the opposite takes place. The sense of the energy flow is reversed, it is directed *downwards*, towards $h = 0$ on the h-axis. This forces the tangent vector $\bar{f}(\bar{x},\bar{u})$ to be directed towards the interior of E_C , forming an obtuse angle γ with $\nabla E(\bar{x})$.

Observe that $f_0(\bar{x},\bar{u}) = \nabla E(\bar{x})^T \cdot \bar{f}(\bar{x},\bar{u})$ is a point-function on Δ and because of its continuity the points where f_0 has a definite sign are not isolated. Hence we can specify regions in Δ covered with the *fields* or *zones* accumulative H^+ : $f_0(\bar{x},\bar{u}) > 0$, neutral H^0 : $f_0(\bar{x},\bar{u}) = 0$ and strictly dissipative H^- : $f_0(\bar{x},\bar{u}) < 0$, on which the energy balance (2.4.17) is respectively positive, zero or negative, and on which the trajectories leave, slide or enter the interior of each current E_C in Δ they cross. Consequently the points of H^+, H^0, H^- are called the *exit points* and *entry points* correspondingly to the E_C level concerned. At the same time, the constant sign of f_0 decides about the direction of the energy flux. The flux on some interval $[t_0,t) \in \mathbb{R}$ may be positive, negative or zero. Whatever happens inside $[t_0,t)$ if, given t_0 , we have

$$h(h^0,t) \leq h(h^0,t_0) , \tag{2.4.20}$$

an energy flow of (2.1.4) is called *dissipative on* $[t_0,t)$: otherwise it is *accumulative on this interval*. The strong inequality in (2.4.20) produces *strict dissipativeness*. The relation (2.4.20) is called the *dissipation inequality*. The system (forces) must have been dissipative (accumulative) if the produced energy flows are dissipative (accumulative).

The same flow is called *monotone dissipative* (*monotone strictly dissipative, monotone accumulative*) on $[t_0,t)$ if for any $\tau \in [t_0,t)$ we have $f_0(\bar{x}(\tau)) \leq 0$, $(f_0(\bar{x}(\tau)) < 0$, $(f_0(\bar{x}(\tau)) > 0)$. We conclude that the monotone dissipative arcs $[t_0,t)$ of trajectories are embedded in the dissipative field H^- , while the monotone accumulative in the accumulative field H^+ .

Monotone dissipativeness includes the *conservative* subcase: $f_0(\bar{x}(t)) \equiv 0$. It is often a convenient convention, though the conservative systems will usually be specified separately. In such systems there is no energy flux, thus $h(h_C,\mathbb{R}) = h_C$ for any h_C or $h(h^0,\mathbb{R}) = h^0$ along any flow, and the flow is at rest or energy-equilibrium. Note that the above refers to the total energy.

The f_0 is the rate of energy flow at each $\bar{x}$, but also average across the system at that point. By (2.4.17) we can see which characteristics of the system contribute to the energy flow, and then estimate the power at the points between particular elements or subsystems of the overall mechanical structure.

As we shall see in the next chapters, the energy flow measured by f_0 at each state $\bar{x}$ and across the system is indicative of the system behavior under control and thus it is a feature helping to design suitable control and adaptation programs. Consider the energy flow dissipative on $[t_0,t)$. By (2.4.20), $h(h^0,t_0) = h^0 \geq h(h^0,t)$. The system acts during $[t_0,t)$ as an energy source for its environment (for instance, mechanical energy transferred into heat via dry friction). To do so, the system requires some supply of energy h^0, accumulated prior to the instant t_0, say during $[\tau_0,t_0)$ with $f_0(\bar{x},\bar{u}) > 0$, $\tau \in [\tau_0,t_0)$ where $\bar{f}_0(\bar{x},\bar{u})$ and $|t_0 - \tau_0|$ are large enough to provide recycling during dissipation on $[t_0,t)$. We assume this amount to be the minimal amount necessary and, by (2.4.15), to be the finite *required energy*

$$E_r(\bar{x}^0) = \inf \int_{\tau_0}^{t_0} f_0(\bar{x},\bar{u}) d\tau > 0 , \quad \text{as} \quad (\bar{x}(\tau_0),\tau) \to \bar{x}^0 ,$$

named so by Willems [1]. Granted the required energy, we define the maximal amount of it which may, at t, have been extracted from the system $(f_0(\bar{x},\bar{u}) \leq 0)$ into the environment:

$$E_a(\bar{x}) = \sup_u \int_{t_0}^{t} - f_0(\bar{x},\bar{u}) ds , \quad \bar{x} = \phi_u(\bar{x}^0,s) , \quad s \to t .$$

We call it the *recoverable work* or the *available storage*, again according to Willems [2]. By dissipativeness, $0 \leq E_a(x) < \infty$. Obviously, the system cannot dissipate more energy than has been supplied to it, thus $h^0 = E_r(\bar{x}^0) \geq h(h^0,t)$, for any given $t \in \mathbb{R}^+$. On the other hand, $E_a(\bar{x}(t))$ has been the maximal extracted, thus

$$0 \leq E_a(\bar{x}) \leq h(h^0,t) \leq E_r(\bar{x}^0) < \infty \tag{2.4.21}$$

for any given $t \in \mathbb{R}_0^+$ as long as the flow is dissipative. Hence the energy outflux is estimated by $E_r(\bar{x}^0) - E_a(\bar{x}) = h^0 - E_a(\bar{x})$, which agrees with (2.4.19), with non-positive $f_0(\bar{x})$; check

$$\int_{t_0}^{t} f_0(\bar{x}(s)) ds \leq E_r(\bar{x}^0) - E_a(\bar{x}) . \tag{2.4.22}$$

Since the flow is additive, we may as well introduce the maximal outflux as the *cycle energy* $E_k(\bar{x}) = E_r(\bar{x}^0) - E_a(\bar{x})$. Note that the cycle energy is in general different from the stored energy in the system at t, which is the flow value $h(h^0,t)$ defined by (2.4.19). The cycle energy serves as an estimate for the energy flow.

We justify (2.3.12) and (2.4.12) by recalling the derivation of (2.4.16). We conclude that the autonomous uncontrolled system may not start from the basic equilibrium: $E_r(\bar{x}^0) = 0$. This confirms our discussion in Section 2.2, cf. (2.3.13).

When $E_r(\bar{x}^0) = E_a(\bar{x})$, the cycle energy vanishes, and the dissipative flow is said to be *reversible*. The flow is *irreversible* if $E_k(\bar{x})$ does not vanish except at the equilibrium $\bar{x} = 0$.

Now let us suppose that the flow is accumulative on a given $[t_0,t)$. Our reasoning of all the above may be inverted. The system consumes the energy from the environment on the interval, similarly as the dissipative system had to do prior to the instant t_0. The available storage of the system at $t < \infty$ becomes the *produced storage*

$$E_p(\bar{x}) = \sup_{t_0 < t} \int_{t_0}^{t} f_0(\bar{x},\bar{u})dt > 0 , \qquad (2.4.23)$$

with $f_0(\bar{x}) > 0$, obtained similarly to E_r for the dissipative system. This storage is fixed relatively to some N, introduced in (2.4.15), making $f_0(\bar{x})$ bounded, cf. (2.4.19). Since no autonomous system may start from the basic equilibrium, we estimate now

$$0 < h(h^0,t) \leq E_p(\bar{x}) , \qquad (2.4.24)$$

for any given $t \in \mathbb{R}$ as long as the flow is accumulative. We immediately have the influx estimated by $E_p(x)$ which by (2.4.24) is also the cycle energy.

Let the flow be dissipative on $[t_0,t_1)$ and accumulative on $[t_1,t_2)$. If the dissipative flow has been reversible, that is, $E_r(\bar{x}^0) = E_a(\bar{x}(t_1))$ we may envisage the situation that

$$E_a(\bar{x}(t_1)) = E_p(\bar{x}(t_2)) = E_r(\bar{x}^0) ,$$

that is, the energy flow is *recycled* to its value h^0, that is, $h(h^0,t_0)$ $= h(h^0,t_2)$. When the above occurs for $[t_0,t_n),t_n = nt_2$ with n a positive integer, the energy flow on this interval may be n-times recycled, and we shall say that the flow is *multiply reversible*. In pariticlar, if this multiple reversibility is such as to allow for a constant $L > 0$, yielding

$$h(h^0,t) = h(h^0,t+L) , \qquad \forall t \in \mathbb{R} , \qquad (2.4.25)$$

114

that is, the function h is periodic with the period L , we shall say that the dissipative flow of $[t_0, t_1)$ is L-*periodically reversible*. The flux on the recycled flow on $[t_0, t_2)$ vanishes, cf. (2.4.22), and so it does for the entire interval of definition of the multiply reversible flow.

When the dissipative flow on $[t_0, t_1)$ is reversible, but $E_a(\bar{x}(t_1)) \neq E_p(\bar{x}(t_2))$, the flow is not recycled on $[t_0, t_2)$, but in particular one may still meet with the circumstances where for any $\varepsilon > 0$ there may be found a $\delta > 0$ sufficiently large for $t_2 - t_1 < \delta$ to yield

$$\left| E_a(\bar{x}(t_1)) - E_p(\bar{x}(t_2)) \right| < \varepsilon . \qquad (2.4.26)$$

One may say then that the energy flow on $[t_0, t_2)$ has been *almost recycled*. The concept of the *almost-periodic reversibility* of the flow on $[t_0, t_1)$ is formed in the obvious manner.

The above discussion applies to the dissipative (accumulative) flows on suitable intervals. In most cases, the intervals of dissipativeness and accumulativeness may be partitioned further into intervals with monotone properties. Thus *time-globally we have a collection of monotone dissipative and/or accumulative time intervals* to consider.

When perturbation force appears in the resultant of the applied forces, cf. (1.5.26), (1.7.10), the derivation of the power balance (2.4.17) remains the same, except for adding one more term of the perturbation power: $\sum_i Q_i^R(\bar{q}, \dot{\bar{q}}, t)\dot{q}_i \neq 0$, $\dot{q}_i \neq 0$, with characteristics $\sum_i R_i(\bar{q}, \dot{\bar{q}}, t)\dot{q}_i \neq 0$ which yields

$$\dot{E}(\bar{q}, \dot{\bar{q}}, t) = \bar{Q}^{DA}(\bar{q}, \dot{\bar{q}})^T \dot{\bar{q}} + \bar{Q}^F(\bar{q}, \dot{\bar{q}}, \bar{u})^T \dot{\bar{q}} + \bar{Q}^R(\bar{q}, \dot{\bar{q}}, t)^T \dot{\bar{q}} + \bar{Q}^{DD}(\bar{q}, \dot{\bar{q}})^T \dot{\bar{q}} \qquad (2.4.27)$$

or

$$\dot{E}(\bar{q}, \dot{\bar{q}}, t) = \bar{F}(\bar{q}, \dot{\bar{q}}, \bar{u})^T \dot{\bar{q}} + \bar{R}(\bar{q}, \dot{\bar{q}}, t)^T \dot{\bar{q}} + \bar{D}^A(\bar{q}, \dot{\bar{q}})^T \dot{\bar{q}} + \bar{D}^D(|\dot{\bar{q}}|, \dot{\bar{q}})^T \dot{\bar{q}} . \qquad (2.4.28)$$

Moreover, the axiom of bounded accumulation (2.4.15) is now complemented by the existence of $N' > 0$ such that

$$\left| \bar{Q}^R(\bar{q}, \dot{\bar{q}}, t)^T \dot{\bar{q}} \right| \leq N' < \infty , \qquad (2.4.29)$$

for all $\bar{q}(t), \dot{\bar{q}}(t) \in \Delta$ along a motion from $(\bar{q}^0, \dot{\bar{q}}^0, t_0)$ which represents the *axiom of bounded perturbation*.

With the perturbation force the system becomes nonautonomous with equation (2.1.20) and solutions-motions $\bar{\phi}(\bar{x}^0, t_0, \mathbb{R}^+)$ in $\Delta \times \mathbb{R} \subset \mathbb{R}^{N+1}$. Consequently the conservative frame of reference H has to be augmented into $H \times \mathbb{R}$, together with all the related motions. The augmentation is cylindrical, since $E(\bar{x})$ is not dependent explicitly on t . Hence the levels $E_C \times \mathbb{R}$ are isometric to previous E_C in Δ . Upon $H \times \mathbb{R}$ we now have *energy-state motions* $h(\bar{x}^0, t_0, h^0, \mathbb{R})$ with the *energy flow* $h(h^0, t_0, \cdot)$: $\mathbb{R}^+ \to \mathbb{R}$ being now t_0-dependent and with the power characteristic $f_0(\bar{x}, \bar{u}, t)$ being t-explicit dependent and determined by (2.4.17). Introducing the augmentation, we still have the same dissipation inequality and all its consequences maintained.

As mentioned at the opening of Section 2.3, when the system is subject to uncertainty, see either (2.2.1), (2.2.2) or (2.2.3), in general (2.2.6), the situation is different. First the characteristics are subject to $\bar{w}$, then since the potential forces may be uncertain, so is the energy $E(\bar{x}, \bar{w})$. This means that the surface H may not be obtained directly from the energy in a unique way. It is defined from the nominal value of either

$$E^-(\bar{x}) = \inf E(\bar{x}, \bar{w}) \mid \bar{w} \in W \qquad (2.4.30)$$

or

$$E^+(\bar{x}) = \sup E(\bar{x}, \bar{w}) \mid \bar{w} \in W \qquad (2.4.31)$$

depending upon our task of study, and upon the sign of the corresponding power which we want to achieve by using the controller. Which of the pair E^-, E^+ of nominal H's is used will thus be specified within particular problems studied. Whatever the defining base for H, the power balance (2.4.17) or (2.4.28) is now specified in terms of the uncertain characteristics, and the same goes for f_0, whence we have for (2.4.18)

$$\left.\begin{array}{l} \dot{h} = f_0(\bar{x}, \bar{u}, \bar{w}, t) , \\[2mm] \dot{\bar{x}} = f(\bar{x}, \bar{u}, \bar{w}, t) , \end{array}\right\} \qquad (2.4.32)$$

which is a selector system for some contingent format of the energy-state equations with energy-state motions discussed in $\Delta \times h \times \mathbb{R}$, denoted by $\bar{\phi}(h^0, \bar{x}^0, t_0, t)$, $t \geq t_0$.

The axioms of bounded accumulation (2.4.5) and bounded perturbation (2.4.9) now take the shape of

$$\left| \bar{F}(\bar{q}, \dot{\bar{q}}, \bar{w}, \bar{u})^T \dot{\bar{q}} - D^A(\bar{q}, \dot{\bar{q}}, \bar{w})^T \dot{\bar{q}} \right| \leq N , \quad \forall w \in W , \qquad (2.4.33)$$

$$\left| R(\bar{q}, \dot{\bar{q}}, t, \bar{w}) \dot{\bar{q}} \right| \leq N' , \quad \forall w \in W , \qquad (2.4.34)$$

respectively. Similarly, the dissipation inequality will require

$$h(h^0, t, \bar{w}) \leq h(h^0, t_0, \bar{w}) , \quad w \in W . \qquad (2.4.35)$$

116

Chapter 3
STABILIZATION

3.1 BASIC CONCEPTS OF CONTROL AND SYNTHESIS

In Section 1.2 we have mentioned the division of kinetic studies into two directions: Analysis, and either Synthesis (Design) or Control. *Analysis* deals with the Second Problem of Mechanics and answers our question (ii): how the system may move under given forces? In that, it gives options for the desired objectives of the second direction: *Synthesis or Control*, which answers the First Problem of Mechanics, that is, our question (iii) of Section 1.2: how to design the forces both internal (Synthesis) and input (Control) in order to obtain a desired type of motions which attain a stipulated objective? As has been said, Synthesis deals with designing characteristics of the internal applied forces such as damping, gravity or elastic forces. One may either choose entire functions (eigen, or coupling, or both) or given a general shape of the functions choose their parameters, say, coefficients in a power series representation.

In the latter case we actually come close to Control, which means choosing the variable $\bar{u}(\cdot)$ within the input terms $F_i(\bar{q},\dot{\bar{q}},\bar{u})$, $i = 1,\ldots,n$. Indeed, if such a choice is generated by a feedback program, see (2.1.1), the problem is called a *control synthesis*, which should not be confused with the synthesis discussed above. To avoid such confusion, we shall call the latter a *design synthesis*, which then may be further classified as *structural* and *kinetic*, depending on whether it refers to designing the structural model (number of DOF, system organization: schematic diagram, types of coupling, etc.) or to designing the functions of characteristics as discussed above. In Section 2.3 we gave examples how the design synthesis can compensate, at least in part, for the control synthesis. A little further in this chapter, we shall discuss this problem again.

Solutions to the problems covered by either control or design synthesis may not exist. Within the admissible control programs or classes of characteristics $D_i(\cdot), \Pi_i(\cdot)$ we may not be able to find such $P^i(\cdot), D_i(\cdot),$ $\Pi_i(\cdot)$ which will allow attaining the stipulated objective, that is, secure *controllability* or *synthesability* for such an objective. Since the admissible classes are bounded and the controllers and characteristics subject to constraints, the problems of controllability or synthesability are by no means academic.

Let us consider the following simple example. A missile pursues a plane, both modelled as point masses equal to a unit. We ignore gravitation as well as any other applied forces except the control force. The plane moves along a straight line, say the q-axis: $q(t) = q^0 + 0.1 t^3$, $q^0 > 0$, and the missile follows with the acceleration $\ddot{q}_M(t)$ equal to the controlling thrust force $u(t)$, $|u| \le 1$. Choosing states $x_1 = q_M$, $x_2 = \dot{q}_M$ we have for the state equations $\dot{x}_1 = x_2$, $\dot{x}_2 = u$. The pursuit is made along the x_1-axis and with the maximal thrust force $u = 1$, until collision at $x_1(t) = q(t)$ for some t, that is assuming $x_1(0) = 0$ when $\frac{1}{2}t^2 = q^0 + 0.1 t^3$. This, solved for $t = t_c$, gives the time of collision. We can easily calculate that collision is not possible, that is, the system is not controllable for collision, if $q^0 > 1.85$, as there is no finite t_c. In this case the airplane is simply too far ahead of the missile with the thrust qualified by the constraints $|u| \le 1$. If $q^0 = 1.85$ the collision occurs at $t_c = 3.33$ and if $q^0 < 1.85$, it occurs earlier, that is, $t_c < 3.33$. From the latter we also see that there is a set of initial positions of the plane, with specified boundary $q^0 = 1.85$, which is the so called region of controllability for collision. Outside such a region the controllability for the specified objective is contradicted. Very much the same argument may be used for synthesability, if we complicate the example slightly. Indeed, introducing some characteristics into the motion equations of either the plane or the missile or both, we may want to design them so that this or some other objective is attained.

It is obvious that both concepts, controllability and synthesability, are very much qualified by the type of objective concerned. In this book we concentrate on controllability which supports our main avenue of study, namely Control. For synthesability, the interested reader may peruse Skowronski [32].

The first concept of controllability, which still persists in the control theory of linear systems, had a very simplistic objective: transfer a state motion of the system concerned between two given states $\bar{x}^0$, and the terminal $\bar{x}^f = \bar{x}(t_f)$, $t_f \leq \infty$ under the designed control $\bar{u}(t)$, $t \in [t_0, t_f]$.

The controllability problem for such an objective has been posed by Kalman-Bertram [1] and Kalman [1] for the autonomous (2.1.3) but solved only for its linear version $\dot{\bar{x}} = A\bar{x} + B\bar{u}$, with A,B constant matrices of suitable dimensions. It was done by use of the Laplace transform, to the joy of people who like it easy, but not necessarily to those who like the models to be justified physically. As the models and the objectives grow more and more complicated, so do the demands for controllers and types of controllability. For instance, it is natural to expect that instead of a point-to-point transfer, we would want to start from a set in order to collide with a set, perhaps attained in stipulated time. Then we may require this collision to be qualified by a permanent capture after some time in the target set, or conversely by only a short rendezvous again in stipulated time. We may need a sequence of such collisions without capture perhaps ending with capture. The control scenario may also require avoidance of some antitarget set both in physical and/or in state spaces, alternatively avoidance of some moving obstacles either permanently or only for a specified interval of time.

We may also want to combine the objectives of the above, for example, collide without capture with some targets and avoid some antitargets - and the process again may be qualified by stipulating time. Still more complex objectives are required when dealing with the relative motion of two or more substructures. We may want two systems to track each other in physical or state spaces or to avoid themselves the same way. The tracking or avoidance may again be required to be permanent or only within some time intervals.

Finally, we may want to obtain a specific phase-space portrait of the mechanical system concerned, like stipulated distribution of equilibria and their types, prescribed limit cycles stable or unstable, structural stability (that is, holding the state-space pattern under perturbation), etc. All the above types of objectives can be grouped into a class which is described as *qualitative*.

In general terms a qualitative objective is of the YES or NO type with respect to obtaining a certain qualitative *objective property* Q of the motions of (2.2.6) on some subset of Δ called then the *Q-zone*. Such a property may be virtually anything as long as it is well defined, together with its zone. Then the ability to attain Q under some control program is called the controllability for Q. Due to the limitations of the chosen controller, the region of such controllability can be smaller than the Q-zone.

We may want to attach to Q an extra subobjective, a cost of attaining Q, measured in terms of some functional defined and additive along the motions concerned, and generating scalar values like time, energy, power, miss-distance from a target, etc., which are to be minimized. Such a subobjective is called *quantitative* or more specifically *optimal*. We then talk about *optimal* Q and controllability for optimal Q. In this sense the composite objectives may be qualitative, quantitative, or both. We may wish to attain collision in minimal time, capture a target with minimum energy or effort, avoid an antitarget with the shortest miss-distance, etc. In fact, any of the qualitative objectives cited can be optimized in terms of some cost. The list of applicable specifications of Q and optimal Q is obviously too long to be quoted, and so is the relevant literature.

In the text to follow, we give a few basic modular subobjectives allowing the building of a variety of composite objectives, some of which are also described together with the applied scenarios they fit best. Each time a brief review of relevant literature is given. The works which initiated research in controllability for general nonlinear systems with untruncated nonlinearity are those by A.A. Krassovski [1], Letov [1]-[3], Markus [1] (see also Lee-Markus [1]), N.N. Krassovski [2], Gershwin-Jacobson [1] and later the Leitmann school: Leitmann [1], Stalford-Leitmann [1], Sticht-Vincent-Schultz [1], Grantham-Vincent [1], Leitmann-Skowronski [1],[2], Vincent-Skowronski [1], Skowronski-Vincent [1], Stonier [1],[2], Skowronski [22],[23],[29], see also Skowronski [32],[38],[45], Aeyels [1]. The systems with untruncated nonlinear but additive perturbations were treated by Cheprasov [1], Dauer [1]-[4], Klamka [1],[2], and Aronsson [1], among others. For a more extensive review, see Skowronski [45].

It seems instructive to introduce the concept of controllability for Q of arbitrarily nonlinear systems, first on a very simple case. We take the autonomous (2.1.4) with Q being some elementary but typical objective property, say, asymptotic stability of the basic Dirichlet stable equilibrium, at which the origin of $\mathbb{R}^N$ is located. Moreover, we shall consider the Q-zone locally, that is, covering only Δ_E about the equilibrium concerned. Let us define such a Q-property as uniform asymptotic stability of such equilibrium. More formally we have . . .

DEFINITION 3.1.1. The basic equilibrium of (2.1.4) is *stable* on Δ_E if and only if given Δ_E for each $\varepsilon > 0$ there is $\delta(\varepsilon) > 0$ such that for each $\bar{x}^0 \in \Delta_E$, $\|\phi(\bar{x}^0,0)\| < \delta$ implies $\|\phi(\bar{x}^0,t)\| < \varepsilon$, for all $t \geq 0$.

Here $\|\cdot\|$ denotes any norm in $\mathbb{R}^N$, for instance, measured in terms of the energy $E(\bar{x})$. As (2.1.4) is autonomous, the trajectories are independent on t_0 and so is δ in Definition 3.1.1. Such stability is called t_0-uniform, or briefly *uniform*, on Δ_E. We may now form the definition specifying the present Q.

DEFINITION 3.1.2. The basic equilibrium of (2.1.4) is *asymptotically stable on* Δ_E if and only if it is stable there and for trajectories of Definition 3.1.1, there is $\delta_0 > 0$ such that $\|\phi(\bar{x}^0,0)\| < \delta_0$ implies $\phi(\bar{x}^0,t) \to 0$, as $t \to \infty$.

Given the above objective property Q, we may now define the controllability for it.

DEFINITION 3.1.3. The system (2.1.4) is *controllable for asymptotic stability of the basic equilibrium on a set* $\Delta_0 \subset \Delta_E$ if and only if there is a control program $\bar{P}(\bar{x})$ defined on Δ_0 such that for each $\bar{x}^0 \in \Delta_0$, the corresponding trajectory of (2.1.4) satisfies Definition 3.1.2.

The union of all such Δ_0's (the maximal such Δ_0 in Δ_E) is the *region of controllability* for the asymptotic stability of above, denoted Δ_{AS}.

Now we want to find a program $\bar{P}(\cdot)$ securing the above, and the region Δ_{AS}.

It is obvious that as the trajectories of (2.1.4) approach the basic equilibrium in the scenario of uniform asymptotic stability, the energy decreases from $E(\bar{x}^0)$ towards the basic level $E(0) = 0$. As such, asymptotic stability holds for all $\bar{x}^0 \in \Delta_E$. The decrease is monotone and at each E-level which is crossed by the trajectory we have the entry points of H^- only:

$$\dot{E}(\bar{x}) = \nabla E(\bar{x})^T \bar{f}(\bar{x},\bar{u}) < 0 , \tag{3.1.1}$$

which then is the *necessary condition* for the uniform asymptotic stability between the levels $E(\bar{x}^0) = h^0$ and $E(0) = 0$. The entry points were defined in Section 2.4, $\nabla E(\bar{x})$ denotes $\mathrm{grad}\, E(\bar{x}) = (\partial E/\partial x_1, \ldots, \partial E/\partial x_N)$.

Note that, given u , (3.1.1) defines an open set in Δ_E . Then, the lowest h^0-level at which (3.1.1) is contradicted generates in Δ_E a surface which can be considered an *estimating candidate* for the boundary $\partial\Delta_{AS}$ of the region Δ_{AS} , see Fig. 3.1. Note also that the above is qualified by some control variable $\bar{u}$, assumed already selected, presumably in terms of a *candidate* $\bar{P}(\cdot)$.

We must now confirm these candidates by suitable sufficient conditions. Let us use the following theorem introduced by Markus [2].

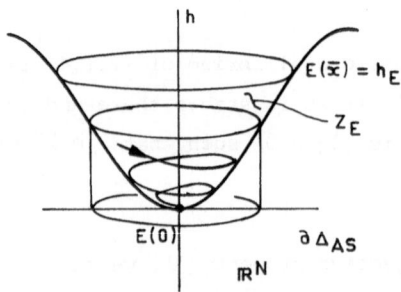

Fig. 3.1

CONDITIONS 3.1.1. If there is a function of class C^1 (continuously differentiable) $V(\cdot) : \mathbb{R}^N \to \mathbb{R}$ and a C^1-program $\bar{P}(\cdot)$ defined on Δ_E such that

(i) $V(\bar{x}) > 0$ for $\bar{x} \neq 0$, $V(0) = 0$;

(ii) $V(\bar{x}) \to \infty$, as $|\bar{x}| \to \infty$;

(iii) $\nabla V(\bar{x})^T \bar{f}(\bar{x},\bar{u}) < 0$ for $\bar{x} \neq 0$,   (3.1.2)

then the system (2.1.4) is controllable for uniform asymptotic stability of its basic equilibrium.

Heuristically the proof follows by contradiction between (i), (ii) and (iii) holding everywhere about the origin. Rigorous proof may be found in Markus [2]. Letting $V(\bar{x}) \overset{\Delta}{=} E(\bar{x})$ we have (i), (ii) holding on Δ_E , and (3.1.1) makes (iii) satisfied as well as necessary and sufficient, with $\bar{u} = P(\bar{x})$ found from the following conclusion.

COROLLARY 3.1.1. *Given* $\bar{x}^0 \in \Delta_E$ *, suppose there is* $\bar{u}*(\cdot) , \bar{u}*(t) \in U$ *, such that*
$$\nabla V(\bar{x})^T \bar{f}(\bar{x},\bar{u}*) = \min_{\bar{u}} \nabla V(\bar{x})^T \bar{f}(\bar{x},\bar{u}) < 0 .$$ (3.1.3)

Then condition (iii) is met with $\bar{u}* = \bar{P}*(\bar{x})$.

We illustrate the above on a local case in-the-small, that is, for amplitudes of $\bar{x}$ near the bottom of the cup Z_E , that is, for a sufficiently small neighborhood of the equilibrium and sufficiently small control range U . In such a case we may *linearize* (2.1.4). Indeed, let $\bar{f}(\cdot)$ be continuously differentiable and such that $\bar{f}(0,0) = 0$, which itself is restrictive. Then we may develop it into a power series and truncate all higher power terms so that (2.1.4) becomes

$$\dot{x} = A\bar{x} + B\bar{u} \tag{3.1.4}$$

with

$$A \overset{\Delta}{=} \left(\frac{\partial f_i}{\partial x_j}\right)\Bigg|_{\bar{x}=0, \bar{u}=0} \ , \qquad B \overset{\Delta}{=} \left(\frac{\partial f_i}{\partial u_k}\right)\Bigg|_{\bar{x}=0, \bar{u}=0} \tag{3.1.5}$$

for $i,j = 1,\ldots,N$, $k = 1,\ldots,r$. Let the control program be defined by $\bar{u} = K\bar{x}$, where K is an $r \times N$ matrix to be designed. Then (3.1.4) becomes

$$\dot{x} = A\bar{x} \ , \tag{3.1.6}$$

with $A \overset{\Delta}{=} (A + BK)$. We let the V-function be the energy expressed as the square form $V = \bar{x}^T P \bar{x}$, where $P = (p_{ij})$ is a symmetric positive definite $N \times N$ matrix. Such $V(\cdot)$ certainly satisfies (i),(ii). We then calculate the derivative $\dot{V} = \dot{\bar{x}}^T P \bar{x} + \bar{x} P \dot{\bar{x}}$. When A is nonsingular, then P and ApA^{-1} have the same eigenvalues. Hence $\dot{V} = \bar{x}^T A^T P \bar{x} + \bar{x}^T P A \bar{x}$. By the symmetry of P , $\bar{x}^T P \dot{\bar{x}} = (P^T \bar{x})^T \dot{\bar{x}}$, implying

$$\dot{V} = x^T [A^T P + PA] \bar{x} = -\bar{x}^T Q \bar{x} \ ,$$

where

$$A^T P + PA = -Q \tag{3.1.7}$$

which is called the *Liapunov Matrix Equation*, with Q also a symmetric matrix (quite often just the unit matrix). It follows that, given suitable A and positive definite Q , we obtain $\dot{V}(\cdot)$ negative definite, thus securing the asymptotic stability required. On the other hand, again given suitable A , the matrix P could be obtained as a positive definite solution to the Liapunov matrix equation, say with $Q = I$, securing the asymptotic stability through the corresponding Liapunov function. The suitable A is called *stable*, and it satisfies the above described role, *if and only if its characteristic roots have negative real parts*. Then the matrix K which secures such a case makes a control program asymptotically stabilizing (3.1.4) about $\bar{x}(t) \equiv 0$.

EXAMPLE 3.1.1. Let

$$A \triangleq \begin{pmatrix} -2 & 0 & 0 \\ 4 & -1 & 0 \\ 0 & 0 & -2 \end{pmatrix} .$$

The Liapunov Matrix Equation is

$$\begin{pmatrix} -2 & 4 & 0 \\ 0 & -1 & 0 \\ 0 & 0 & -2 \end{pmatrix} \begin{pmatrix} P_{11} & P_{12} & P_{13} \\ P_{21} & P_{22} & P_{23} \\ P_{31} & P_{32} & P_{33} \end{pmatrix} + \begin{pmatrix} P_{11} & P_{12} & P_{13} \\ P_{21} & P_{22} & P_{23} \\ P_{31} & P_{32} & P_{33} \end{pmatrix} \begin{pmatrix} -2 & 0 & 0 \\ 4 & -1 & 0 \\ 0 & 0 & -2 \end{pmatrix}$$

$$= \begin{pmatrix} -1 & 0 & 0 \\ 0 & -1 & 0 \\ 0 & 0 & -1 \end{pmatrix}$$

wherefrom

$$\begin{pmatrix} -4p_{11} +4p_{21}+4p_{12} & -3p_{12} +4p_{22} & -4p_{13}+4p_{23} \\ -3p_{21} +4p_{22} & -2p_{22} & -3p_{23} \\ -4p_{31} +4p_{32} & -3p_{32} & -4p_{33} \end{pmatrix} = \begin{pmatrix} -1 & 0 & 0 \\ 0 & -1 & 0 \\ 0 & 0 & -1 \end{pmatrix}$$

and thus $P_{22} = \frac{1}{2}$, $P_{23} = P_{32} = 0$, $P_{33} = \frac{1}{4}$, $P_{13} = P_{31} = 0$, $P_{21} = P_{12} = \frac{2}{3}$, $P_{11} = \frac{19}{12}$, yielding the symmetric matrix

$$P = \begin{pmatrix} \frac{19}{12} & \frac{2}{3} & 0 \\ \frac{2}{3} & \frac{1}{2} & 0 \\ 0 & 0 & \frac{1}{4} \end{pmatrix} .$$

Then the Sylvester conditions for positive definiteness (all minors positive) give

$$|P_{11}| = \frac{19}{12} , \quad \det \begin{pmatrix} P_{11} & P_{12} \\ P_{21} & P_{22} \end{pmatrix} = \frac{25}{72} , \quad \det P = \frac{1}{12} \cdot \frac{25}{24}$$

all positive. Hence P is positive definite and symmetric, thus yielding $\dot{V} < 0$. Brief calculation gives $V(\bar{x})$:

$$V(\bar{x}) = (x_1 \ x_2 \ x_3) \begin{pmatrix} \frac{19}{12} & \frac{2}{3} & 0 \\ \frac{2}{3} & \frac{1}{2} & 0 \\ 0 & 0 & \frac{1}{4} \end{pmatrix} \begin{pmatrix} x_1 \\ x_2 \\ x_3 \end{pmatrix} = \frac{19}{12} x_1^2 + \frac{4}{3} x_1 x_2 + \frac{1}{2} x_2^2 + \frac{1}{4} x_3^2$$

$$= \frac{25}{36} x_1^2 + \left(\frac{2\sqrt{2}}{3} x_1 + \frac{1}{\sqrt{2}} x_2 \right)^2 + \frac{1}{4} x_3^2 > 0 ,$$

which secures the asymptotic stability of the origin as required. □

EXAMPLE 3.1.2. To illustrate the nonlinear case we consider the classical Lienard system with the scalar motion equation

$$\ddot{q} + d(q)\dot{q} + k(q) = u \tag{3.1.8}$$

where $d(\cdot), k(\cdot)$ are C^1-functions, with the assumption of the restitution law (2.3.18) stretching for all $\mathbb{R}^2 : k(q)q > 0$, $\forall q \neq 0$, that is, the energy cup $\Delta_E = \Delta$. The energy is immediately obtained as

$$E(q,\dot{q}) = \frac{1}{2} \dot{q}^2 + \int k(q)dq \tag{3.1.9}$$

generating the power $\dot{E}(q,\dot{q}) = \frac{\partial E}{\partial q} \dot{q} + \frac{\partial E}{\partial \dot{q}} \ddot{q}$, and substituting (3.1.8), (3.1.9), we obtain

$$\dot{E}(q,\dot{q}) = u\dot{q} - d(q)\dot{q}^2 . \tag{3.1.10}$$

Lienard [1] assumed $\int k(q)dq \to \infty$ as $|q| \to \infty$, which with (3.1.9) implies that conditions (i) and (ii) hold. To make $\dot{E}(q,\dot{q})$ negative definite, we choose $P(q,\dot{q})$ from (3.1.3) such that

$$\max_u \ (u\dot{q}) < d(q)\dot{q}^2 , \quad \dot{q} \neq 0 ,$$

or

$$u \ \text{sgn} \ \dot{q} < d(q)|\dot{q}| , \quad \dot{q} \neq 0 , \tag{3.1.11}$$

which determines $P(\cdot)$, depending upon the sign of $d(q)$. The latter may be negative, producing negative damping, that is, contributing to the positive power balance (3.1.10) which opposes our objective. The controller $P(\cdot)$ must be so designed as to counterbalance the damping power $d(q)\dot{q}^2$ in order to make (3.1.10) negative. □

After the above introduction, we may return now to the uncertain system (2.2.6) and the general property Q. We do this not wishing to generalize, but solely for technical reasons, namely in order to avoid repetitive defining of controllabilities separately for each of the later discussed objectives. Indeed, although some conclusions follow directly from the general definitions given below, they will be left for discussion on particular Q's later in the text.

First, we want to establish whether, given the objective Q, the *system is capable of attaining it at all*, that is, possibly with cooperation of the opposing uncertainty. This means there must exist a control program $\bar{P}(\cdot)$ such that for some, possibly friendly values w, we can attain Q. The above generates a pair of functions $\bar{u}(\cdot),\bar{w}(\cdot)$ producing a motion, possibly just one motion, exhibiting Q. The above is the so called *weak mode of control* (sensitive to $\bar{w}(\cdot)$) securing controllability for Q under suitable $\bar{w}(\cdot)$. We classify it as follows.

DEFINITION 3.1.4. The system (2.2.6) is *controllable for the objective* Q *on some* $\Delta_0 \subset \Delta$ *at* t_0 if and only if there is a program $P(\cdot)$ such that for each $\bar{x}^0 \in \Delta_0$ there is a motion $\bar{\phi}(\cdot) \in K(\bar{x}^0,t_0)$ exhibiting Q. When $\bar{P}(\cdot),Q$ are independent of t_0, the system is *uniform* controllable for Q on Δ_0. When Q is exhibited before, after or during a stipulated interval of time T_Q we have controllability for Q *before, after* or *during* T_q.

The notions below follow immediately. The set Δ_0 is called *controllable for* Q *at* t_0 and the union of such sets (maximal Δ_0 in Δ) is called the *region of controllability for* Q *at* t_0, denoted Δ_{qt_0}. When $\bar{P}(\cdot),Q$ are independent of t_0, we have the *region of uniform controllability for* Q, denoted Δ_q. Finally when the appearance of Q is referred to T_q, we have the region of uniform controllability for Q *before, after* (ultimate controllability) or *during* T_q, denoted $\Delta_q(T_q)$. All three regions are subsets of the Q-zone and any subset of Δ_{qt_0}, Δ_q or $\Delta_q(T_q)$ is correspondingly controllable. When the region covers Δ the corresponding controllability is called *complete*.

When $\bar{P}(\cdot)$ of Definition 3.1.4 is given, we acknowledge this fact by referencing the region accordingly: $\Delta_{qt_0}(\bar{P})$, $\Delta_q(\bar{P})$ or $\Delta_q(T_q,P)$, calling it the *recovery region* for Q under $P(\cdot)$. The size of such a region, or in particular its diameter, determines what, following Schmitendorf [2], we call the *degree* of controllability under $\bar{P}(\cdot)$, defined by

$$\rho(T_q) = \inf\{\|\bar{x}^0\| \mid \bar{x}^0 \in \partial\Delta_q(\bar{P},T_q)\} \tag{3.1.12}$$

which is the radius of the largest ball $\|\cdot\|$ contained in $\Delta_q(\bar{P},T_q)$. Here $\|\cdot\|$ means any norm in $\mathbb{R}^N$. Obviously

$$\Delta_q(\bar{P},T_q) \subset \Delta_q(\bar{P}) \subset \Delta_q. \tag{3.1.13}$$

Next, we want the controllability robust against all options of uncertainty, that is, irrespectively of what the uncertainty does. This is called a *strong mode of control*, specified as follows.

DEFINITION 3.1.5. The system (2.2.6) is *strongly controllable for* Q on Δ_0 *at* t_0 if and only if there is a program $\bar{P}(\cdot)$ such that for each $\bar{x}^0 \in \Delta_0$, motions $\bar{\phi}(\bar{x}^0, t_0, \cdot)$ exhibit Q for all $\bar{\phi}(\cdot) \in K(x^0, t_0)$. When $\bar{P}(\cdot), Q$ are independent of t_0 , the system is strongly *uniform controllable for* Q *on* Δ_0 , and when Q is qualified by a stipulated time interval T_Q , so is the controllability.

Again Δ_0 is called strongly controllable for Q at t_0 and the union of such sets (maximal such Δ_0) is called the *region of strong controllability for* Q at t_0 , denoted Δ_{Qt_0} . In case of t_0-independence we have the region of *uniform* strong controllability for Q , denoted Δ_Q . We shall also abbreviate the name as *strong region for* Q . Finally when Q occurs *before, during* or *after* T_Q , we add these qualifying notions to the name of the region and notation, that is, $\Delta_Q(T_Q)$. Here again when the region covers the entire Δ , the corresponding strong controllability is *complete*.

By obvious implications applying between the defined types of controllability, we have

$$\Delta_Q(T_Q) \subset \Delta_Q \subset \Delta_{Qt_0} \ , \tag{3.1.14}$$

$$\Delta_{Qt_0} \subset \Delta_{qt_0} \tag{3.1.15}$$

and

$$\Delta_Q \subset \Delta_q \ . \tag{3.1.16}$$

The control program $P(\cdot)$, briefly the *controller*, of Definition 3.1.5 is called *robust controller*, later also a *winning* controller. When $\bar{P}(\cdot)$ is given we have the *strong recovery region* $\Delta_Q(P)$, possibly qualified by $T_Q : \Delta_Q(P, T_Q)$. Moreover, in a similar fashion as for controllability, its size (diameter) gives the degree of strong controllability determined by (3.1.12) but with $\partial \Delta_Q(P, T_Q)$ replacing $\partial \Delta_q(P, T_q)$.

Since the strong controllability with given $\bar{P}(\cdot)$ implies the strong controllability, we have

$$\Delta_Q(\bar{P}, T_Q) \subset \Delta_Q(\bar{P}) \subset \Delta_Q \ . \tag{3.1.17}$$

EXAMPLE 3.1.3. Consider a damped oscillator with the scalar motion equation

$$\ddot{q} + u\dot{q} + wk(q) = 0 \tag{3.1.18}$$

and let $u = P(q, \dot{q})$ be the control program satisfying the positive viscous damping requirements (2.4.7) and (2.3.12):

$$P(q,\dot{q})\dot{q} > 0 , \quad \dot{q} \neq 0 \quad \text{and} \quad P(q,0) \equiv 0 . \tag{3.1.19}$$

Suppose Q is defined to be a node (non-oscillatory) type behavior of the trajectories of (3.1.14) near its basic equilibrium. From the elementary theory of oscillations $u \geq 2\sqrt{w}$ yields such a node, while its contradiction $u < 2\sqrt{w}$ yields a focus. With the bound on uncertainty $W : w \in [-1,1]$ the worst the uncertainty can do occurs when $w(t) \equiv 1$. Then the system is strongly controllable for the node on $\mathbb{R}^2$ for $P(q,\dot{q}) \geq 2$, which means that apart from $P(q,\dot{q})\dot{q} > 0$, $\dot{q} \neq 0$ and $P(q,0) \equiv 0$, the damping coefficient used must be for all $q,\dot{q}$ bounded below by 2. These are the only restrictions upon the choice of the robust program to generate the node at the basic equilibrium. Consequently there is much freedom left for designing $P(\cdot)$ to serve other purposes in addition to the node, say, for instance shortest time approach to the equilibrium, etc. $\square$

For many objectives Q the region of strong controllability Δ_Q is essential in applications but not easy to determine. We shall discuss various methods leading to the latter when referring to particular Q's, but it is convenient to introduce now a barrier type estimate of Δ_Q which can be used in all the cases.

Since only $\Delta_Q \cap \Delta$ is of interest, we may as well simplify notation assuming $\Delta_Q \subset \Delta$ and thus bounded, with finite diameter. This also means that the degree of strong controllability is assumed of finite value. Let $\mathcal{S}$ be a surface that separates Δ into two disjoint open sets Δ^P and $C\Delta^P \triangleq \Delta - \Delta^P$, the first enclosing Δ_Q, called *interior*, the other called *exterior*, see Fig. 3.2. The surface $\mathcal{S}$ will act as a repellor from Δ^P for at least one motion of (2.2.6)'.

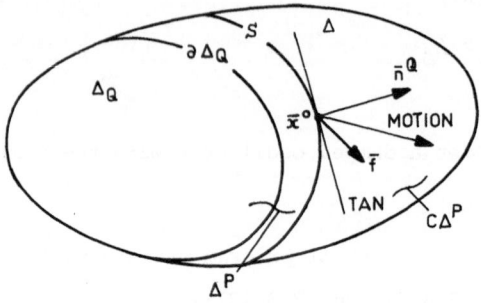

Fig. 3.2

DEFINITION 3.1.6. The surface $\mathcal{S}$ is *weakly $\bar{P}(\cdot)$-nonpermeable*, if and only if given the robust $\bar{P}(\cdot)$ generating Q and given $\bar{x}^0 \in \mathcal{S}$ there is $\bar{w}^*(\cdot)$ such that for some t_0, and corresponding motion $\bar{\phi}(\bar{x}^0, t_0, \cdot) \in K(\bar{x}^0, t_0)$, we have

$$\bar{\phi}(\bar{x}^0, t_0, \mathbb{R}^+) \cap \Delta^P = \phi \tag{3.1.20}$$

for all $\bar{u} \in \bar{P}(\bar{x}, t)$.

Obviously $\bar{P}(\cdot)$ will still generate other motions which will not be prevented from entering Δ_Q and exhibiting Q. But the "spell" of $\bar{P}(\cdot)$ is broken, in that from $\mathcal{S}$ on, it is unable to dominate the scene completely.

Indeed, the above contradicts the strong controllability for Q and by definition of Δ_Q the weakly $\bar{P}(\cdot)$-nonpermeable surface $\mathcal{S}$ must belong to the complement $C\Delta_Q \triangleq \Delta - \Delta_Q$. When $\mathcal{S}$ is smooth, with non-zero gradient $\bar{n}^Q \triangleq (n_1^Q, \ldots, n_N^Q)$ directed away from Δ^P, we conclude from Definition 3.1.6 that $\bar{x}^0 \in \mathcal{S}$ implies that the angle $\nmid (\bar{n}^Q, \bar{f})$ is smaller than $90°$, which means that for each $\bar{x} \in \mathcal{S}$

$$(\bar{n}^Q)^T \cdot \bar{f}(\bar{x}, \bar{u}, \bar{w}^*, t) \geq 0 \tag{3.1.21}$$

for all $\bar{u} \in \bar{P}(\bar{x}, t)$, $t \geq t_0$, cf. Section 2.4, Fig. 2.11. Such (3.1.21) establishes the necessary condition for weak $\bar{P}(\cdot)$-nonpermeability of $\mathcal{S}$. Any $\mathcal{S}$ on which (3.1.21) holds may be considered a legitimate candidate for a weakly $\bar{P}(\cdot)$-nonpermeable surface. Such a candidate must be confirmed by some sufficient conditions.

CONDITIONS 3.1.2. A surface $\mathcal{S}$ separating Δ into two disjoint open sets $\Delta^P, C\Delta^P$ is weakly $\bar{P}(\cdot)$-nonpermeable, if given $\bar{P}(\cdot)$ there is $\bar{w}^*(\cdot)$ and a C^1-function $V_B(\cdot) : D_B \to \mathbb{R}$, $D_B \triangleq N(\mathcal{S}) \cap \Delta^P$, where $N(\mathcal{S})$ is a neighborhood of $\mathcal{S}$, such that for all $\bar{x} \in D_B$,

   (i) $V_B(\bar{x}) < V_B(\bar{\xi})$, $\forall \bar{\xi} \in \mathcal{S}$,

   (ii) $\nabla V_B(\bar{x})^T \bar{f}(\bar{x}, \bar{u}, \bar{w}^*, t) \geq 0$, $\tag{3.1.22}$

for all $\bar{u} \in P(\bar{x}, t)$, $t \geq t_0$.

The above conditions follow immediately from the contradiction between (i) and (ii) arising when the motion concerned leaves $\mathcal{S}$ into Δ^P. The same argument proves the conditions with the inequalities of both (i) and (ii) reversed. Observe, moreover, that when $\mathcal{S}$ is specified by a V_B-level with $\nabla V_B(\bar{x}) = \bar{n}^Q \neq 0$, by the argument leading to (3.1.21), condition (i) is automatically satisfied and condition (ii) becomes also necessary. Thus we have . . .

COROLLARY 3.1.2. *When the surface $\mathcal{S}$ in Conditions 3.1.2 is defined as a V_B-level, condition (ii) of Conditions 3.1.2 is necessary and sufficient for weak $\bar{P}(\cdot)$-nonpermeability of $\mathcal{S}$.*

Obviously there may be many weakly $\bar{P}(\cdot)$-nonpermeable surfaces in $C\Delta_C$. As Δ_C has a finite diameter, there exists in $C\Delta_C$ a weakly $\bar{P}(\cdot)$-nonpermeable $\mathcal{S}$ which is "closest" to $\partial\Delta_C$ in the sense that there is no other such $\mathcal{S}$ between it and the boundary $\partial\Delta_C$.

DEFINITION 3.1.7. A weakly $\bar{P}(\cdot)$-nonpermeable surface which is "closest" to $\partial\Delta_Q$ in the above sense (or the boundary itself) is called the *weak $\bar{P}(\cdot)$-barrier* for Q and denoted B_q^P.

Such a weak $\bar{P}(\cdot)$-barrier consists of points closest to $\partial\Delta_Q$ wherefrom the strong role of $\bar{P}(\cdot)$ is contradicted, that is, a leak may appear, with the "leaking" motion exhibiting no Q.. Consequently B_q^P is defined as the lower bound of (3.1.22) and estimates Δ_Q from above. In ideal circumstances $B_q^P = \partial\Delta_Q$. Further properties of such a barrier as well as its use in determining Δ_Q are left to our discussion on particular objective properties Q later in the text.

EXERCISES 3.1

3.1.1 Investigate the stability of the basic equilibrium (0,0) for the system
$$\dot{x}_1 = x_1^2 - x_2^2 ,$$
$$\dot{x}_2 = -2x_1x_2 ,$$
using $V = 3x_1x_2^2 - x_1^3$.

3.1.2 Using $V = 5x_1^2 + 2x_1x_2 + 2x_2^2$ show that the basic equilibrium (0,0) of the system
$$\dot{x}_1 = x_2$$
$$\dot{x}_2 = -x_1 - x_2 + (x_1 + 2x_2)(x_2^2 - 1)$$
is asymptotically stable in the region $|x_2| < 1$.

3.1.3 Find the region of asymptotic stability for the scalar system

$\dot{x} = \frac{1}{2} x(x-2)$.

3.1.4 Discuss the stability of the basic equilibrium $(0,0)$ of the following systems using the indicated test function. Find the regions of attraction wherever appropriate:

(i) $\dot{x}_1 = x_2$, $\qquad\qquad$ $V = 3x_1^2 + 6x_1x_2 + 3x_2^2$

$\dot{x}_2 = -2x_1 - 3x_2$,

(ii) $\dot{x}_1 = e^{x_1} - x_2$, $\qquad\quad$ $V = x_1^2 + x_2^2$

$\dot{x}_2 = x_1$,

(iii) $\dot{x}_1 = \sin(x_1 - x_2)$, $\qquad$ $V = x_1 x_2$,

$\dot{x}_2 = \sin(x_2 - x_1)$,

(iv) $\ddot{q} = q^3$, $\qquad\qquad\qquad$ $V = q\dot{q}$,

(v) $\dot{x}_1 = x_1 + 4x_2$, $\qquad\quad$ $V = 8x_1^2 + 11x_1x_2 + 5x_2^2$

$\dot{x}_2 = -2x_1 - 5x_2$,

(vi) $\dot{x}_1 = -x_1 + 6x_2$, $\qquad$ $V = 3x_1^2 + 14x_1x_2 + 8x_2^2$

$\dot{x}_2 = 4x_1 + x_2$.

3.1.5 Show that trajectories of the system

$$\dot{x}_1 = -x_2 - x_1\sqrt{x_1^2 + x_2^2}$$
$$\dot{x}_2 = x_1 - x_2\sqrt{x_1^2 + x_2^2}$$

form a center at $(0,0)$, that is, are stable in the linear approximation, while in fact $(0,0)$ is an asymptotically stable equilibrium (focus). For full discussion, use the critical points classification known from your differential equations course.

3.1.6 Find a test function $V(\cdot)$ for asymptotic stability of $(0,0)$ for the system

$$\dot{x}_1 = x_1(x_2 - b) , \qquad \dot{x}_2 = x_2(x_1 - a)$$

and confirm that the region of attraction is defined by $(x_1/a)^2 + (x_2/b)^2 < 1$.

3.1.7 The motion equations of a gyroscope without external forces are

131

$$J_1\dot{\omega}_1 + (J_3 - J_2)\omega_2\omega_3 = 0$$

$$J_2\dot{\omega}_2 + (J_1 - J_3)\omega_3\omega_1 = 0$$

$$J_3\dot{\omega}_3 + (J_2 - J_1)\omega_1\omega_2 = 0$$

where J_i principal moments of inertia and ω_i angular velocities about principal axes, $i = 1,2,3$. Let $\omega_1 = \omega_0 + x_1$, $\omega_2 = x_2$, $\omega_3 = x_3$ ($\omega_0 = $ const). Show that the origin is stable if $J_1 < J_2 < J_3$, using

$$V = J_2(J_2 - J_1)x_2^2 + J_3(J_3 - J_1)x_3^2 + [J_2x_2^2 + J_3x_3^2 + J_1(x_1^2 + 2x_1\omega_0)]^2.$$

3.1.8 Consider the system

$$\dot{x}_1 = x_2,$$
$$\dot{x}_2 = -2x_2 - 2x_1 - 3x_1^2.$$

Investigate stability of the basic equilibrium $(0,0)$ with $V = x_2^2 + x_1^2$. Determine the other equilibrium and apply a suitable transformation of coordinates to move $(0,0)$ to this equilibrium. Write the new equations of motion in $0'\xi_1\xi_2$ and using $V = (\xi_2^2/2) + \xi_1^2 + \xi_1^3$ investigate the stability about the new origin.

3.1.9 Using the Liapunov Matrix Equation, prove asymptotic stability of the origin for

(i)
$$\begin{pmatrix} \dot{x}_1 \\ \dot{x}_2 \end{pmatrix} = \begin{pmatrix} 0 & -1 \\ 3 & -6 \end{pmatrix} \begin{pmatrix} x_1 \\ x_2 \end{pmatrix};$$

(ii)
$$\begin{pmatrix} \dot{x}_1 \\ \dot{x}_2 \\ \dot{x}_3 \end{pmatrix} = \begin{pmatrix} 2 & 1 & 1 \\ 0 & 2 & 0 \\ 0 & -1 & 2 \end{pmatrix} \begin{pmatrix} x_1 \\ x_2 \\ x_3 \end{pmatrix};$$

(iii)
$$\begin{pmatrix} \dot{x}_1 \\ \dot{x}_2 \\ \dot{x}_3 \end{pmatrix} = \begin{pmatrix} -2 & -7 & -5 \\ 0 & -5 & -3 \\ 0 & 0 & -6 \end{pmatrix} \begin{pmatrix} x_1 \\ x_2 \\ x_3 \end{pmatrix}.$$

3.2 STABILITY OBJECTIVES

In general, we need more specified definitions of stability than those
given in Section 3.1, if we deal with the uncertain nonautonomous system
(2.2.6)' instead of (2.1.4). The stability of a set defined below covers
stability of an equilibrium, ultimate boundedness of motions and orbital
and structural stabilities - the main objectives of stabilization.

Suppose M is a subset of Δ and let $M(t)$ denote a t-section of
$M : M(t) \times \mathbb{R} \triangleq M$. If there is a compact (bounded and closed) $\Delta_M \subset \mathbb{R}^N$
such that $M(t) \subset \Delta_M$ for all t, then M is bounded, and thus compact if
closed. Assume the latter. Then let $\rho(\bar{x},\bar{y})$ be the distance between
points $\bar{x},\bar{y} \in \mathbb{R}^N$. For some t, the distance between a point $\bar{x}(t) \in \mathbb{R}^N$
and the set $M(t)$ is determined by

$$\rho(\bar{x}(t),M(t)) \triangleq \inf \rho(\bar{x},\bar{y}) \mid \bar{y} \in M(t) . \tag{3.2.1}$$

We may define now the first objectives of stabilization. Let Δ_0 be a
subset of Δ , and recall that by their defining properties the trajectories
of (2.2.6) and motions of (2.2.6)' are solutions of the corresponding
selector equations (2.2.5) and (2.2.5)' with $\bar{u},\bar{w}$ substituted.

DEFINITION 3.2.1. $\bar{M}$ is said to be *stable* with respect to motions of
(2.2.5)' on $\Delta_0 \times \mathbb{R}$, if and only if for each $\varepsilon > 0$, $t_0 \in \mathbb{R}$ there is
$\delta > 0$ such that $\bar{x}^0 \in \Delta_0$, $\rho(\bar{x}^0,\bar{M}(t_0)) < \delta$ imply $\rho(\bar{\phi}(\bar{x}^0,t_0,t),M(t)) < \varepsilon$,
for all $t \geq t_0$. When δ_0 does not depend upon t_0 , $\bar{M}$ is *uniformly stable*
on Δ_0 . $\bar{M}$ is *unstable* if and only if it is not stable. See Fig. 3.3.

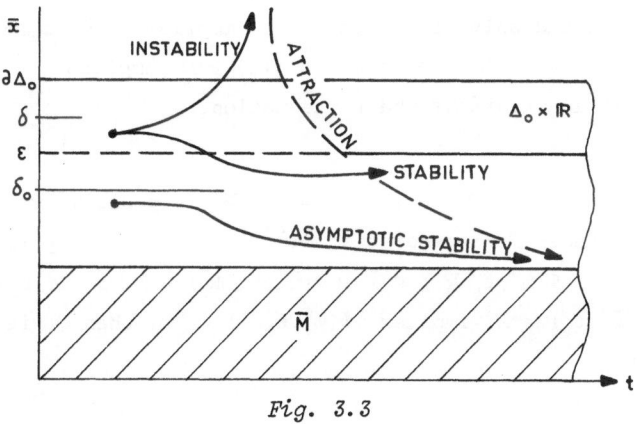

Fig. 3.3

DEFINITION 3.2.2. $\bar{M}$ is an *attractor (repellor)* for motions of (2.2.5)' on $\Delta_0 \times \mathbb{R}$ if and only if for each $\varepsilon > 0$, $t_0 \in \mathbb{R}$ there is $\delta_0 > 0$ such that $\bar{x}^0 \in \Delta_0$ and $\rho(\bar{x}^0, \bar{M}(t_0)) < \delta_0$ implies $\rho(\bar{\phi}(\bar{x}^0, t_0, t), \bar{M}(t)) \to 0$, as $t \to \infty$ ($t \to -\infty$). When δ_0 does not depend upon t_0 we have *uniform attraction (repulsion)*.

Note that $\bar{M}$ being an attractor does not imply stability, as the motion may escape to infinity and return before approaching $\bar{M}$. This is the reason behind calling the attraction *quasi-asymptotic stability*, see Antosiewicz [1], Yoshizawa [1]. The set of all $\bar{x}^0 \in \Delta$ wherefrom motions are uniformly attracted to $\bar{M}$ is called the *region of uniform attraction*, denoted Δ_{AT}.

DEFINITION 3.2.3. $\bar{M}$ is *asymptotically stable* with respect to motions of (2.2.5)' on $\Delta_0 \times \mathbb{R}$ if and only if it is a stable attractor on this set. It is *uniform asymptotically stable* if and only if it is a uniformly stable uniform attractor. $\bar{M}$ is *completely unstable* if and only if it is an unstable repellor.

The set of all $\bar{x}^0$'s in Δ at which $\bar{M}$ is uniform asymptotically stable is called the *region* of such stability, denoted Δ_{as}.

So far $\bar{M}$ was an arbitrary compact set in Δ, which may be specified by a motion or motions of (2.2.5)' or not. The first case leads to stability of an equilibrium.

DEFINITION 3.2.4. Given a control program $\bar{P}(\cdot)$, a set $M \subset \Delta \times \mathbb{R}$ is *positively (negatively) invariant* under $\bar{P}(\cdot)$-generated motion of the selector equation (2.2.5)' if and only if $(\bar{x}^0, t_0) \in M$ implies $\bar{\phi}(\bar{x}^0, t_0, \mathbb{R}^+) \subset M$ ($\bar{\phi}(\bar{x}^0, t_0, \mathbb{R}^-) \subset M$). If both, that is, $(\bar{x}^0, t_0) \in M$ implies $\bar{\phi}(\bar{x}^0, t_0, \mathbb{R}) \subset M$, it is *invariant* under the motion.

DEFINITION 3.2.5. Given $\bar{P}(\cdot)$, a set $M \subset \Delta \times \mathbb{R}$ is positively (negatively) *strongly invariant* under $\bar{P}(\cdot)$-generated $K(\bar{x}^0, t_0)$ if and only if $(\bar{x}^0, t_0) \in M$ implies $\Phi(\bar{x}^0, t_0, \mathbb{R}^+) \subset M$ ($\Phi(\bar{x}^0, t_0, \mathbb{R}^-) \subset M$). If both are true, that is, $(\bar{x}^0, t_0) \in M$ implies $\Phi(\bar{x}^0, t_0, \mathbb{R}) \subset M$, then M is called strongly invariant.

The same two definitions hold uniformly in t_0 for the autonomous (2.2.5) and (2.2.6). Let $\tilde{M}$ be a map of M into Δ. The semi-invariances (positive, negative) and the invariance are defined by $\bar{x}^0 \in \tilde{M}$ implying $\bar{\phi}(\bar{x}^0, \mathbb{R}^+) \subset \tilde{M}$, $\bar{\phi}(\bar{x}^0, \mathbb{R}) \subset \tilde{M}$, respectively, while the strong invariances are defined correspondingly by $\bar{x}^0 \in \tilde{M}$ implying $\Phi(\bar{x}^0, \mathbb{R}^+) \subset \tilde{M}$ and $\Phi(\bar{x}^0, \mathbb{R}) \subset \tilde{M}$.

As already mentioned, trajectories of (2.2.6) and motions of (2.2.6)' are solutions of (2.2.5), (2.2.5)' respectively, with $\bar{u}, \bar{w}$ substituted. Then the invariant sets for (2.2.5), (2.2.5)' are correspondingly those for the dynamical systems (2.1.4), (2.1.15).

A positively (negatively) invariant or invariant set is termed *minimal* if and only if it is closed and does not contain any other set of the same type. The same applies to strongly invariant sets.

The steady state trajectories of (2.1.4) and steady state sets of (2.1.15), see Section 2.1, are minimal invariant. In Section 3.4 some will become minimal strongly invariant for (2.2.6) and (2.2.6)'. In this section we refer to equilibria only.

For (2.1.15) the minimal invariant M will thus be defined as the t-axis extension of an equilibrium $\bar{x}^e$ specified by (2.1.16): $M = \{\bar{x}^e\} \times \mathbb{R}$. For (2.2.6)' with unknown $\bar{w}$, the relation (2.1.16) becomes $\bar{f}(\bar{x}^e, \bar{u}, \bar{w}, t) \equiv 0$ with given $\bar{u} = \bar{u}^*$, leaving $\bar{x}^e$ unknown but bounded within the set

$$x^e(u^*) \overset{\Delta}{=} \{\bar{x} \in \Delta \mid \bar{f}(\bar{x}, \bar{u}, \bar{w}, t) = 0, \forall t \geq t_0, \bar{u} = \bar{u}^*, \bar{w} \in W\} . \qquad (3.2.1)$$

On the other hand, the reference equilibrium must be chosen compatible with the energy reference frame of Section 2.3. The latter is based on the nominal $E(\bar{x})$ obtained by substituting an extremizing $\bar{w}^*$ into $E(\bar{x}, \bar{w})$, with extrema of such $E(\bar{x})$ coinciding with the equilibria, see Section 2.3. Consequently we obtain the equilibrium from the selector (2.2.5)' with $\bar{w}^*$ substituted, that is, from the adjusted (2.1.16) which now, given $\bar{u}$, becomes

$$\bar{f}(\bar{x}^e, \bar{u}, \bar{w}^*, t) \equiv 0 . \qquad (3.2.2)$$

Obviously we can shift the equilibria by changing $\bar{u}$ and/or by using suitable compensation in $\bar{\Pi}(\bar{q}, w^*)$, either in gravity or in the elastic component. Since we know W, such a shift may cancel the influence of $\bar{w}$ and go beyond x^e, limited only by the constraints U or the capacity of the compensation. The balancing role of control and compensation in influencing the equilibria becomes visible when we express (3.2.2) in terms of characteristics. Then from (2.1.10) and (2.3.12), (2.3.13) we obtain for specific $\bar{u}, \bar{w}$

$$\dot{\bar{q}} = 0 , \quad \bar{\Pi}(\bar{q},w^*) = \bar{F}(\bar{q},\dot{\bar{q}},\bar{u},w^*) + \bar{R}(\bar{q},\dot{\bar{q}},\bar{w}^*,t) . \tag{3.2.3}$$

For the autonomous system: $\bar{R} \equiv 0$ this means that the controller balanced the potential characteristics, which usually are subject to design synthesis. On the other hand, the distribution of control load (forces or momenta) among actuators acting on particular DOF's is also a part of the design of the mechanical system. It plays a particularly important role in the control of flexible structures, recall Section 1.8. If $\nu_1,\dots,\nu_r$ are the *weight-parameters* of such a distribution, the relation

$$\Phi(\nu_1,\dots,\nu_r,F_1(\cdot),\dots,F_n(\cdot)) = 0 \tag{3.2.4}$$

specifies a *distribution law* for the actuators. For instance, in the case of linear distribution with $r = n$ (robotic actuators), we shall have for given $\bar{u},\bar{w}$,

$$\sum_i \nu_i F_i(\bar{q},\dot{\bar{q}},\bar{u},\bar{w}) = 0 . \tag{3.2.5}$$

With fixed $\Pi_i(\cdot)$, the equilibria move with varying $\bar{u}$ in the Configuration Space $\mathbb{R}^n$ defined by $\dot{\bar{q}} = 0$. By (3.2.3), (2.3.3) such motion is located on the surface

$$\Phi(\nu_1,\dots,\nu_r, \frac{\partial V}{\partial q_1} ,\dots, \frac{\partial V}{\partial q_n}) = 0 \tag{3.2.6}$$

or vectorially with $\nu = (\nu_1,\dots,\nu_r)^T = \text{const}$ on

$$\Phi(\bar{\nu},\nabla V(\bar{q})) = 0 , \tag{3.2.7}$$

in Δ_q . Assume for the moment that we are not interested in design synthesis (compensation in $\Pi \cdot (\cdot)$), that is, that the gradient $\nabla V(\bar{q})$ is given. Then by (3.2.7), the position of the vector $\bar{\nu}$ with respect to the gradient may specify the motion of the equilibrium. For instance, in the case of the linear distribution of control loads (3.2.5), we have

$$\nabla V(\bar{q})^T \cdot \bar{\nu} = 0 \tag{3.2.8}$$

making the geometric locus of controlled equilibria into a set of points in Δ_q where $\bar{\nu}$ is orthogonal to the gradient, that is, tangent to the E-levels. The reader may note here that when such motion of the controlled equilibrium crosses the uncontrolled equilibrium: $\nabla V(\bar{q}) = 0$, the distribution vector $\bar{\nu}$ becomes arbitrary, thus undefined.

It is convenient to place the origin of $\mathbb{R}^N$ in the reference equilibrium whose neighborhood is of interest, making it $\bar{x}(t) \equiv 0$ with the minimal invariant set taken as the t-axis: $M = \{0\} \times \mathbb{R}$. Definitions 3.2.1-3.2.3 apply immediately with $\rho(\bar{x}(t),M(t)) = \|\bar{x}(t)\|$, where $\|\cdot\|$ is any norm in $\mathbb{R}^N$.

DEFINITION 3.2.6. The motion $\bar{x}(t) \equiv 0$ of (2.2.5)' is *uniformly stable on* Δ_0 if and only if for each $\varepsilon > 0$ there is $\delta > 0$ such that $\bar{x}^0 \in \Delta_0$, $\|\bar{x}^0\| < \delta$ implies that $\|\phi(\bar{x}^0, t_0, t)\| < \varepsilon$, $t \geq t_0$. It is *uniform asymptotically stable on* Δ_0 if and only if there is $\delta_0 > 0$ such that $\|\bar{x}^0\| < \delta_0$ implies $\bar{\phi}(\bar{x}^0, t_0, t) \to 0$, $t \to \infty$.

The above definition extends from $\bar{x}(t) \equiv 0$ to any motion of (2.2.5)' as long as it is given by using the technique introduced for relative state representation in Section 2.1, see (2.1.6), and the subsequent discussion. The latter is immediately augmentable to our nonautonomous system (2.2.5)'. Indeed, if $\bar{x}_m = \bar{\phi}(x_m^0, t_0, t)$, $t \geq t_0$ is the given motion, setting $\bar{x} = \bar{y} + \bar{x}_m(t)$, (2.2.5)' becomes $\dot{y} = \bar{f}(\bar{y} + \bar{x}_m(t), \bar{u}, \bar{w}, t) - \bar{f}(\bar{x}_m(t), \bar{u}, \bar{w}, t)$ $\triangleq \bar{g}(\bar{y}, \bar{u}, \bar{w}, t)$, with $\bar{g}(0, \bar{u}, \bar{w}, t) \equiv 0$, and given $\bar{u}, \bar{w}$, the stability of $\bar{y}(t) \equiv 0$ is covered by Definition 3.2.6.

As mentioned in Section 2.1 and what we shall see later in detail, the reference curve relative to which we investigate the stability of motions, or other behavior for that matter, may not be a motion at all. We may, using a suitable control program for $u_m(t)$, make it a motion and then, using the transformation $\bar{x} \to \bar{y}$ of above, investigate the equilibrium of the relative system. Alternatively we may treat the curve directly as *M* and apply Definitions 3.2.1 - 3.2.3, as suggested in the first place.

Sufficient conditions for controllability and strong controllability for stability and asymptotic stability can be obtained by adapting the classical arguments used in proving the stabilities, see Yoshizawa [1].

CONDITIONS 3.2.1. The system (2.2.6)' is controllable for uniform stability of $\bar{x}(t) \equiv 0$ on Δ_0, briefly *stabilizable on* Δ_0 *about* $\bar{x}(t) \equiv 0$, if there is a program $\bar{P}(\cdot)$ defined on $\Delta_0 \times \mathbb{R}$ and a C^1-function $V(\cdot) : \Delta_0 \times \mathbb{R} \to \mathbb{R}$, $V(0, t) \equiv 0$, such that for all $(\bar{x}, t) \in \Delta_0 \times \mathbb{R}$,

   (i) $a(\|\bar{x}\|) \leq V(x, t) \leq b(\|\bar{x}\|)$, where $a(\cdot), b(\cdot)$ are positive, continuously increasing functions:

   (ii) for each $\bar{u} \in \bar{P}(\bar{x}, t)$, $\bar{w} \in W$,

$$\frac{\partial V(\bar{x}, t)}{\partial t} + \nabla_x V(\bar{x}, t)^T \bar{f}(\bar{x}, \bar{u}, \bar{w}, t) \leq 0 . \qquad (3.2.9)$$

To prove the above, first we show stability. As $V(\cdot)$ is continuous and $V(0, t_0) = 0$. there is $\delta(\varepsilon, t_0) > 0$ such that $\|\bar{x}^0\| < \delta$ implies

$$V(\bar{x}^0, t_0) < a(\varepsilon) . \qquad (3.2.10)$$

Suppose instability, that is, that $\phi(t)$ crosses $\|\bar{x}\| = \varepsilon$. Then, there is $t_1 > t_0$ such that $\phi(\bar{x}^0, t_0, t_1) = \bar{x}^1$, $\|\bar{x}^1\| = \varepsilon$. On the other hand, condition (ii) implies $V(\bar{x}^1, t_1) \leq V(\bar{x}^0, t_0)$, thus by (3.2.10), $V(\bar{x}^1, t_1) < a(\varepsilon)$ contradicting (i), whence proving stability. Now by choosing the above $\delta(\varepsilon) > 0$ such that $b(\delta) < a(\varepsilon)$ for all t_0, by the same argument we may show that $\|\bar{x}^0\| < \delta(\varepsilon)$, $t_0 \in \mathbb{R}$, then $\|\phi(\bar{x}^0, t_0, t)\| < \varepsilon$, $t \geq t_0$ which proves uniform stability.

CONDITIONS 3.2.2. The system (2.2.6)' is strongly controllable for the uniform stability of $\bar{x}(t) \equiv 0$ on Δ_0, briefly *strongly stabilizable on* Δ_0 *about* $\bar{x}(t) \equiv 0$, if Conditions 3.2.1 hold but (ii) is replaced by the following:

(ii)' for each $\bar{u} \in \bar{P}(\bar{x}, t)$,

$$\frac{\partial V(\bar{x}, t)}{\partial t} + \nabla_x V(\bar{x}, t)^T \bar{f}(\bar{x}, \bar{u}, \bar{w}, t) \leq 0 \qquad (3.2.11)$$

for all $\bar{w} \in W$.

Indeed, replacing the motion $\phi(\cdot)$ in the proof of Conditions 3.1.1 by some $\phi(\cdot) \in K(\bar{x}^0, t_0)$ of (2.2.6)', we obtain the same contradiction, which then proves the hypothesis for all motions from $K(\bar{x}^0, t_0)$.

The above conditions have an immediate conclusion which gives the possibility of defining a suitable robust controller. Denote $L(\bar{x}, \bar{u}, \bar{w}, t) \overset{\Delta}{=} \partial V/\partial t + \nabla_x V(\bar{x}, t) \cdot \bar{f}(\bar{x}, \bar{u}, \bar{w}, t)$.

COROLLARY 3.2.2. *Given* $(\bar{x}^0, t) \in \Delta_0 \times \mathbb{R}$, *if there is a pair* $(\bar{u}*, \bar{w}*) \in U \times W$ *such that*

$$L(\bar{x}, \bar{u}*, \bar{w}*, t) = \min_{\bar{u}} \max_{\bar{w}} L(\bar{x}, \bar{u}, \bar{w}, t) \leq 0,$$

then condition (3.2.11) is met with $\bar{u}* \in \bar{P}*(\bar{x}, t)$.

The Corollary follows immediately from the fact that

$$\min_{\bar{u}} \max_{\bar{w}} L(\bar{x}, \bar{u}, \bar{w}, t) \geq \min_{\bar{u}} L(\bar{x}, \bar{u}, \bar{w}, t)$$

for all $\bar{w} \in W$. The obtained robust controller will be called *energy dissipative*.

For the objective of asymptotic stability, we shall go directly to the strong mode, then the weak mode follows.

CONDITIONS 3.2.3. The system (2.2.6)' is strongly controllable for the uniform asymptotic stability of $\bar{x}(t) \equiv 0$ on Δ_0, briefly *strongly asymptotically stabilizable about* $\bar{x}(t) \equiv 0$, if Conditions 3.2.2 hold with (ii)' replaced by the following:

(ii)" for each $\bar{u} \in \bar{P}(\bar{x},t)$,

$$\frac{\partial V}{\partial t} + \nabla_x V(\bar{x},t)^T \bar{f}(\bar{x},\bar{u},\bar{w},t) \leq -c(\|\bar{x}\|) , \qquad (3.2.12)$$

for all $\bar{w} \in W$, where $c(\cdot)$ is a continuous positive definite function on Δ_0.

Indeed by Conditions 3.2.2, $\bar{x}(t) \equiv 0$ is uniformly stable, for all motions $\bar{\phi}(\bar{x}^0,t_0,\cdot) \in K(\bar{x}^0,t_0)$, $(x^0,t_0) \in \Delta_0 \times \mathbb{R}$, thus we must only show that it is a uniform attractor. We do it adapting the arguments of Yoshizawa [1].

Let us define $\Delta_0 : \|\bar{x}\| < \alpha$, $\alpha > 0$. The uniform stability implies that there is $\delta_1 > 0$ such that $\|\bar{x}^0\| < \delta_1$, $t_0 \in \mathbb{R}$ yields $\|\phi(\bar{x}^0,t_0,t)\| < \alpha$, for all $t \geq t_0$, and all $\bar{\phi}(\bar{x}^0,t_0,\cdot) \in K(\bar{x}^0,t_0)$. Indeed by the same argument as for (3.2.10), we have $V(\bar{x}^0,t_0) < a(\alpha)$ which is contradicted upon crossing $\partial\Delta_0$ by some motion from Δ_0.

By the same uniform stability for each $\varepsilon > 0$, there is $\delta(a) > 0$ such that $\|\bar{x}^0\| < \delta$, $t_0 \in \mathbb{R}$ implies $\|\phi(\bar{x}^0,t_0,t)\| < \varepsilon$ for all $\phi(x^0,t_0,\cdot) \in K(\bar{x}^0,t_0)$ and all $t \geq t_0$. We will now show that $\|\bar{x}^0\| < \delta_1$, $t_0 \in \mathbb{R}$ implies $\|\phi(\bar{x}^0,t_0,t)\| < \delta(a)$ as some t, for some $\phi(\cdot) \in K(\bar{x}^0,t_0)$. Suppose that $\delta(\varepsilon) \leq \|\phi(\bar{x}^0,t_0,t)\| < \alpha$, for all $t \geq t_0$, $\phi(\cdot) \in K(\bar{x}^0,t_0)$. Since there exists a $c > 0$ such that $\dot{V}(\bar{x},t) \leq -c$ on $\delta \leq \|x\| < \alpha$, integrating we have

$$V(\phi(\bar{x}^0,t_0,t),t) \leq V(\phi(x^0,t_0,t_0)) - c(t-t_0) . \qquad (3.2.13)$$

For $t \geq t_0 + T$, $T = [b(\delta_1) - a(\delta)]/c$ we have $V(x^0,t_0) - c(t-t_0) < a(\delta)$, because $V(x^0,t_0) \leq b(\delta_1)$. By (3.2.13), $V(\phi(x^0,t_0,t),t) < a(\delta)$ which contradicts $V(\phi(\bar{x}^0,t_0,t),t) \geq a(\delta)$. Thus, at some t_0 such that $t_0 \leq t_1 \leq t_0 + T$, we must have $\|\phi(\bar{x}^0,t_0,t_1)\| < \delta(\varepsilon)$ for all $\phi(\bar{x}^0,t_0,\cdot) \in K(\bar{x}^0,t_0)$. Hence, if $t \geq t_0 + T$, we have $\|\phi(\bar{x}^0,t_0,t)\| < \varepsilon$ with T depending only upon ε, which generates uniform attraction, and completes the proof.

Similarly as before, we have the immediate conclusion making it possible to design a dissipative $P(\cdot)$.

COROLLARY 3.2.3. *Given* $(\bar{x},t) \in \Delta_0 \times \mathbb{R}$, *if there is a pair* $(\bar{u}^*,\bar{w}^*) \in U \times W$ *such that*

$$L(\bar{x},\bar{u}^*,\bar{w}^*,t) = \min_{\bar{u}} \max_{\bar{w}} L(\bar{x},\bar{u},\bar{w},t) \leq -c(\|\bar{x}\|) \qquad (3.2.14)$$

then condition (3.2.12) is met with $\bar{u}^* \in \bar{P}^*(\bar{x},t)$.

EXAMPLE 3.2.1. Before analysing the above conditions in the full context of our model, let us briefly illustrate their work on a simple pendulum of Example 1.1.1, see Fig. 1.1. The state equations concerned are

$$\left. \begin{aligned} \dot{x}_1 &= x_2 \\ \dot{x}_2 &= -(g/\ell) \sin x_1 - d|x_2|x_2 + u \end{aligned} \right\} \qquad (1.1.6)$$

and we consider a neighborhood of the basic equilibrium $(0,0)$, that is, the region $\Delta_{qE} : -\pi < x_1 < \pi$ which allows us to reduce (1.1.6) to

$$\left. \begin{aligned} \dot{x}_1 &= x_2 \\ \dot{x}_2 &= (g/\ell)(x_1 - \tfrac{1}{6} x_1^3) - d|x_2|x_2 + u \end{aligned} \right\} . \qquad (3.2.15)$$

The test function selected will be the total energy

$$E(x_1,x_2) = \frac{x_2^2}{2} + \left(\frac{x_1^2}{2} - \frac{x_1^4}{24} \right)(g/\ell)$$

with its time derivative = power:

$$\dot{E}(x_1,x_2) = ux_2 - d|x_2|x_2^2 . \qquad (3.2.16)$$

Letting $V(\bar{x},t) \equiv E(x_1,x_2)$ we obviously have condition (i) satisfied within the whole cup about $(0,0)$ and in the absence of uncertainty the asymptotic stabilizability of $(0,0)$ depends on whether $\dot{E}(x_1,x_2)$ can be made negative definite, say,

$$ux_2 - d|x_2|x_2^2 \leq -c|x_2|$$

with $d > 0$, $c > 0$. This gives the control program determined by the condition

$$u \, \mathrm{sgn} \, x_2 \leq dx_2^2 - c$$

determining the choice of a controller. More economically, following Corollary 3.2.3, we need

$$\min_u (u \, \mathrm{sgn} \, x_2) \leq dx_2^2 - c \qquad (3.2.17)$$

to satisfy condition (ii)" for our system. Note that so far there is no uncertainty. This means

$$-dx_2^2 + c \leq u \leq dx_2^2 - c$$

which defines the set valued dissipative program

$$P(\cdot) \; : \; u(t) \; \in \; [-dx_2^2 + c \, , \, dx_2^2 - c] \; .$$

Here $c > 0$ is best chosen while actually substituting $P(\bar{x})$ into the equations for the purpose of numerical simulation. It may be guessed now, but will become clearer later, that c actually estimates the rate of change of $E(\bar{x}(t))$ along the motion concerned.

Let us suppose now that the damping coefficient d is for some reason uncertain and allowed negative, say, $W : -1 \leq d \leq 1$, which will naturally oppose our asymptotic stabilization. Since the uncertainty is only in damping, it does not affect $E(\bar{x})$ itself and we still have the reference frame H untouched, the cup Z_E being well defined. The power (3.2.16) is however already affected and the control condition (3.2.17), which implies condition (ii)' now becomes

$$\min_{u} \; (u \, \text{sgn} \, x_2) \; \leq \; \min_{d} \; (dx_2^2) \; - \; c \tag{3.2.18}$$

which simply means

$$\min_{u} \; (u \, \text{sgn} \, x_2) \; \leq \; -(x_2^2 + c) \; ,$$

making the robust dissipative controller defined by

$$P(\bar{x}) \; : \; \begin{cases} u \leq -(x_2^2 + c) \, , & \text{for} \quad x_2 \geq 0 \, , \\ u \geq x_2^2 + c \, , & \text{for} \quad x_2 < 0 \, . \end{cases} \tag{3.2.19}$$

So far the system considered is autonomous although uncertain. Introducing external perturbation $R(t)$, (3.2.15) becomes

$$\left. \begin{aligned} \dot{x}_1 &= x_2 \\ \dot{x}_2 &= -(g/\ell)(x_1 - \tfrac{1}{6} x_1^3) - d|x_2|x_2 + u + R(t) \end{aligned} \right\} \tag{3.2.20}$$

generating a slight change in our reasoning. Although the frame of reference H and thus the energy cup Z_E still stays the same together with the basic equilibrium concerned, the power (3.2.16) is now expanded by an extra term:

$$\dot{E} = R(t)x_2 + ux_2 - d|x_2|x_2^2 \tag{3.2.21}$$

wherefrom (3.2.18) becomes

$$\min_{u} \; (u \, \text{sgn} \, x_2) \; \leq \; -R(t) \, \text{sgn} \, x_2 + \min_{d} \; (dx_2^2) \; - \; c \tag{3.2.22}$$

which means that the dissipative program $P(\cdot)$ is non-stationary and defined by

$$\left. \begin{aligned} P(\bar{x},t) \; : \quad u(t) &\leq -(x_2^2 + c + R(t)) \, , & x_2 \geq 0 \\ u(t) &\geq x_2^2 + c - R(t) \, , & x_2 < 0 \end{aligned} \right\} \tag{3.2.23}$$

or, taking only equalities, has two branches

$$u(t) = \pm[x_2^2 + c \pm R(t)] ,$$ (3.2.24)

for $x_2 \geq 0$, $x_2 < 0$ respectively.

EXAMPLE 3.2.2. Let us now extend Example 3.2.1 to the case of the coupled pendula of Example 1.1.2, see Fig. 1.6. Equations (1.1.3) with the substituted characteristics and written in the state format become

$$\left.\begin{aligned}
\dot{x}_1 &= x_3 \\
\dot{x}_2 &= x_4 \\
\dot{x}_3 &= -d_1 |x_3| x_3 - (g/\ell)(x_1 - \tfrac{1}{6} x_1^3) - a^2 k(x_1 - x_2) + u_1 \\
\dot{x}_4 &= -d_2 |x_4| x_4 - (g/\ell)(x_2 - \tfrac{1}{6} x_2^3) + a^2 k(x_1 - x_2) + u_2 .
\end{aligned}\right\}$$ (3.2.25)

We maintain the same asymptotic stabilization objective as in the previous Example 3.2.1 and use the same conditions with $V(\bar{x}, t) \equiv E(\bar{x})$, which now is

$$E(\bar{x}) = \tfrac{1}{2}(x_3^2 + x_4^2) + [\tfrac{1}{2}(x_1^2 + x_2^2) - \tfrac{1}{24}(x_1^4 + x_2^4)](g/\ell) + \tfrac{1}{2} a^2 k(x_1 - x_2)^2 .$$ (3.2.26)

This obviously preserves the reference energy surface H and the cup Z_E in the same shape as in Example 3.2.1, satisfying condition (i) of Conditions 3.2.3. Calculating the power,

$$\dot{E}(\bar{x}) = (u_1 x_3 - d_1 |x_3| x_3^2) + (u_2 x_4 - d_2 |x_4| x_4^2) .$$ (3.2.27)

Assuming again uncertain d_i, $i = 1,2$ bounded by $W: -1 \leq d_i \leq 1$, we have, by Remark 3.2.3, Condition (ii)" implied by the condition corresponding to (3.2.18) which is now calculated as

$$\min_{u_i} (u_i \, \text{sgn} \, x_{2+i}) \leq \min_{d_i} (d_i x_{2+i}^2) - \tfrac{c}{2} , \quad i = 1,2 .$$ (3.2.28)

Because of the additive character of (3.2.26) and (3.2.27), the control condition (3.2.28) applies generally for $i = 1,\dots,n$, and illustrates the procedure of obtaining algorithms for robust controllers in our later cases. Substituting the bounds of W, we have

$$\min_{u_i} (u_i \, \text{sgn} \, x_{2+i}) \leq -[x_{2+i}^2 + (c/2)] , \quad i = 1,2$$ (3.2.29)

making the vector program $\bar{P}(\cdot)$ defined by

$$P_i(\bar{x}) : \begin{cases} u_i(t) \leq -(x_{2+i}^2 + c/2) , & \text{for } x_{2+i} < 0 , \\ u_i(t) \geq (x_{2+i}^2 + c/2) , & \text{for } x_{2+i} \geq 0 . \end{cases} \qquad (3.2.30)$$

Needless to say that in the nonautonomous case, when perturbations $R_i(t)$, $i = 1,2$ are added to the right hand sides of equations (1.1.3) and thus also to the dynamic equations of (3.2.25), we have by the same argument as in Example 3.2.1, the nonstationary dissipative $\bar{P}(\bar{x},t)$ defined by

$$P_i(\bar{x},t) : u(t) = \pm[x_{2+i}^2 + c/2 \pm R(t)] \qquad (3.2.31)$$

for $x_2 \geq 0$, $x_2 < 0$ respectively. □

EXAMPLE 3.2.3. We shall illustrate the asymptotic stabilization on a slightly longer case study.

A retrieval of a tethered satellite under rotation is an unstable procedure, see Misra-Modi [1]. Banerjee-Kane [1] and Xu-Misra-Modi [1] apply thruster-augmented control to such a procedure to stabilize it. A simplified version of the model used by the latter two works was produced by Fudjii-Ishijima [1], ignoring flexibility and mass of the tether as well as the atmospheric influence, but emphasising the fundamental aspects of the case and applying Liapunov control technique with full nonlinearity. Below we follow this work. The system is shown in Fig. 3.4. The satellite is considered a gravitational body S which is to be asymptotically stabilized, the auxiliary subsatellite mass m is sufficiently small so that CG of the shuttle remains in its nominal orbit after deploying the tether, and no motion of m influences the shuttle. The tether is very long but

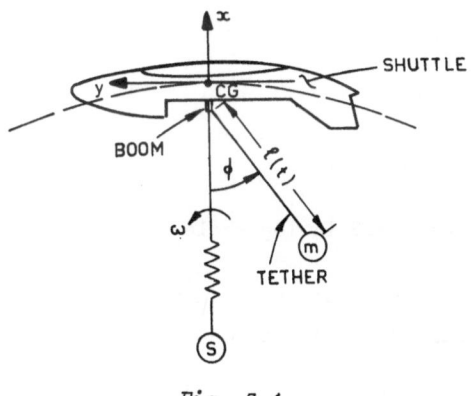

Fig. 3.4

mass-less. The controller acts along the tether through its tension F and no other control force or energy dissipation exists perpendicularly to the tether. Let ℓ_m be the desired length of the tether for doing its assignment (deployment or retrieval of S) and let $\ell^0 = \ell(t_0)$ be the initial length. Then with the angular velocity ω of the satellite S in its orbit, we take the dimensionless time $\tau = \omega t$ and form the dimensionless deflections $\tilde{\ell} = \ell/|\ell^0 - \ell_m|$ and controls $\tilde{u} = u/(m\,\omega^2|\ell^0 - \ell_m|)$. Following Fudjii-Ishijima [1] we may now write the motion equations as

$$
\left. \begin{array}{l}
\overset{\circ\circ}{\tilde{\ell}} - (\overset{\circ}{\tilde{\ell}})^2 - 2\tilde{\ell}\overset{\circ}{\phi} - 3\tilde{\ell}\cos^2\phi = -\tilde{u} , \\[2mm]
\overset{\circ\circ}{\phi} + 2(\overset{\circ}{\tilde{\ell}}/\ell)(1+\overset{\circ}{\phi}) + 3\sin\phi\cos\phi = 0 .
\end{array} \right\}
\tag{3.2.32}
$$

Here the circles above $\tilde{\ell}$ and ϕ denote differentiation with respect to τ : $(\overset{\circ}{-}) = d(-)/d\tau$. Considering the state $\bar{x} \overset{\Delta}{=} (\tilde{\ell},\phi,\overset{\circ}{\tilde{\ell}},\overset{\circ}{\phi})^T$, our goal is to asymptotically stabilize $\bar{x}(t) \equiv 0$. We do it with the test function

$$
V = \frac{1}{2}\left[a_1\overset{\circ}{\tilde{\ell}}^2 + (e^{a_2(\tilde{\ell}-\tilde{\ell}_m)} - 1) + b_1\left(\frac{\overset{\circ}{\phi}}{\sqrt{3}\sin(\pi/4)}\right)^2 + b_2\left(\frac{\sin\phi}{\sin(\pi/4)}\right)^2\right]
\tag{3.2.33}
$$

where the second term refers to deployment and retrieval, while ϕ is the normalized modulo $\pi/4$ and $\overset{\circ}{\phi}$ with the latter multiplied by $\sqrt{3}$. From (3.2.32) and (3.2.33) we obtain

$$
-\overset{\circ}{V} = a_1\overset{\circ}{\tilde{\ell}}(\tilde{\ell}\overset{\circ}{\phi}^2 + 2\tilde{\ell}\overset{\circ}{\phi} + 3\tilde{\ell}\cos^2\phi - \tilde{u}) + a_2(\tilde{\ell}-\tilde{\ell}_m)\overset{\circ}{\tilde{\ell}}e^{a_2(\tilde{\ell}-\tilde{\ell}_m)^2}
$$
$$
+ \frac{2}{3}b_1\overset{\circ}{\phi}(-2\tilde{\ell}\overset{\circ\circ}{\phi}/\tilde{\ell}) - 3\sin\phi\cos\phi(-2\overset{\circ}{\tilde{\ell}}/\tilde{\ell}) + 2b_2\overset{\circ}{\phi}\sin\phi\cos\phi .
$$

Then assuming $b_1 = b_2$ we choose

$$
\tilde{u} = \tilde{\ell}\overset{\circ}{\phi}^2 + 2\tilde{\ell}\overset{\circ}{\phi} + 3\tilde{\ell}\cos^2\phi + (a_2/a_1)(\tilde{\ell}-\tilde{\ell}_m)\exp a_2(\tilde{\ell}-\tilde{\ell}_m)^2
$$
$$
- (4b_1/3a_1)\overset{\circ}{\phi}(\overset{\circ}{\phi}+1)/(\tilde{\ell}+c)
$$

where $c(\bar{x})$ is an arbitrary suitable function. Substituting $c = (k/a_1)\tilde{\ell}\overset{\circ}{\tilde{\ell}}v$, $k > 0$ gives

$$
\overset{\circ}{V} = -k\tilde{\ell}\overset{\circ}{\tilde{\ell}}^2 v ,
$$

and then

$$
V(\tau) = V(0)\exp\left(-\int_0^\tau k\tilde{\ell}\overset{\circ}{\tilde{\ell}}^2 d\tau\right) ,
$$

with the plot shown in Fig. 3.5 in log scale.

The decoy of $V(\cdot)$ is slow after 100 min. because the control program is not adaptive. We shall discuss the adaptive control in Chapter 7.

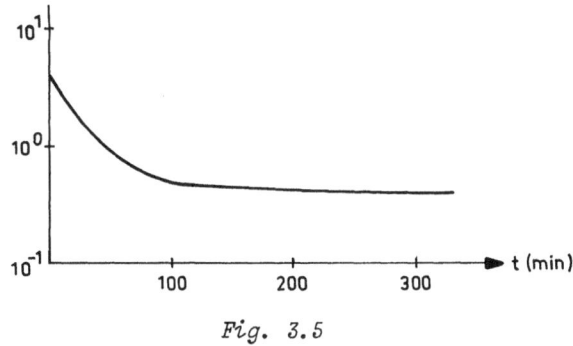

Fig. 3.5

Fudjii-Ishijima [1] give some simulation results assuming the circular orbit for the shuttle of 6220 km with orbital speed 7.065×10^{-2} rad/min. With the dimensions of $\ell = 1$ km , $\dot{\ell} = 12.66$ km/min , $\ell_m = 100$ km , $a_1 = a_2 = 5$, $b_1 = 1$, $k = 100$, and initial values $\phi^0 = 0$, $\dot{\phi}^0 = 0$, $\dot{\ell}_m^0 = 0$, $\phi_m^0 = 0$, $\dot{\phi}_m = 0$ the deployment is shown in Fig. 3.6.

Sketch (a) shows the position in Oxy while sketch (b) shows the length of the tether. The retrieval case is shown in Fig. 3.7, with $\ell = 100$ km , $\ell_m = 1$ km , and $\dot{\ell}^0 = 0$, $\phi^0 = 0$, $\dot{\phi}^0 = 0$, $\dot{\ell}_m = 0$, $\phi_m = 0$, $\dot{\phi}_m = 0$, $a_1 = 0.5$, $a_2 = 2$, $b_1 = 5$, $k = 100$.

The deployment destabilizes at the distance about 50 km, before stabilizing at 100 km, while the retrieval destabilizes twice at about 75 km and 25 km before the terminal stabilization.

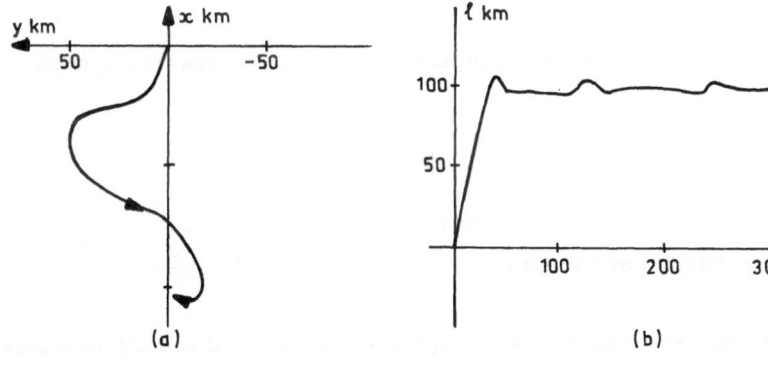

Fig. 3.6

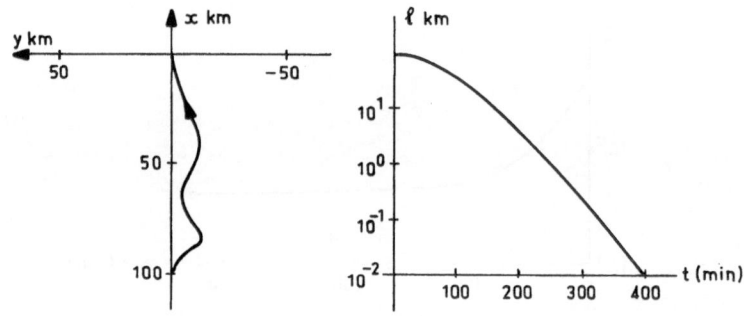

Fig. 3.7

□

EXERCISES 3.2

3.2.1 Show that the set $M = \{0\}$ is minimal invariant for trajectories of $\dot{x} = -x^2$, $t \geq 0$ with $x^0 \geq 0$.

3.2.2 Design a controller to stabilize the system

$$\dot{x}_1 = -x_1 + 6x_2 + u, \quad \dot{x}_2 = 4x_1 + x_2$$

about $(0,0)$.

3.2.3 Show that the program $u_1 = -4x_1^3$, $u_2 = -4x_2^3$ asymptotically stabilizes the system

$$\dot{x}_1 = x_1^3 - x_2^3 - x_1x_2^2 - x_2x_1^2 + u_1$$
$$\dot{x}_2 = x_1^3 + x_2^3 + x_1x_2^2 + x_2x_1^2 + u_2$$

about $(0,0)$. Find the region of such stabilization.

3.2.4 Find the control program asymptotically stabilizing the system

$$\left. \begin{array}{c} \ddot{q}_1 + \dot{q}_1 + \dot{q}_2 + q_1 = u_1 \\ \ddot{q}_2 + \dot{q}_2\dot{q}_1^2 + \dot{q}_2 + q_2 = u_2 \end{array} \right\}$$

about the basic equilibrium assuming that the trajectories start from the basic energy cup.

3.2.5 For the scalar system $\dot{x} = (1/2)u(x - 2)$, $|u| \leq 2$, find feedback controllers making the origin asymptotically stable and find a region of such stability.

3.2.6 Discuss the same problem as above for the system

$$\dot{x}_1 = x_2 - x_1^3 , \quad \dot{x}_2 = -2x_1 - x_2 + u .$$

3.2.7 Consider the system $\dot{x} = Ax + Bu$ with

$$A = \begin{pmatrix} \cos \alpha & -\sin \alpha \\ \sin \alpha & \cos \alpha \end{pmatrix} , \quad \alpha = \text{const}$$

and $|u_i| \leq \hat{u}_i = \text{const}$, and design the matrix B with appropriate dimension. Then verify that the free system: $u(t) \equiv 0$ becomes

$$\dot{r} = r , \quad \dot{\theta} = \theta + \alpha$$

in polar coordinates and show that the circles are invariant. How the trajectories in (r,θ) of the forced system behave under dissipative controllers. Write a program securing asymptotic stability of $r = 0$.

3.2.8   (i)   Consider the system

$$\dot{x}_1 = x_1 (x_2^2 - 1) + u_1 ,$$

$$\dot{x}_2 = x_2 (x_1^2 - 1) + u_2 .$$

Using $V = \frac{1}{2} (x_1^2 + x_2^2)$, estimate the behavior of trajectories with respect to the V-level curves for $u_1 \equiv 0$, $u_2 \equiv 0$. Show that the disc $V(x_1,x_2) \leq \frac{1}{2}$ is positively invariant. Sketch the trajectories.

  (ii)   For the system of (i), find a feedback control program which makes $(0,0)$ globally asymptotically stable.

  (iii)   Write the system of (i) in time discretized format, find the discrete Liapunov function and discuss the stabilization of (ii) (optional).

3.2.9 Consider the system

$$\dot{x}_1 = -3x_1 + x_2 + u_1 ,$$

$$\dot{x}_2 = -x_1 - 2x_2 + u_2 ,$$

and show that for its free version $u_i(t) \equiv 0$, $i = 1,2$, the origin $(0,0)$ is globally asymptotically stable. Destabilize it.

3.2.10 For the system

$$\dot{x}_1 = x_2 + x_1 (u - 1) ,$$

$$\dot{x}_2 = -x_1 + x_2(u-1)$$

with controller $u = x_1^2 + x_2^2$ show that the region of asymptotic stabilizability (retrieving region) is defined by $x_1^2 + x_2^2 \leq 1$.

3.2.11 Take $V = x_1^2 + x_2^2$ as a test function for the Van der Pol system

$$\dot{x}_1 = x_2 - \beta[(x_1^3/3) - x_1]$$

$$\dot{x}_2 = -x_1$$

and prove the following:

(i)   If $\beta < 0$, then $(0,0)$ is asymptotically stable and the circle $x_1^2 + x_2^2 \leq 3$ is the region of such stability for any β .

(ii) If $\beta = 0$, then $(0,0)$ is stable but not asymptotically stable.

(iii) If $\beta > 0$, then $(0,0)$ is not stable.

3.2.12 Find the trajectories of $\dot{x}_1 = x_2$, $\dot{x}_2 = -x_1$ and show that the equilibrium is stable using $V = \tfrac{1}{2} r^2$, $r = \sqrt{x_1^2 + x_2^2}$.
Show that there is no region about $(0,0)$ where the test function forms a threshold.

3.3 STABILIZATION RELATIVE TO ENERGY LEVELS

The concepts of stability, in particular that of asymptotic stability, are obviously local as they relate to an equilibrium or $\bar{x}(t) \equiv 0$. Consequently they may only be used when a boundedness in a particular energy cup or a neighborhood of such equilibrium has been already secured.

Another shortcoming of asymptotic stability is that in real scenarios, more often than not, we need a real (finite) time approach to a target set. Both of the above inconvenient features are avoided in the objective property of ultimate boundedness of the motions, which thus often complements the stability objective. It leads to stabilization below or above a certain E-level either in a cup or above one, or enclosing several cups. In fact it may also refer to the cup in-the-large.

A motion of (2.2.5)' will be called *bounded* if and only if there is a bounded set Δ_β in Δ such that $\bar{\phi}(\bar{x}^0,t_0,t) \in \Delta_\beta$, for all $t \geq t_0$. The motion is called *bounded on some* $\Delta_0 \subset \Delta$ if and only if given Δ_0 for any

$t_0 \in \mathbb{R}$ there is Δ_β such that $(\bar{x}^0, t_0) \in \Delta_0 \times \mathbb{R}$ implies $\bar{\phi}(x^0, t_0, t) \in \Delta_\beta$ for all $t \geq t_0$. Finally the boundedness on Δ_0 is *uniform* if Δ_β does not depend on t_0. It will be convenient to describe the sets Δ_0, Δ_β in terms of some norm $\|\cdot\|$ in $\mathbb{R}^N$, specifically a test function $V(\cdot)$. The narrowing of generality for our purposes is negligible. In this sense, demanding boundedness of a motion we must require $\beta > 0$ such that $\|\bar{\phi}(\bar{x}^0, t_0, t)\| < \beta$ for all $t \geq t_0$.

A different type of boundedness is obtained when we refer to terminal behavior of all motions.

DEFINITION 3.3.1. The motions of (2.2.5)' are *ultimately bounded on a set* $\Delta_0 \times \mathbb{R} \subset \Delta \times \mathbb{R}$ *for bound* $\partial \Delta_B$, if and only if there is a bounded subset $\Delta_B \subset \Delta$ and for each $t_0 \in \mathbb{R}$ there is a constant $T_B > 0$ such that $(\bar{x}^0, t_0) \in \Delta_0 \times \mathbb{R}$ implies $\bar{\phi}(\bar{x}^0, t_0, t) \in \Delta_B$ for all $t \geq t_0 + T_B$. The above boundedness is *uniform* if T_B does not depend upon t_0.

Observe that ultimate boundedness of motions on $\Delta_0 \times \mathbb{R}$ does not imply that the motions from this set are bounded. In fact, one can easily envisage the scenario that a motion escapes to infinity before entering Δ_B at the time $t_0 + T_B$. So boundedness and ultimate boundedness are independent properties. Obviously for the autonomous systems, both are automatically uniform. For the ultimate boundedness as much as before for the boundedness, we find it convenient to specify Δ_B in terms of a norm in $\mathbb{R}^N$ and adjust Definition 3.3.1 to demanding that there is an independent bound $B > 0$, and for each $t_0 \in \mathbb{R}$ there is a constant T_B such that $(x^0, t_0) \in \Delta_0$ implies $\|\bar{\phi}(\bar{x}^0, t_0, t)\| < B$, for all $t \geq t_0 + T_B$. When $\partial \Delta_B$ in general, or B in our case, is given a priori and $\Delta_B \subset \Delta_0$, we have the following property.

DEFINITION 3.3.2. The set $\Delta_B \subset \Delta$ is a *real-time* or *finite attractor* for motions of (2.2.5)' *from some set* $\Delta_0 \times \mathbb{R} \supset \Delta_B \times \mathbb{R}$, if and only if the motions are uniform ultimately bounded on $\Delta_0 \times \mathbb{R}$ for bound $\partial \Delta_B$ (or B).

When T_B is stipulated, we call Δ_B a *stipulated time attractor*, when Δ_0 is stipulated, we call Δ_B a *practical attractor*, see Fig. 3.8.

Observe from the above definitions that both sets Δ_0 and Δ_B are positively invariant. The largest Δ_0 of Definition 3.3.2 is called the *region of finite attraction*, denoted Δ_{BO}. The following sufficient conditions imply the controllability for boundedness of motions.

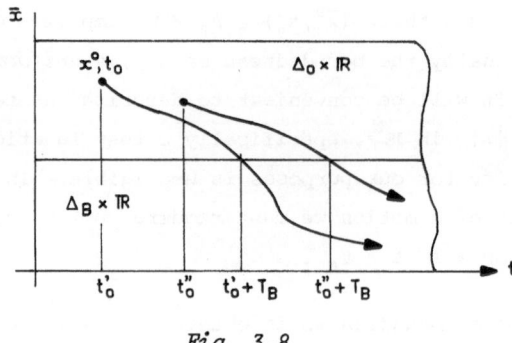

Fig. 3.8

CONDITIONS 3.3.1. The system (2.2.6)' is *strongly controllable for uniform boundedness of motions* on $\Delta_0 \times \mathbb{R}$ if there is a constant $r > 0$, possibly large, a program $P(\cdot)$ and a C^1-function $V(\cdot)$ both defined on $C\Delta_r \times [0,\infty)$, $C\Delta_r \triangleq \Delta_0 - \Delta_r$, $\Delta_r : \|\bar{x}\| < r$, such that

(i) $\qquad a(\|\bar{x}\|) \leq V(\bar{x},t) \leq b(\|\bar{x}\|)$

where $a(\cdot), b(\cdot)$ are continuous functions and $a(\|\bar{x}\|) \to \infty$, as $\|\bar{x}\| \to \infty$;

(ii) for each $\bar{u} \in P(\bar{x},t)$ we have

$$\frac{\partial V(\bar{x},t)}{\partial t} + \nabla_x V(\bar{x},t)^T \cdot \bar{f}(\bar{x},\bar{u},\bar{w},t) \leq 0, \quad \forall \bar{w} \in W. \qquad (3.3.1)$$

Controllability is secured if (ii) holds for some given $\bar{w}$.

The conditions may be proved by following the same argument as in Yoshizawa [1]. Define $\Delta_0 : \|\bar{x}\| \leq \alpha$, and let $\alpha > r$. By the properties of $a(\cdot), b(\cdot)$ we may find $\beta > 0$ such that $a(\beta) > b(\alpha)$. Suppose now that the boundedness is contradicted along some motion: $\bar{\phi}(\bar{x}^0,t_0,t_1)\| = \beta$ at some $t_1 > t_0$. Since $\|\bar{x}^0\| \leq \alpha$ there exist two values t_2, t_3 such that $t_0 \leq t_2 < t_3 \leq t_1$, $\|\bar{\phi}(\bar{x}^0,t_0,t_2)\| = \alpha$, $\|\bar{\phi}(\bar{x}^0,t_0,t_3)\| = \beta$ implying $\alpha < \|\bar{\phi}(\bar{x}^0,t_0,t)\| < \beta$, $t \in (t_2,t_3)$. Then we must have $V(\bar{\phi}(t_2),t_2) \leq b(\alpha)$, $V(\bar{\phi}(t_3),t_3) \geq a(\beta)$, which contradicts (ii), proving the boundedness for all motions in $K(\bar{x}^0,t_0)$. In the case of the weak mode of control (controllability only) the contradiction refers to the single motion at each $(\bar{x}^0,t_0)$.

CONDITIONS 3.3.2. The system (2.2.6)' is *strongly controllable* on $\Delta_0 \times \mathbb{R}$ *for uniform ultimate boundedness* of motions for bound B if the Conditions 3.3.1 hold with (ii) replaced by

(ii)' for each $\bar{u} \in P(\bar{x},t)$ there is a continuous positive function $c(\|\bar{x}\|)$, such that

$$\frac{\partial V(\bar{x},t)}{\partial t} + \nabla_x V(\bar{x},t)^T \cdot \bar{f}(\bar{x},\bar{u},\bar{w},t) \le -c(\|\bar{x}\|) \ , \quad \forall \ \bar{w} \in W \ . \qquad (3.3.2)$$

Controllability is secured when (ii)' holds for a given $\bar{w}$.

The conditions may be verified by the same argument as applied to proving Conditions 3.2.3, see also Yoshizawa [1]. Finally we arrive at for-the-level stabilization which involves the real-time-attraction objective.

Given the set Δ_B in some Δ_0, denote $C\Delta_B \overset{\Delta}{=} \Delta_0 - \Delta_B$ and let an open $D \supset \overline{C\Delta_B}$, $D \cap \{0\} = \phi$. Then introduce a C^1-function $V(\cdot) : D \times \mathbb{R} \to \mathbb{R}$, with

$$v_0 = \inf V(\bar{x},t) \mid \bar{x} \in \partial\Delta_0 \ , \quad t \in \mathbb{R} \ ,$$
$$v_B = \inf V(\bar{x},t) \mid \bar{x} \in \partial\Delta_B \ , \quad t \in \mathbb{R} \ . \qquad (3.3.3)$$

CONDITIONS 3.3.3. The system (2.2.6)' is strongly controllable on $\Delta_0 \times \mathbb{R}$ for real-time attraction to Δ_B, or *strongly stabilizable for the level* $\|\bar{x}\| = B$, if there is a program $\bar{P}(\cdot)$ and a C^1-function $V(\cdot)$, both defined on $D \times \mathbb{R}$ such that for all $(\bar{x},t) \in D \times \mathbb{R}$,

(i) $v_B \le V(\bar{x},t) \le v_0$;

(ii) for each $\bar{u} \in \bar{P}(\bar{x},t)$, there is $T_B > 0$ such that

$$\frac{\partial V(\bar{x},t)}{\partial t} + \nabla_x V(\bar{x},t)^T \cdot \bar{f}(\bar{x},\bar{u},\bar{w},t) \le -\frac{v_0 - v_B}{T_B} \ , \quad \forall \ \bar{w} \in W \ . \qquad (3.3.4)$$

Stabilization is secured when (ii) holds for a given $\bar{w}$. Stabilization and strong stabilization in stipulated time is secured when (ii) holds for a given T_B.

In order to verify the above Conditions, we need to show that:

(1) Δ_0 is strongly positively invariant;
(2) $C\Delta_B \times \mathbb{R}$ is left by the motions of $K(x^0,t_0)$ in real-time;
(3) Δ_B is strongly positively invariant.

We begin the first argument by assuming the contrary to (1), that is, that at least one motion of (2.2.6)' from $\Delta_0 \times \mathbb{R}$ crosses $\partial\Delta_0$, upon which there is $t_1 > t_0$ such that by (i), $V(\bar{\phi}(\bar{x}^0,t_0,t_1),t_1) \ge v_0 \ge V(\bar{x}^0,t_0)$, contradicting (ii). To show (2) we integrate (3.3.4) along an arbitrary motion from $C\Delta_B \times \mathbb{R}$ obtaining the time estimate

$$t - t_0 \le T_B \frac{V(x^0,t) - V(\bar{x},t)}{v_0 - v_B} \ . \qquad (3.3.5)$$

From (i) we have $V(\bar{x},t) - v_B \geq 0$, $V(\bar{x}^0,t_0) - v_0 \leq 0$ or
$V(\bar{x}^0,t_0) - V(\bar{x},t) \leq v_0 - v_B$, yielding $t - t_0 \leq T_B$ which means that for
$t > t_0 + T_B$ the motions from $C\Delta_B \times \mathbb{R}$ must leave this set and by the
proved feature (1) must enter Δ_B. The latter is shown to be positively
invariant (feature (3)) by the same argument as for (1), that is, by (i),
(ii) contradicting on D, which completes our verification.

With the same notation and by the same argument as for Corollary 3.2.2
we obtain the following:

COROLLARY 3.3.1. *Given* $(x^0,t_0) \in C\Delta_B \times \mathbb{R}$, *if there is a pair* $(\bar{u}^*,\bar{w}^*)$
$\in U \times W$ *such that*

$$L(\bar{x},\bar{u}^*,\bar{w}^*,t) = \min_{\bar{u}} \max_{\bar{w}} L(\bar{x},\bar{u},\bar{w},t) \leq - \frac{v_0 - v_B}{T_B} , \qquad (3.3.6)$$

then (ii) of Conditions 3.3.3 is met with $u^* \in \bar{P}^*(\bar{x},t)$.

The above allows to determine the dissipative robust $P^*(\cdot)$ as illustrated in Examples 3.2.1, 3.2.2.

In practical terms, by the level $\partial\Delta_B : \|\bar{x}\| = B$, we shall mean
either the amplitude-level with $\|\bar{x}\| \overset{\Delta}{=} |\bar{x}|$, or a V-level determined by
some stationary function $\tilde{v}(\cdot) : \Delta \to \mathbb{R}$ such that $V(\bar{x},t) \equiv \tilde{V}(\bar{x}) = \|\bar{x}\|$,
with $\partial\Delta_B : \tilde{V}(x) = B$. In the latter case, specifying as well
$\partial\Delta_0 : \tilde{V}(\bar{x}) = v_0$ we obtain (ii) of Conditions 3.3.3 trivially satisfied and
we may show that the following is true, see Skowronski [4], [9], [12], [20]
and Skowronski-Ziemba [4].

COROLLARY 3.3.2. *Given* $\partial\Delta_0, \partial\Delta_B$ *determined in terms of* $\tilde{v}$-*levels, (ii) of*
Conditions 3.3.3 becomes necessary and sufficient.

Indeed, consider (3.3.4) in terms of $\tilde{v}(\cdot)$:

$$\nabla\tilde{V}(\bar{x})^T \cdot \bar{f}(\bar{x},\bar{u},\bar{w},t) \leq - \frac{v_0 - v_B}{T_B} \qquad (3.3.7)$$

for all $\bar{w} \in W$ on D and write it as $\tilde{L}(\bar{x},\bar{u},\bar{w},t) \leq -\delta v/T_B$ where
$\delta v = v_0 - v_B$ is the difference between the $\tilde{v}$-levels of $\partial\Delta_0$ and $\partial\Delta_B$
required for reaching $\partial\Delta_B$ from all points between $\partial\Delta_0$ and $\partial\Delta_B$. When
$\tilde{L}$ is smaller than the rate $|-\delta v/T_B|$, that is, $\tilde{L} > -(\delta v/T_B)$, the level
$\partial\Delta_B$ is not reachable. The sufficiency of (3.3.7) follows from implying
(3.3.4).

Turning now to the energy reference frame, we may have two objectives in stabilization for a given E-level: to keep the motions of (2.2.6) or (2.2.6)' either below it or above it. The latter may apply for instance in the case of stabilization about an unstable equilibrium (a saddle), the first is obviously more common.

Let us refer first to *stabilization below* an E-level. According to (2.4.30) we use $\widetilde{V}(\bar{x}) \equiv E^-(\bar{x})$, whence (3.3.7) is

$$f_0(\bar{x},\bar{u},\bar{w},t) = \nabla E^-(\bar{x})^T \cdot \bar{f}(\bar{x},\bar{u},\bar{w},t) \leq - \delta h/T_B , \qquad (3.3.7)'$$

for all $\bar{w} \in W$. It means that the complement $\Delta_0 - \Delta_B$ is covered by the field of entry points H^- , with $\delta h \overset{\Delta}{=} h_{CO} - h_{CB} > 0$ specifying the *outflux* in $E^-(\bar{x})$ between $\partial\Delta_0$ and $\partial\Delta_B$, see (2.4.22). This secures reaching $E^-(\bar{x}) = B$ from all h^0 in $\Delta_0 \times h$, see (2.4.19).

We may now use the obtained conditions for controlling the behavior of our general mechanical system over Δ . We shall derive several properties which will specify the control pattern of such behavior. With controllability for ultimate boundedness we shall be able to assess the behavior in-the-large, see Section 2.3, then with stabilization below a level and about equilibrium, we shall specify local patterns in energy cups. Anything between these two regions will be the subject of the global study in Section 3.4.

First of all we observe that granted our power limiting axioms of Section 2.3, the motions of (2.2.6)' are uniform ultimately bounded in-the-large for an arbitrary control program $\bar{P}(\cdot)$, as much as the influence of such a controller may be ignored. See Fig. 3.9.

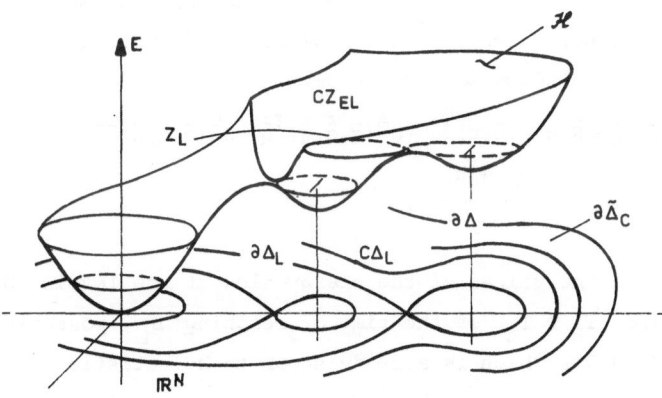

Fig. 3.9

Indeed, consider the cup in-the-large CZ_{EL} , and directly let $V(\bar{x},t) \equiv E^-(\bar{x})$, for the use in Conditions 3.3.2 on CZ_{EL} . By virtue of (2.4.21) we have (i) of Conditions 3.3.1 and thus the first part of Conditions 3.3.2 satisfied. This is independent of any $P(\bar{x},t)$. It remains to check upon (ii)' of Conditions 3.3.2, which reads now

$$\nabla E^-(\bar{x})^T \cdot \bar{f}(\bar{x},\bar{u},\bar{w},t) \le -c(\|\bar{x}\|) , \quad \forall \, \bar{w} \in W \tag{3.3.8}$$

or in terms of particular characteristics:

$$\left. \begin{array}{l} \bar{F}(\bar{q},\dot{\bar{q}},\bar{u},\bar{w})^T \dot{\bar{q}} + \bar{R}(\bar{q},\dot{\bar{q}},t,\bar{w})^T \dot{\bar{q}} - \bar{D}^A(\bar{q},\dot{\bar{q}},\bar{w})^T \dot{\bar{q}} - \bar{D}^D(\bar{q},\dot{\bar{q}},\bar{w})^T \dot{\bar{q}} \\ \qquad \le -c(\|q\|,\|\dot{q}\|) \end{array} \right\} \tag{3.3.9}$$

for all $\bar{w} \in W$, cf. (2.4.28). The axiom of bounded accumulation (2.4.15), augmented by (2.4.29), gives

$$|\bar{F}(\bar{q},\dot{\bar{q}},\bar{u},\bar{w})^T \dot{\bar{q}} + \bar{R}(\bar{q},\dot{\bar{q}},\bar{w},t)^T \dot{\bar{q}} - \bar{D}^A(\bar{q},\dot{\bar{q}},\bar{w})^T \dot{\bar{q}}| < N' < \infty , \tag{3.3.10}$$

and (2.4.8), (2.4.9) secure the growth of $D^D(\cdot)$ with the increase of the amplitudes $|\bar{q}|, |\dot{\bar{q}}|$ on $C\Delta_L$. This means there exists some energy level $\tilde{Z}_C$ within the cup in-the-large Z_{EL} such that we will be able to find positive $c(\cdot)$ to satisfy (3.3.9) and thus cover $\Delta - \tilde{\Delta}_C$ with entry points of H^-, for any $\bar{u} \in U$, that is, disregarding any program $\bar{P}(\cdot)$. Consequently we have $\tilde{Z}_C : E^-(\bar{x}) = B$ such that the motions of (2.2.6)' are uniformly ultimately bounded in $\tilde{\Delta}_C$, and the latter fact is independent of control. In the above, $\tilde{\Delta}_C$ is a subset of Δ with boundary $\partial\tilde{\Delta}_C$ being a map of $\tilde{Z}_C$ into Δ .

Granted the above we can now assume this $\tilde{\Delta}_C$ as Δ_0 and control the system for stabilization under the level Z_L , that is, secure strong controllability on $\tilde{\Delta}_C \times \mathbb{R}$ for real-time attraction to Δ_L . This can be done satisfying Conditions 3.3.3 under the controller designed from (3.3.6). Observe that letting $V(\bar{x},t) \equiv E^-(\bar{x})$ again, we can now use Corollary 3.3.2 to produce the required result. The necessary and sufficient condition for strong stabilization below Z_L now becomes

$$\left. \begin{array}{l} \bar{F}(\bar{q},\dot{\bar{q}},\bar{u},\bar{w})^T \dot{\bar{q}} + \bar{R}(\bar{q},\dot{\bar{q}},\bar{w},t)^T \dot{\bar{q}} - D^A(\bar{q},\dot{\bar{q}},\bar{w})^T \dot{\bar{q}} - D^D(\bar{q},\dot{\bar{q}},\bar{w})^T \dot{\bar{q}} \\ \qquad \le -\dfrac{\tilde{h}_C - h_L}{T_L} , \end{array} \right\} \tag{3.3.11}$$

where $\tilde{h}_C, h_L$ are the values of the energy flow at the levels $\tilde{Z}_C, Z_L$ respectively and $T_B = T_L$ is the time of reaching Δ_L , possibly stipulated. Observe that (3.3.9), which is already shown to be satisfied on $C\Delta_L$, secures the negative value of the left hand side of (3.3.11), that is, the fact that $C\Delta_L$ is covered by entry points of H^-. We only have to qualify this "negativeness" or the "speed" of entry by the controller to secure

154

(3.3.11), that is, to make the outflux of energy between $\tilde{Z}_C$ and Z_L resulting from satisfied (3.3.9) with the rate $c(\|\bar{x}\|)$, such as to produce $\tilde{h}_C - h_L$ in time interval T_L . The controller of this type is obviously dissipative, cf. (2.4.20). For the positive damping, see (2.4.7), it is rational to expect $-D^D(\bar{q},\dot{\bar{q}},\bar{w})^T\dot{\bar{q}} \leq 0$ for all $\bar{w} \in W$ and thus

$$\bar{F}(\bar{q},\dot{\bar{q}},\bar{u},\bar{w})^T\dot{\bar{q}} + \bar{R}(\bar{q},\dot{\bar{q}},\bar{w},t)^T\dot{\bar{q}} - D^A(\bar{q},\dot{\bar{q}},\bar{w})^T\dot{\bar{q}} \leq - \frac{\tilde{h}_C - h_L}{T_L}$$

for all $\bar{w} \in W$, which secures (3.3.11). Following Corollary 3.3.1, the *control condition* becomes

$$\min_{\bar{u}} \max_{\bar{w}} [\bar{F}(\bar{q},\dot{\bar{q}},\bar{u},\bar{w})^T\dot{\bar{q}}]$$

$$\leq - \frac{\tilde{h}_C - h_L}{T_L} - \max_{\bar{w}} [\bar{R}(\bar{q},\dot{\bar{q}},\bar{w},t)^T\dot{\bar{q}} - \bar{D}^A(\bar{q},\dot{\bar{q}},\bar{w})^T\dot{\bar{q}}] . \qquad (3.3.12)$$

Observe that all the terms of (3.3.12) except one are dot products of n-vectors, so (3.3.12) is implied by simultaneous holding of the following inequalities related to components of the dot product:

$$\min_{\bar{u}} \max_{\bar{w}} [F_i(\bar{q},\dot{\bar{q}},\bar{u},\bar{w}) \, \text{sgn} \, \dot{q}_i]$$

$$\leq - \frac{\tilde{h}_C - h_L}{n|\dot{q}_i|T_L} - \max_{\bar{w}} \{[R_i(\bar{q},\dot{\bar{q}},\bar{w},t) - D_i^A(\bar{q},\dot{\bar{q}},\bar{w})] \, \text{sgn} \, \dot{q}_i\} , \qquad (3.3.12)'$$

$$i = 1,\ldots,n .$$

Obviously $\tilde{h}_C$ in (3.3.12)' may be replaced by its upper estimate h_Δ which is well determined and thus convenient. Calculation of $u_i(t)$ from (3.3.13) requires specifying the functions $F_i(\cdot)$, $R_i(\cdot)$, $D_i(\cdot)$, so we must rest the general case here, until passing over to case investigations. On the other hand, a few general comments are still in order.

Observe that when the controlled motion approaches the hypersurface $\dot{q} = 0$, part of (3.3.12)' blows up to infinity, making the program demand control values above saturation level. In order to illustrate the case, let us assume what often happens, namely that $F_i(\cdot)$ is linear in $\bar{u}_i$ and that every DOF has its own actuator (if not, we can treat the redundant components as nominal: $u_i \equiv 0$) "geared positively", that is, $\text{sgn} \, F_i = \text{sgn} \, u_i$. Thus we have $F_i(\bar{q},\dot{\bar{q}},\bar{u},\bar{w}) = B_i(\bar{q},\dot{\bar{q}},\bar{w})u_i$, $B_i > 0$. Moreover, let $\bar{w}*$ be the maximizing value of $\bar{w}$ in (3.3.12)'. Then the latter is implied by the program

155

$$\min_{\bar{u}} (u_i \, \text{sgn} \, q_i) \leq \begin{cases} \dfrac{-1}{B_i(\bar{q},\dot{\bar{q}},\bar{w}^*)} \left\{ \dfrac{h_\Delta - h_L}{n|\dot{q}_i|T_L} + [R_i(\bar{q},\dot{\bar{q}},\bar{w}^*,t) - D_i^A(\bar{q},\dot{\bar{q}},\bar{w}^*)]\text{sgn} \, \dot{q}_i \right\} \\ \qquad\qquad\qquad\qquad \text{for } |\dot{q}_i| \geq \beta_i \ , \\[2mm] \dfrac{1}{B_i(\bar{q},\dot{\bar{q}},\bar{w}^*)} \left\{ [R_i(\bar{q},\dot{\bar{q}},\bar{w}^*,t) - D_i^A(\bar{q},\dot{\bar{q}},\bar{w}^*)]\text{sgn} \, \dot{q}_i \right\} \\ \qquad\qquad\qquad\qquad \text{for } |\dot{q}_i| < \beta_i \ , \end{cases}$$

$$i = 1,\ldots,n \ , \qquad\qquad (3.3.13)$$

with $\bar{\beta} = (\beta_1,\ldots,\beta_n) = \text{const}$ calculated from

$$P_i(\bar{q},\bar{\beta},\bar{w}^*,t) = \hat{u}_i \ , \quad i = 1,\ldots,n \ ,$$

where $\hat{u}_1,\ldots,\hat{u}_n = \text{const}$ are the saturation values. Alternatively β_i, $i = 1,\ldots,n$ may be obtained experimentally during the simulation procedure by taking them as minimal distances of points where the control action is secured from the surface $\dot{\bar{q}} = 0$. Indeed, observe that below β_i the controller makes the system either a damped or a conservative oscillator which as a rule crosses the axes $\dot{q}_i = 0$ instantaneously at all regular points, and does not blow the trajectories to infinity along the velocity surface. All that is missing over the (hopefully small) strip $|\dot{q}_i| < \beta_i$, $i = 1,\ldots,n$, is the active role of the controller towards its control objective. In fact, with the accumulation power $(R_i - D_i^A)\dot{q}_i$ bounded, see Section 2.3, switching the controller off: $u_i(t) \equiv 0$, over the small strip $|\dot{q}_i| < \beta_i$, would do negligible harm.

On the other hand, the level $\tilde{h}_C$ in (3.3.12)' can be replaced by the initial value $h^0 = h(\bar{x}^0,t_0,t_0)$ actually needed for reaching Z_L from this initial energy level. In this case we make the rate of outflux related to position, and denote it

$$c(\bar{x}^0) \stackrel{\Delta}{=} \frac{E^-(\bar{x}^0) - h_L}{T_L} \ . \qquad\qquad (3.3.14)$$

This may or may not be convenient, but is certainly more accurate and thus a more economic (power saving) design formula for the controller. The above (3.3.14) is a special case of the positive definite function $c(\|\bar{x}\|)$ used before with the norm specified by $E^-(\cdot)$. The case when the representation (3.3.14) is convenient is illustrated in Examples 3.2.1, 3.2.2, 3.3.1 where (3.3.14) cancels the velocity $\dot{\bar{q}}$ in the denominator of (3.3.12), thus allowing us to avoid the procedure of introducing $\beta_1,\ldots,\beta_n$.

The controller calculated from (3.3.12)' secures strong stabilization under the level Z_L, that is, makes Δ_L a strong real-time attractor with all motions of (2.2.6)' from $\Delta \times \mathbb{R}$ being in Δ_L after $T_L < \infty$, the latter either to be found or stipulated. This means that all points of $\partial\Delta_L$ are entry points of H^- and that Δ_L is strongly positively invariant. This also means that no asymptotic $t \to \infty$ approach to any set in $C\Delta_L$ is possible and that Δ_L must enclose all steady state positive limit sets, including but not narrowed to equilibria.

Note that the controller based on (3.3.12)' is designed to work on $C\Delta_L \times \mathbb{R}$ only. What happens to the motions after they enter Δ_L depends upon the shape of H with its extremal points $\equiv$ equilibria, the appearance of H^-, H^0, H^+ in Δ_L and upon distribution of the steady state sets mentioned, see Skowronski [32], [38]. It is the subject of what we have called global investigations and which, in the general case of $\mathbb{R}^N$, is very difficult. We shall explain some aspects of it in Section 4.1. At present, using the stabilization tools available, we can discuss how to design another branch of the control program, now acting in $\Delta_L \times \mathbb{R}$ with the following objective. We want to bring the motions below the energy level corresponding to a threshold $z_{CE}^\vee$ which separates one or more cups $z_E^\vee$. Then, upon entering one of such cups, we want to make the motions strongly stabilized about the corresponding equilibrium.

Since the equilibria are all located in the Configuration Space of displacements: $\dot{\bar{q}} = 0$, and the energy cups are symmetric about them, the surface H in Z_L forms a $2n-1$ dimensional canyon with the bottom along $\bar{q}$ and parabolic walls ascending from it along $\dot{\bar{q}}$, see Fig. 3.10. Consequently, aiming at a specific cup $z_E^\vee$ we may either avoid or be forced to pass over other, higher located cups. The real scenario depends upon the location of the cups in the configuration space. As the origin may be moved

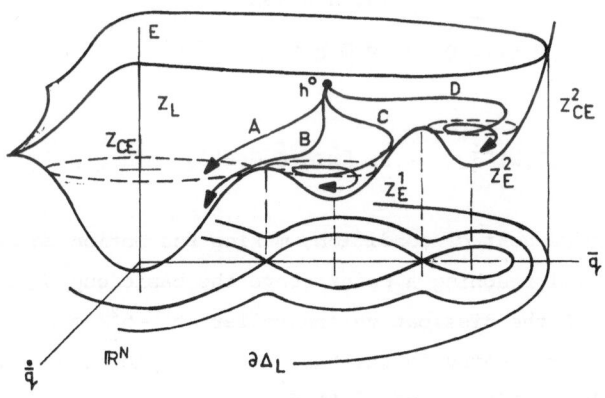

Fig. 3.10

with suitable transformation of variables, it would not narrow generality to concentrate on the basic cup about $\bar{x}(t) \equiv 0$ for the sake of being specific.

Consider h^0 indicated in Fig. 3.5 as a starting energy value over a starting position $\bar{x}^0 = (\bar{q}^0, \dot{\bar{q}}^0)$ in Δ_L. By the same argument as for strong stabilization under Z_L, with the use of Corollary 3.3.2 and the same dissipative control condition (3.3.14), we may arrive at the highest threshold level below h^0, say z_{CE}^{ν}. It results in securing the corresponding Δ_{CE}^{ν} as a real-time attractor. Here the dissipative controller has been used, not on $C\Delta_L \times \mathbb{R}$, but on some subset of $\Delta_L \times \mathbb{R}$ with the generated outflux of energy $\delta h = h^0 - h_{CE}^{\nu} > 0$ which has to be substituted into (3.3.14). Note, however, that stabilizing the motions at the above level did not indicate where the motion actually landed. The controller used relates to energy levels only and is thus incapable of bringing the motion to a specific position on z_{CE}^2. If the motion covered the paths A or B in Fig. 3.10, the same controller may achieve the task of getting it into the basic cup and stablizing it about $\bar{x}(t) \equiv 0$. The latter follows immediately from the fact that such a controller would satisfy Corollary 3.2.3, producing entry points of H^- all the way to the basic equilibrium.

However, when the motion covers the paths C or D with the energy dissipative controller kept in action, it must enter cups z_E^1 or z_E^2 respectively, and get stuck there. How to avoid the case is not a stabilization problem, but rather a problem of controllability for collision or capture in a specific target set discussed in Chapters 5 and 6. One of the methods which we can mention at this stage is, in the case of following path D, to adjust the controller calculated from (3.3.14) so that upon reaching z_{CE}^2 it preserves the energy along the motion instead of dissipating it: $h(h^0, t_0, t) \equiv h^0$ for all t concerned. This is obtained by substituting $\delta h = h^0 - h_{CE}^2 = 0$ which gives

$$\min_{\bar{u}} \max_{\bar{w}} f_0(\bar{x}, \bar{u}, \bar{w}, t) = 0, \quad \forall \; \bar{w} \in W \tag{3.3.15}$$

or

$$\min_{\bar{u}} \max_{\bar{w}} [\bar{F}(\bar{q}, \dot{\bar{q}}, \bar{u}, \bar{w})^T \dot{\bar{q}}] = -\max_{\bar{w}} [\bar{R}(\bar{q}, \dot{\bar{q}}, \bar{w}, t)^T \dot{\bar{q}} - D(\bar{q}, \dot{\bar{q}}, \bar{w})^T \dot{\bar{q}}] \tag{3.3.16}$$

for the *conservative control* condition, making the motion to slide along the level z_{CE}^2 until reaching a point above the basic cup z_E. Then some energy outflux with the dissipative controller $h^0 - h_{CE}^2 > 0$ is used again, producing the required entry to the basic cup and yielding, for $h^0 - 0 > 0$, the required stabilization about $\bar{x}(t) \equiv 0$.

158

The case of following the path C is identical, except we use (3.3.16) slightly lower, that is, at z_{CE}^1.

In the case of both paths C and D, it is easy to see that the motion could have entered the corresponding cups. If it did happen, we obviously need exit points for the field H^+, thus the energy *accumulative controller*, in order to get back to the threshold. The technique is identical to *stabilization above* some E-level, which is the second option of using Conditions 3.3.3 and Corollary 3.3.2, as indicated earlier in this section.

Let us now briefly refer to stabilization above some E-level. We choose now $\tilde{V}(\bar{x}) \equiv h_\Delta - E^+(\bar{x})$, cf. (2.4.31), with v_0, v_B specified by the constants h^0, h_B: $v_0 = h_\Delta - h^0$, $v_B = h_\Delta - h_B$, such that $h_B > h^0$ (accumulation of energy). With such a choice (i) of Conditions 3.3.3 is satisfied and (3.3.7) becomes

$$f_0(\bar{x}, \bar{u}, \bar{w}, t) = \nabla E^+(\bar{x})^T \cdot \bar{f}(\bar{x}, \bar{u}, \bar{w}, t) \geq \delta h / T_B \qquad (3.3.7)''$$

for all $\bar{w} \in W$, with $\delta h = h_B - h^0 > 0$ specifying the influx in $E^+(\bar{x})$ between $\partial \Delta_0$ and $\partial \Delta_B$, see (2.4.22). This secures reaching $E^+(\bar{x}) = B$ from below and staying above it for all $t \geq t_0 + T_B$. Now, the Corollary 3.3.1 requires

$$\max_{\bar{u}} \min_{\bar{w}} \nabla E^+(\bar{x})^T \cdot \bar{f}(\bar{x}, \bar{u}, \bar{w}, t) \geq \delta h / T_B \; ,$$

or by a similar argument to that for (3.3.14),

$$\left. \begin{array}{l} \max\limits_{\bar{u}} \min\limits_{\bar{w}} \; [F_i(\bar{q}, \dot{\bar{q}}, \bar{u}, \bar{w}) \, \text{sgn} \, \dot{q}_i] \\[2mm] \geq \dfrac{\delta h}{n |\dot{q}_i| T_B} - \min\limits_{\bar{w}} \{[R_i(\bar{q}, \dot{\bar{q}}, \bar{w}, t) - D_i(\bar{q}, \dot{\bar{q}}, \bar{w})] \, \text{sgn} \, \dot{q}_i\} \end{array} \right\} \qquad (3.3.17)$$

which allows us to calculate the accumulative controller for the influx δh. Similarly as with the dissipative controller, such a controller is useful in several regions of Δ_L, easy to establish. One of them is about a saddle or for crossing over a threshold: $h_B = h_{CE}^{\vee}$, as mentioned, see Gabrielyan [1].

EXAMPLE 3.3.1. Let us return again to Example 1.1.1 and show first ultimate boundedness in-the-large and then stabilization below two energy levels z_L and z_{CE} surrounding the basic cup z_E.

To enclose more equilibria than needed in our previous Example 3.2.1 for the single cup investigated there, we shall consider the state equations (1.1.6) truncated at the third term: $(g/\ell) \sin x_1 \cong a x_1 - b x_1^3 + c x_1^5$, which

gives three equilibria Dirichlet stable $x_1^e = 0$, $(1/2c)^{\frac{1}{2}}(-b \pm \sqrt{b^2 - 4ac})^{\frac{1}{2}}$
and two unstable $x_1^e = -(1/2c)^{\frac{1}{2}}(-b \pm \sqrt{b^2 - 4ac})^{\frac{1}{2}}$, see Example 2.3.1. The
latter two produce the single threshold specified by the energy

$$E(x_1,x_2) = \frac{1}{2} x_2^2 + \frac{1}{2} ax_1^2 - \frac{1}{4} bx_1^4 + \frac{1}{6} cx_1^6$$

with the first unstable equilibrium substituted. We then obtain the threshold as the line in Δ

$$-\frac{1}{2} x_2^2 + \frac{1}{2} ax_1^2 - \frac{1}{4} bx_1^4 + \frac{1}{6} cx_1^6 = h_{CE} \qquad (3.3.18)$$

called the *conservative separatrix*, see Fig. 3.11. Letting $(g/\ell) = 1$,
$a = c = 1$, $b = -2.1$, the equilibrium concerned is $x_1^e = 0.85$ which,
substituted to (3.3.17), gives $h_{CE} = 0.15$. Since we deal with the single
threshold, (3.3.18) represents Z_L and the boundary $\partial \Delta_L$. Both the ultimate
boundedness of motions in-the-large and strong stabilization below h_{CE} are
then obtained with the dissipative controller (3.2.14) for a given d , and
(3.2.16), (3.2.19) for unknown $d \in [-1,1]$, the latter successively in the
autonomous and nonautonomous versions fo (1.1.6).

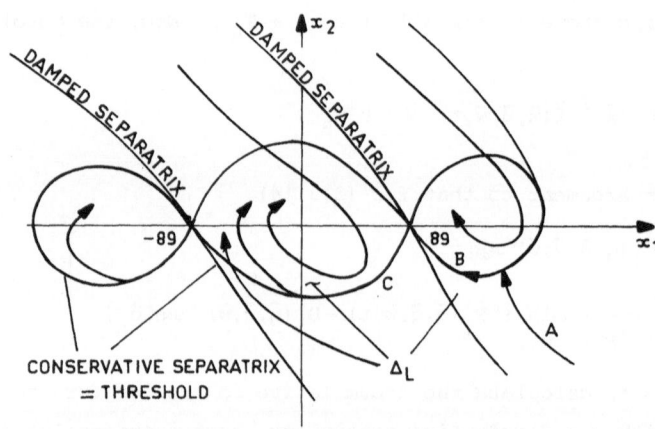

Fig. 3.11

In our simple example, a trajectory of (1.1.6) arriving at Δ_L must
already enter one of the cups. The question which of them - without
control - depends on the point of arrival embedded in one of the regions of
attraction specified by the damped separatrices: trajectories (1.1.8)
passing through $x_1^e = 0.85$. They will be termed in later chapters the
regions of controllability for capture. It is however obvious as well,
that with a conservative controller that secures the motion along the
conservative separatrix, we may prevent entering a particular cup and force
the system into another, see for instance the controlled trajectory $\overset{\frown}{ABC}$ in

Fig. 3.6. Parts A and C are steered by a dissipative controller (3.2.14) while B is controlled by its conservative counterpart calculated from $\dot{E}(x_1,x_2) = 0$ as by (3.2.10) $u = d|x_2|x_2$, in the autonomous version. $\square$

EXERCISES 3.3

3.3.1 Stabilize the system

$$\dot{x}_1 = -x_1 + 2x_2 - u,$$
$$\dot{x}_2 = -x_2 - u,$$
$$\dot{x}_3 = -u,$$

under the V-level: $x_1^2 + x_2^2 + x_3^2 = 1$. HINT: Choose the program $u = x_2$.

3.3.2 For the system $\ddot{q} - |q|\dot{q} + q + q^3 = u$, $q \in \mathbb{R}^n$, find a dissipative controller satisfying Corollary 3.3.1 which stabilizes below the E-level specified by $h = 1$. Calculate the time T_B as a function of $\bar{x}^0 = (q^0,\dot{q}^0)^T$. If the control values ought to be constrained, what is the saturation value $\hat{u}$ needed for the objective concerned?

3.3.3 Consider the system (1.1.13) of Example 1.1.2, Fig. 1.6, find the equilibria, energy threshold values and calculate how much influx is needed in order to reach the threshold about the basic cup from particular initial states $(\bar{q}^0,\dot{\bar{q}}^0)$ with this cup. Find z_L, Δ_L and determine the outflux needed to reach the highest Dirichlet stable equilibrium (highest local minimum of $E(\cdot)$).

3.3.4 Stabilize the system $\ddot{q} + w\dot{q} + aq - bq^3 = u$, $q \in \mathbb{R}$, $|u| < \hat{u}$, $w \in [-1,1]$; $a,b,\hat{u} = \text{const} > 0$, about the Dirichlet unstable equilibrium $q^e = \sqrt{a/b}$, $\dot{q}^e = 0$ above a designed secure E-level.

3.3.5 Write the system $\ddot{q} + q^2\dot{q} - q + wq^3 = u$, $q \in \mathbb{R}$ in the phase space form and then find a controller stabilizing the trajectories below some E-level surrounding one of the stable equilibria. In the above $|u| \le \hat{u}$, $w \in [1,2]$.

3.3.6 Consider the system $\ddot{q} + wq = u$, $x \in \mathbb{R}$, with the program

$$u = \begin{cases} k, & \text{for } \dot{q} < 0, \\ -k, & \text{for } \dot{q} > 0, \end{cases}$$

and $w \in [1,2]$. By completing the square show that the trajectories form elipses centered at $q^e = \pm k/w$, $\dot{q}^e = 0$ for $\dot{q} \gtrless 0$, respectively. Then verify stabilization at the basic energy level reached in finite time. Calculate the total energy outflux and the time.

3.3.7 Let the origin of $\mathbb{R}^2$ be the source of a potential with intensity proportional to the square distance $V(q_1,q_2) = k/(q_1^2 + q_2^2)$, and consider a point mass with a given initial kinetic energy $T = \frac{1}{2}[(\dot{q}_1^0)^2 + (\dot{q}_2^0)^2]$. Ignoring other forces, find a controller under which the mass is charged with the full potential V_{max} (that is, the point collides with the source) subject to the repelling force $w : -1 \leq w \leq 1$.

HINT: $V = \left(V_{max} - \dfrac{k}{q_1^2 + q_2^2} \right) + \dot{q}_1^2 + q_2^2$.

3.4 HOW TO FIND A LIAPUNOV FUNCTION

We have already used two test functions $V(\cdot)$, both in terms of the total energy, and have seen that the design of controllers as well as the size of controllability regions depends upon the choice of $V(\cdot)$, which is by no means unique. Even if we are able to specify controllability for some objective by necessary and sufficient conditions, such conditions are still subject to the choice of $V(\cdot)$. It is thus always an open question whether a better function may not be found: generating some more accurate controller, a larger region of controllability, etc.

The search for Liapunov functions began before the turn of the century, by Liapunov, himself showing that the energy in terms of the square form $V = \bar{x}^T P \bar{x}$ is the function for linearized problems. In spite of such long development of the Liapunov formalism, there is no general rule for finding a suitable Liapunov function on all occasions, but there are a number of fairly general methods available. A very good account of the early search for Liapunov functions is given by Antosiewicz [1], Barbashin [1], [2], Szegö [1] and Grayson [1]. The history of the region estimates, and new results in the formalism as such, are reviewed up-to-date in Genesio-Tartaglia-Vicino [1], with special cases discussed in Chin [1], and more recently in Chiang-Hirsch-Wu [1]. We shall refer to the region estimates again in Section 5.4, with some references listed there.

From the variety of methods available, three seem to be most popular. The first two, referring to total energy, appear particularly useful for mechanical systems and will be briefly described below - they are the *first integral method* and the *variable gradient method*. The third, the so called *Zubov's method*, requires solving of a nonlinear partial differential equation, see Szegö [1], and thus developed along the practical line of numerical design of the Liapunov function, see Margolis-Vogt [1], Davison-Kurak [1]. The latter proves useful in determining the regions of stability and controllability discussed in this text in Section 5.4.

METHOD OF FIRST INTEGRAL

The first suggestion and early development of the method belong to A.M. Liapunov himself. Later references may be found in the quoted reviews.

It seems that our choice of Liapunov function is at its best if we can find a function which is somehow inherent in the physical specification of the system, possibly characterizing its dynamic behavior - like our energy frame of reference. It usually is a *storage* (potential) *function* with levels preserving energy, entropy, any type of cost connected with a motion, etc., and with a flow of storage specifying the motion.

A natural way of finding such a storage function is to select from the state equations their part which is exactly integrable. Then the first integral of such a subsystem, storage conservative, will give a family of levels of the potential function concerned. There are two versions of such a first-integral method. The first is used when the integrable sub-system may be immediately selected in an obvious way, and the second when this is not the case, and some rearranging of the state equations is needed for attaining the above integrability.

In the *first* version it is assumed that we may separate a Lipschitz continuous function $\bar{\psi}(\bar{x},t)$ from the selector (2.2.5)':

$$\bar{f}(\bar{x},\bar{u},\bar{w},t) = \bar{\psi}(\bar{x},t) + \bar{\phi}(\bar{x},\bar{u},\bar{w},t) \qquad (3.4.1)$$

such that the corresponding *reference system*

$$\dot{\bar{x}} = \bar{\psi}(\bar{x},t) \qquad (3.4.2)$$

is exactly integrable, that is, that there is a smooth scalar function $V(\cdot) : \Delta \times \mathbb{R} \to \mathbb{R}$ which is a first integral of (3.4.2), meaning $V(\bar{x},t)$ = const or

$$dV(\bar{x},t) = \nabla_{\bar{x},t} V(\bar{x},t) \cdot d(\bar{x},t) = 0 \qquad (3.4.3)$$

along the solutions of (3.4.2) in $\Delta \times \mathbb{R}$, that is,

$$\frac{\partial V(\bar{x},t)}{\partial t} + \nabla_{\bar{x}} \, V(\bar{x},t) \cdot \bar{\psi}(\bar{x},t) = 0 \ . \tag{3.4.4}$$

The checking or designing conditions for the procedure are the Cauchy-Euler necessary and sufficient conditions for exactness:

$$\frac{\partial g_i}{\partial x_j} = \frac{\partial g_j}{\partial x_i} \ , \qquad i,j = 1,\ldots,N+1 \tag{3.4.5}$$

where $\bar{g}(\bar{x},t) = (g_1,\ldots,g_{N+1}) \overset{\Delta}{=} \nabla_{\bar{x},t} \, V(\bar{x},t)$, $x_{N+1} \overset{\Delta}{=} t$. Then we find the Liapunov function from (3.4.3) by the usual procedure of calculating the line integral along solutions of an exact equation

$$V(\bar{x},t) = V(\bar{x}^0,t_0) + \int_{x^0,t_0}^{x,t} \bar{g}(\xi,\tau) \cdot d(\xi,\tau) \ . \tag{3.4.6}$$

With the direct selection of state variables, see Section 2.1, our choice of $\bar{\psi}(\cdot)$ means $\bar{\psi} = (\dot{\bar{q}} \, , \, -\bar{\Pi}(\bar{q}))^T$ which makes (3.4.2) the energy conservative reference frame introduced in Section 2.3, with the energy flow (2.4.19) discussed in Section 2.4 and corresponding to (3.4.6). For the relative state variables or for investigating a relative behavior of systems, we may have to use different $\bar{\psi}(\cdot)$, together with different Liapunov functions, as will be seen in later chapters.

Examples 3.1.2, 3.2.1, 3.2.2 and 3.3.1 illustrate the version described above of the first integral method. Let us add one more typical case.

EXAMPLE 3.4.1. Consider the scalar Duffing's equation

$$\ddot{q} + d(q,\dot{q}) + kq - \beta q^3 = u \ , \tag{3.4.7}$$

or equivalently

$$\begin{aligned} \dot{x}_1 &= x_2 \\ \dot{x}_2 &= -d(x_1,x_2) - kx_1 + \beta x_1^3 + u \end{aligned} \tag{3.4.8}$$

with our standard assumptions on damping $d(q,\dot{q})$ and potential characteristic $y = kq - \beta q^3$. We obtain the reference system by cutting off the damping and input terms which leaves the conservative $\psi(\bar{x},t) \equiv kx_1 - \beta x_1^3$. Then (3.4.2) becomes

$$\frac{dx_2}{dx_1} = \frac{-kx_1 + \beta x_1^3}{x_2} \tag{3.4.9}$$

which integrates exactly by separable variables:

$$\int x_2 \, dx_2 + \int (kx_1 - \beta x_1^3) = \text{const}$$

or

$$E(x_1, x_2) = \frac{x_2^2}{2} + k_1 \frac{x_1^2}{2} - \beta \frac{x_1^4}{4} = \text{const} \qquad (3.4.10)$$

representing the E-levels in $0x_1x_2$ and yielding $E(x_1, x_2) = 0$ along the trajectories of (3.4.9) as required, see Example 3.2.1. Observe that the system has three equilibria $x_2 = 0$, $x_1 = 0, \pm\sqrt{k/\beta}$ with only $(0,0)$ being Dirichlet stable and the corresponding energy cups reaching up to thresholds $h = k^2/2\beta$.

Taking $V(\bar{x}, t) \equiv E(\bar{x})$, the energy of the system, we may study the trajectories locally, that is, below $h = k^2/2\beta$, but also beyond the cup, that is, about the thresholds themselves - with the use of the accumulative control program as indicated in Section 3.3. On the other hand, note that choosing the standard function $V = \frac{1}{2}(x_1^2 + x_2^2)$ without our method, that is, without reference to energy or some other physically meaningful function, we shall obtain the correct results only up to the energy threshold

$$\partial\Delta_E : \frac{x_2^2}{2} + k_1 \frac{x_1^2}{2} - \beta \frac{x_1^4}{4} = \frac{k^2}{2\beta} , \qquad (3.4.11)$$

that is, locally. Anywhere outside (3.4.11), another test function must be found and some conditions securing the continuity of investigated trajectories must be determined and satisfied on the overlap between the operating regions of the two functions. In the above sense, our choice of energy is global (non-local) and avoids a number of technical difficulties in using the Liapunov formalism.   □

The *second* version of the method of first integral applies when the separation (3.4.1) is not possible. Following Kinnen-Chen[1], Chin[1], we then rearrange (2.2.5)' in such a way as to obtain a separable part of it which integrates exactly. To do so, let us "autonomize" (2.2.5)' in $\mathbb{R}^{N+1}$ as

$$\left.\begin{array}{l} \dfrac{dx_i}{d\tau} = f_i(\bar{x}, \bar{u}, \bar{w}, t) \\[3mm] \dfrac{dt}{d\tau} = 1 \end{array}\right\} \qquad (3.4.12)$$

and introduce the $(N+1)$-vector function $\bar{g}(\bar{x}, \bar{u}, \bar{w}, t)$ with components

$$\left.\begin{array}{l} g_i(\bar{x}, \bar{u}, \bar{w}, t) = f_1(\bar{x}, \bar{u}, \bar{w}, t) + \cdots + f_{i-1}(\bar{x}, \bar{u}, \bar{w}, t) \\[3mm] \qquad - f_{i+1}(\bar{x}, \bar{u}, \bar{w}, t) - \cdots - f_N(\bar{x}, \bar{u}, \bar{w}, t) - 1 . \end{array}\right\} \qquad (3.4.13)$$

Then the equation (2.2.5)' gives

$$g_{N+1}(\bar{x},\bar{u},\bar{w},t) + \sum_{i=1}^{N} g_i(\bar{x},\bar{u},\bar{w},t)\dot{x}_i = 0$$

whence also

$$\left[\sum_{i=1}^{N} f_i(\bar{x},\bar{u},\bar{w},t) \right] dt + \sum_{i=1}^{N} g_i(\bar{x},\bar{u},\bar{w},t)dx_i = 0$$

or in the (N+1)-vector format

$$\bar{g}(\bar{x},\bar{u},\bar{w},t) \cdot d(\bar{x},t) = 0 . \qquad (3.4.14)$$

For example, for the 2D system $\dot{x}_1 = f_1(x_1,x_2)$, $\dot{x}_2 = f_2(x_1,x_2)$ we have $g_1 = -f_2$, $g_2 = f_1$ which yields $g_1\dot{x}_1 + g_2\dot{x}_2 = g_1 f_1 + g_2 f_2 = -f_2 f_1 + f_1 f_2 = 0$.

The equation (3.4.14) is equivalent to (2.2.5)' producing the same motions. Obviously, in general, (3.4.14) is not integrable. To remedy it, we design some (N+1)-vector function $\bar{\ell}(\bar{x},\bar{u},\bar{w},t)$ which, subtracted from $\bar{g}(\bar{x},\bar{u},\bar{w},t)$ of (3.4.14) with given $\bar{u},\bar{w}$, makes the resulting equation

$$[\bar{g}(\bar{x},\bar{u},\bar{w},t) - \bar{\ell}(\bar{x},\bar{u},\bar{w},t)]^T \cdot d(\bar{x},t) = 0 \qquad (3.4.15)$$

exactly integrable. Because of this, there is a $V(\bar{x},t)$ for which, denoting $x_{N+1} \overset{\Delta}{=} t$, we have

$$\frac{\partial V(\bar{x},t)}{\partial x_i} = g_i(\bar{x},\bar{u},\bar{w},t) - \ell_i(\bar{x},\bar{u},\bar{w},t) , \quad i = 1,\ldots,N+1 \qquad (3.4.16)$$

and

$$\frac{\partial(g_i - \ell_i)}{\partial x_j} = \frac{\partial(g_j - \ell_j)}{\partial x_i} , \quad i,j = 1,\ldots,N+1 . \qquad (3.4.17)$$

This means that $[\bar{g} - \bar{\ell}] = \nabla_{\bar{x},t} V(\bar{x},t)$. Then by (3.4.15),

$$\bar{g}(\bar{x},\bar{u},\bar{w},t)^T \cdot d(\bar{x},t) = \bar{\ell}(\bar{x},\bar{u},\bar{w},t)^T \cdot d(\bar{x},t) . \qquad (3.4.18)$$

On the other hand, along motions of (2.2.5)' or (3.4.14):

$$\dot{V}(\bar{x},t) = \bar{g}(\bar{x},\bar{u},\bar{w},t)^T \frac{d(\bar{x},t)}{d\tau} = \bar{g}(\bar{x},\bar{u},\bar{w},t)^T [\bar{f}(\bar{x},\bar{u},\bar{w},t),1] .$$

Substituting (3.4.18), we obtain

$$\dot{V}(\bar{x},t) = \bar{\ell}(\bar{x},\bar{u},\bar{w},t)^T [\bar{f}(\bar{x},\bar{u},\bar{w},t),1] . \qquad (3.4.19)$$

If $\bar{\ell}$ is designed with $\ell_{N+1} = 1 : \bar{\ell} = (\bar{\ell}',1)^T$, we have

$$\dot{V}(\bar{x},t) = \frac{\partial V(\bar{x},t)}{\partial t} + \bar{\ell}'(\bar{x},\bar{u},\bar{w},t)^T \cdot \bar{f}(\bar{x},\bar{u},\bar{w},t) . \qquad (3.4.20)$$

Integrating (3.4.19) along motions of (2.2.5)', we obtain

$$V(\bar{x},t) = \int_{t_0}^{t} ds + \sum_{i=1}^{N} \int_{x_i^0}^{x_i} \ell_i'(\bar{\xi},\bar{u},\bar{w},t)d\xi_i . \qquad (3.4.21)$$

Technically we start with designing $\bar{\ell}$ which satisfies conditions (3.4.17), and then we check whether $\dot{V} < 0$ and $V > 0$. The procedure is illustrated in the following example.

EXAMPLE 3.4.2. Consider the system

$$
\left.
\begin{array}{l}
\dot{x}_1 = ux_1^3 + wx_2^2 \qquad\qquad = f_1 \\[6pt]
\dot{x}_2 = -x_1 - x_2 + ux_1^3 - wx_2^3 = f_2
\end{array}
\right\}
$$

with $u, w \in [-1, 1]$. From (3.4.13), $g_1 = -f_2$, $g_2 = f_1$. First we search for $\ell_i(x_1, x_2, u, w)$, $i = 1, 2$ such that (3.4.17) holds. One of the possible ways is to require both partial derivatives vanishing

$$
\frac{\partial(g_1 + \ell_1)}{\partial x_2} = 1 + 3wx_2^2 + \frac{\partial \ell_1}{\partial x_2} = 0 ,
$$

$$
\frac{\partial(g_2 + \ell_2)}{\partial x_1} = 3ux_1^2 + \frac{\partial \ell_2}{\partial x_2} = 0 ,
$$

wherefrom $\ell_1 = -x_2 - wx_2^3$, $\ell_2 = -ux_1^3$. Next, if we choose the minimizing $u = -1$ and assume the maximizing uncertainty $w \equiv 1$, see Corollary 3.2.3, we obtain (3.4.19) in the form

$$
\ell_1 f_1 + \ell_2 f_2 = -x_2^4 - x_1^4 - x_1^6 < 0
$$

as required. Then we calculate

$$
V(x_1, x_2) = \int_0^{x_1} (g_1 - \ell_1)d\xi_1 + \int_0^{x_2} (g_2 - \ell_2)
$$

$$
= \frac{x_1^2}{2} + \frac{x_1^4}{4} + \frac{x_2^4}{4} > 0 ,
$$

again as required. $\qquad\qquad\qquad\qquad\qquad\qquad\qquad\qquad\qquad$ □

VARIABLE GRADIENT METHOD

The early development of the method belongs to Schultz-Gibson [1]. Later stages are reviewed as quoted at the opening of this section.

Suppose that we want the order of our previous procedure reversed: start with assuming a suitable time derivative (3.4.19) and check upon (3.4.17) and the sign of $V(\cdot)$ obtained by integration (3.4.21). In more general terms, the method is expressed as follows.

Consider (3.4.14) and assume the gradient $\bar{g}(\bar{x}, \bar{u}, \bar{w}, t)$ to be unknown but of the format

$$g_i = \alpha_{ii}t + \sum_{j=1}^{N} \alpha_{ij}x_j , \quad i = 1,\ldots,N .$$

Then we choose α_{ii}, α_{ij} such that $\bar{g}$ is a gradient of a scalar function, that is, that (3.4.17) holds, while $\bar{u},\bar{w}$ are chosen based upon Corollary 3.2.3. The procedure is purely technical and thus best explained in a simple but typical example. Some different aspects of the method are described in Prusty [1] and Byrne-Wall [1].

EXAMPLE 3.4.2. Consider the system

$$\dot{x}_1 = -3ux_2 - x_1^5 ,$$
$$\dot{x}_2 = -2x_2 + wx_1^5 ,$$

with $u \in [1,2]$, $w \in [0,1]$. We set up

$$\frac{\partial}{\partial x_1} = \alpha_{11}x_1 + \alpha_{12}x_2 , \qquad \frac{\partial V}{\partial x_2} = \alpha_{21}x_1 + \alpha_{22}x_2 .$$

With minimizing $u = 1$ and maximizing $w = 1$, we demand (3.2.7) of Corollary 3.2.3:

$$\dot{V} = -(\alpha_{11}x_1 + \alpha_{12}x_2)(3x_2 + x_1^5) + (\alpha_{21}x_1 + \alpha_{22}x_2)(x_1^5 - 2x_2)$$
$$= -x_1^2(\alpha_{11}x_1^4 - \alpha_{21}x_1^4) + x_1x_2(-3\alpha_{11} - \alpha_{12}x_1^4 - 2\alpha_{21} + \alpha_{22}x_1^4)$$
$$- x_2^2(3\alpha_{12} + 2\alpha_{22}) \leq -c(\|\bar{x}\|) .$$

Since α_{ij} are of our choice, we let $\alpha_{12} = \alpha_{21} = 0$, $\alpha_{11} = \frac{1}{3}\alpha_{22}x_1^4$, which gives $\dot{V}(x_1,x_2) = -\alpha_{22}(\frac{1}{3}x_1^6 + 2x_2^2)$ satisfying the requirement provided $\alpha_{22} > 0$. With this assumption on α_{22} we also have

$$g_1 = \frac{\partial V}{\partial x_1} = \frac{1}{3}\alpha_{22}x_1^5 , \qquad g_2 = \frac{\partial V}{\partial x_2} = \alpha_{22}x_2 .$$

Then checking (3.4.17),

$$\frac{\partial}{\partial x_2}(\frac{1}{3}\alpha_{22}x_1^5) = \frac{\partial}{\partial x_1}(\alpha_{22}x_2) = 0 ,$$

provided $\alpha_{22} = \text{const}$. Finally we calculate

$$V(x_1,x_2) = \int_0^{x_1} \frac{1}{3}\alpha_{22}\xi_1^5 d\xi_2 + \int_0^{x_2} \alpha_{22}\xi_2 d\xi_2 = \frac{1}{18}\alpha_{22}(x_1^6 + 9x_2^2)$$

which is positive definite for $\alpha_{22} > 0$. Observe that if $\alpha_{22} < 0$ then $V(x_1,x_2) < 0$ but also $\dot{V} > -c(\|\bar{x}\|)$ which satisfies Conditions 3.2.3 with $-V(x_1,x_2)$. As regards control, the latter gives accumulative control.

$\square$

EXERCISES 3.4

3.4.1 Using the first integral method, find a test function $V(\cdot)$ and secure the stability or asymptotic stability of the origin for the following systems. Find the regions.

(i) $\dot{x}_1 = x_2 + 3x_2$,

$\dot{x}_2 = -3x_2 - 4x_1^5 - 4x_1^3 - x_1 + u$;

(ii) $\ddot{q} + \dot{q} \sin^2 q + 4q^3 = u$;

(iii) $\dot{x}_1 = x_2$,

$\dot{x}_2 = -ux_2 - 4x_1^2 - 3x_1$;

(iv) $\dot{x}_1 = x_2$,

$\dot{x}_2 = -ux_2^2 - 3x_1 + 4x_1^3 - wx_1^5$.

3.4.2 Consider the system $\ddot{q} + 8|q|\dot{q} + 4q = 3$, $q \in \mathbb{R}$, and establish a test function $V(\cdot)$ for stabilizing the trajectories about the equilibrium $q^e = 3/4$, $\dot{q}^e = 0$.

3.4.3 Using the Variable Gradient Method, find stabilizing test functions for the following systems.

(i) $\dot{x}_1 = -x_1^2 x_2 - x_1^5$,

$\dot{x}_2 = x_1^3 - x_2^3$;

(ii) $\dot{x}_1 = x_2$,

$\dot{x}_2 = -x_2 - x_2 f(x_1) - x_1 f(x_1) - x_1 x_2 \dfrac{df}{dx_1}$;

with $f(\cdot)$ being a positive C^1 function for all x_1 ;

(iii) $\dot{x}_1 = x_1 x_2^2 + x_1 \sin^2 x_2$,

$\dot{x}_2 = -x_2 (x_1^2 + 1)$;

(iv) $\ddot{q} + q(1 + q\dot{q}) + \dot{q}(q^2 + \dot{q}^2) = 0$;

(v) $\dot{x}_1 = x_1 x_2$,

$\dot{x}_2 = -2x_1^2 - x_2$;

(vi) $\dot{x}_1 = x_1 (x_2 - 1)$,

$\dot{x}_2 = -2(x_1^4 + \ell x_2)$, $\ell \geq 0$;

(vii) $\dot{x}_1 = x_1 (x_2 - x_2^2)$,

$\dot{x}_2 = -x_1^2 - 2x_3^2 x_2$,

$\dot{x}_3 = -x_2^2 x_3 + x_2^2 x_3$;

(viii) $\dot{x}_1 = x_2$,

$\dot{x}_2 = -3x_2 - 2x_1 + 4x_1^3$.

Chapter 4
GLOBAL PATTERN OF STEADY STATES

4.1 AUTONOMOUS LIMIT SETS

What happens to the motions of (2.2.6) or (2.2.6)' after they have entered Δ_L or when they have started from there? How is the conservative reference frame, in particular the energy cups and thresholds, distributed in Δ_L for a specified system? Assuming that we aim at stabilizing the system about steady state sets, given a controller, how are the particular attractors and real-time attractors distributed in Δ_L, both inside and outside the energy cups? Can we find controllers forcing motions from one cup or region of attraction to another?

These, and others similar, are the questions we would like to answer in our global study of Δ_L. However, as it already may have been guessed from Section 3.3, the answers are rarely possible in the general case. Nevertheless, there are some generally valid properties of the global state or phase space pattern which can facilitate the case studies. We shall attempt in this section to sum them up and illustrate their use in examples.

The first question to ask is about the location and size of the energy cups. It has been largely answered in Section 2.3 for the free, conservative system considered our reference, and in Section 3.2 for the controlled system, possibly also nonautonomous. On this background we shall now look at the properties and location of the steady state sets.

The reader may recall here our discussion on steady state trajectories and sets in Section 2.1 and on invariant sets in Section 3.2, as well as the fact that the steady state trajectories and sets are minimal invariant. We shall elaborate the point slightly further now, referring first to the autonomous system (2.2.6) with the selector equation (2.2.5) and steady state trajectories.

The set of points $\bar{y} \in \Delta$ such that we can find a sequence of time instants $\{t_n\} \subset \mathbb{R}$ generating $\bar{\phi}(\bar{x}^0, t_n) \to \bar{y}$ as $t \to \infty$ ($t \to -\infty$) is called the positive (negative) *limit set* $\Lambda^{\pm}$ for trajectories of (2.2.5):

$$\Lambda^{\pm} = \{\bar{y} \in \Delta \mid \exists \ \{t_n\}, \ t_n \to \pm\infty \Rightarrow \bar{\phi}(\bar{x}^0, t_n) \to \bar{y}\} \ , \qquad (4.1.1)$$

see Fig. 4.1. The point $\bar{y}$ of above is called a *limit point* of the trajectory. It may be shown that $\Lambda^{\pm}$ is closed, invariant (Nemitzky-Stepanov [1]) and represents a boundary of a trajectory (Bhatia-Szegö [1]):

$$\overline{\bar{\phi}(\bar{x}^0, \mathbb{R}^{\mp})} = \bar{\phi}(\bar{x}^0, \mathbb{R}^{\pm}) \cup \Lambda^{\pm}(\bar{x}^0) \ . \qquad (4.1.2)$$

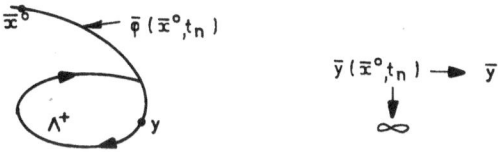

Fig. 4.1

Then if we can find a trajectory which is its own limit set: $\bar{\phi}(\bar{x}^0, \mathbb{R}^{\pm}) = \Lambda^{\pm}(\bar{x}^0)$, it also must be its own closure $\Lambda^{\pm}(\bar{x}^0) = \overline{\bar{\phi}(\bar{x}^0, \mathbb{R}^{\pm})}$. It is heuristically obvious that such an invariant set is minimal. More formally, it follows from (4.1.2) that, in particular, if a set is a closure of a trajectory: $\bar{\phi}(\tilde{M}, \mathbb{R}^{\pm}) = \tilde{M}$ or $\bar{\phi}(\tilde{M}, \mathbb{R}) = \tilde{M}$, then it is minimal invariant under (2.2.5), and conversely any set that is minimal invariant under (2.2.5) represents the closure of a trajectory, see Skowronski [32]. This is then also equivalent to

$$\Lambda^{\pm}(\tilde{M}) = \tilde{M} \ , \qquad (4.1.3)$$

which may serve as a definition of minimal invariant sets. The best example here is a periodic orbit which consists of its own limit points only and for which one can prove that

$$\bar{\phi}(\bar{x}^0, \mathbb{R}) = \phi(\bar{x}^0, L) = \Lambda^+(\bar{x}^0) = \Lambda^-(\bar{x}^0) \ . \qquad (4.1.4)$$

The orbit is a closed set and forms its own limit sets, it is thus minimal invariant. In fact (4.1.4) defines the periodic orbit as there are no other trajectories with this property. In a plane, a periodic limit orbit becomes a *limit cycle*. Obviously the orbits may enclose one another within or outside an energy cup. A very simple example illustrates the case.

171

EXAMPLE 4.1.1. Consider kinematics of a point-mass given in polar coordinates by the equation

$$\dot{r} = r(u-r)(w-r)^2 \ , \qquad \Big\}$$
$$\dot{\theta} \equiv 1 \ , \qquad\qquad \Big\} \qquad\qquad (4.1.5)$$

with the state variables $x = r$, $t = \theta$. Given the test function $V(x,t) \equiv r$, $w = 5$, and the control program defined by $u(t) \equiv 2$, we see that for $0 < r < 2$, we have $\dot{r} > 0$ which makes the negative limit cycle $\Lambda^- : r(t) \equiv 0$ unstable, and the positive limit cycle $\Lambda^+ : r(t) \equiv 2$ asymptotically stable, as on the other side, that is, for $2 < r < 5$, we have $\dot{r} < 0$. By the latter, we also have the negative limit cycle $\Lambda^- : r(t) \equiv 5$ for trajectories from $r < 5$, see Fig. 4.2.

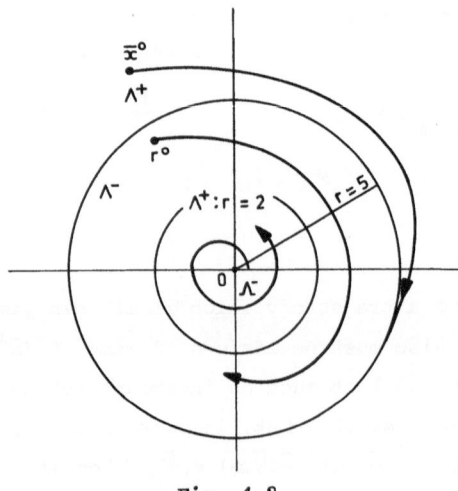

Fig. 4.2

Observe that changing the controller $u = P(r)$ we may change the pattern significantly. For $u = r$ we have the r^0-family of cycles $r(t) = r^0 = const$. Taking $u(t) = const$ and increasing it along $(0,5]$ we move Λ^+ towards the Λ^- surrounding it, and finally for $u(t) \equiv 5$ and still $w = 5$, we reduce the two cycles to a single asymptotically stable $\Lambda^+ : r(t) \equiv 5$. In turn, making $w \in [2,5]$, we may collapse the outside Λ^- to Λ^+ sooner than at $r = 5$, but on the other hand $w > 5$ can make Λ^+ chase the Λ^- concerned until the controller saturates, thus making the unification of the two cycles impossible. □

Obviously an equilibrium is trivially a periodic orbit, with a single limit point comprising closure of a trajectory and being its own limit set, see (2.1.7). We illustrate the case again with a simple example.

EXAMPLE 4.1.2. Consider the system

$$\begin{aligned} \dot{x}_1 &= \sin x_1 \\ \dot{x}_2 &= u(1 - x_2^2) \end{aligned} \right\} \qquad (4.1.6)$$

with the controller $u = x_2$ and equilibria $(0,0), (\pi,1), (\pi,-1), (-\pi,1),$ $(-\pi,-1)$ which are limit points. Indeed, no trajectory from the open rectangle $(-\pi,\pi) \times (-1,1)$ can reach the lines $x_1 = \pm\pi$, $x_2 = \pm 1$ in finite time (no point of the boundary leaves or enters the boundary), so the rectangle and its boundary are invariant. Consequently the closure is also invariant. It is easily seen that $\Lambda^{\pm}$ are located as shown in Fig. 4.3.

□

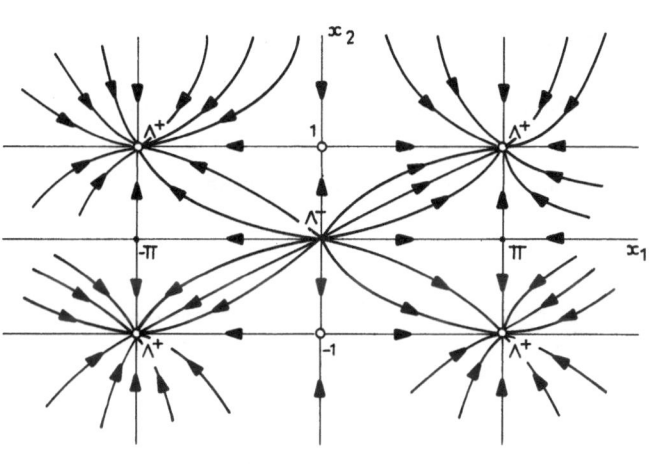

Fig. 4.3

It now remains to see whether and where in Δ_L we can find the minimal invariant sets. Our hypothetical answer is: they are in any subset of Δ_L in which the trajectories of (2.2.5) from Δ_L are uniformly ultimately bounded for bound $\partial\Delta_B$. Indeed the latter implies that Δ_B is positively invariant. Then, since the system is autonomous, it is also invariant. One can then also show (Bhatia-Szegö [1]) that the closure $\bar{\Delta}_B$ is invariant. Nemitzky-Stepanov [1] proved that every nonempty invariant set which is bounded and closed contains a minimal invariant set which is also closed and bounded. This confirms our hypothetical answer. Moreover, by Birkhoff [4], every trajectory of a bounded and closed, minimal invariant set is at least recurrent which means, in our case, a steady state trajectory. Consequently the uniform ultimately bounded Δ_B localizes the steady state trajectories which by (4.1.3) are limit sets. The real-time attractor is a stipulated Δ_B of the above, and we must specify such sets a-priori if we want to draw a state space pattern over specific regions in Δ. Consequently, the *real-time attractors* either identified for given controllers or formed by a suitably chosen control action, *localize the steady state limit trajectories*. They may be found or posed in local cups as equilibria or any other kind, the latter also outside the cups: enclosing several cups, as long as they stay in Δ_L .

When the limit trajectories are expected inside an energy cup, the real-time attractor obtained by strong stabilization below the E-level of h_B may suffice. Elsewhere, enclosing several equilibria or still in a cup, but well above the equilibrium, we need a combination of two E-levels: stabilization below some h_B' and above some h_B'' after the time interval $T_B = \max(T_B', T_B'')$, see Fig. 4.4. Then we may need the La Salle's theorem on limit sets to determine the steady state, see La Salle – Lefschetz [1]. Before stating the theorem, we need some new concepts.

A positively (negatively) invariant set or invariant set is termed *maximal* if and only if it is closed and not a proper subset of a set of the same type. The same applies to strongly invariant sets.

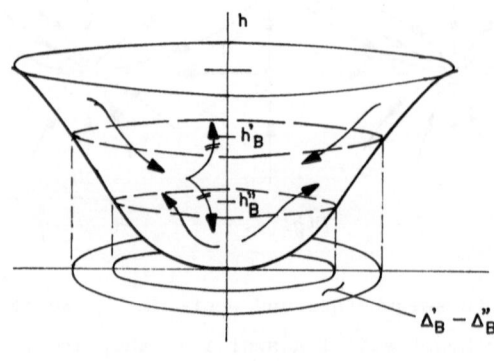

Fig. 4.4

We assume $\Delta_B' - \Delta_B''$ bounded and introduce an open connected set D such that $D \cap (\Delta_B' - \Delta_B'') \neq \phi$. By the same argument as in La Salle – Lefschetz [1] we obtain the following proposition.

PROPOSITION 4.1.1. *Let* $V(\cdot) : \bar{D} \rightarrow \mathbb{R}$ *be a* C^1-*function such that the boundaries* $\partial\Delta_B'$, $\partial\Delta_B''$ *on* D *are specified by* V-*levels, and let* $\tilde{M}$ *be the maximal invariant set in the set of all points* $\bar{x} \in \Delta_B' - \Delta_B''$ *where* $\dot{V}(\bar{x}) = 0$. *If there is a control program* $\bar{P}(x)$ *such that for each* $\bar{u} \in \bar{P}(\bar{x})$, *either*

$$\left. \begin{array}{l} V(\bar{x}) > 0 , \quad \dot{V}(\bar{x}) \leq 0 , \\[2mm] V(\bar{x}) < 0 , \quad \dot{V}(\bar{x}) \geq 0 , \end{array} \right\} \qquad (4.1.7)$$

or

174

for all $\bar{x} \neq 0$, $\bar{w} \in W$, *then the system* (2.2.6) *is strongly asymptotically stabilizable about* $\tilde{M}$.

The above proposition has an immediate corollary which follows by the same argument as that for Corollary 3.2.2 but used twice, first with substituted $E^-(\bar{x})$ and second with substituted $-E^+(\bar{x})$.

COROLLARY 4.1.1. *Given* $\bar{x}^0 \in D$, *if there are* $\bar{u}* \in U$ $\bar{w}* \subset W$ *such that either* $V(\bar{x}) > 0$ *with*

$$\nabla V(\bar{x})^T \cdot \bar{f}(\bar{x}, \bar{u}*, \bar{w}*) = \min_{\bar{u}} \max_{\bar{w}} \left[\nabla V(\bar{x})^T \cdot \bar{f}(\bar{x}, \bar{u}, \bar{w}) \right] \leq 0$$

or $V(\bar{x}) < 0$ *with*

$$\nabla V(\bar{x})^T \cdot \bar{f}(\bar{x}, \bar{u}*, \bar{w}*) = \max_{\bar{u}} \min_{\bar{w}} \left[\nabla V(\bar{x})^T \cdot \bar{f}(\bar{x}, \bar{u}, \bar{w}) \right] \geq 0$$

$$(4.1.8)$$

then condition (4.1.7) *is met with* $u* \in \bar{P}*(\bar{x})$.

Clearly Proposition 4.1.1 and Corollary 4.1.1 are sufficient conditions. We use $V(\bar{x}) \equiv E^-(\bar{x})$ for descending from $\partial \Delta_B'$ and $V(\bar{x}) \equiv -E^+(\bar{x})$ for ascending from $\partial \Delta_B''$. Then, proposing the same controllers on $\Delta_B' - \Delta_B''$ as those used for reaching the real time attractors Δ_B', Δ_B'', we see that they satisfy (4.1.8) and thus (4.1.7). Indeed, the dissipative controller (3.3.7)' acting from above across finite $|\delta h| < \infty$ for $T_B \rightarrow \infty$ generates

$$\min_{\bar{u}} \max_{\bar{w}} f_0(\bar{x}, \bar{u}, \bar{w}) \leq 0 \qquad (4.1.9)$$

while the accumulative controller (3.3.7)" acting from below for the same reason produces

$$\max_{\bar{u}} \min_{\bar{w}} f_0(\bar{x}, \bar{u}, \bar{w}) \geq 0, \qquad (4.1.10)$$

proving our point. By the same argument as for Corollary 3.3.2, with a choice of $V(\cdot)$ and $\bar{P}(\cdot)$, condition (4.1.7) becomes also *necessary*, and thus so are (4.1.9) and (4.1.10), cf. (3.1.21), Corollary 3.1.2.

Continuing further, (4.1.9) and (4.1.10) imply that on the conservative field H^0 between H^- and H^+ we must have

$$\min_{\bar{u}} \max_{\bar{w}} f_0(\bar{x}, \bar{u}, \bar{w}) = \max_{\bar{u}} \min_{\bar{w}} f_0(\bar{x}, \bar{u}, \bar{w}) = 0. \qquad (4.1.11)$$

The above necessary condition determines a reasonable candidate for the set $\tilde{M}$ of the Proposition 4.1.1. We confirm the candidate by showing it to be maximal invariant between h_B' and h_B''. Let us start with the invariance.

Indeed, (4.1.9) with $E^-(\bar{x}) > 0$ prevents trajectories, starting at such $\widetilde{M}$, from entering the field H^- above it and (4.1.10) with $-E^+(\bar{x}) < 0$ prevents such trajectories from entering H^- below it, see Figs. 4.4 and 2.9. Then it is also seen that such $\widetilde{M}$ is maximal invariant, since trajectories from outside $\widetilde{M}$ start either from H^- or H^+ which excludes the preventive properties of the above.

Note here that such $\widetilde{M}$ defined by (4.1.11) may not necessarily be filled up by a single steady state trajectory. In the general case of $\mathbb{R}^N$ it will be an N-1 dimensional manifold.

Technically, in order to identify $\widetilde{M}$ with approximation to minimal $|\delta h|$ we search for lowest h'_B and highest h''_B, which means $\min E^-(\bar{x})$ and $\max E^+(\bar{x})$, subject to constraints $\dot{E}(\bar{x}) = 0$. This is illustrated below.

EXAMPLE 4.1.3. Consider the Lienard-type system

$$
\left.
\begin{aligned}
\dot{x}_1 &= x_2 \\
\dot{x}_2 &= -x_2(x_1^2 + x_2^2 - w) - (x_1 + x_1^3) + u
\end{aligned}
\right\} \tag{4.1.12}
$$

with the single equilibrium $(0,0)$ which is Dirichlet unstable (saddle) generated by the energy

$$
E(x_1, x_2) = \frac{x_2^2}{2} + \frac{x_1^2}{2} + \frac{x_1^4}{4}, \tag{4.1.13}
$$

so that the local cup extends over Δ. The energy is not related to w, thus we let $V(\bar{x}, t) \equiv E(x_1, x_2)$, whence

$$
\dot{E}(x_1, x_2) = ux_2 - (x_1^2 + x_2^2 - w)x_2^2. \tag{4.1.14}
$$

Suppose now that w is known, say $w(t) \equiv 4$. Since the problem is two dimensional, the geometric locus $H^0 : \dot{E}(x_1, x_2) = 0$ for conservative trajectories is a $\bar{x}^0$-family of lines and since we are in the energy cup these lines are closed, for each $\bar{x}^0$ generating a periodic orbit. Hence specifying the controller by

$$
ux_2 - (x_1^2 + x_2^2 - 4)x_2^2 = 0 \tag{4.1.15}
$$

or $u = x_2$, we obtain at suitable $\bar{x}^0$ a single, isolated asymptotically stable limit cycle $\Lambda^+ : x_1^2 + x_2^2 = 5$. Indeed the latter fills up $\widetilde{M}$, since for $x_1^2 + x_2^2 < 5$ there is $\dot{E} > 0$ making the equilibrium an unstable Λ^- and the region covered by exit points of H^+, while for $x_1^2 + x_2^2 > 5$ there is $\dot{E} < 0$ making it covered by entry points of H^-, see Fig. 4.5.

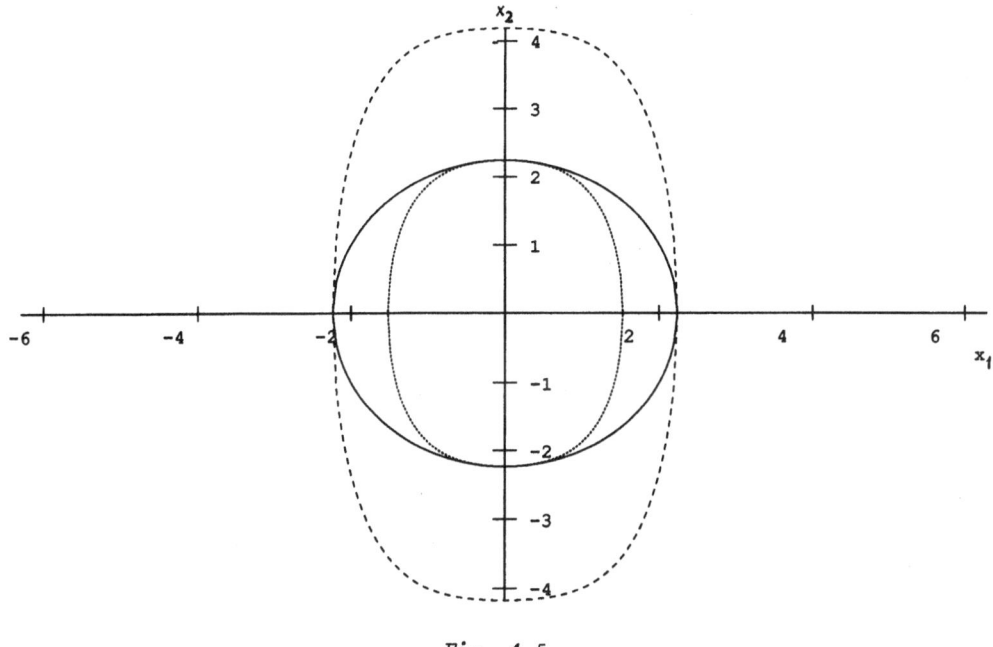

Fig. 4.5

With w unknown, say $w \in [2,6]$, we have

$$\min_{u} (ux_2) + \max_{w} [-(x_1^2 + x_2^2 - w)x_2^2] \leq 0$$

or

$$\min_{\bar{u}} (u \operatorname{sgn} x_2) \leq \min_{\bar{w}} (x_1^2 + x_2^2 - w)|x_2|$$

or

$$\min_{\bar{u}} (u \operatorname{sgn} x_2) \leq (x_1^2 + x_2^2 - 6)|x_2| \qquad (4.1.16)$$

as the dissipative control condition above the cycle Λ^+, and by a similar argument,

$$\max_{u} (u \operatorname{sgn} x_2) \geq (x_1^2 + x_2^2 - 2)|x_2| . \qquad (4.1.17)$$

Observe that the controller $u = x_2$ satisfies the above inequalities in their sets of holding, that is, for $x_1^2 + x_2^2 > 7$ and $x_1^2 + x_2^2 < 3$ respectively, with (4.1.15) between (4.1.16) and (4.1.17), implying (4.1.11). Since $E^-(\bar{x}) \equiv E^+(\bar{x}) \equiv E(\bar{x})$, in order to find the estimates h_B', h_B'' we take successively $\min E(\bar{x})$ subject to $x_1^2 + x_2^2 = 7$ and $\max E(\bar{x})$ subject to $x_1^2 + x_2^2 = 3$. Substituting (4.1.13), the Lagrange multipliers method gives $h_B' = 15/4$, $h_B'' = 7/2$.

Note that the method of defining the unique cycle $\Lambda^+ : x_1^2 + x_2^2 = 5$ would not have been possible in $\mathbb{R}^N$ even for specified $\bar{w}$, while the estimation method by the two levels h'_B, h''_B works for unknown $\bar{w}$ in both $\mathbb{R}^2$ and $\mathbb{R}^N$, $N > 2$, cases. $\quad\quad\quad\quad\quad\quad\quad\quad\quad\quad\quad\quad\quad\quad\quad$ □

Obviously the function $V(\cdot)$ of Proposition 4.1.1 does not have to be energy, as the following brief example shows. Consider

$$\dot{x}_1 = x_1 + 4x_3 - x_1(x_1^2 + 8x_2^2 + 8x_3^2)$$
$$\dot{x}_2 = x_2 - x_3 - x_2(4x_2^2 + 8x_3^2)$$
$$\dot{x}_3 = -x_1 + x_2 + x_3 + 2x_3(x_1^2 + 4x_2^2 - 2x_3^2)$$

and find two ellipsoids centered at the origin such that the inner ellipsoid contains only exits and the other only entries. We set up $V = x_1^2 + 4x_2^2 + 4x_3^2$ wherefrom

$$\dot{V} = x_1[x_1 + 4x_3 - x_1(x_1^2 + 8x_2^2 + 8x_3^2)] + 4x_2[x_2 - x_3 - x_2(4x_2^2 + 8x_3^2)]$$
$$+ 4x_3[-x_1 + x_2 + x_3 + 2x_3(x_1^3 + 4x_2^2 - 2x_3^2)] \ .$$

Introducing $V(\cdot)$ as distance ρ^2 with $\sin^2\theta = 4x_3^2/\rho^2$, $\cos^2\theta = (x_1^2 + 4x_2^2)/\rho^2$ we have $\dot{V} = \rho^2(1 + \frac{1}{2}\sin^2 2\theta) - \rho^4$ and thus $\dot{V} > 0$ for $\rho = 1/2$, $\dot{V} < 0$ for $\rho = 3/2$, generating the two ellipsoids. For a better estimate, we find ρ, θ when $\dot{V} = \rho^2(1 + \frac{1}{2}\sin^2 2\theta) - \rho^4 = 0$ or $\sin^2 2\theta = 2\rho^2 - 2$ which defines the limit surface. Note that $\sin^2 2\theta$ varies from 0 to 1, which gives $1 \le \rho \le \sqrt{3/2} = 1.225$.

In all the above we did not say much about the character of the steady state limit trajectory, except that by Birkhoff's result it is recurrent, and by Hilmy, it fills up an N-1 dimensional manifold. Unfortunately, little more can be said in the general case.

A trajectory is called positively (negatively, both) *internally stable*, that is, stable with respect to the set of its own points, if and only if for each $\varepsilon > 0$ there is $\delta > 0$ such that any $\bar{x}^1$ on this trajectory satisfying $\rho(\bar{x}^0, \bar{x}^1) < \delta$ implies $\rho(\bar{\phi}(\bar{x}^0, t), \bar{\phi}(\bar{x}^1, t)) < \varepsilon$ for all $t \in \mathbb{R}^+ (\mathbb{R}^-, \mathbb{R})$. Markov [1] proved that if a trajectory is recurrent and positively internally stable, it is almost periodic. This result was later confirmed by Nemitzky [1], [2] under the stronger condition for internal stability.

The existence of a periodic trajectory is in general an open problem. It is traditionally discussed in terms of the so called fixed-point theorems (fixed point under the map of the state space into itself), and has a rich literature, see Nimitzky-Stepanov [1], Cesari [1], Sansone-Conti [1], Yoshizawa [1].

The existence of a periodic limit trajectory appears to be slightly easier to show. The problem is solved for $N < 4$, cf. Nemitzky [2], Shirokorad [1], Cartwright [1], Minc [1], Tribieleva [1], etc. The account is certainly not exhaustive. Then, in the N-dimensional case, the existing attempts go along several avenues: the method of integral manifolds (thori), see Blinchevski [1], the method of point-transformations, see Neimark [1], and oscillatory regimes, see Nemitzky [3], Skowronski [3],[7], Skowronski-Shannon [1], global convergence, Skowronski [4]-[7]. An entirely different approach via frequency domain was used in Noldus [1]. The work on limit orbits is well reviewed in Nemitzky [4],[5] and Pliss [1]. For the current review, see Skowronski [45].

In terms of control design we may, however, have some solution to the problem of a periodic limit trajectory. Using necessary conditions, such as (4.1.11) or similar, applied to a closed trajectory in Δ_L, we may consider such a trajectory a candidate for the limit orbit, and then confirm the candidate by Proposition 4.1.1. Alternatively to (4.1.11), one can use a condition built upon the estimate of an energy flux, see Section 2.4, along the proposed closed curve, see also Kauderer [1].

PROPERTY 4.1.1. If there is a closed trajectory $\bar{\phi}(\bar{x}^0, \mathbb{R})$ in Δ_L and the time needed for a single tracing along it equals $T < \infty$, then there are $(u^*, w^*) \in U \times W$ such that

$$\int_0^T f_0(\bar{x}, \bar{u}^*, \bar{w}^*) dt = \sum_i \oint_{\bar{\phi}(\mathbb{R})} [F_i(\bar{q}, \dot{\bar{q}}, \bar{u}^*, \bar{w}^*) - D_i(\bar{q}, \dot{\bar{q}}, \bar{w})] dq_i = 0 . \quad (4.1.18)$$

It means that for the extremizing value $\bar{w}^*$ taken as in (4.1.11) we may design a program $\bar{P}(\cdot)$ such that the trajectory concerned is either conservative or interchangeably controlled by balanced dissipative or accumulative controllers to attain (4.1.18), see also our derivation of (4.1.11). The first case may be easier to design.

We shall do it by generalizing Example 4.1.3. Consider (2.1.4) and choose the program from the control condition

$$\sum_i F_i(\bar{q}, \dot{\bar{q}}, \bar{u}, \bar{w}^*) \dot{q}_i = \sum_i D_i(\bar{q}, \dot{\bar{q}}, \bar{w}) \dot{q}_i \quad (4.1.19)$$

satisfying (4.1.18) and thus making the system integrable, with the energy integral $E(\bar{x}) = h^0$, and the trajectory embedded in E-level. The *program* is thus *conservative*. By the same argument as for deriving (4.1.11), below and above the E-level concerned, we shall have the balance (4.1.19) cancelled and the program becomes accumulative and dissipative, respectively. This, by Proposition 4.1.1 and Corollary 4.1.1, makes the trajectory a limit steady state trajectory, periodic since closed. Let us illustrate the case with a single DOF example.

The trajectory $kq^2 + \dot{q}^2 = 2h^0$ is the energy integral of the conservative linear system $\ddot{q} + kq = 0$, $k > 0$, and at the same time it is an ellipse, thus closed. The condition (4.1.18) holds along such an ellipse in an obvious way. Consider now the system

$$\ddot{q} + d(q,\dot{q}) + kq = u$$

and choose $u = d(q,\dot{q})$ for all $q,\dot{q}$ satisfying $kq^2 + \dot{q}^2 = 2h^0$. By the format of the latter, this balance is distorted for any point below and above the h^0-level generating $\dot{E} > 0$, $\dot{E} < 0$, respectively.

The question which reamins open is how many limit trajectories are enclosed below some E-level h_B, in a neighborhood of an equilibrium, or between two such levels, h_B' and h_B'', or elsewhere? Obviously, if the $\Delta_B' - \Delta_B''$ concerned is covered by a region of attraction to a single limit set, the problem is solved. In general, however, assuming a suitable controller, we need to investigate $\Delta_B' - \Delta_B''$ further to discover a steady state behavior of the trajectories and to assume them attractors from subsets of $\Delta_B' - \Delta_B''$. There are general methods for such an investigation on differentiable manifolds, see Meyer [1], Conley-Easton [1], Wilson Jr [1], Wilson Jr-Yorke [1], but they are not directly applicable to our type of study. Consequently our study must often be numerical only, in the absence of other methods. It is substantially facilitated by knowing some general properties of distribution of attractors and real-time attractors, their mutual relationship and the relations between the region in-the-large and the local regions. We shall comment on this below.

Let $\widetilde{M}$ and $\widetilde{M}'$ be two nonvoid disjoint attractors from Δ_0, Δ_0' respectively. Each is bounded, closed and connected. By uniqueness, irrespective of the controller used, no trajectory from $\bar{x}^0$ may be attracted to both $\widetilde{M}$ and $\widetilde{M}'$ which means that $\Delta_0 \cap \Delta_0' = \phi$. Conversely, since $\widetilde{M} \subset \Delta_0$ we have $\widetilde{M}' \subset \Delta_0'$, and the fact that $\Delta_0 \cap \Delta_0' = \phi$ yields $\widetilde{M} \cap \widetilde{M}' = \phi$. Hence, the result given below follows.

PROPERTY OF DISJOINTNESS I: Regions of attraction of any two connected attractors are disjoint if and only if these attractors are disjoint.

It follows that the pair Δ_0, Δ_0' is a partition of $\Delta_0 \cup \Delta_0'$ and thus given the same controller, trajectories attracted $\widetilde{M}$ and $\widetilde{M}'$ form classes defined by this property or *equivalence classes* with respect to attraction. This leads to the following.

PROPERTY OF IDENTIFICATION I: Given a control program, trajectories attracted to a distinct stable attractor form an equivalence class among those that start in the union of regions of attraction.

Let $\widetilde{M}$ and $\widetilde{M}'$ be uniformly stable, thus also uniformly asymptotically stable. This feature is essential for determining the state space pattern of attraction via numerical integration of the state equations. Since by Property of Identification I the trajectories attracted to $\widetilde{M}$ form equivalence class, it suffices to find one of them in order to determine all trajectories from the region of attraction Δ_{AT} concerned. The boundary $\partial\Delta_{AT}$ is found by the so called "retrograde" integration, that is, solving the state equations for $\tau = -t$ from initial conditions located at some Liapunov level (V - functional level) close to the attractor. The "retro-trajectory" obtained will, for $q \to \infty$ or $t \to -\infty$, coil up on some V-level which generates $\partial\Delta_{AT}$, see Chapter 5 for details.

To illustrate the property of identification and the above, let us consider the Van der Pol's equation given by

$$\left. \begin{array}{l} \dot{x}_1 = x_2 \\ \dot{x}_2 = -\epsilon(1 - x_1^2)x_2 - x_1 + u \end{array} \right\} \qquad (4.1.20)$$

where ϵ is a parameter. For the case $\epsilon = 2.5$, $u \equiv 0$, the origin $x = 0$ is an asymptotically stable equilibrium.

Figure 4.6 shows the evolution of the region of attraction to $\bar{x}(t) \equiv 0$ obtained by employing the iterative retrograde integration. The closed

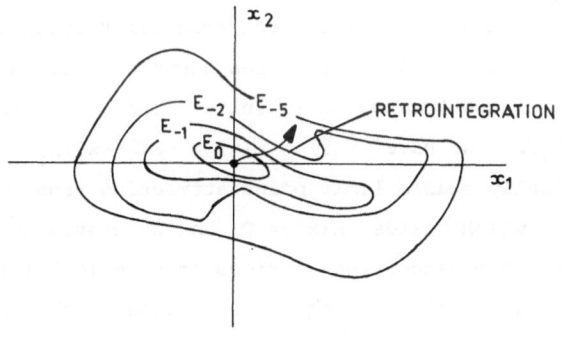

Fig. 4.6

E-levels E_{-1}, E_{-2} and E_{-5} are the retro-images of the initial E-level E_0 respectively obtained at time $t_1 = -1$, $t_2 = -2$, $t_5 = -5$. The curve E_{-5} coincides with the total region of attraction. Any points of E_{-5} can be mapped forward into E_0 after time t_5. Two such trajectories with initial conditions $\bar{x}^0 = (-1.0, 2.5)$ and $\bar{x}^0 = (1.8, 0.0)$ are shown in Fig. 4.7(a) with time history in Fig. 4.7(b). For more details of the method employed, the reader is directed to the monograph by Hsu [1] and later works by Flashner-Guttalu [1], Guttalu-Flashner [1] and Guttalu-Skowronski [1]. The method is based on cell-to-cell mapping ($E_0 \rightarrow E_{-5}$) introduced by Hsu [1] and on the Property of Identification I.

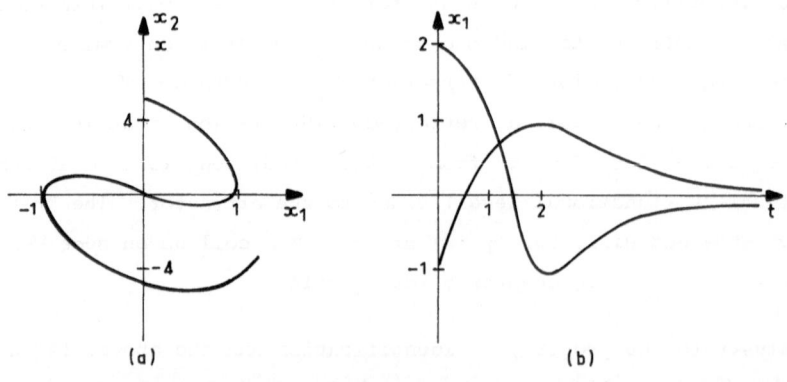

(a)   (b)

Fig. 4.7

Two closed sets are called *non-overlapping* if and only if their interiors do not intersect. When Property of Disjointness I holds, two closures $\bar{\Delta}_{AT}$, $\bar{\Delta}'_{AT}$ do not overlap. However, such sets may still partition their union, now closed, and the Property of Identification I holds. Then given a controller for each $\tilde{M}$, there is a set S called the *separatrix*: $S \overset{\Delta}{=} \partial \Delta_{AT} = \bar{\Delta}_{AT} - \Delta_{AT}$ fencing the class of trajectories identified. For the Van der Pol equation discussed, the unstable limit cycle is such a separatrix between the class of trajectories asymptotically attracted to the origin and the class attracted to infinity. Indeed, taking $V(\cdot)$ $= E(\bar{x}) = (x_2^2/2) + (x_1^2/2)$ we have $\dot{V}(\bar{x}) = \dot{E}(\bar{x}) = -\varepsilon(1 - x_1^2)x_2^2$ which makes $(0,0)$ an asymptotically stable limit point attracting from some neighborhood up to $x_1^2 = 1$, which yields $\dot{E}(\bar{x}) = 0$ and generates Λ^- producing the said separatrix. The damped separatrices of Example 1.1.1 are another good illustration of separating the equivalence classes of trajectories attracted to different attractors. Still another case may be shown in terms of the Dufting's equation already discussed. As above, we take the controller as given and substituted already, and adapt the simulation results of Guttalu-Flashner [1] in the following example.

EXAMPLE 4.1.4. Consider the following subcase of the Duffing equation:

$$\left.\begin{aligned} \dot{y}_1 &= y_2 \ , \\ \dot{y}_2 &= -dy_2 - ky_1 - y_1^3 + u \ , \end{aligned}\right\} \qquad (4.1.21)$$

where d, k and β are damping and elastic parameters. The equilibria are located at $(0,0)$ and $(\pm\sqrt{-k/\beta}, 0)$. By the transformation $y_2 = \sqrt{-k/\beta}\, x_1$ $y_2 = \sqrt{-k/\beta}\, x_2$, we have

$$\left.\begin{aligned} \dot{x}_1 &= x_2 \ , \\ \dot{x}_2 &= -dx_2 - kx_1(1 - x_1^2) \ , \end{aligned}\right\} \qquad (4.1.22)$$

whose equilibria are now at $(0,0)$ and $(\pm1,0)$. We choose the parameters $k = -2$, and $d = 3$, $u = 0$ for free Duffing's system for which $(0,0)$ is a saddle and both $(+1,0)$ and $(-1,0)$ are asymptotically stable equilibrium points. Figure 4.8 shows the regions of attraction of the two stable equilibrium points computed using the algorithm described before, following Flashner-Guttalu [1]. They are equal to regions of asymptotic stability. The manifolds of the separatrices at the origin divide the phase plane into two separate regions of attraction as shown in the figure which illustrates the Property of Disjointness I. It is easy to observe that the trajectories in each region of attraction are of the same character, confirming the Property of Identification I. □

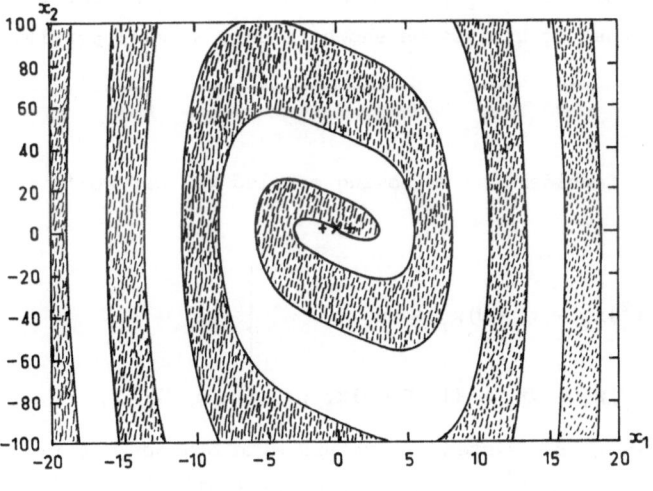

Fig. 4.8

183

In higher dimensional state spaces, defining the *separatrix surface* is not an unique or easy task, as it must be done solely on the grounds of the numerical procedure. Then the help of the Properties of Identification I and Disjointness becomes essential. We illustrate the case with two well known spatial examples, first the 3D Lorenz equation and then 4D double Van der Pol's equation.

EXAMPLE 4.1.5. The Lorenz equation is

$$\left. \begin{array}{l} \dot{y}_1 = \sigma(y_2 - y_1) \ , \\ \dot{y}_2 = ry_1 - y_2 - y_1 y_3 \ , \\ \dot{y}_3 = y_1 y_2 - by_3 \ , \end{array} \right\} \qquad (4.1.23)$$

where σ, r and b are positive constants which define the physical charac-teristics of the system, see Thompson-Stewart [1]. Some of these parameters may be considered given control variables. With the transformation $y_1 = \sqrt{(b(r-1))}x_1$, $y_2 = \sqrt{(b(r-1))}x_2$ and $y_3 = (r-1)x_3$, the above system becomes

$$\left. \begin{array}{l} \dot{x}_1 = \sigma(x_2 - x_1) \ , \\ \dot{x}_2 = rx_1 - x_2 - (r-1)x_1 x_3 \ , \\ \dot{x}_3 = b(x_1 x_2 - x_3) \ , \end{array} \right\} \qquad (4.1.24)$$

which has equilibria at $\bar{x}^{e1} = (0,0,0)$, $\bar{x}^{e2} = (1,1,1)$ and $\bar{x}^{e3} = (-1,-1,1)$. For the case $\sigma = 10$, $r = 2$ and $b = 8$, $\bar{x}^{e1}$ is unstable while $\bar{x}^{e2}$ and $\bar{x}^{e3}$ are both asymptotically stable. The two attractors $\bar{x}^{e2}$ and $\bar{x}^{e3}$ are disjoint. Through the Property of Disjointness I, we can separately calcu-late the regions of attraction of $\bar{x}^{e2}$ and $\bar{x}^{e3}$. However, by the Property of Identification I, it suffices to compute just one trajectory. Figures 4.9 show the projections of the region of attraction of $\bar{x}^{e2}$ onto the three principal planes $0x_1 x_2$, $0x_1 x_3$ and $0x_2 x_3$. For the sake of clarity, the subset of attraction shown is only a small portion of the entire region. □

EXAMPLE 4.1.6. Consider the following coupled Van der Pol's equations, see Hsu [1],

$$\left. \begin{array}{l} \dot{x}_1 = x_2 \ , \\ \dot{x}_2 = \mu(1 - x_1^2)x_2 - (1 + \nu)x_1 + \nu x_3 \ , \\ \dot{x}_3 = x_4 \ , \\ \dot{x}_4 = \mu(1 - x_3^2)x_4 + \nu x_1 - (1 + \eta + \nu)x_3 \ , \end{array} \right\} \qquad (4.1.25)$$

where μ , ν and η are system parameters with obvious meaning. Here, we treat a case for which the origin is asymptotically stable. For the values

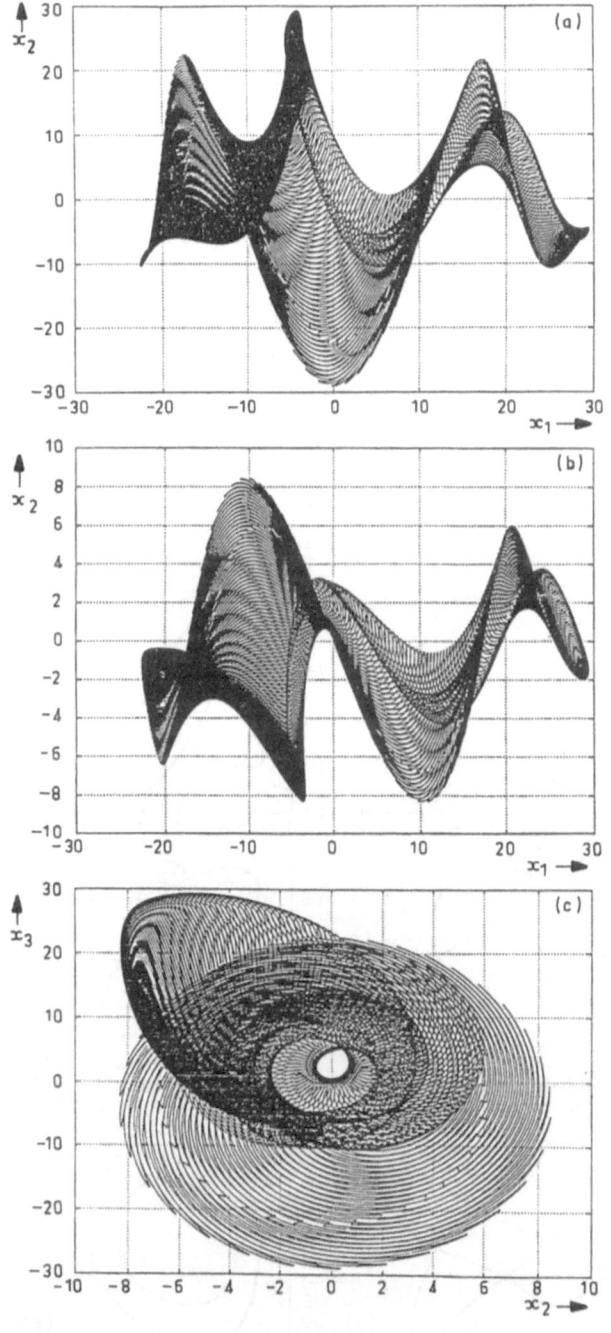

Fig. 4.9

of the parameters $\mu = -0.40$, $\eta = 0.2$, and $\nu = 0.1$, this system possesses an asymptotically stable attractor at $\bar{x}^e = 0$ and two unstable
periodic solutions (limit cycles) which bound the domain of attraction of
$\bar{x}^e = 0$. Figures 4.10 show the projections of the regions of attraction of
the origin on the $0x_1x_2$, $0x_1x_3$, $0x_1x_4$, $0x_3x_2$, $0x_2x_4$ and $0x_3x_4$
planes. The two unstable limit cycles are identified in this figure by the
symbols Λ_1^- and Λ_2^-. For the values of the parameters $\mu = 0.40$, $\eta = 0.2$
and $\nu = 0.1$, the local stability character of the three attractors are
reversed. The two asymptotically stable attractors (limit cycles) are
disjoint and their domains of attraction have been found to be disjoint.
The conceptual properties of disjointness and identification of attractors
have been quite useful in calculating this higher dimensional example.

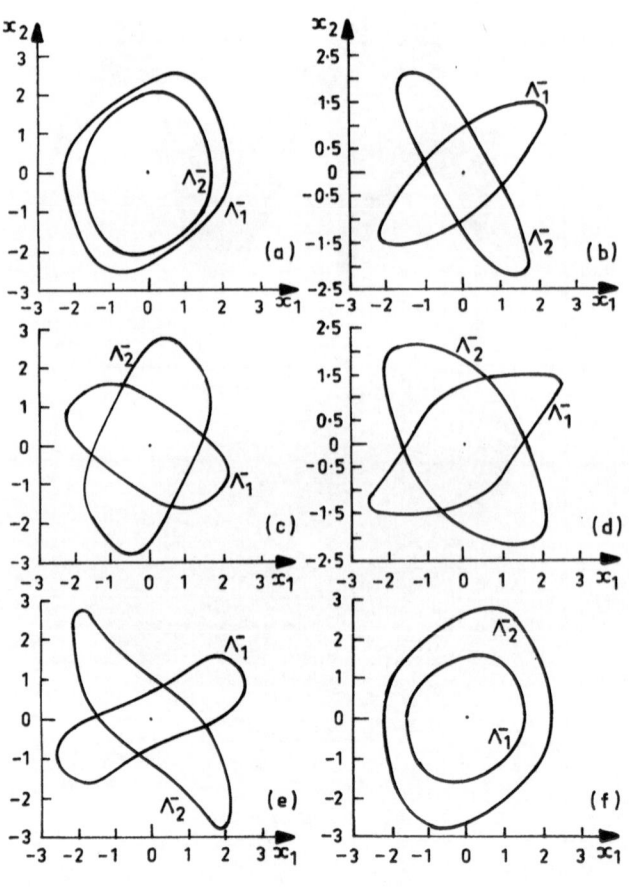

Fig. 4.10

We turn now to real-time attractors. We already stated that each such attractor will contain at least one positive limit set and thus it should contain at least one attractor. On the other hand, it is obvious that since all trajectories attracted to Δ_B in real time from some $\Delta_0 \subset \Delta_{BO}$ stay in Δ_B for $t \to \infty$, there are no attractors in $\Delta_0 - \Delta_B$. Consequently we have the following property.

PROPERTY OF EMBEDDING. It is always possible to find Δ_0 covering a real time attractor Δ_B which either encloses an attractor $\tilde{M}$ or is disjoint from it, in which case $\Delta_0 \cap \Delta_{AT} = \phi$.

We may illustrate the application of the above embedding with the case in Example 4.1.4, see Fig. 4.8. Here, the real time attractors may be drawn as an open target set about each of the attractors with $\Delta_{BO} = \Delta_{AT}$. Consequently $\partial\Delta_B$ coincides here with separatrices, except at the entrance to Δ_{AT}'s marked by arrows in Fig. 4.8.

From the Property of Embedding, we may also conclude that $\Delta_{BO} - \Delta_B$ contains no attractors. Hence either $\Delta_{BO} = \Delta_{AT}$ as in Example 4.1.4, or $\partial\Delta_{BO}$ separates attractors and the distribution of Δ_{BO} determines what may be used to locate attractors in Δ.

By the uniqueness of trajectories, irrespective of the controller involved, no trajectory from $\bar{x}^0$ may tend to both Δ_B and Δ_B', provided they are disjoint, see Fig. 4.11. Hence $\Delta_B \cap \Delta_B' = \phi$ implies $\Delta_{BO} \cap \Delta_{BO}' = \phi$. On the other hand, if $\Delta_B \subset \Delta_{BO}$, then $\Delta_{BO} \cap \Delta_{BO}' = \phi$ implies $\Delta_B \cap \Delta_B' = \phi$. Hence we have the following property.

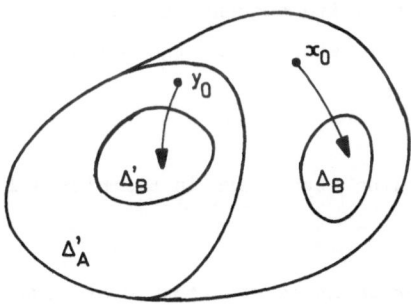

Fig. 4.11

PROPERTY OF DISJOINTNESS II. Regions of real time attraction are pairwise disjoint if the corresponding real time attractors, selected from a finite sequence, are disjoint. The converse holds if for at least one member of any pair Δ_B , Δ_B' we have $\Delta_B \subset \Delta_{BO}$.

Given two Δ_B , Δ_B' with their regions Δ_{BO} , Δ_{BO}' , there must be $T_M = \max (T_B , T_B')$ such that for $t > T_M$ all trajectories from $\Delta_{BO} \cup \Delta_{BO}'$ stay in $\Delta_B \cup \Delta_B'$. The argument obviously extends to any number of Δ_B's , whence we have the following property.

PROPERTY OF $\cup\Delta_B$. Given a control program, the union of a finite sequence of real time attractors attracts in real time from the union of the corresponding regions.

The above is well illustrated in Fig. 4.8. Now let the trajectories start at $\Delta_{BO} \cap \Delta_{BO}'$. By uniqueness, all of them must be real time attracted to $\Delta_B \cap \Delta_B'$ after $T_M = \min (T_B , T_B')$. Again this is true for any finite number of attractors and we have the following property.

PROPERTY OF $\cap\Delta_B$. Given a control program, the intersection of a finite sequence of real time attractors attracts in real time from the intersection of corresponding regions.

From the Property of Disjointness II, it follows that, given the finite sequence of disjoint real time attractors Δ_B^k , $k = 0,1,2,\cdots,\ell < \infty$, the sequence partitions the union $\cup_k \Delta_B^k$ and thus we obtain the following identification property.

PROPERTY OF IDENTIFICATION II. Given a control program, trajectories attracted to a member of a finite sequence of real time attractors form an equivalence class among those starting in the union of regions of real time attraction of the sequences.

The discussion regarding attractors in Examples 4.1.4 - 4.1.6 applies here directly.

Following the same argument as used for Identification I, we may prove Identification II for closures $\overline{\Delta}_{BO}$ and introduce the *real time* or *finite separatrix* $S_F \triangleq \overline{\Delta}_{BO} - \Delta_{BO}$.

Both attractors and real time attractors are final in the sense that the solutions cannot be attracted elsewhere. It is of some interest to study the relationship between *real time* and *transient attractors*. Two types of transient attractors may be discussed in terms of our topic. The type that forces the motions to enter Δ_B after they have left the transient attractor Δ_{BT} defined below, which steers or *conducts*, denoted Δ_{BP}, and the type which does not do that, for example, allows motions to drift everywhere including a bounce off some Δ_B after Δ_{BT}. We shall successively discuss these two types. A similar notion to conductor appears in Russian literature under the name of "bridge", see Krassovski and Subbotin [1].

We delete trivial cases by assuming that $\Delta_B \cap \Delta_{BP} \neq \phi$. Then by continuity of trajectories in $\bar{x}^0$, no part of the sets Δ_B, Δ_{BP} is empty. It follows that, given a control program, there must be at least one trajectory from Δ_{BO} that passes through Δ_{BP} before entering Δ_B, see Fig. 4.12(a). Hence there must also exist a nonempty set $\Delta_{BPO} \subset \Delta_{BO}$ such that trajectories from Δ_{BPO} and only those trajectories exhibit the above property. We shall then say that Δ_{BP} is a *conductor* from Δ_{BPO} to Δ_B, the set Δ_{BPO} being the *region of conduction*. Assuming $\Delta_{BP} \subset \Delta_{BO}$ we have $\Delta_{BP} \subset \Delta_{BPO}$ and conduction becomes a specified version of transient attraction, namely that which takes trajectories from Δ_{BT} into Δ_B.

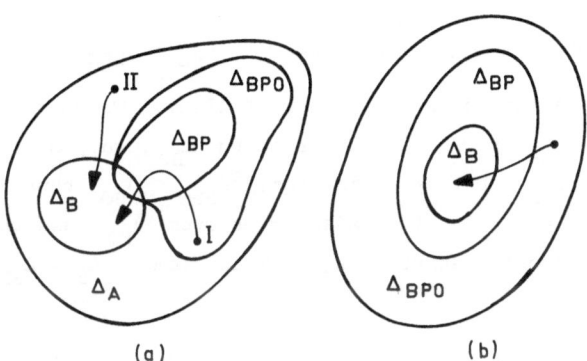

Fig. 4.12

The significance of conducting transient attractors becomes obvious when considering them as geometric loci of transient responses of the system to some external inputs which finally steer the system trajectories to a steady state limit enclosed in some Δ_B.

(a) $\Delta_{AP} \cap \Delta_B \subset \Delta_{BP}$

Otherwise, the definition of Δ_{BPO} is contradicted, that is, it would be possible to find trajectories from Δ_{BPO} which do not pass over Δ_{BP} (type II in Figure 4.12(a)).

(b) $\Delta_B \cap \Delta_{BP} \subset \Delta_{BPO}$

This follows from $\Delta_{BP} \subset \Delta_{AP}$ and the conductor property (a) above. It means that some trajectories from Δ_{BPO} start at Δ_{BP}. Then also

(c) $\Delta_B \subset \Delta_A \Rightarrow \Delta_{BP} \cap \Delta_{BO} \neq \phi$.

Indeed, we have $\Delta_B \cap \Delta_{BP} \neq \phi$ which, together with $\Delta_B \subset \Delta_{BO}$ yields the hypothesis by virtue of the definitions. This property means again that there is at least one trajectory entering Δ_B from Δ_{BP} thus yielding $\Delta_{BPO} \neq \phi$.

(d) Δ_{AP} is connected.

Indeed, should we have two disjoint members of Δ_{BPO} and the connected Δ_B, the trajectories would have to cross $\Delta_{BO} - \Delta_{BPO}$ on their way to Δ_B, which contradicts the definition of Δ_{BPO}.

(e)   Given Δ_B and Δ_{BP} , $\Delta_B \subset \Delta_{BP} \iff \Delta_{BO} = \Delta_{BPO}$.

This is a rather specific case, see Figure 4.12(b). There may not be trajectories entering Δ_B at t_B without crossing Δ_{BP}, say, of the type II. This yields $\Delta_{BO} \subset \Delta_{BP}$, but by definition, $\Delta_{BPO} \subset \Delta_{BO}$ thus $\Delta_{BO} = \Delta_{BPO}$. Conversely, if $\Delta_{BO} = \Delta_{BPO}$, all trajectories from Δ_{BO} pass through Δ_{BP} and only such trajectories start in Δ_{BO} , which means $\Delta_B \subset \Delta_{BP}$.

The conductor properties are easily seen in the examples quoted in this section already. They may seem trivial for the case of the 2D state space. However, they become useful in higher dimensions and when the control objective ceases to be just a single, well defined property, as it has been so far. The conductors and non-conducting transient attractors (leading to nowhere) will become important for objectives with conflict, that is, in our qualitative dynamic game to be discussed in Chapter 8. We leave our further discussion of transient attractors to that chapter.

EXERCISES 4.1

4.1.1 Find a positive limit set $\Lambda^+(\bar{x})$ of the system

$$\ddot{q} + (q^2 + \dot{q}^2 - 1)u + q = 0$$

with the program $u = \dot{q}$ and determine the region of attraction to such a Λ^+. How many periodic points are in this region? Which subsets of Δ are positively or negatively invariant and how do they relate to Λ^+?

4.1.2 Prove that no trajectory of a dynamical system in Δ reaches a critical point in finite time.

4.1.3 Prove that $\Lambda^+(\bar{x})$, $\bar{x} \in \mathbb{R}^N$ of a dynamical system in Δ is closed and invariant.

4.1.4 Show that the equation

$$\ddot{q} + u(q^2 + \dot{q}^2 - 1)\dot{q} + q^3 = 0 , \quad u > 0 ,$$

has at least one periodic motion. Discuss uniqueness of such a motion.

4.1.5 Return to Exercise 2.1.9 and consider the 1-periodic dynamical system (*) for which we have

$$f_i(\phi,\theta) = f_i(\phi+1,\theta) = f_i(\phi,\theta+1) , \quad i = 1,2 .$$

Show that the state trajectories consist of:
  (i)   the critical point $\{0\}$;
  (ii) the periodic orbit coinciding with the unit circle;
  (iii) spiralling trajectories through each point (r,θ) where $r \neq 0$, $r \neq 1$.
Then show that for the region $0 < r < 1$ the unit circle is Λ^+ while the origin is Λ^-. For $r > 0$ the unit circle is still Λ^+ while $\Lambda^- = \phi$.

4.1.6 Consider the system

$$\dot{x}_1 = x_1 - x_2 + (-4x_1^2 - 5x_2^2)x_1$$

$$\dot{x}_2 = x_1 + x_2 + (-3x_1^2 - 4x_2^2)x_2$$

with the test function $V = \frac{1}{2}(x_1^2 + x_2^2)$ and show that there are two circles centered at $(0,0)$, the outer $r = \sqrt{x_1^2 + x_2^2} = 1$ covered by entry points and the inner $r = 1/3$ covered by exit points for the trajectories, bounding the periodic orbit $r = 1/2$, which is the limit set Λ^+.

4.1.7 Study the trajectories of the system

$$\dot{x}_1 = (x_1 - x_2)u$$
$$\dot{x}_2 = (x_1 + x_2)u$$

with $V = \frac{1}{2}(x_1^2 + x_2^2)$ and show that under the controller $u = \frac{1}{4}x_1^2 + x_2^2 - 1$ there is a limit cycle Λ^+ defined by the ellipse $\frac{1}{4}x_1^2 + x_2^2 = 1$. How can the Λ^+ be controlled by changing the above specified control program?

4.1.8 Consider the forced Van der Pol system

$$\ddot{q} + u(q^2 - 1)\dot{q} + q = w \sin \omega t$$

with $q(t) \in \mathbb{R}$, $w \in [0,1]$, $|u| \leq \hat{u}$ and ω being the perturbation frequency.

(i) For the system without uncertain perturbation ($w \equiv 0$) find the controller ($u < 0$) which generates dissipation for $|q^0| < 1$ towards $\Lambda^+ = (0,0)$ and accumulation for $|q^0| > 1$, and the controller ($u > 0$) which produces the opposite accumulation for $|q^0| < 1$ from $\Lambda^- = (0,0)$ to Λ^+ located outside the circle of radius $\sqrt{3}$.

(ii) For the uncertainty at its upper bound $w \equiv 1$ find the controller ($u < 0$) generating the stable limit cycle Λ^+ with radius 2. Show that when ω is equal to the frequency of the unperturbed case $w \equiv 0$, the perturbation function has maximum effect.

(iii) For a given initial value $(q^0, \dot{q}^0)$ discuss two separate trajectories of (i) and (ii) respectively and the reachability cone achieved by them.

4.1.9 Discuss the possible responses of the system of Problem 4.1.4 with $w = 1$ and small values of $u < 0$.

4.2 VIBRATING PANEL UNDER DISSIPATIVE CONTROL

The following is a case study of the *global pattern of vibrations* of an elastic panel under a dissipative controller acting persistently, thus generating the dissipation inequality (2.4.20) along all trajectories from Δ_L. It makes the system *passive*. We want to treat this study as an illustrative example to our discussion in Sections 3.2, 3.3 and 4.1.

We consider a half-cylindrical resilient panel of infinite length, hinged at the ends along generatrices upon an absolutely rigid foundation, with the mass of a unit-length lumped at the center of gravity, that is, in the middle of the span, therefore reducing the model to the single-degree-of-freedom case. It suffices to study the motion of the cut through the mass along the span, Fig. 4.13. The panel admits finite (not small) amplitude oscillations about several of its possible equilibria.

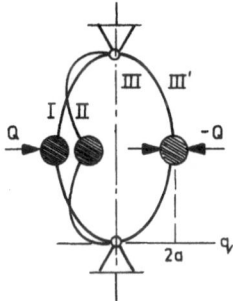

Fig. 4.13

The control program is dissipative and selected in such a way as to reduce the non-potential forces to linear positive damping, thus after substitution, resulting in $F(\bar{q},\dot{\bar{q}},\bar{u},\bar{w}) + D(\bar{q},\dot{\bar{q}},\bar{w}) \overset{\Delta}{=} 2\ell\dot{q}$, $\ell > 0$. The restoring characteristic is, in the general shape, $\Pi(q) = \alpha q + \beta q^3$, as in Section 2.3 with all the axioms holding almost everywhere in Δ_L. We shall specify it below, see Fig. 4.14. Thus the equation under study is generally

$$\ddot{q} + 2\ell\dot{q} + \alpha q + \beta q^3 = 0 , \tag{4.2.1}$$

with the initial conditions $(q^0)^2 + (\dot{q}^0)^2 \geq 0$, and sufficient required initial energy

$$\tfrac{1}{2} (\dot{q}^0)^2 + V(q^0) \geq \int_0^{q^0} \Pi(q)\,dq , \tag{4.2.2}$$

see Section 2.4.

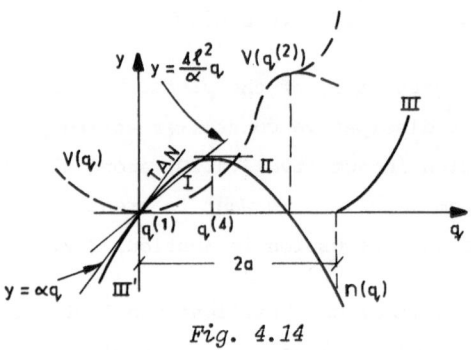

Fig. 4.14

The restoring characteristic represents a spring force in the panel and will have a few branches depending upon the specified ranges of deflections. The spring force values are those of the static load Q which executes the deflection. There are four various static equilibrium ranges (modes) specifying the mentioned ranges of deflection, see Fig. 4.14:

I : no inflection, buckling opposite to the direction of Q;

II : with one or several inflection points;

III, III' : no inflections, buckling in the direction of Q, $-Q$ respectively.

For the non-engineering reader, it may be instructive to go once again through the role of the static characteristic in this perhaps slightly more complicated example. Engineers may choose to ignore the remarks immediately below.

Suppose the panel is put on a static testing stand where we shall check its response to constant loads Q successively growing but maintaining each time the instantaneous balance between Q and the spring force $-\Pi$ in the panel, that is, maintaining the static equilibrium. Observe that for the sake of this balance, Q has the opposite sign to the spring force, thus the same sign as Π. Now glance at Fig. 4.14. Under increasing Q, the deflection q increases and these circumstances continue within I, that is, until some value of q, say $q^{(4)}$, where the stage II begins: the panel has bent in, and with q increasing, smaller and smaller Q is necessary to balance $-\Pi$. Eventually, for some value $q^{(2)}$ the required Q vanishes, the balance, if at all, is kept by the reaction in the supporting bearings, the equilibrium is Dirichlet unstable and any further increase of q requires $-Q$, that is $-(-\Pi)$, to keep the balance. Then $|-Q|$ grows rapidly with growing q and the system has a strong tendency to jump from the unstable equilibrium to a different structural form guaranteeing stability of the structure. Then the phase III begins, where Q grows safely again up to breaking the panel. The latter situation occurs also when, having started from $q = 0$, we proceed in the opposite direction, that is, for $q < 0$, range III', which requires $-Q$ at once.

It is rather amazing how many biological and social or political phenomena may be modelled on this very type of static characteristic, and thus later dynamically analysed along with our present example. Indeed, "overplaying one's hand" is proverbial in many languages, and there is a whole "catastrophe theory" along these lines developed not long ago. We hope the reader may also appreciate the importance of limiting the amplitude in oscillations about any of the mentioned static equilibria, in particular, about $q^{(2)} = 2a$ and what is even more essential, the importance of controllability either to $q = 0$ or to $q = q^{(2)}$ just as desired.

Having described the static test, we may now substantiate introduction of the following branches of Π.

  (A) $\Pi_A(q) = \alpha q + \beta_A q^3$, $q \in (-\infty, 0]$, $\alpha, \beta_A > 0$ due to III'.

  (B) $\Pi_B(q) = \alpha q + \beta_B q^3$, $q \in (0, 2a]$,   where a is the buckling
        distance and $\alpha > 0$, $\beta_B < 0$, comprising I and II.

  (C) $\Pi_C(q) = \alpha q + \beta_C q^3$, $q \in (2a, \infty)$, $\alpha, \beta_C > 0$, due to III.

(B) is rather the "eventful" branch, thus requiring special attention. Obviously (A) and (B) must be made to join smoothly:

$$\Pi_A(0) = \Pi_B(0) = 0 , \qquad \Pi_A'(0) = \Pi_B'(0) . \qquad\qquad (4.2.3)$$

In anticipation of II, Π_B must be made soft and start at 0 with the tangent below $y = \alpha_B q$. This requires $\Pi_B(q)/q > \Pi'(q)$, $q = 0^+$, which means $\beta_B/\alpha_B < 1$ as assumed. Let $-\beta_B = \beta' > 0$, then Π_B has two zero values at $q^{(1)} = 0$, $q^{(2)} = (\alpha/\beta')^{\frac{1}{2}}$ and the maximum $q^{(4)} = (\alpha/3\beta')^{\frac{1}{2}}$. The points $q^{(1)}, q^{(2)}$ are the extremal arguments of the potential energy

$$V_B(q) = \int \Pi_B(q)\,dq ,$$

and from $V_B''(q) = \Pi_B'(q) = \alpha - 3\beta'q^2$ it follows that $q^{(1)}$ is the stable and $q^{(2)}$ the unstable equilibrium (maximum if $\beta'/\alpha > 1/\sqrt{3}$, inflection if $\beta'/\alpha = 1/\sqrt{3}$) which agrees with our description of the static test, and justifies the choice of Π_B. Similar justification is readily obtainable for Π_A, Π_C.

It is instructive (and applicable in Synthesis) to discuss the influence of coefficients, in particular, of the "eventful" branch Π_B. To this aim, we rewrite

$$\Pi_B(q) = \alpha(q - \beta'/\alpha\, q^3)$$

in order to discuss the conclusive ratio $\beta'/\alpha > 0$, call it *softness* of Π.

Obviously, the growth of β'/α decreases $q^{(2)} = (\alpha/\beta')^{\frac{1}{2}}$, but particular stages of this "softening" may be of interest. Clearly, for $\beta' = 0$ (in fact, excluded), there is no $q^{(2)}$. For $0 < \beta'/\alpha < 1/4a^2$, $q^{(2)}$ is too far: $q^{(2)} > 2a$, which makes it lose its physical sense. For $\beta'/\alpha = 1/4a^2$ we have $q^{(2)} = 2a$ yielding continuity of Π with a corner (undefined derivative) at the point. Now if $\beta'/\alpha = 1/4a^2$ and still grows, $q^{(2)}$ tends to q^1 and $\Pi_B(q^4)$ tries to escape to infinity eo-ipso out of our model.

In the meantime, we have observed already that $\beta'/\alpha > 1/\sqrt{3}$ makes $V_B(q^{(2)}) = \alpha^2/4\beta'$ the maximum thus a distinct threshold, while $\beta'/\alpha = 1/\sqrt{3}$ makes it an inflection only, cf. Fig. 4.14. Observe moreover that

$$V_B(2a) - V_B(0) \gtreqless 0 \quad \text{for} \quad \beta'/\alpha \lesseqgtr 1/2a^2 . \tag{4.2.4}$$

We foreshadow that in order for the model to perform the prescribed role, the consistency with physical conditions requires the softness to remain $1/2a^2 > \beta'/\alpha > 1/4a^2$.

We may turn now to the state-plane pattern. The trajectory equation can also be written:

$$\frac{d\dot{q}}{dq} = -\frac{2\ell\dot{q} + \alpha q + \beta' q^3}{\dot{q}}, \quad q \leq 0, \quad q > 2a ; \tag{4.2.5}$$

$$\frac{d\dot{q}}{dq} = -\frac{2\ell\dot{q} + \alpha q - \beta' q^3}{\dot{q}}, \quad 0 < q \leq 2a , \tag{4.2.6}$$

where we have assumed $\beta' = \beta_A = \beta_C$ in order to simplify matters. First, we must define the reference system. The total energy is

$$H_{A,C}(q,\dot{q}) = \frac{1}{2}\dot{q}^2 + \frac{\alpha}{2}q^2 + \frac{\beta'}{4}q^4 , \tag{4.2.7}$$

$$H_B(q,\dot{q}) = \frac{1}{2}\dot{q}^2 + \frac{\alpha}{2}q^2 - \frac{\beta'}{4}q^4 . \tag{4.2.8}$$

Consequently, the topographic surfaces, now lines, are

$$E_C : \dot{q}^2 + \alpha q^2 \pm \frac{1}{2}\beta' q^4 = 2h_C = h_C' , \tag{4.2.9}$$

successively for (A), (C) and (B). Here again we shall base our analysis on (B). Indeed, the branches (A), (C) are hard, and thus generate no thresholds. Whatever the shape of $\tilde{H}$ over (A)-, (C)-arguments, to each $Z_C : E_B(q,\dot{q}) = h_C$ there will correspond some level over (A), (C) and the threshold of E_B determines the topography. Calculation of the threshold

is immediate

$$E_B(q^{(2)},0) = \frac{\alpha}{2}\left(\frac{\alpha}{\beta'}\right) - \frac{\beta'}{4}\left(\frac{\alpha}{\beta'}\right)^2 = \frac{\alpha^2}{4\beta'} .$$

We can now find $E_{CE} = \partial\Delta_E$ simply by picking up those E_C over (B) and (A) - (C) that correspond to the threshold

$$E_{CE} : \dot{q}^2 + \alpha q^2 \pm \frac{1}{2}\beta'q^4 = \frac{\alpha^2}{2\beta'} . \qquad (4.2.10)$$

The line was called the *separatrix*. It includes the plane separating set and $q^{(2)}$. In fact, for $\dot{q} \in (0,2a]$, the result is the same, as it might have been if defined by requiring some E_C to pass over $q^{(2)}$.

Incidentally, the local character of the conservative trajectories about $q^{(1)}$ and $q^{(2)}$ is obtainable via the known plane Poincaré analysis, cf. any quoted textbook. The result is that $q^{(1)}$ the so-called *saddle-point*, which is obvious from Fig. 4.15, as well as from our description of Δ_E in Chapter 2.

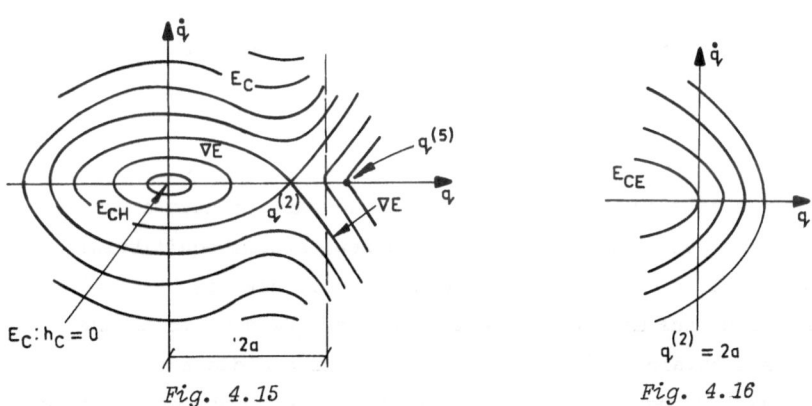

Fig. 4.15   Fig. 4.16

We turn our attention now to the region to the right of E_{CE}, $q > 2a$ in which the conservative trajectories cross the q-axis only once and do not cross the $\dot{q}$-axis. The picture is obtained owing to the alleged shape of H to the right of $q^{(2)}$. According to our theoretic description in Section 2.3, this shape depends upon the driving function $V(q)$. We have observed already that $V(q^{(2)})$ is either a maximum (if $\beta'/\alpha > 1/\sqrt{3}$) or an inflection (if $\beta'/\alpha = 1/\sqrt{3}$). In the first case, $E(q,\dot{q})$ decreases for $q > q^{(2)}$; in the second, it is monotonically increasing up to the boundary of Δ. In the latter case, the panel is simply too stiff to perform any jumps. The second minimum of V being unobtainable, thus the second cup degenerated, we have only one basic cup about $q^{(1)}$, and as the reader expects, the state-plane pattern is then local, thus of no interest at the moment. It is also inconsistent with the proverb "too far East is West" applicable before, for nothing really happens when we push too far - the panel just breaks.

Thus let us consider the first case with the maximum

$$V_B(q^{(2)}) = \frac{\alpha^2}{4\beta'} .$$

For $q > q^{(2)}$, $V(q)$ decreases and so does $E(q,\dot{q})$ making H fall down the hill from the saddle $E_B(q,\dot{q}) = \alpha^2/4\beta'$ along the well's river $V(q)$ through, till $E(q,\dot{q}) = 0$ and still further . . . to a physical nonsense, of an indefinite sink of energy, see Section 2.3. Fortunately, much sooner than that possibility, the system will respond by a jump to the phase III and the shape (C) of the characteristic, securing the endangered axiom (2.4.21), and, safeguarding the dissipativeness in-the-large which would not survive otherwise, see Section 3.3. Clearly, the least sensible value of energy is zero which determines a topographic line that cuts the q - axis at, say q^5; the latter from $0^2 + \alpha q^2 - \frac{1}{2} \beta' q^4 = 0$ which gives $q^{(5)} = (2\alpha/\beta')^{\frac{1}{2}}$. Thus in order not to lose all the energy before the jump, we must have $q^5 > 2a$, which immediately conditions either the buckling distance $a < (\alpha/2\beta')^{\frac{1}{2}}$ or the softness $\beta'/\alpha < 1/2a^2$. The latter agrees with the first condition of (4.2.4). Indeed, in order for the system to pass from vibrations about the unstable equilibrium over to vibrations about the new form of stable equilibrium (stabilize itself), it is obviously necessary to have a non-negative difference between potential energies at the new, stable equilibrium $q = 2a$ and the basic equilibrium $q^{(1)}$. The least we must have is $V(0) = V(2a)$ which happens to be the most convenient solution. We have to assume $\beta'/\alpha = 1/2a^2$ and obtain $q^{(5)} = 2a$. The jump is avoided but the aim is achieved; let us say the jump simply occurs under the right circumstances.

The reference system of this case is illustrated in Fig. 4.17. The point $q^{(5)} = 2a$ becomes the new stable equilibrium. The branch $\Pi_C(q)$ now in force for $q > 2a$ is hard and the topographic pattern does not differ from that discussed for $q < 0$. By the same argument as for (A), the threshold $V_B(q^{(2)})$ applies here and we simply expand the E_{CE} to this region. We obtain two cups on the same (basic) energy level.

The levels Z_C and the lines E_C everywhere are now those defined by E_B . Note that in this case we have left quite a freedom to the designer's synthesis of our panel, by requiring $\beta'/\alpha = 1/2a^2$. He may choose either to adjust the material in order to get the appropriate softness, or to adjust the construction to get a proper buckling. Finally, he may do both

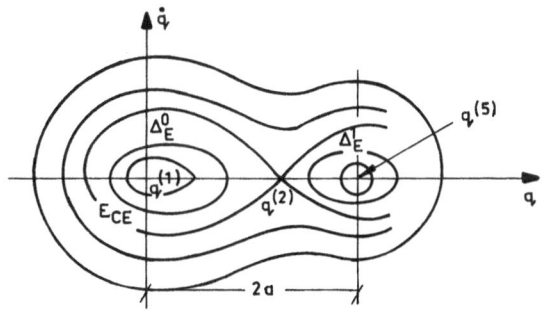

Fig. 4.17

adjustments, actually playing with them as one opposes the other. However, nothing is really discouraging in the fact that the above case may not be achieved. Indeed, the requirement $V(0) = V(2a)$ linked with $\beta'/\alpha = 1/2a^2$ was the least we had to have from the reasonable range $V(2a) \geq V(0)$ with $\beta'/\alpha < 1/2a^2$, cf. (4.2.4). Now, making the domains of (B) and (C) gradually more and more overlapping, that is, increasing the negative difference between $2a$ and $q^{(5)}$ by the decrease of $\beta'/\alpha \rightarrow 1/4a^2$, we move $q^{(2)} \rightarrow 2a$ and approach the second border-case allowed, namely $\beta'/\alpha = 1/4a^2$. We recall here our former argument that $\beta'/\alpha < 1/4a^2$ does not make sense physically.

For $q^{(2)} = 2a$ the restitutive characteristic is continuous, there is no jump, cf. Fig. 4.14, we have only one cup, cf. Fig. 4.16, and the study is local.

We may now confirm the selection of $\beta'/\alpha \in (1/4a^2 , 1/2a^2)$ for the sake of maintaining a reasonable "jump under control". In fact, it is quite clear that $1/4a^2 , 1/2a^2$ are the *bifurcative* values of the softness coefficient.

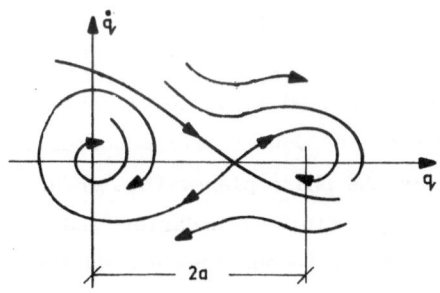

Fig. 4.18

It will be convenient for our later discussion to stick to the case of two stable equilibria with $q^{(5)} = 2a$ ($\beta'/\alpha = 1/2a^2$), described first, cf. Fig. 4.17. Let us discuss the damped oscillations in this case, with the non-local problems in attention. The damping is positive, thus

$$\dot{E}(q,\dot{q}) = -\ell\dot{q}^2 < 0 \quad \text{for} \quad \dot{q} \neq 0$$

since $\ell > 0$. It is an immediate conclusion that $q^{(1)}, q^{(5)}$ are asymptotically stable. As a rule, since $\dot{E}(q,\dot{q}) < 0$, $\dot{q} \neq 0$, no trajectory that had entered Δ_E^0, Δ_E^1 may leave these sets which are thus positively invariant with ultimate boundedness. Consequently, they are final attractors, each with an attractor enclosed. It would be a huge underestimation, however, to assume Δ_E^0, Δ_E^1 as the regions of attraction Δ_{AT}^0, Δ_{AT}^1 respectively. These reach much higher energy levels than the threshold

$$E(q^{(2)},0) = V(q^{(2)}) = \frac{\alpha^2}{4\beta'}$$

and thus include more than the set $\Delta_L = \Delta_E^0 \cup \Delta_E^1$ (Δ_L is defined by the highest threshold; in our case, there is only one threshold), see Fig. 4.19

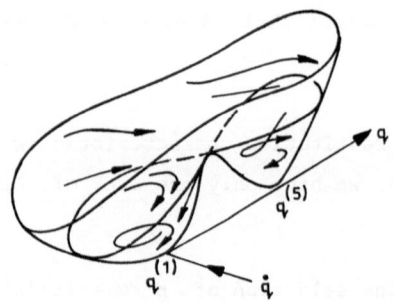

Fig. 4.19

From Section 4.1, we have that

$$\Delta_{AT}^0 \cap \Delta_{AT}^1 = \phi, \qquad \Delta_{AT}^0 \cup \Delta_{AT}^1 = \Delta,$$

that is, Δ_{AT}^0, Δ_{AT}^1 is a partition of Δ. By the Property of Identification I, it suffices to find one trajectory from Δ_{AT}^0 or Δ_{AT}^1 to describe the other. And we have been instructed to do so via the retrogression. On the other hand, we not only need to identify trajectories from Δ_{AT}^0 or Δ_{AT}^1, but also to locate them in the phase plane, thus we need to know the size of these sets, obtainable by defining boundaries, and that means estimating them by damped separating sets, in our case damped separatrices, cf. Fig 4.18. From Section 4.1, we know that they belong in part to the equivalence class of trajectories approaching the corresponding equilibrium, and from Section 2.3 that they enclose the unstable equilibria - we then retrace (negative time) a trajectory that happened to touch the unstable equilibrium, which in the plane case indeed separates Δ_{AT}^0 from Δ_{AT}^1, see Section 3.2. The question is, how can the retracing be done without solving the equation? We do it with the power series approximation. We move the origin to $q^{(2)}$ using $(q,\dot{q}) \to (\xi,\dot{q})$ such that $\xi = q - q^{(2)}$. Since $q^{(2)} = (\alpha/\beta')^{\frac{1}{2}}$, we have

$$\Pi_2(\xi) = \alpha\xi - \beta'\xi^3 ,$$

and

$$\frac{d\dot{q}}{d\xi} = \frac{-2\ell\dot{q} - \Pi_2(\xi)}{\dot{q}}$$

or

$$\dot{q}\left(\frac{d\dot{q}}{d\xi} + 2\ell\right) = -\Pi_2(\xi) .$$

Let $z = -\xi$,

$$-\dot{q}(z)\left(\frac{d\dot{q}}{dz} + 2\ell\right) = \alpha z - \beta'z^3 .$$

We develop $\dot{q}(\xi)$ at $\xi = 0$ into $\dot{q}(\xi) = a_1\xi + a_3\xi^3 + \cdots$ or $-\dot{q}(z) = -a_1 z - a_3 z^3 - \cdots$. On substitution,

$$(-a_1 z - a_3 z^3)(2\ell - a_1 - 3a_3 z^2) = \alpha z - \beta'z^3 .$$

Multiplying and comparing coefficients of like powers, we obtain

$$a_1 = \ell \pm (\ell^2 + \alpha)^{\frac{1}{2}} , \qquad a_3 = \frac{\beta'}{2\ell \pm 4(\ell^2 + \alpha)^{\frac{1}{2}}} ,$$

that is,

$$\dot{q}(\xi) = \ell \pm \xi(\ell^2 + \alpha)^{\frac{1}{2}} - \xi^3 \frac{\beta'}{2\ell \pm 4(\ell^2 + \alpha)^{\frac{1}{2}}} \qquad (4.2.11)$$

for an approximate equation of two branches of the damped separatrix in a neighborhood of $\xi = 0$, that is, $q^{(2)}$.

We want to distinguish some locating points for the damped separatrix, and its crossing the $\dot{q}$-axis may as well do the job. Substituting $\xi = -q^{(2)} = -(\alpha/\beta')^{\frac{1}{2}}$, we obtain

$$\dot{q}^{(2),(3)} = \pm\left(\frac{\alpha}{\beta'}\right)^{\frac{1}{2}}\left[\frac{\alpha}{2\ell \pm 4(\ell^2 + \alpha)^{\frac{1}{2}}} - (\ell \pm (\ell^2 + \alpha)^{\frac{1}{2}})\right] , \qquad (4.2.12)$$

see Fig. 4.18.

201

The conservative (frame-of-reference) counterpart of $\dot{q}^{(2)}, \dot{q}^{(3)}$ is obtainable by substituting $q = 0$ into (4.2.10). The result is

$$\dot{q}_{CE}(0) = \pm \left(\frac{1}{2} \frac{\alpha^2}{\beta^{\mathrm{r}}} \right)^{\frac{1}{2}} .$$

On the other hand, assuming $\ell = 0$ in (4.2.12), we obtain

$$\dot{q}_{\ell=0}^{(2)} = \left(\frac{3}{4} \frac{\alpha^2}{\beta^{\mathrm{r}}} \right)^{\frac{1}{2}} .$$

The difference between $\dot{q}_{CH}(0)$ and $\dot{q}_{\ell=0}^{(2)}$ shows the approximation error of (4.2.12), which is considerable indeed! We could have done better staging the return $(\xi, \dot{q}) \rightarrow (q, \dot{q})$ directly at (4.2.11) and then calculating $\dot{q}^{(2)}$. However, even then, there still would have been an error owing to:

(i) the fact that (4.2.11) holds locally about $q^{(2)}$; and

(ii) cut off terms of the series.

Our point is that there always will be some error. Its size does not matter as long as we do not make practical conclusions, so we may as well take the present one. Once there is an error, it is important to be on the safe side. Using (4.2.12) for Δ_{AT}^1 and $\dot{q}_{CE}(0)$ for Δ_{AT}^0, we are safe. Envisage Fig. 4.17 and Fig. 4.18 superposed.

For any other $\ell > 0$, $\dot{q}^{(2)}$ calculated from (4.2.12) is smaller than $\dot{q}_{\ell=0}^{(2)}$, it decreases with ℓ increasing but may not come below $\dot{q}_{CE}(0)$ for obvious reasons.

4.3 SELF-SUSTAINED MOTION. AIRCRAFT FLUTTER

Describing the energy-in-the-well for passive systems, we have let the minima of cups be open sinks for outflowing energy. Envisage now that they become, at least partially, sources. The falling down energy-state-trajectories somewhere on their way shall meet influxes which come from these minima open now to supply. The influxes may raise the energy levels as high as eventually allowed under the limited accumulation axiom, the limit measured by the amount of energy available, while the system remains autonomous. The system has the option of internal energy transfer through some mechanism which usually closes the energy in a periodic manner, with frequency corresponding to a natural frequency of the system. The influx of energy makes the basic equilibrium unstable. In the linear model of the system with a single equilibrium and no other steady states, we would have the amplitudes of motions growing indefinitely. Fortunately, the real structure is nonlinear, and the increase in amplitude is accompanied by the

increase in nonlinearity which usually damps the amplitude, leading to a
steady state regime. The latter, owing to the periodic energy supply, is
periodic. Real situations in which the above scenario occurs are those with
negative damping, see Section 2.3, frequently related to coupling between
a particular DOF, thus appearing in systems with at least two DOF. We refer
to the cases of negative damping in autonomous systems as to *self-accumulation*, as opposed to hetero-accumulation induced by external perturbations
in the nonautonomous systems.

The physical criterion for testing the self-accumulation in a free
system is the fact that it is absent at rest, see the axioms on negative
damping (2.4.12). Another sufficient feature is the presence of a periodic
limit trajectory, as described above. If the dynamics of the structure
generates such an option, a perturbation from the basic equilibrium could
move the system to a stable, possibly destructive high amplitude periodic
limit trajectory. A helicopter on the ground but autonomous, that is, not
subject to ground resonance, is a good example. A small perturbation
from the rest position will only cause a transient disturbance, but a large
perturbation, like a heavy wind gust, may cause a jump towards a stable
limit cycle, see Tongue-Flowers [1]. Other air or fluid induced self-
accumulation examples are aircraft flutter, discussed in this section, and
oil or steam whirl-whip auto-vibrations of rotating machines which we
discuss in Section 5.6, see Muszynska-Franklin-Bently [1].

The controller may take the role of either positive or negative damping, depending upon the objective. It is, however, more likely to be the
first case as the auto-accumulation is usually an undesired phenomenon in
mechanical structures. The action-scheme of the auto-accumulating system
is represented in Fig. 4.20, which is an extended part of the scheme in
Fig. 2.1. The elements: source or plant, could be linear but at least the
program, which is now solely state (feedback) dependent, must be nonlinear.
Obviously, all connections are multiple.

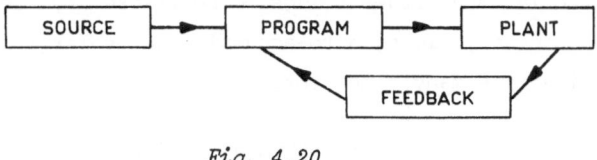

Fig. 4.20

We shall explain the mechanism of the negative damping in the typical example of dry friction between a weight and a surface moving on drums, mentioned already in Section 1.5.

Consider a mass m attached by a massless spring to a frame. The mass rests upon a horizontal rough surface which is gradually brought to move with the uniform speed v directed as shown in Fig. 4.21. Neglecting the initial disturbance, we might assume that the mass would take a position of equilibrium determined by a friction force D and the equal but opposite tensile force Π in the spring. Experiment shows, however, that this position is unstable and that horizontal vibrations of the mass will be built up. To explain the phenomenon, we must take into account that the Coulomb (dry) friction is not constant but diminishes slightly with the increase of the relative velocity. If, owing to some outside disturbance, vibrations of the mass have started already, the force D, always in the direction of the velocity v, will become larger when the mass moves in the direction of v and smaller when it moves in the opposite direction. In the first case, D produces positive work while in the second, the work is negative. Over a cycle of vibrations, a net work is positive, yielding an accumulation of energy.

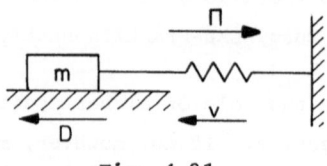

Fig. 4.21

Dry friction is responsible for swinging of the Froude-pendulum where the self-excitation phenomenon was first observed; it is responsible for crushes of steel cutting tools in machines and playing a violin, to quote physical examples. Identical models, however, carry us as far away as to the dynamic description of ... revolutions, in social science modelling. Another phenomenon producing auto-accumulation with a very similar explanation is that which generates the flutter of aeroplane wings, shimmy of car wheels, caused the Tacona-Narrows bridge to be wrecked by the wind, etc. The number of phenomena yielding auto-accumulation is large and the number of possible applications many times more so.

Returning to the first example seems as good as to any, thus we may describe briefly a double pendulum of W. Froude relying upon the works of

204

Bogusz [1] and Serebriakova [1]. Let us have two joined Froude's pendula
with equal masses $m = 1$ hanging from a shaft that rotates with a constant
speed ω. The dry friction of the shaft against the sleeve of the suspension causes the auto-accumulation precisely according to the rules described.
This accumulation may not only sustain the oscillations that started with
the initial perturbation, but actually start the pendula swinging or even
rotating – put them into periodic motion . . . it all depends upon ω. We
let the angles of swing of the pendula be the generalised coordinates
q_1, q_2, with the relative angular velocities $\omega - \dot{q}_1$, $\omega - \dot{q}_2$, $\dot{q}_2 - \dot{q}_1$.
Usually there will not be any coupling except frictional, so the restoring
forces may be represented by the functions

$$\Pi_1(q_1,q_2) = \sin q_1, \qquad \Pi_2(q_1,q_2) = \sin q_2,$$

and the positive viscous damping, including the controller, successively by
the functions $u\dot{q}_1$, $u\dot{q}_2$, $u \in R^+$, which are linear because the damping is
small. Further, we let the negative damping D_1^A, D_2^A be symmetric and with
only the coupling terms nonlinear, such that

$$D_1^A = -\ell_1^A(\omega - \dot{q}_1) - \ell_2^A\widetilde{D}(\dot{q}_2 - \dot{q}_1),$$
$$D_2^A = -\ell_1^A(\omega - \dot{q}_2) + \ell_2^A\widetilde{D}(\dot{q}_2 - \dot{q}_1),$$

where $\ell_1^A, \ell_2^A \in R^+$, and $\widetilde{D}(\cdot)$ is a smooth positive valued function. Then
the motion equations are

$$\begin{cases} \ddot{q}_1 + u\dot{q}_1 - \ell_1^A(\omega - \dot{q}_1) - \ell_2^A\widetilde{D}(\dot{q}_2 - \dot{q}_1) + \sin q_1 = 0, \\ \ddot{q}_2 + u\dot{q}_2 - \ell_1^A(\omega - \dot{q}_2) + \ell_2^A\widetilde{D}(\dot{q}_2 - \dot{q}_1) + \sin q_2 = 0. \end{cases} \qquad (4.3.1)$$

If each of the pendula carries out a circular motion, we may say that
the system has a solution which is periodic in q_1 and q_2, thus in $\bar{q}$.
The control conditions for such a case are derived in Wazewski [5]:

$$\omega \geq \frac{2(u + \ell_1^A)}{\ell_1^A}[\pi(u + \ell_1^A) + 2] + \ell_2^A/\ell_1^A.$$

The role of control in balancing the negative damping is quite visible from
the above, and the choice of u is essential to design.

This particularly refers to auto-synchronism which is a phenomenon
occurring often together with auto-accumulation, in fact, as a part of it.
The phenomenon is now quite successfully employed in smoothing and silencing
almost everywhere - eccentric vibrators employed in sieves, mills, conveyors,
electronic equipment and all types of amplifiers ought to work synchronously.
It is often autosynchronization which silences your home airconditioner or
your washing machine. On the other hand, an uncontrolled autosynchronism

can produce unexpected vibrations with sizeable amplitude in various con-
structural systems and thus present the designer or user with a dangerous
surprise.

The discovery of autosynchronism goes back to Huygens (1629-1695) who
observed that two clocks, slightly out of synchronism when fixed on a wall,
tend to run at the same speed when hung on a thin wooden board. A similar
observation in acoustics was made by Rayleigh who experimented on two
organ pipes of slightly different frequencies coupled through a resonator.
More than two centuries elapsed since Huygens, before synchronism was
discovered in electric circuits by E.V. Appleton (1922). Van der Pol (1927)
followed with the more detailed study of the phenomenon in terms of the
normal form of differential equations. The present literature is quite
vast as the phenomenon is one of the best known among other nonlinear
phenomena. Minorsky [2], of all researchers so far, made the most significant contribution to the topic.

To illustrate the self-sustained vibrations case, we shall take the
example of an aircraft wing flutter. We begin, following Kauderer [1],
with a reduced case represented by a single (resultant) DOF laboratory
model of the noncircular (rectangular) plate shown in Fig. 4.22 between
linear springs, with the resultant elasticity coefficient k and subject
to an airflow with the controlled velocity v_0 . The plate moves vertically,
with the deflection q(t) measured from its equilibrium position obtained
by the springs in unstressed mode, and with the velocity $v(t)$. Then the
velocity of the air flow relative to the motion is $v = v_0 + v$. It generates the aerodynamic forces of *lift* $L = \frac{1}{2} \rho S v^2 C_L(\alpha)$, and *drag*
$D = \frac{1}{2} \rho S v^2 C_D(\alpha)$ marked in Fig. 4.22, with the air density ρ , area of the

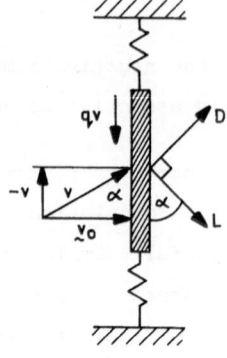

Fig. 4.22

plate S and $C_L(\alpha)$, $C_D(\alpha)$ being lift and drag coefficients, usually smooth so that a power series can be developed, and for small angles of attack α reduced to linear form. In most of the practical cases, the aerodynamic forces are given experimentally.

From Fig. 4.22 it is seen that

$$v = v_0\sqrt{1 + (v/v_0)^2}, \quad v/v_0 = \cotan\alpha .$$

The forces mapped on the line of motion become

$$-kq + \tfrac{1}{2}\rho Sv_0^2 \sqrt{1 + (v/v_0)^2} [C_L - (v/v_0)C_D]$$

whence the motion equation is

$$mv\frac{dv}{dq} = -kq + \tfrac{1}{2}\rho Sv_0^2 \sqrt{1 + (v/v_0)^2} [C_L - (v/v_0)C_D] \qquad (4.3.2)$$

where m is the mass of the plate. Denoting $k/m \overset{\Delta}{=} \omega^2$, $(\rho Sv_0^2/2m) \overset{\Delta}{=} d$ to obtain the characteristics, we have

$$v\frac{dv}{dq} - d\sqrt{1 + (v/v_0)^2} [C_L - (v/v_0)C_D] - kq = 0 . \qquad (4.3.3)$$

Observe that the damping characteristic depends only on the velocity $v = \dot{q}$, and is partially negative, that is, it can generate accumulation of energy for some values of v. The characteristic is usually given by experimental data wherefrom we know that the values concerned appear in the interval $-0.45v_0 \le v \le 0.45v_0$. We conclude that the origin is unstable and thus we may expect a periodic limit cycle about it. Such a Λ^+ must intersect the lines $v = \pm 0.45v_0$ which bound the conservative region H^- above from the region of accumulation H^+ below the cycle, see Section 4.1. Owing to the fact that the characteristics are odd functions, the regions H^+, H^- should be symmetric with respect to the origin.

The realistic flutter case works on a similar principle to that used for the above experimental stand, but must have at least two DOF to produce the elastic and inertial coupling which constitute the wind-motor generating the auto-accumulation. We usually see the wing as a hinged-free beam, hinged at the fuselage side and allowed the two DOF required, namely lateral vibration of the stiffness axis and torsional vibrations about that axis, see Fig. 4.23.

At steady state motion, which is of interest, the airplane proceeds with a constant subsonic speed v, which can be controlled, with a given angle of attack α and a given wing setting angle θ. Owing to the symmetry of the profile and other less significant factors, any initial

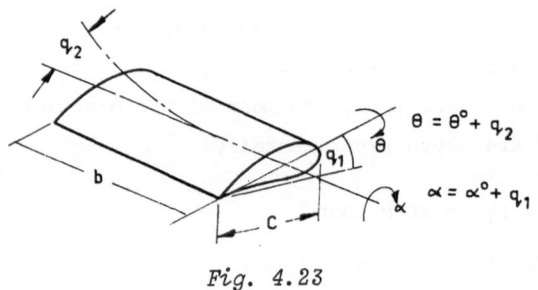

Fig. 4.23

torsional disturbance $\theta^0, \alpha^0 \neq 0$ from an equilibrium may produce the said vibrations which in specific circumstances could have an auto-accumulative character.

The corresponding block scheme in Fig. 4.24 is a specified case of that in Fig. 4.20. The plant, which is now the wing, is presented in terms of its two DOF with the aerodynamic coupling always present and the inertial coupling often appearing.

Similarly as in the previous laboratory experiment, the source is the wind flowing about the profile and the control program is determined by the aerodynamic actions, strongly nonlinear functions of the angle of attack and speed. The latter is considered the control variable, entering the program implicitly. Indeed, given a specific flight path in $Oxyz$ and with a certain amount of positive damping supplied by the structure itself (following suitable design of characteristics), the control of auto-accumulation (that is, flutter) will depend upon the speed $v = const$ kept during the flight, away from some critical value that may force the system trajectories out of an energy cup where the damping provides enough shelter against auto-accumulation. For details, we refer the reader to Den Hartog [1], Bednarz-Giergiel [1] and Skowronski [32].

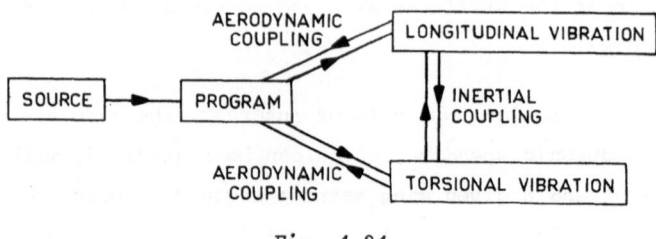

Fig. 4.24

4.4 NONAUTONOMOUS STEADY STATES

Let us now make a few comments on how the objectives of Section 4.1 can be augmented to the case of nonautonomous systems (2.2.6)', and thus its selector equation (2.2.5)', which must be used when the structure is subject to external perturbations $\bar{R}(\bar{x},t,w)$, related to time explicitly.

The main difference between the trajectories of autonomous systems considered before and the motions of the nonautonomous systems refers to initial conditions, namely the fact that the motions start from $(\bar{x}^0,t_0)$ $\epsilon \ \Delta \times \mathbb{R}$ and are unique only in this set, while the trajectories have the same uniqueness property in Δ , that is, starting from $\bar{x}^0 \ \epsilon \ \Delta$ do not cross for arbitrary $t_0 \ \epsilon \ \mathbb{R}$. However, as long as various notions, specifying objectives for nonautonomous systems in indefinite time, can be considered t_0- uniform, the geometric extension of the corresponding sets from Δ to $\Delta \times \mathbb{R}$ remains cylindrical and thus these notions may still be discussed in Δ . Indeed, let $\bar{\phi}(\bar{x}^0,\mathbb{R})$ be the t_0- family of motions $\bar{\phi}(\bar{x}^0,t_0,\mathbb{R})$ projected from $\Delta \times \mathbb{R}$ into Δ, and let $\Delta_c \subset \Delta$ be any set to be intersected at an arbitrary $t \ \epsilon \ \mathbb{R}$. We have $[\bar{\phi}(\bar{x}^0,\mathbb{R}) \times \mathbb{R}] \cap (\Delta_c \times \mathbb{R})$ $= [\bar{\phi}(\bar{x}^0,\mathbb{R}) \cap \Delta_c] \times \mathbb{R}$ which makes $\bar{\phi}(\bar{x}^0,\mathbb{R}) \cap \Delta_c \neq \phi$ necessary and sufficient for such intersection in $\Delta \times \mathbb{R}$.

We have already defined in a similar way the equilibria of (2.2.5)' and its steady state sets. In the same manner, definition (4.1.1) of the limit set holds for (2.2.5)' together with the conclusions which follow, including that of the closure of a trajectory - this time a motion:

$$\Lambda^+ \ = \ \cap_{s<\infty} \overline{\cup_{s\leq t} \bar{\phi}(\bar{x}^0,t_0,t)} \ . \tag{4.4.1}$$

We may illustrate the case with Example 4.1.1. Suppose the variable θ is not measured in angular units but in time, then Fig. 4.2 may be redrawn as shown in Fig. 4.25.

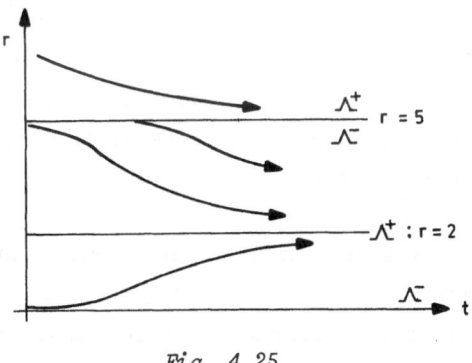

Fig. 4.25

The E-levels h_B', h_B'' have already been discussed for a nonautonomous system in Section 3.3, together with the corresponding dissipative and accumulative controllers. They form cylindrical surfaces in $\Delta \times \mathbb{R}$. The invariant and minimal invariant sets $\tilde{M}$ now become such sets M of Section 3.3. By analogy we also have the maximal invariant set $M \triangleq \hat{M} \times \mathbb{R}$, with Proposition 4.1.1 applicable without changing the choice of $V(\cdot)$, since our $E^{\pm}(\bar{x})$ do not depend explicitly upon t, and both the Proposition and Corollary 4.1.1 have been considered for objectives which are t_0-uniform. Obviously the control conditions will now be specified in terms of $f_0(\bar{x}, \bar{u}, \bar{w}, t)$ instead of $f_0(\bar{x}, \bar{u}, \bar{w})$, which is as well, if we use the controllers of Section 3.3.

For autonomous systems we did not have any general criterion for periodicity of the limit trajectories in higher dimensional spaces. The advantage of introducing a periodic perturbation $\bar{R}(\bar{x}, \bar{w}, t)$ is that the resultant steady state is also periodic, at least after some transient interval of time, see Yoshizawa [1]. Moreover, periodic perturbation allows us to treat (2.2.5)' as a dynamical system in Δ, see Skowronski [32], [45], which means the methods discussed for the autonomous case are directly applicable. This is also significant for numerical simulation allowing the use of the cell-to-cell mapping in retro-integration, applied before. For periodic systems it is a period-to-period map. Regions of attraction and real-time-attraction are now to be interpreted as those corresponding to periodic points of the dynamical system concerned. We illustrate the case with the following example adapted from Guttalu-Flashner [1].

EXAMPLE 4.4.1. Consider the forced Duffing's equation

$$\left. \begin{array}{l} \dot{x}_1 = x_2 \, , \\[2mm] \dot{x}_2 = -ux_2 - kx_1 - \beta x_1^3 + a\cos\omega t \, , \end{array} \right\} \tag{4.4.2}$$

where $u = \text{const} = 0.2$, eigenfrequency $k = 0$, hard-spring coefficient $\beta = 1$, forcing amplitude $a = 0.3$ and perturbation frequency $\omega = 1$. The system possesses two disjoint asymptotically stable attractors (harmonic) marked by a solid dot and "+" in Fig. 4.26, and unstable motion, marked "×", all with period 2π. ☐

Referring now to the global study of distributing attractors, if $M = M(t) \times \mathbb{R}$ is a stable attractor from $\Delta_{AT} \triangleq \Delta_{AT}(t) \times \mathbb{R}$, for each t we can prove the Property of Disjointness I, while the Property of Identifica-

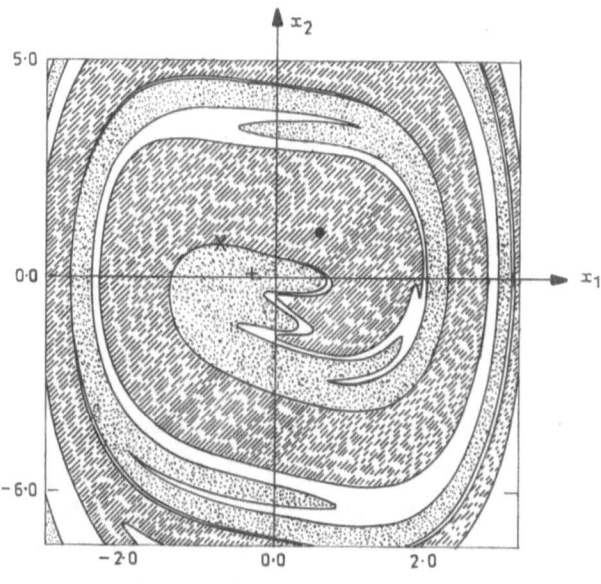

Fig. 4.26

tion I holds for all $t \geq t_0$. On the other hand, the finite time pattern
of real time attractors requires attention, as we may not automatically
assume that the study in Δ suffices.

Define now $\mathbb{R}_B(t_0) \triangleq \{t \in \mathbb{R} \mid t \geq t_B = t_0 + T_B\}$, $t_0 \in \mathbb{R}$. Then $\Delta_B \times \mathbb{R}$
is a real time attractor from $\Delta_{BO} \times \mathbb{R}$ in $\Delta \times \mathbb{R}$. Note that $\Delta_B \times \mathbb{R}_B$
does not have to be a subset of $\Delta_{BO} \times \mathbb{R}$, they may even be disjoint, as
long as $(\bar{x}^0, t_0) \in \Delta_{BO} \times \mathbb{R}$ implies $\bar{\phi}(\bar{x}^0, t_0, \mathbb{R}_B) \subset \Delta_B \times \mathbb{R}$. For any t_0 ,
$\Delta_B \times \mathbb{R}_B$ and $\Delta_B' \times \mathbb{R}_B'$ yield

$$(\Delta_B \times \mathbb{R}_B) \cap (\Delta_B' \times \mathbb{R}_B') = (\Delta_B \cap \Delta_B') \times (\mathbb{R}_B \cap \mathbb{R}_B')$$

from which $(\Delta_B \times \mathbb{R}_B) \cap (\Delta_B' \times \mathbb{R}_B') = \phi$ if and only if $\Delta_B \cap \Delta_B' = \phi$ or
$\mathbb{R}_B \cap \mathbb{R}_B' = \phi$. Following the definition of real time attraction we assume
$\mathbb{R}_B \cap \mathbb{R}_B' \neq \phi$ ($t_{B'} \geq t_B \Rightarrow \mathbb{R}_B \cap \mathbb{R}_B' = \mathbb{R}_{B'}$ and $t_B \geq t_{B'} \Rightarrow \mathbb{R}_{B'} \cap \mathbb{R}_B = \mathbb{R}_B$).
Thus disjointness of Δ_B , Δ_B' decides the disjointness of $\Delta_B \times \mathbb{R}_B$ and
$\Delta_B' \times \mathbb{R}_B'$. Since the attraction concerned is t_0- uniform, the above applies
to $\Delta_B \times \mathbb{R}$, $\Delta_B' \times \mathbb{R}$ as well. Correspondingly, we have $(\Delta_{BO} \times \mathbb{R})$
$\cap (\Delta_{BO}' \times \mathbb{R}) = (\Delta_{BO} \cap \Delta_{BO}') \times \mathbb{R}$, and we may use the Property of Disjointness
II *in Δ equivalently for* $\Delta \times \mathbb{R}$. Using the same argument, we may show the
equivalent use of the Property of Embedding and the Properties of $\cup \Delta_B$, $\cap \Delta_B$
as well as the Property of Identification II.

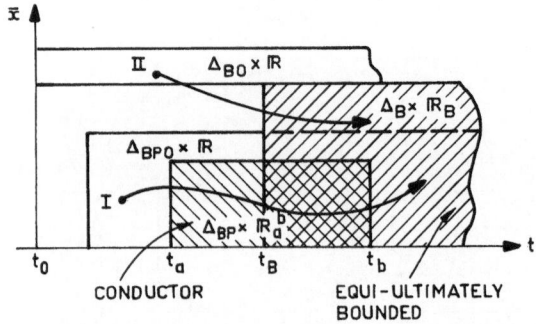

Fig. 4.27

Turning now to nonautonomous conductors, given $t_a, t_b \in \mathbb{R}$, $t_a \leq t_b$, let $[t_a, t_b] \triangleq \mathbb{R}_a^b$, and consider two sets $\Delta_B \times \mathbb{R}$, $\Delta_{BP} \times \mathbb{R}_a^b$. We observe here a different feature, namely $(\Delta_B \times \mathbb{R}_B) \cap (\Delta_{BP} \times \mathbb{R}_a^b) = (\Delta_B \cap \Delta_{BP}) \times (\mathbb{R}_B \cap \mathbb{R}_a^b)$. It follows that, given t_0, $(\Delta_B \times \mathbb{R}_B) \cap (\Delta_{BP} \times \mathbb{R}_a^b) \neq \phi$ if and only if both $\Delta_B \cap \Delta_B \neq \phi$ and $\mathbb{R}_B \cap \mathbb{R}_a^b \neq \phi$. Hence the latter *two conditions together* are equivalent to conducting by $\Delta_{BP} \times \mathbb{R}_a^b$, see Fig. 4.27.

Observe that $\mathbb{R}_B$ depends upon t_0 and $\cup_{t_0} \mathbb{R}_B(t_0) = \mathbb{R}$ as $t_0 \in \mathbb{R}$. Hence there is always a t_0 such that for the motion generated, we have $\mathbb{R}_B \cap \mathbb{R}_a^b \neq \phi$, and again the study in Δ suffices, so that the Properties of Conductors apply in the nonautonomous case.

Chapter 5
COLLISION AND CAPTURE

5.1 COLLISION RELATED OBJECTIVES

Objective properties Q related to various ways of crossing a target
are classical but still basic, at least for systems which are already
stabilized. Ever since its inception about 1960 when the first work
appeared (see Kalman-Bertram [1], Kalman [1]), the notion of controllability
has been connected with reaching a target. Initially, the problem was posed
in a rather simplified and applicationally rigid form of attempting a transfer between two given states $\bar{x}^0$ and $\bar{x}(t_f) \in \Delta$ under specified control.

It later developed into the transfer between sets of initial conditions
and specified target sets in Δ, then an optimal transfer, and as such had
been the subject for classical, linear control theory for many decades,
starting from Kalman's presentation at the I IFAC Congress in 1960, see
Barnett [1], through a wide literature continuing up to the present, too
vast to be quoted.

Leaving aside also the linearized models (perburbed linear), mainly by
Aronsson [1], Dauer [1]-[4], Lukas [1],[2], Davison-Silverman-Varayia [1],
Lobry [1], Mirza-Womack [1], Nguen [1], Klamka [1],[2], and a few others,
we refer to the results without an attempt to linearize, like Markus [1],
Cheprasov [1], Gershwin-Jacobson [1], Liu-Leake [1], Heinen-Wu [1], and later
the Leitmann school, Sticht-Vincent-Schultz [1], Grantham-Vincent [1],
Vincent-Skowronski [1], Stonier [1], Skowronski-Vincent [1], Skowronski
[29],[38],[44],[45].

Without underestimating the mentioned objective of transfer between
some initial states and a target, the present demands on control theory

require much more. The conventional concepts of collision (meeting a target) or avoidance (not meeting it at all) must be looked at in a more flexible manner. Say, collision with or without penetration of the target, with or without permanent capture in such a target, perhaps with rendezvous only, and if so, before or after a stipulated time or ultimate capture after stipulated time. There is obviously another set of such objectives for avoidance, see Chapter 7. The objectives mentioned are *modular*, that is, one may be able to form real complex objectives using them as components. Let us start with the most basic concept, namely collision.

It will be assumed that the particular scenario in Cartesian space *Oxyz* has been transformed to the state space variables, see Section 1.6, and thus we deal with the state space targets $T = T_q \times T_{\dot{q}} \subset \Delta$. Let T be a given set in Δ which is bounded and closed (that is, compact) but not necessarily connected. It might be made to move in time, and then considered $T(t) \times \mathbb{R}$ in $\Delta \times \mathbb{R}$, or made stationary, back as T , by redefining the state into relative variables measuring the instantaneous distance $\rho(\bar{x}(t), T(t))$, $t \geq t_0$. In the latter sense taking a stationary target serves the moving target scenarios as well.

Modelling in relative coordinates somehow obscures the physical meaning of the state representation but has a distinctive advantage of reducing the dimensions of the state space and thus making the computation faster. The latter is of particular importance for on-line computation on small and thus slow computers on board vehicles, for robots working in difficult conditions (space, under water), etc., where fast computation may be vital. Moreover, the relative coordinates approach may be necessary anyway, say, discussing coordination of several mechanical structures like aircraft or spacecraft keeping formation, coordination of robotic arms, etc., in general for all tracking problems.

For directly chosen state variables, the target may also mean either a curve or a trajectory of a system in Δ as well as a family of such curves or trajectories. We shall then talk of stipulated *path tracking* or *model reference following*. This can be seen as being alternative to the relative variables method of control.

The classical notion of *collision* is understood as at least an instantaneous contact of a motion of (2.2.6)' with T . More formally, we express it by the following definition.

DEFINITION 5.1.1. Given T and $(\bar{x}^0, t_0) \in \Delta \times \mathbb{R}$, a motion $\bar{\phi}(\bar{x}^0, t_0, \mathbb{R}^+)$ of (2.2.6)' collides with T if and only if

$$\bar{\phi}(\bar{x}^0, t_0, \mathbb{R}^+) \cap T \neq \phi . \tag{5.1.1}$$

This means that there is some, generally unspecified, constant $T_c < \infty$ such that for $t_c = t_0 + T_c$ we have

$$\bar{\phi}(\bar{x}^0, t_0, t_c) = \bar{x}^c \in T . \tag{5.1.2}$$

When T_c is stipulated a-priori, we say that the motion collides *at stipulated time*.

Considering the collision of Definition 5.1.1 or its stipulated time version to be the objective property Q discussed in Section 3.1, we have the controllability and strong controllability for such collisions specified by Definitions 3.1.4 and 3.1.5, respectively, together with the notion of regions of such controllabilities. Let the region of controllability be denoted by Δ_c and of strong controllability by Δ_C. Note that the relations (3.1.14) - (3.1.16) hold for collision as well. The regions corresponding to collision at stipulated time will be denoted $\Delta_c(T_c), \Delta_C(T_c)$.

Let us make here a blanket assumption for all the future text that *any newly defined modular subobjective property will generate controllabilities specified by Definitions 3.1.4, 3.1.5 with all their following notions.* To save space, this statement will not be repeated each time we introduce such subobjectives, unless we would like to indicate exceptions to the rule. Obviously each time the regions will have different subscripts related to the specified Q and thus notation will be given, as well as any other properties related specifically to the case concerned. For instance, for collision it must be obviously stated that, by definition, the target belongs to all the regions trivially: $T \subset \Delta_C \subset \Delta_c$, $T \subset \Delta_C(T_c) \subset \Delta_c(T_c)$.

In practical situations, we often come across the demand of collision *before or after stipulated time*, rather than at the given time instant, the latter being difficult to achieve.

DEFINITION 5.1.2. Given T, $T_c^+ < \infty$ and $(\bar{x}^0, t_0) \in \Delta \times \mathbb{R}$, a motion $\bar{\phi}(\bar{x}^0, t_0, \mathbb{R}^+)$ of (2.2.6)' *collides* with T *before the time* T_c^+ if for some $t' \leq t_c = t_0 + T_c^+$ we have

$$\bar{\phi}(\bar{x}^0, t_0, t') = \bar{x}^c \in T . \tag{5.1.3}$$

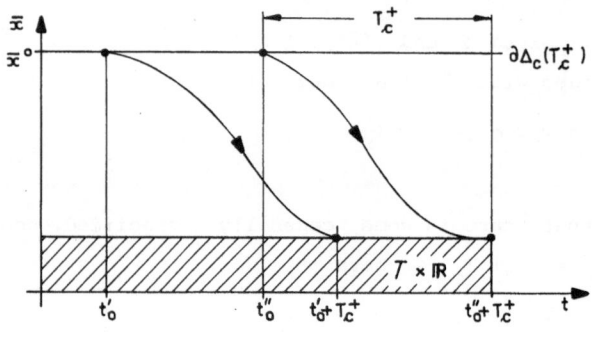

Fig. 5.1

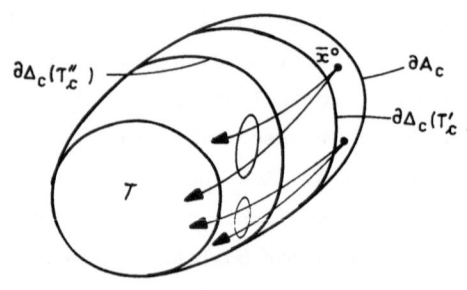

Fig. 5.2

The regions of controllability are denoted $\Delta_c(T_c^+)$ and $\Delta_C(T_c^+)$. We shall concentrate on some features of the latter, though the same conclusions are applied to the former. The strong controllability before T_c^+ implies the strong controllability for collision, hence we have

$$\Delta_C(T_c^+) \subset \Delta_C .$$
$$(5.1.4)$$

Each motion collides with T during a time interval which is smaller or equal to the upper bound T_c^+ which is common for all motions from $\Delta_C(T_c^+)$, see Fig. 5.1. In this sense, the boundary $\partial\Delta_C(T_c^+)$ forms a surface in Δ_C which can be called isochronal. It can be seen from the Fig. 5.1 that with the same rate of attraction to $T \times \mathbb{R}$ the surface $\partial\Delta_C(T_c^+)$ is cylindrical along the t-axis. Indeed if a motion with the same rate of attraction were to start above the x^0-level indicated, it would come to ∂T later than after T_c^+. The cylindrical shape in $\Delta \times \mathbb{R}$ means that this surface projected into $\Delta_C \subset \Delta \subset \mathbb{R}^N$ forms an *isochronal level* in this set, see Fig. 5.2. The isochronal levels are *nested* about T, that is, they enclose each other successively with growing T_c^+.

Indeed, by the same argument, it follows from Definition 3.1.4 that for any pair T'_c, T''_c such that $T''_c < T'_c$,

$$\Delta_c(T''_c) \subset \Delta_c(T'_c) \quad . \tag{5.1.5}$$

For stipulated $P(\cdot)$, the strong recovery region for our collision before T^+_c is denoted $\Delta_c(\bar{P},T^+_c)$ and is measured by (3.1.12) adjusted to this set. Moreover,

$$\Delta_c(\bar{P},T^+_c) \subset \Delta_c(\bar{P}) \subset \Delta_c \quad . \tag{5.1.6}$$

Another subobjective is the *ultimate collision*, that is, collision not before but after stipulated $T^-_c < \infty$. We shall use the following definition.

DEFINITION 5.1.3. Given T, $T^-_c < \infty$ and $(\bar{x}^0,t_0) \in \Delta \times \mathbb{R}$, a motion $\bar{\phi}(\bar{x}^0,t_0,\mathbb{R}^+)$ of (2.2.6)' *collides ultimately* with T after T^-_c, if and only if for some $t' \geq t^-_c = t_0 + T^-_c$ we have

$$\bar{\phi}(\bar{x}^0,t_0,t') = \bar{x}^c \in T \quad . \tag{5.1.7}$$

The corresponding regions of controllability are denoted $\Delta_c(T^-_c)$, $\Delta_C(T^-_c)$ and recovery regions by $\Delta_c(\bar{\ },T^-_c)$, $\Delta_C(\bar{\ },T^-_c)$. The properties (3.1.13) - (3.1.17) and (5.1.4) - (5.1.7) apply after obvious adjustment, for instance (5.1.5) requires reversed inclusion.

If we now want the collision at a stipulated instant of time, we must have $T^+_c = T^-_c = T_c$ and apply (5.1.2).

Collision before T^+_c may apply successively to various targets T_j, $j = 1,2,\ldots$, generating a sequence of subobjectives, usually ending with ultimate collision. We then talk about sequential collision $T^+_{c1},T^+_{c2},\ldots,T^-_c$.

It is often of obvious practical interest to exclude cases when the motion can bounce off the boundary ∂T. Then we demand *penetration* of T rather than collision with it. The problem is the same as collision with the interior, an open set $\text{int } T$.

DEFINITION 5.1.4. Given T, and $(\bar{x}^0,t_0) \in \Delta \times \mathbb{R}$, a motion $\bar{\phi}(\bar{x}^0,t_0,\mathbb{R}^+)$ of (2.2.6)' *penetrates* T, if and only if

$$\bar{\phi}(\bar{x}^0,t_0,\mathbb{R}^+) \cap \text{int } T \neq \phi \quad . \tag{5.1.8}$$

Similarly as in collision, it means that there is $T_p < \infty$ (unspecified) such that for $t_p = t_0 + T_p$, we have

$$\bar{\phi}(\bar{x}^0, t_0, t_p) = \bar{x}^P \in \text{int } T ,\qquad (5.1.9)$$

and if the constant T_p is stipulated, we consider the *penetration at stipulated* T_p, see Fig. 5.3.

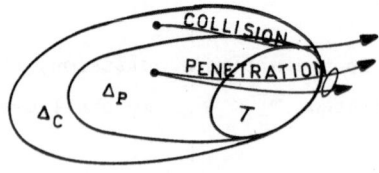

Fig. 5.3

The controllability regions are denoted Δ_p, Δ_p and for the case of stipulated time $\Delta_p(T_p)$, $\Delta_p(T_p)$. The properties (3.1.13) – (3.1.17) and (5.1.4) hold here as well. Moreover, since penetration implies collision, we also have

$$\Delta_p \subset \Delta_c , \qquad \Delta_p \subset \Delta_c \qquad\qquad (5.1.10)$$

and for the same T_p,

$$\Delta_p(T_p) \subset \Delta_p , \qquad \Delta_p(T_p) \subset \Delta_p . \qquad\qquad (5.1.11)$$

Identically as for collision, we may introduce penetration before T_p^+ and after T_p^- (ultimate). The discussion of $\Delta_p(T_p^+)$, $\Delta_p(T_p^+)$ is the same as for $\Delta_c(T_c^+)$ outlined above. Penetration is simply a collision with an open set. Considering the sequential collision, particular subobjectives may appear with or without penetration. In general, when discussing an objective Q, we may combine these two modular properties or separate them, and obviously use Definitions 3.1.4, 3.1.5 for the controllabilities of the resulting objective. Let us look closely at the case of collision without penetration or, which is the same, *collision with rejection*. To attain such an objective, a motion of (2.2.6)' would have to satisfy the collision (5.1.3) but contradict the penetration (5.1.8), that is, to satisfy

$$\bar{\phi}(\bar{x}^0, t_0, \mathbf{R}^+) \cap \text{int } T = \phi .$$

Denote $C\Delta_p \overset{\Delta}{=} \Delta_c - \Delta_p$. It is obviously the region of strong controllability for rejected collision. Observe that any motion leaving $C\Delta_p$ would have to do the impossible, miss T or penetrate it. Hence we obtain the following.

PROPERTY 5.1.1.   The region $C\Delta_P$ is positively strongly invariant under
$P(\cdot)$.

Then we also have

PROPERTY 5.1.2. $T \subset \Delta_P$ if and only if $C\Delta_P = \phi$.

Indeed, no motion from $C\Delta_P$ can reach T without penetration, which
proves necessity. Sufficiency is obvious as $\Delta_C - \Delta_P = \phi$ implies
$\Delta_C \subset \Delta_P$ and we have $T \subset \Delta_C$.

Now let $\partial^R T \triangleq \partial T \cap C\Delta_P$ be the rejecting part of the target boundary
∂T termed *non-useable*. Immediately we obtain the contra positive to the
above sufficiency:

PROPERTY 5.1.3. $C\Delta_P \neq \phi \Rightarrow \partial^R T \neq \phi$

and conclude the obvious that with rejected collision, not all of the
boundary ∂T is used for penetration. It also means that $\partial^R T \subset C\Delta_P$. On
the other hand, int $T \subset \Delta_P$ whence int $T \cap C\Delta_P = \phi$ and we have

PROPERTY 5.1.4.   There is no protruding of T into $C\Delta_P$ except for ∂T ,
cf. Fig. 5.4.

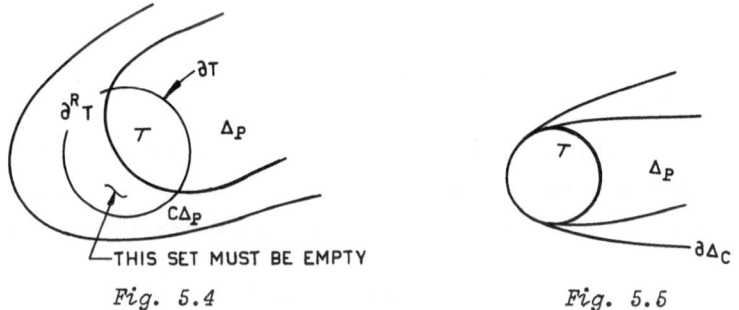

Fig. 5.4   Fig. 5.5

Thus we may have two cases:

(i)   non-protruding, $\partial^R T = \phi$, that is, all of ∂T is
      useable for penetration;

(ii) protruding, $\partial^R T \neq \phi$, that is, some part of ∂T is
      non-useable.

In the first case, by Property 5.1.2, we have $C\Delta_P = \phi$, that is, there is
no rejected reaching whence all reaching motions penetrate.

Let us consider now the protruding case (ii) with $\partial^R T \neq \phi$, see Fig. 5.5. Suppose that Δ_p is closed, that is, $C\Delta_p$ is "open below", then $\partial^R T \subset$ int $\overline{C\Delta_p}$ and there may be a point internal to both Δ_p and $C\Delta_p$, which contradicts the obvious int $T \subset \Delta_p$. Hence we have

PROPERTY 5.1.5. $C\Delta_p \neq \phi$ implies that Δ_p is open.

Granted that Δ_p is open and that only ∂T can have common points with $C\Delta_p$, it follows that only ∂T of the entire T can protrude into the closed $C\Delta_p$. Hence

PROPERTY 5.1.6. $C\Delta_p \neq \phi$ implies $\partial^R T \subset \partial\Delta_p$.

In turn, by definition, rejected collision implies that all motions from $C\Delta_p$ terminate in $\partial^R T$ and thus, by the above, we have

PROPERTY 5.1.7. Motions from closed $C\Delta_p \neq \phi$ terminate in $\partial\Delta_p$ and through this set in $\partial^R T$.

This means that $\partial\Delta_p$ is strongly positively invariant under $\bar{P}(\bar{x},t)$ generating the rejected collision.

The next modular subobjective property, often met with in control scenarios, is *capture* in the target T, and *rendezvous* with T which is a temporary capture. These two subobjectives serve much better than collision alone in all tracking and path or model following situations. It might be worth mentioning here that in some control theory presentations, the word capture is used in the sense of our collision. Our meaning of capture is a permanent holding in T after a specific time, again possibly stipulated.

In the above sense, we may obviously have collision with or without capture. Controllability for capture in Liapunov terms was introduced in Vincent-Skowronski [1], and we shall refer to this work. The study on rendezvous will represent a modified version of that in Skowronski [38].

DEFINITION 5.1.5. Given the target T, and $(\bar{x}^0,t_0) \in \Delta \times \mathbb{R}$, some motion $\bar{\phi}(\bar{x}^0,t_0,\mathbb{R}^+)$ is *captured* in T if and only if there is $T_c < \infty$ such that

$$\bar{\phi}(\bar{x}^0,t_0,t) \in T, \quad \forall t \geq t_c = t_0 + T_c. \tag{5.1.12}$$

The same motion performs a *rendezvous* with T if and only if there is $T_Z \triangleq [t_{Z1},t_{Z2}] \subset \mathbb{R}$, $t_{Z2} < \infty$ such that

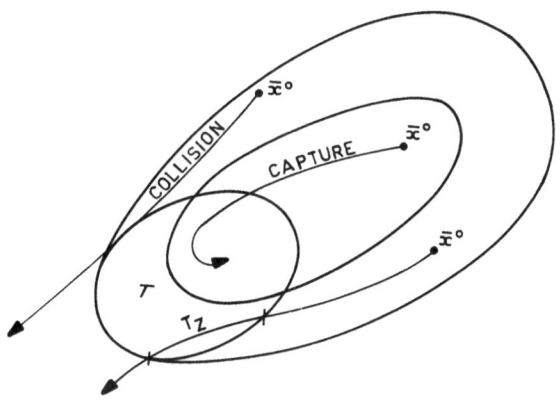

Fig. 5.6

$$\bar{\phi}(\bar{x}^0, t_0, t) \in T, \quad \forall \, t \in T_Z$$

$$\text{and} \quad \bar{\phi}(\bar{x}^0, t_0, (t_{Z2}, \infty)) \cap T = \phi \, . \qquad\qquad (5.1.13)$$

When T_C, T_Z are stipulated, we refer to *capture after* T_C, and *rendezvous during* T_Z, respectively, see Fig. 5.6.

For $t_{Z2} \to \infty$, rendezvous becomes capture, and for $t_{Z2} \to t_{Z1}$, it becomes collision.

EXAMPLE 5.1.1. Consider a wheel of a railroad car passing over a low spot on the rail, as shown in Fig. 5.7. As the wheel traverses the low spot, the vertical displacement of the wheel is a combination of the variable depth of the low spot and the additional deflection of the rail due to the weight of the wheel interacting with the low spot. If we let $s = f(\ell)$ represent the variable depth of the low spot and $q(t)$ the additional vertical deflection of the rail as the wheel passes by, we can write the motion equation of the wheel in the vertical direction as

$$m \frac{d^2(q+s)}{dt^2} + kq = 0 \, ,$$

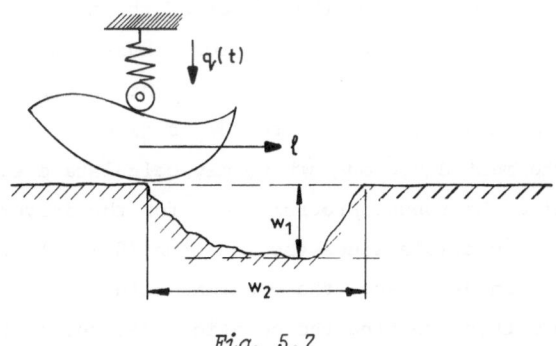

Fig. 5.7

221

where m is the mass of the wheel and k is the vertical load that produces
a unit deflection (coefficient of elasticity). Using the chain rule of
differentiation, $\ddot{s}$ can be written in terms of the speed v of the train,
producing

$$m(\ddot{q} + v^2 \frac{d^2 s}{d\ell^2}) + kq = 0 \ ,$$

or

$$m\ddot{q} + kq = -mv^2 \frac{d^2 s}{d\ell^2} \quad .$$

Defining the shape of the low spot in terms of a simplified Fourier series
we have

$$s(\ell) = \frac{w_1}{2} (1 + \cos \frac{2\pi\ell}{w_2})$$

with the uncertain length of the low spot w_2 and uncertain depth of that
spot at its midpoint w_1 . Remembering that $\ell = vt$, we can write the
motion equations as

$$m\ddot{q} + kq = mv^2 \frac{w_1}{2} \cdot \frac{4\pi^2}{(w_2)^2} \cos (2\pi vt/w_2) \tag{5.1.14}$$

which represents forced vibrations, that is, the track is forced to vibrate
by the wheel passing over the low spot. Given initial conditions
$q(0) = 0$, $\dot{q}(0) = 0$, the motion equation has a solution

$$q(t) = -\frac{w_1/2}{(1 - w_2/vL)} (\cos (2\pi vt/w_2) - \cos (2\pi t/L)) \tag{5.1.15}$$

with the period L . It follows that the additional vertical deflection q
is directly proportional to the depth of the spot w_1 . Moreover, it also
depends upon the ratio w_2/vL which ought to be separately analysed. The
period L of the vibrations is related to the radius of the wheel, which
is thus a parameter of the system. The deflection increases with the ratio
of wheel size to length of the low spot. The vector $(w_1,w_2)^T = \bar{w} \in W$
gives the uncertainty under which the system operates, while the speed v
of the train, appearing in both the ratio w_2/vL and in the bracket, is
the control variable programmed by the driver of the train. He wants to
avoid an excessive amplitude of the deflection-vibrations, *capturing* q(t)
within specific bounds which are safe and represent the target concerned:
$q_{min} \leq q(t) \leq q_{max}$. Granted that the set W is given by statistical
evidence, taking the most dangerous $\bar{w}*$ we may calculate a suitable v from
(5.1.15) thus getting the robust program $P(\cdot)$ for the driver. Such a
direct calculation is possible since the equation (5.1.14) has an exact
solution. In a more general case, the program would have to come from the
Liapunov formalism without solving the equation, see the next section.

Owing to the shape of (5.1.15) the program obtained from it always theoretically exists, but we must also check the obtained values of v against the possibilities of the motor of the train.

It may occur that it is not necessary to control the system for an indefinite duration after the impact of hitting the low spot hole. It all depends on how often we expect the low spots and how many of them exist. Calculating (5.1.15) over a specific duration T_z subject to the above discussion produces the rendezvous during T_z generated by the same controller. $\qquad\qquad\qquad\qquad\qquad\qquad\qquad\qquad\qquad$ $\square$

The regions for controllability and strong controllability for capture and rendezvous will be denoted by Δ_c, Δ_c and Δ_z, Δ_z respectively with obvious adjustment when referring to stipulated time intervals T_c, T_z. Clearly, capture is an ultimate rendezvous with $T_z \to \infty$. Moreover both, capture and rendezvous, imply collision. Thus we have

$$\Delta_c \subset \Delta_z \subset \Delta_c \; ; \quad \Delta_c \subset \Delta_z \subset \Delta_C \qquad\qquad (5.1.16)$$

and the same inclusions for corresponding regions with stipulated time intervals. From (5.1.16) we conclude that $\Delta_c \cap T \neq \phi$, but not necessarily $T \subset \Delta_c$. Hence, if the system is strongly controllable for collision with capture in T, then there is a robust $\bar{P}(\cdot)$ such that for all $\bar{w}(\cdot)$ the motions are in T after some time interval. However, it does not follow that the system state can be maintained in an arbitrary subset of T. It does not follow either that the state can be maintained even in a given subset of $T \cap \Delta_c$, since for a certain point of this set it may be possible that the state passes through the point but that there is a $\bar{w}(\cdot)$ such that no $\bar{P}(\cdot)$ is able to return the state to a neighborhood of the point. However, it does follow that there is at least one nonempty subset of T located in $T \cap \Delta_c$ which is positively strongly invariant under $\bar{P}(\cdot)$. Any such subset is called a *capturing subtarget* T_c in T, and there may exist an indefinite number of them, see Fig. 5.6. Thus, in practical terms we refer to a given *candidate set* T_c which is required to be such a subtarget, and must be confirmed by some sufficient conditions. Note, however, that for fixed T_c the choice of T_c affects the size of Δ_c, see Section 5.2.

The concept of strong controllability for collision with capture, and robust controllers in order to achieve it, using Liapunov formalism, were developed in Skowronski [27],[29], and Skowronski-Vincent [1]. Since capture implies collision, there is no need to consider collision with capture

as a combined objective unless one has an opponent who can deny capture
but not necessarily collision. The uncertainty $\bar{w}$ can play the role of
such an opponent. The above cited investigations give necessary and
sufficient conditions for strong controllability for capture, including
collision, but if the necessary conditions cannot be satisfied, offer the
flexible alternative that collision can be achieved even if, due to the
opposing uncertainties, capture cannot. In particular, the uncertain
perturbations may deny capture (*collision without capture*) or they may be
capable of driving the motion out of T in finite time and keeping it out
(*collision with escape*) no matter what the controller does. Such an escape
leaves us with rendezvous for the time duration, which becomes shorter as
the option becomes stronger.

EXAMPLE 5.1.2. Let us consider a single point-mass system with the kinematics defined by

$$\dot{q} = u \cos q + w \sin q \qquad\qquad (5.1.17)$$

with $u \in [-1,1]$, $w \in [-1,1]$, and the target T : $q_1 = 0$, see Fig. 5.8.

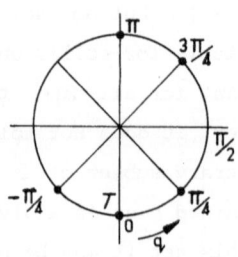

Fig. 5.8

The objective is to collide with T,
possibly with capture. For
$q^0 \in [0, \pi/4)$ the robust control is
achieved with $u = -1$ yielding
$\dot{q}(t) < 0$. The latter implies that
for any point in this interval the
system is pushed towards 0 no matter
what the uncertainty w does, even if
it does the worst, that is, uses
$w \equiv 1$. When $q = 0$ is passed, the
collision is achieved, and if we had wanted the collision only, we let the
system be free: $u \equiv 0$, which makes the trajectory reach $q = -\pi$ under
the opposition $w \equiv 1$ still attempting to push the point-mass away from
the target. In fact, such control is convenient, if we want the sequential
collision with T, achievable from $q = -\pi$ by pushing the mass further
down to $q = 0$. The story is different if we wanted capture in T. Then
after colliding with T the first time, we would switch to $u \equiv 1$ which
generates $\dot{q}(t) > 0$ thus stopping the mass at $q = 0$ or making it to
return there from any $q^0 \in (-\pi/4, 0]$, against $w \equiv 1$. For $q^0 < -\pi/4$
we are not able to attain the latter even with $u \equiv 1$. The uncertainty
wins, making capture impossible, but leaving us with the option of sequential collisions mentioned before. The strong region for capture is thus

$\Delta_{\mathfrak{C}}$: $-\pi/4 \leq q \leq \pi/4$ while the strong region for collision is the whole q-axis, cf. (5.1.16).

The above example illustrates also the role of the weak versus strong modes of control. As we recall from Section 3.1 the weak mode, that is, controllability secures the objective with the "cooperation" of uncertainty, meaning that it is possible to select a pair $\bar{u}, \bar{w}$ which generates Q, in our case collision with capture in T : $q = 0$. Indeed, in the above example, the choice of such a pair is possible for both collision and capture along with all q so the controllability is complete, and we are free to search for the robust controllers, as was done in the first part of the example. $\square$

5.2 OPTIMAL CONTROLLABILITY

Suppose the system (2.2.6)' is strongly controllable at $(\bar{x}^0, t_0)$ $\epsilon \Delta \times \mathbb{R}$ for Q and moves along some of the motions of $K(\bar{x}^0, t_0)$. Let us introduce a functional $\mathfrak{V}(\cdot)$ which assigns a unique real number to such a motion. The number will be called the *cost* of the transfer from $(\bar{x}^0, t_0)$ to $\Phi(\bar{x}^0, t_0, t_f)$ while achieving Q and it must be consistent with Q during the time concerned for the functional to be called a (quantitative) *performance index*. The cost will be assumed dependent upon the corresponding robust $\bar{P}(\cdot)$ and possibly t_f . Although t_f must exist for each $\bar{\phi}(\cdot)$ concerned, it may not be specified. Then we delete it from under the bracket, writing $\mathfrak{V}(\bar{x}^0, t_0, \bar{\phi}(\cdot), \bar{P})$. The cost is identified at some event

$$(\bar{x}^\sigma, t_\sigma) : \quad (\bar{x}^\sigma, t_\sigma, \bar{\phi}(\cdot), \bar{P}(\cdot), \bar{w}(\cdot), t_\sigma) = 0 , \qquad (5.2.1)$$

for all admissible $\bar{P}(\cdot)$, and all $\bar{\phi}(\cdot) \epsilon K(\bar{x}^\sigma, t_\sigma)$, $\bar{x}^\sigma = \bar{\phi}(\bar{u}, \bar{x}^0, t_0, t_\sigma)$, the t_σ being either t_0 or t_f depending upon Q . For example, the cost of collision is smaller the closer we are to a target and zero at the target. The cost of avoidance behaves in the opposite manner. The cost is additive along the motion. For a more detailed and very illustrative description of the above concepts, we recommend Leitmann [8].

The function $\mathfrak{V}(\bar{x}^0, t_0, \bar{\phi}(\cdot), \bar{P}, \bar{w}(\cdot), \cdot) : t \rightarrow \mathfrak{V}(t) \epsilon \mathbb{R}$ is called the *cost flow*, and it plays the same role in optimisation as the energy flow in qualitative behavior. It is assumed to be C^1 and written $\mathfrak{V}(\bar{u}, \bar{x}^0, t_0, \cdot)$ or even $\mathfrak{V}(\cdot)$ when the initial state and time are obvious or irrelevant. A qualitative strong controllability with some adjoint quantitative objective will consist of two problems.

Problem 1. *Given* $(\bar{x}^0, t_0) \in \Delta \times \mathbb{R}$, *secure the strong controllability for Q and then among winning programs find* $\bar{P}_*(\cdot)$ *such that*

$$\mathfrak{V}(\bar{x}^0, t_0, \bar{\phi}(\cdot), \bar{P}_*(\cdot), w(\cdot)) \leq \min_{\bar{P}} \max_{\bar{w}(\cdot)} \inf_{\bar{\phi}(\cdot)} \mathfrak{V}(\bar{x}^0, t_0, \bar{\phi}(\cdot), w(\cdot)), \quad (5.2.2)$$

for all admissible $\bar{P}(\cdot), \bar{w}(\cdot)$ *and all* $\bar{\phi}(\cdot) \in K(\bar{x}^0, t_0)$.

The right hand side of (5.2.2) is called the *upper value* of the game against uncertainty and denoted $\mathfrak{V}^+$ = constant . Here we adapt the philosophy of the game against nature or worst-case-design, outlined in Section 2.2, according to which $\bar{w}$ aims to make it worst for us, that is, attempts to maximize our cost while we try to minimize it. The minimizing program $\bar{P}_*$ is called *optimal*.

The second problem reflects the reverse viewpoint of the uncertainty which aims at preventing our goal.

Problem 2. *Given* $(\bar{x}^0, t_0) \in \Delta \times \mathbb{R}$, *the uncertainty* w *wants to contradict strong controllability for Q and if that is not possible, at least use* $\bar{w}*(\cdot)$ *such that*

$$\mathfrak{V}(\bar{x}^0, t_0, k(\cdot), P(\cdot), \bar{w}*(\cdot)) \geq \max_{\bar{w}(\cdot)} \min_{\bar{P}} \sup_{\bar{\phi}(\cdot)} \mathfrak{V}(\bar{x}^0, t_0, \bar{\phi}(\cdot), \bar{P}(\cdot), \bar{w}(\cdot)),$$
$$(5.2.3)$$

for all admissible $\bar{P}(\cdot), \bar{w}(\cdot)$ *and all* $\bar{\phi}(\cdot) \in K(\bar{x}^0, t_0)$.

Here the right hand side of (5.2.3) is a constant $\mathfrak{V}^-$ and is called the *lower value* of the game against uncertainty. Again $\bar{w}*(\cdot)$ is called *optimal* for the uncertainty. If $\mathfrak{V}^- = \mathfrak{V}^+$, we say that there exists a *Game Value* $\mathfrak{V}*(\bar{x}^0, t_0) = \mathfrak{V}^- = \mathfrak{V}^+$. Let us write $\mathfrak{V}(\bar{x}^0, t_0, \bar{P}(\cdot), \bar{w}(\cdot))$ for the cost, to acknowledge the fact that for each pair $\bar{P}(\cdot), \bar{w}(\cdot)$, we admit many motions $\bar{\phi}(\bar{u}, \bar{x}^0, t_0, \cdot) \in K(\bar{x}^0, t_0)$, each generating a cost, possibly different. The pair $\bar{P}_*(\cdot), \bar{w}*(\cdot)$ such that the cost $\mathfrak{V}(\bar{x}^0, t_0, \bar{P}_*(\cdot), \bar{w}*(\cdot))$ $= \mathfrak{V}*(\bar{x}^0, t_0)$ is called *optimal* and so are the corresponding motions from some $K*(\bar{x}^0, t_0)$. Note that all the values $\mathfrak{V}(\bar{x}^0, t_0, \bar{P}_*(\cdot), \bar{w}*(\cdot))$ must be equal, for given $(\bar{x}^0, t_0)$.

PROPERTY 5.2.1. *A game against uncertainty has the value, if and only if there are* $P_*(\cdot)$ *and* $\bar{w}*(\cdot)$ *such that given* $(\bar{x}^0, t_0)$,

$$\mathfrak{V}(\bar{x}^0, t_0, \bar{P}_*(\cdot), \bar{w}(t)) \leq \mathfrak{V}(\bar{x}^0, t_0, \bar{P}_*(\cdot), \bar{w}*(t)) \leq \mathfrak{V}(\bar{x}^0, t_0, \bar{P}(\cdot), \bar{w}*(\cdot))$$
$$(5.2.4)$$

for all $\bar{P}(\cdot)$ *and* $\bar{w}(\cdot)$.

The proof follows immediately from the fact that in view of (5.2.2), (5.2.3), $\mathfrak{V}^- = \mathfrak{V}^+$ implies (5.2.4). The converse holds since always min max $\mathfrak{V}(\bar{P}(\cdot), \bar{w}(\cdot)) \geq$ max min $\mathfrak{V}(\bar{P}(\cdot), \bar{w}(\cdot))$. The condition (5.2.4) is called *saddle condition*, and is frequently used to determine the optimal program, in a similar way to (5.2.2) of Corollary 3.2.2. The game value $\mathfrak{V}*(\bar{x}^0, t_0) = \mathfrak{V}(\bar{x}^0, t_0, \bar{P}_*(\cdot), \bar{w}_*(\cdot))$ is unique, that is, the $\mathfrak{V}*$'s along all $\bar{\phi}(\cdot)$'s are equal.

DEFINITION 5.2.1. A system (2.2.6)' is *optimally controllable* at $(\bar{x}^0, t_0)$ *for* Q if and only if it is controllable for Q there and the corresponding $\bar{P}(\cdot), \bar{w}(\cdot)$ are optimal. Then $\bar{x}^0$ is optimally controllable for Q and Δ_q^* denotes the *region* of such controllability.

DEFINITION 5.2.2. A system (2.2.6)' is *optimally strongly controllable* at (x^0, t_0) *for* Q if and only if it is strongly controllable at this point for Q and among the winning $\bar{P}(\cdot)$ there is $\bar{P}*(\cdot)$ such that combined with some $\bar{w}*(\cdot)$ they form an optimal pair. Then $\bar{x}^0$ is optimally strongly controllable for Q and Δ_Q^* is the corresponding region.

The remaining concepts follow the pattern in Section 3.1. Note that optimal strong controllability solves Problem 1. On the other hand, the optimal controllability for the contradiction of Q, denoted $\mathcal{Q}$ makes the strong controllability for Q impossible, thus solving Problem 2 in full, rather than in its optimizing part only.

Definitions 5.2.1 and 5.2.2 yield

PROPERTY 5.2.2. $\qquad \Delta_q^* \subset \Delta_q \qquad \Delta_Q^* \subset \Delta_Q$.

Let us investigate the necessary conditions. We said that the cost flow plays the correspondent role to the energy flow. To show it, we need a *cost surface* $\Sigma : x_0 = \mathfrak{V}*(\bar{x}, t)$ in the space $\mathbb{R}^{N+2}$ of vectors $(\hat{x}, t)$, where $\hat{x} = (x_0, x_1, \dots, x_N)^T$, cf. Fig. 5.9.

We let $\mathfrak{V}*(\bar{x}, t)$ be a C^1-function, single sheeted due to the property that the value of the game is unique. It is often called the *game potential*, cf. Krassovski-Subbotin [1], forming a reference frame for a conflict as much as the energy surface does for the qualitative control. Similarly to the energy, we form a family of levels: $\Sigma*(c) : \mathfrak{V}*(\bar{x}, t)$ = constant = c , which, when projected isometrically onto $\Delta \times \mathbb{R}$, produce a continuous family of surfaces S_c filling up $\Delta \times \mathbb{R}$, ordered by the parameter c . The motions cross the iso-cost levels exactly the same way as they crossed the constant-energy levels E_c in Sections 2.3, 2.4.

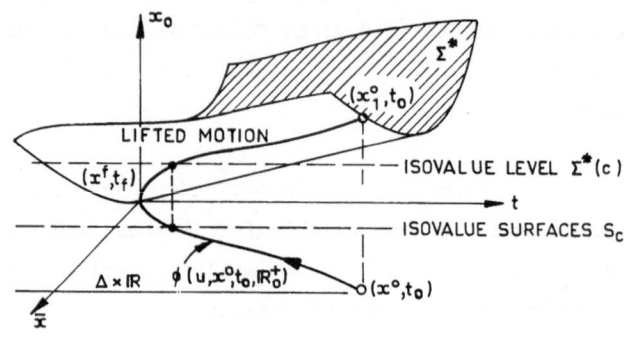

Fig. 5.9

Let the investigated slice of Σ^* be regular (no extremal points) and such that, given S_c , separates its neighborhood into two disjoint sets

EXT : $\mathfrak{V}^*(\bar{x},t) > c$, (5.2.5)

INT : $\mathfrak{V}^*(\bar{x},t) < c$. (5.2.6)

As Σ^* is smooth, so is S_c and at any $(\bar{x},t) \in S_c$ there is a gradient

$$\nabla_{x,t} \, \mathfrak{V}^*(\bar{x},t) \overset{\Delta}{=} \left[\frac{\partial \mathfrak{V}^*}{\partial t} , \frac{\partial \mathfrak{V}^*}{\partial t} , \dots , \frac{\partial \mathfrak{V}^*}{\partial x_N} \right] \neq 0$$

directed towards EXT. Consider now the point $(\bar{x},t)$ on a motion with the selector $\bar{f}(\bar{x},\bar{u},\bar{w},t) \equiv 0$ and let this point be regular on S_c , see Fig. 5.10 and compare Fig. 2.11. There obviously will be $K(\bar{x},t)$ of such motions through the same point $(\bar{x},t)$. From (5.2.4) we have at this point

$$\mathfrak{V}(\bar{x},t,\bar{P}_*(\cdot),\bar{w}(\cdot)) \leq \mathfrak{V}^*(\bar{x},t) ,$$
(5.2.7)

for all $w(\cdot)$, and

$$\mathfrak{V}(\bar{x},t,\bar{P}(\cdot),\bar{w}^*(\cdot)) \geq \mathfrak{V}^*(\bar{x},t) ,$$
(5.2.8)

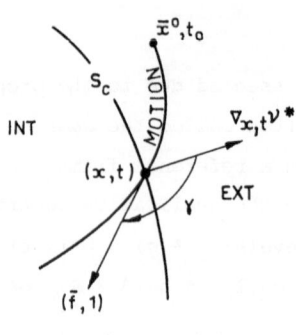

Fig. 5.10

for all $\bar{P}(\cdot)$, cf. (4.1.9), (4.1.10). From (5.2.7) we conclude that no motion with the selector $\bar{f}(\bar{x},\bar{u}_*,\bar{w},t)$ may penetrate EXT no matter which $\bar{w}(\cdot)$ is used. This contradicts $\dot{\mathfrak{V}}^*(t) > 0$ along such motions, thus yielding $\dot{\mathfrak{V}}^*(t) \leq 0$, that is,

$$\frac{\partial \mathfrak{V}^*}{\partial t} + \nabla_x \mathfrak{V}^*(\bar{x},t)^T \cdot \bar{f}(\bar{x},\bar{u}_*,\bar{w},t) \le 0 ,$$
(5.2.9)

for all $\bar{u}_* \in \bar{P}_*(\bar{x},\bar{\lambda},t)$, and all $\bar{w}(\cdot)$. In turn, from (5.2.8) we conclude that no motion with $\bar{f}(\bar{x},\bar{u},\bar{w}^*,t)$ may penetrate INT no matter which $\bar{P}(\cdot)$ is used, yielding

$$\frac{\partial \mathfrak{V}^*}{\partial t} + \nabla_x \mathfrak{V}^*(\bar{x},t) \cdot \bar{f}(\bar{x},\bar{u},\bar{w}^*,t) \ge 0$$
(5.2.10)

for all $\bar{u} \in \bar{P}(\bar{x},t)$. It follows that the conditions (5.2.9), (5.2.10) are necessary for the *saddle* (5.2.4). The surface S_c with the above properties has been called semipermeable by Isaacs [1]. We shall call it *cost-nonpermeable*, for reasons which will become clear later, cf. Definition 3.1.3 and Section 8.4. There obviously may be many such surfaces in Δ .

Further implications lead to the following three cases.

1. Motions with selectors $\bar{f}(\bar{x},\bar{u}_*,\bar{w},t)$ enter INT, possibly with sliding upon S_c, see $\gamma \ge 90°$ in Fig. 5.10. We call $(\bar{x},t)$ an *entry point* with contact, or if $\gamma > 90°$ *strict entry*. Then the motions are *cost dissipative* or strictly dissipative, respectively, which is the objective of our control.

2. Motions with selectors $\bar{f}(\bar{x},\bar{u},\bar{w}_*,t)$ leave INT possibly with sliding upon S_c, see $\gamma \le 90°$. Then we call our point $(\bar{x},t)$ an *exit point* with contact, or if $\gamma < 90°$ *strict exit*. The latter means that the motions are *cost accumulative* as desired by the uncertainty opposition.

3. The motions with selectors $\bar{f}(\bar{x},\bar{u}_*,\bar{w}^*,t)$ stay on S_c and with (5.2.9), (5.2.10) applied together we have, cf. (4.1.11),

$$\frac{\partial \mathfrak{V}^*}{\partial t} + \min_{\bar{u}} \max_{\bar{w}} [\nabla_x \mathfrak{V}^*(\bar{x},t) \cdot \bar{f}(\bar{x},\bar{u},\bar{w},t)] = 0$$
(5.2.11)

for all $\bar{u} \in \bar{P}(\bar{x},t)$, and all $\bar{w} \in W$, which defines S_c .

The surface is thus *cost conservative* or cost positively invariant, and the forces forming $\bar{f}(\bar{x},\bar{u}_*,\bar{w}^*,t)$ may be called *cost potential forces*. It is thus immediately suggestive of the fact that the equation (2.2.5)' with the right hand side $\bar{\psi}(\bar{x},t) \stackrel{\Delta}{=} \bar{f}(\bar{x},t,\bar{u}_*,\bar{w}^*)$ should be integrable and that $\mathfrak{V}^*(\bar{x},t) = $ constant is a first integral, see Sections 3.4 and 4.1, so that the analogy to energy is complete.

Suppose now that our assumption of regularity (non-zero gradient) does not hold in the investigated part of $\Sigma^*(t)$, and let $\bar{x}^e$ be an extremal point in $\Sigma^*(t)$, such that $\Sigma^* \overset{\Delta}{=} \Sigma^*(t) \times \mathbb{R}$. Moreover, let us take the cost level $S_c = \{\bar{x}^e\} \times \mathbb{R}$ with the gradient vanishing. It is either a local minimum or maximum (threshold). Consider the minimum first. The neighborhood is a cup EXT with INT collapsed to an empty set. As the gradient vanishes, we work directly from (5.2.7) and (5.2.8) which obviously still hold. As before, motions with $\bar{f}(\bar{x}^e, \bar{u}_*, \bar{w}, t)$ cannot penetrate EXT, but since INT = ϕ the best the uncertainty can do is to force the motions to rest at $\{\bar{x}^e\} \times \mathbb{R}$, that is,

$$\bar{f}(\bar{x}^e, \bar{u}_*, \bar{w}, t) \underset{t}{\equiv} 0 \qquad\qquad (5.2.12)$$

for all $\bar{u}_* \in \bar{P}_*(\bar{x}, \bar{\lambda}, t)$, $\bar{w} \in W$, $t \in \mathbb{R}^+$, which thus replaces (5.2.9) as a necessary condition for the minimum. By symmetric argument with EXT = ϕ we obtain

$$\bar{f}(\bar{x}^e, \bar{u}, \bar{w}*, t) \underset{t}{\equiv} 0$$

for all $\bar{u} \in \bar{P}(\bar{x}, \bar{\lambda}, t)$, all $\bar{P}(\cdot)$ and $\bar{w} \in W$, as a necessary condition for $\bar{x}^e$ being a local maximum. As a result, the motions with $\bar{f}(\bar{x}, \bar{u}_*, \bar{w}*, t)$ cannot leave $\{\bar{x}^e\} \times \mathbb{R}$ at all, whence

$$\bar{f}(\bar{x}^e, \bar{u}_*, \bar{w}*, t) \underset{t}{\equiv} 0 , \qquad\qquad (5.2.13)$$

is necessary for $\bar{x}^e$ being an equilibrium, cf. (5.2.8).

Apart from the extrema $\bar{x}^e$, the surface Σ may have discontinuities in differentiability - sets of points where the gradient is not defined. The corresponding projection into $\mathbb{R}^N$ forms what Isaacs calls a *singular surface*, an (N-1)-dimensional manifold on which regular behavior of motions fails. These surfaces may attract, repel or be neutral as much as the extrema could.

As a practical conclusion of our analogy between the energy and cost surfaces, we can have that if the cost is considered energy, the stabilization of Sections 3.2, 3.3 and 4.1 applies directly. When the cost is something else, it applies as well but must be translated into the cost language. Then we talk about *strong cost stabilization* rather than strong stabilization. On the other hand, when the objectives are mixed, qualitative and quantitative, e.g. we have both energy and other cost stabilization, the name is *strong optimal stabilization*. The translations are immediate and left to the reader.

The necessary conditions (5.2.11) are useful in many ways, but to us mainly to produce semipermeable candidates for strong controllability, not necessarily optimal, for various objectives - in particular capture, see the next section. The practical means of applying (5.2.11) is through the so called Pontriagin Principle, which we outline here very briefly for further use.

Let us specify the cost as the integral

$$\mathfrak{V} = \int_{t_0}^{t_f} f_0^V(\bar{x},t,\bar{u},\bar{w})dt \qquad\qquad (5.2.14)$$

and write the cost flow as

$$\bar{x}_0(t) = \bar{x}_0(t_0) + \int_{t_0}^{t} f_0^V(\bar{x},\tau,\bar{u},\bar{w})d\tau , \quad \bar{x}_0(t_0) = 0 ,$$

which in terms of the vector $\hat{\bar{x}} = (x_0,x_1,\ldots,x_N)^T$, see Fig. 5.9, produces the *cost-state equations*

$$\dot{\hat{\bar{x}}} = \hat{\bar{f}}(\hat{\bar{x}},t,\bar{u},\bar{w}) \qquad\qquad (5.2.15)$$

where $\hat{\bar{f}} = (f_0^V,\bar{f})^T$, with cost-state motions $\hat{\phi}(\cdot)$ in $\Delta \times \{x_0\} \times \mathbf{R}$, governed by the augmented selectors $\hat{\bar{f}}$. Then (5.2.11) becomes

$$\frac{\partial\mathfrak{V}*}{\partial t} + \min_{\bar{u}} \max_{\bar{w}} [\nabla_{\hat{\bar{x}}}\mathfrak{V}*(\bar{x},t)\cdot\hat{\bar{f}}(\hat{\bar{x}},t,\bar{u},\bar{w})] = 0 , \qquad\qquad (5.2.11)'$$

where

$$\nabla_{\hat{\bar{x}}}\mathfrak{V}*(\bar{x},t) = \left(\frac{\partial\mathfrak{V}*}{\partial x_0},\ldots,\frac{\partial\mathfrak{V}*}{\partial x_N}\right) \quad \text{with} \quad \frac{\partial\mathfrak{V}*}{\partial x_0} = 1 .$$

Following Isaacs, we call it the Main Equation. Introduce now the vector $\bar{n}$ normal to S_c and let $\hat{\bar{n}} = (n_0,n_1,\ldots,n_{N+1})^T$ be defined by $n_0 = \partial\mathfrak{V}*/\partial x_0 = 1$, $n_i = \partial\mathfrak{V}*/\partial x_i$, $i = 1,\ldots,N$, $n_{N+1} = \partial\mathfrak{V}*/\partial t$. We shall refer to it as *adjoint* or *costate vector*. Then (5.2.11)' becomes

$$\min_{\bar{u}} \max_{\bar{w}} \hat{\bar{n}}\cdot\hat{\bar{f}}(\hat{\bar{x}},t,\bar{u},\bar{w}) = 0 \qquad\qquad (5.2.16)$$

with the geometric interpretation in Fig. 5.10 and the corresponding discussion, see Fig. 2.11. Let us introduce now the Pontriagin's Hamiltonian $H(\hat{\bar{x}},t,\bar{u},\bar{w},\hat{\bar{n}}) \overset{\Delta}{=} \hat{\bar{n}}(t)\cdot\hat{\bar{f}}(\bar{x},t,\bar{u},\bar{w})$ and rewrite (5.2.16) as

$$\min_{\bar{u}} \max_{\bar{w}} H(\hat{\bar{x}},t,\bar{u},\bar{w}) = 0 , \qquad\qquad (5.2.17)$$

while the selecting costate equations (5.2.15) and the so called *adjoint equations* producing $\hat{\bar{n}}$ are

$$\dot{x}_i = \frac{\partial H}{\partial n_i} , \quad i = 0,1,\ldots,N ; \qquad\qquad (5.2.18)$$

$$\dot{n}_i = -\frac{\partial H}{\partial x_i} = \sum_{j=0}^{N} n_j \frac{\partial f_i}{\partial x_i} , \quad i = 0,1,\ldots,N . \qquad\qquad (5.2.19)$$

231

The conditions (5.2.17), (5.2.18), (5.2.19) form what is known as the Pontriagin *Min-Max Principle*. When the cost is time: $f_0^V(\bar{x},t,\bar{u},\bar{w}) \equiv 1$, all the above simplifies. Indeed, $x_0(t) = x_0(t_0) + (t_f - t_0)$, $\hat{\bar{f}} = (1,f)$ with

$$H = n_0 + \sum_{i=1}^{N} n_i f_i(\bar{x},t,\bar{u},\bar{w}) + \frac{\partial \mathcal{V}^*}{\partial t} \ . \tag{5.2.20}$$

5.3 CONDITIONS FOR COLLISION

There are two basic problems in controlling the system for collision-type objectives: to *design a feedback controller* and to *define the region* where it may successfully apply, that is, the region of both controllability and strong controllability. The problems are usually solved in practical terms by assuming reasonably justified candidates for $\bar{P}(\cdot)$ and Δ_q, Δ_Q and then verifying them against some sufficient conditions for the controllabilities concerned. The candidates may come from an experienced guess of the designer or user, or from necessary conditions for the controllability. Controllers can also be derived from conditions which are both necessary and sufficient, like Corollaries in Sections 3.2, 3.3 and 4.1.

Let us begin with collision. First, sufficient conditions for the corresponding controllability of a non-truncated nonlinear system followed from the work by Markus [1] already quoted in Section 3.2. Then Gershwin-Jacobson [1] were historically the first to introduce conditions for controllability for stipulated time collision. Both attempts used the Liapunov formalism, but it is in Blaquière-Gerard-Leitmann [1] where the formalism is rigorously posed. In more practical terms, the case was proved by Sticht-Vincent-Schultz [1]. Later Stonier [1] and Skowronski [29] gave alternative versions of these conditions. All the above dealt with autonomous systems without uncertainty. The perturbed uncertain system has been discussed by Skowronski [32]. We shall follow that line, beginning with conditions under which smoothness of the test function $V(\cdot)$ is not required.

Let $\Delta_0 \subset \Delta$ enclosing T be the proposed candidate for a strong region. Denote $CT \overset{\Delta}{=} \Delta_0 - T$, and introduce an open envelope $D \supset \overline{CT}$ and a function $V(\cdot) : D \times \mathbb{R} \to \mathbb{R}$ with

$$\left. \begin{array}{l} v_0^+ = \sup V(\bar{x}) \mid \bar{x} \in \partial\Delta_0 \ , \\[2mm] v_T = \inf V(\bar{x}) \mid \bar{x} \in \partial T \ . \end{array} \right\} \tag{5.3.1}$$

CONDITIONS 5.3.1. The system (2.2.6)' is *strongly controllable on* Δ_0 *for collision with* T, if there is $\bar{P}(\cdot)$ defined on D and a function $V(\cdot) : D \times R \rightarrow R$ such that

    (i) $V(\bar{x},t) > v_T$ for $x \notin T$, $t \in \mathbb{R}$;

    (ii) $V(\bar{x},t) \leq v_0^+$ for $x \in CT$, $t \in \mathbb{R}$;

    (iii) for each $\bar{u} \in P(\bar{x},t)$ there is a constant
        $T_c(\bar{x}^0,t_0,\bar{\phi}(\cdot)) > 0$ such that

$$V[\bar{\phi}(\bar{x}^0,t_0,t_0+T_c),t_0+T_c] - V(\bar{x}^0,t_0) \leq -(v_0^+ - v_T) \qquad (5.3.2)$$

        for all $\bar{\phi}(\bar{x}^0,t_0,\cdot) \in K(\bar{x}^0,t_0)$, $(\bar{x}^0,t_0) \in CT \times \mathbb{R}$.

As motions from $\bar{x}^0 \in T$ collide trivially, to prove the conditions we assume $\bar{x}^0 \in CT$. Then denote $t_c = t_0 + T_c$, $\bar{x}^c = \bar{\phi}(\bar{x}^0,t_0,t_c)$. From (iii) we have $V(\bar{x}^c,t_c) + [v_0^+ - V(\bar{x}^0,t_0)] \leq v_T$. By (ii) there is a constant $a(\bar{x}^0,t_0) > 0$ such that the above becomes $V(\bar{x}^c,t_c) + a \leq v_T$ which by (i) means collision with T.

Condition (5.3.2) means that the drop in V-level ($\equiv$ E-level) must be larger than the V-measured distance from the boundary $\partial\Delta_0$ to the target T. The same may be more conveniently expressed in terms of V-outflux and values $\bar{w}(t) \in W$ rather than functions, provided $V(\cdot)$ is smooth, see Skowronski [29]. Similar conditions on flux have been derived by Leipholz [2] and Olas [1].

CONDITIONS 5.3.2. The system (2.2.6)' is strongly controllable on Δ_0 for collision with T under Conditions 5.3.1 with C^1-function $V(\cdot)$ and (5.3.2) replaced by

$$\int_{t_0}^{t_0+T_c} \left[\frac{\partial V(\bar{x},t)}{\partial t} + \nabla_{\bar{x}} V(\bar{x},t)^T \cdot \bar{f}(\bar{x},\bar{u},\bar{w},t) \right] dt \leq -(v_0^+ - v_T) \qquad (5.3.3)$$

for all $\bar{w} \in W$.

To prove this, we again take the non-trivial $\bar{x}^0 \notin T$, and suppose some motion from $CT \times \mathbb{R}$ avoids collision. Taking $V(\cdot)$ along such a motion, that is, calculating the flow of $V(t)$, condition (i) gives

$$V(\phi(\bar{x}^0,t_0,t),t) = V(\bar{x}^0,t_0) + \int_{t_0}^{t} \dot{V}(\bar{x},\tau) d\tau > v_T \qquad (5.3.4)$$

for all $t \in \mathbb{R}^+$. As $V(\bar{x}^0,t_0) \leq v_0^+$, (5.3.4) yields

$$\int_{t_0}^{t_f} \dot{V}(\bar{x},t)\,dt > -(v_0^+ - v_T) \qquad\qquad (5.3.5)$$

which contradicts (5.3.3) proving our hypothesis.

If T_c is stipulated, both Conditions 5.3.1 and 5.3.2 remain the same. We also have the following conclusion.

COROLLARY 5.3.1. *When* Δ_0, T *are defined by* v-*levels:*

$$\partial\Delta_0 : V(\bar{x},t) \underset{t}{\equiv} v_0^+ , \qquad \partial T : V(\bar{x},t) \underset{t}{\equiv} v_T , \qquad\qquad (5.3.6)$$

then either (5.3.2) or (5.3.3) becomes necessary *as well as* sufficient.

Indeed, if not (5.3.3), then we have (5.3.5) wherefrom there is $(\bar{x}^0, t_0) \in CT \times \mathbb{R}$ yielding $V(\bar{x}^0, t_0) = v_0^+$ such that

$$V(\bar{x},t) = V(\bar{x}^0, t_0) + \int_{t_0}^{t} \dot{V}(\bar{x},\tau)\,d\tau > v_T$$

for all $t \geq t_0$, contradicting collision.

Note that specifying $\partial\Delta_0$, ∂T by V-levels is almost always practical. If (5.3.6) does not fit the required sizes of $\hat{\Delta}_0, \hat{T}$, we use safe estimates taking the V-levels such that $T \subset \hat{T}$, $\Delta_0 \supset \hat{\Delta}_0$. Obviously such estimates are as good as our choice of $V(\cdot)$.

The controller for collision may be found from the following corollary to Conditions 5.3.2.

COROLLARY 5.3.2. *Given* $(\bar{x}^0, t_0) \in CT \times \mathbb{R}$, *if there is a pair* $(\bar{u}^*, \bar{w}^*) \in U \times W$ *generating motion* $\bar{\phi}(\bar{x}^0, t_0, \cdot)$ *such that*

$$V(\bar{\phi}^*(t),t) - V(\bar{x}^0, t_0) = \min_{\bar{u}} \max_{\bar{w}} \int_{t_0}^{t_c} L(\bar{x},\bar{u},\bar{w},t)\,dt \leq -(v_0^+ - v_T) \quad (5.3.7)$$

then (5.3.3) is met with $\bar{u}^* \in \bar{P}(\bar{x},t)$.

For notation, see Corollary 3.2.2. The above follows by the same argument as Corollary 3.2.2 and the observation that the min max problem (5.3.7) can be seen as the extremizing problem with free terminal end for which there is a solution for any $T_c < \infty$.

If the calculation of (5.3.7) is inconvenient, we can use the following obvious but stronger alternative.

COROLLARY 5.3.3. *Given* $(\bar{x}, t) \in CT \times \mathbb{R}$, *if there is a pair*
$(\bar{u}^*, \bar{w}^*) \in U \times W$ *such that*

$$L(\bar{x}, \bar{u}^*, \bar{w}^*, t) = \min_{\bar{u}} \max_{\bar{w}} L(\bar{x}, \bar{u}, \bar{w}, t) \le - \frac{v_0^+ - v_T}{T_c} \ , \tag{5.3.7)'}$$

then (5.3.3) is met with $\bar{u}^* \in \bar{P}_*(\bar{x}, t)$.

Implementation of the above conditions obviously depends upon the selection of $V(\cdot)$ and $\bar{P}(\cdot)$. Methods for obtaining $V(\cdot)$ have been outlined in Section 3.4. It may also be seen that a fairly general method for obtaining $\bar{P}(\cdot)$ follows from the corollaries in Chapter 3 and Corollaries 5.3.2, 5.3.3. It has been introduced and developed by the Leitmann school in the Seventies. The stabilizing robust controllers in Examples 3.2.1, 3.2.2, cf. (3.2.25), (3.2.26), have been obtained the same way. The following example illustrates the case of using the Leitmann-Gutman controller.

EXAMPLE 5.3.1. Let $\bar{x} = \bar{f}^0(\bar{x})$ be a dynamical system with the asymptotically stable equilibrium at the origin of $\mathbb{R}^N$ achieved by the Liapunov function $V^0(\cdot)$ taken as the energy $E^0(\bar{x})$ of the above system, with $\nabla V^0(\bar{x})^T \cdot \bar{f}^0(x) \le -c$ on some Δ_0 enclosing the origin. Suppose the system is perturbed by an uncertainty $\bar{w} \in W$ and must be controlled against it, that is, we investigate a case of the typical Leitmann system (2.2.9):

$$\dot{\bar{x}} = \bar{f}^0(x) + B(\bar{x})(\bar{u} + \bar{w}) \ ,$$

with $B(\bar{x})$ a continuous $N \times r$ matrix function, and using the uncertainty band $W = \{\bar{w} \mid \|\bar{w}\| \le \rho(x)\}$, see (2.2.10) and Gutman-Leitmann [1]. Taking the same energy $E^0(\bar{x})$ as our test function $V(\bar{x})$, we shall satisfy (iii) if

$$\min_{\bar{u}} \max_{\bar{w}} \int_{t_0}^{t_c} \nabla E^0(\bar{x})^T B(\bar{x})(\bar{u} + \bar{w}) \ \le \ -(v_0^+ - v_T)$$

which is satisfied if

$$\min_{\bar{u}} \max_{\bar{w}} \nabla E^0(\bar{x})^T \cdot B(\bar{x})(\bar{u} + \bar{w}) \ \le \ \frac{1}{T_c^+} (v_0^+ - v_T) \ .$$

The latter condition is implied by the Leitmann-Gutman controller, briefly LG-controller, see Leitmann [2], defined by

$$\bar{u}(t) \ = \ \begin{cases} -\rho(\bar{x}) \ \dfrac{\alpha(\bar{x})}{\|\alpha(\bar{x})\|} \ , & \alpha(\bar{x}) \ne 0 \\[4mm] 0 \ , & \alpha(\bar{x}) = 0 \end{cases}$$

where $\alpha(\bar{x}) = \nabla E^0(\bar{x})^T B(\bar{x})$. Simple calculation (see Gutman [3], Gutman-Palmor [1]) shows that (5.3.12) substituted into the Liapunov derivative makes $\dot{V}(\bar{x}) \leq \nabla E^0(\bar{x})^T \cdot \bar{f}^0(\bar{x}) < 0$ thus implying (iii). Note that $P(\cdot)$ defined by (5.3.12) is smooth and single valued for $\alpha \neq 0$ but it is a set for $\alpha = 0$. It means that for the latter case $\bar{u}$ chooses its values in a set and the choice depends upon unknown $\bar{w}$. However, as $\dot{V}(\bar{x}) < 0$, $\alpha = 0$ for any $\bar{u}$, say $\bar{u} = 0$, for a short time $V(\cdot)$ decreases. We are secure in our task as long as $\bar{u}$ is not switched off for too long. $\quad\square$

There is an alternative to defining $P(\cdot)$ from Corollary 5.3.2. It follows from the inverse Liapunov method proposed by Liu-Leake [1], see also Peczkowski-Liu [1], and is based on the following general property.

Let $\bar{y}, \bar{z}$ be two real N-vectors with $\bar{y} \neq 0$. Given a scalar α, we have $\bar{y} \cdot \bar{z} = \alpha$ if and only if there is a real skew symmetric $N \times N$ matrix C such that

$$\bar{z} = [(\alpha/\|\bar{y}\|^2)I + C]\bar{y} \tag{5.3.8}$$

where I is the unit matrix. The above gives an explicit but nonunique solution to the implicit algebraic equation $\bar{z} \cdot \bar{y} = \alpha$, $\bar{y} \neq 0$. When $\bar{y} = 0$, α must be zero for a solution to exist, and $\bar{z}$ is then arbitrary. Clearly C is not unique so it produces a class of values of $\bar{z}$, Liu-Leake proved that this class is an equivalence class, so that any $\bar{z}$ is a legitimate representative of such a class. The relation (5.3.8) has been used by Skowronski [13] for design synthesis by determining $\bar{f}$ in (2.2.5)'. We may use it as a control condition, see Skowronski [32]. Observe that, applying (5.3.8) to the inner product in (5.3.3) of Conditions 5.3.2, we obtain that any

$$\bar{f}(\bar{x}, \bar{u}, \bar{w}, t) = \left\{ \frac{-(v_0^+ - v_T)}{T_c[V(\bar{x})]^2} I + C \right\} \nabla V(\bar{x}) \tag{5.3.8'}$$

considered for all $\bar{w} \in W$ implies (5.3.3), thus satisfying Conditions 5.3.2, with a free choice of the skew symmetric C. The choice generates an equivalence class of the above $\bar{f}$'s and thus, if $\bar{u}$ is calculated from (5.3.8)', an equivalence class of functions $\bar{u}(\cdot)$.

EXAMPLE 5.3.2. Consider the system

$$\dot{x}_1 = x_2$$
$$\dot{x}_2 = -3x_2 - x_1 + u,$$

with the objective to collide with the target-strip $T : 0 \leq x_2 \leq 2$, $\forall x_1$,

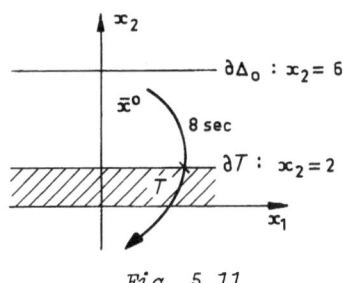

Fig. 5.11

shown in Fig. 5.11, starting from some $\Delta_0 : x_2 \leq 6$, $\forall\, x_1$. Choosing $V(\bar{x}) = \frac{1}{2} x_2^2$ we obtain $v_0^+ = 18$ and $v_T = 2$, while requiring $T_c = 8$ sec. Then (5.3.8)' yields

$$
\begin{pmatrix} x_2 \\ -3x_2 - x_1 + u \end{pmatrix} = \left\{ \frac{-2}{x_2^4/4} \begin{pmatrix} 1 & 0 \\ 0 & 1 \end{pmatrix} + \begin{pmatrix} 0 & m \\ -m & 0 \end{pmatrix} \right\} \begin{pmatrix} 0 \\ x_2 \end{pmatrix} = \begin{pmatrix} mx_2 \\ -8/x_2^3 \end{pmatrix}
$$

wherefrom $x_2 = mx_2$ or $m = 1$ and $-3x_2 - x_1 + u = -8/x_2^3$. Hence the control program is

$$
u = \begin{cases} (-8/x_2^3) + 3x_2 + x_1 , & |x_2| \geq \beta , \\ 0 , & |x_2| < \beta , \end{cases}
$$

with β calculated as in (3.3.8).   □

Note here that choosing $P(\cdot)$ either from (5.3.7)' or (5.3.8)' provides overkill, as what we really need for collision is only the outflux estimated by (5.3.3), and not a monotonically negative rate of change $\dot{E}(\bar{q},\dot{\bar{q}})$. However, taking (5.3.3) directly as a control condition gives an integral format of the controller which is inconvenient and inaccurate (free constant of integration).

The following two sets of conditions serve the case of the stipulated time collision.

CONDITIONS 5.3.3. The system (2.2.6)' is strongly controllable on Δ for collision with T before T_c^+ , if Conditions 5.3.2 hold for $T_c(\bar{x}^0, t_0, \bar{\phi}(\cdot)) \leq T_c^+$.

CONDITIONS 5.3.4. The system (2.2.6)' is strongly controllable on Δ_0 for ultimate collision in T if given T_c^- , Conditions 5.3.2 hold for $T_c(\bar{x}^0, t_0, \bar{\phi}(\cdot)) \geq T_c^-$.

The above follows by obvious adjustment to the proof of Conditions 5.3.2. The Corollaries 5.3.1, 5.3.2, 5.3.3 are immediately adjustable, and so are the methods for defining $P(\bar{x},t)$.

The conditions for penetration are obtained by slight adjustments to Conditions 5.3.2.

CONDITIONS 5.3.5. The system (2.2.6)' is strongly controllable on Δ_0 for penetration of T , if there is $\bar{P}(\cdot)$ defined on D and a C^1-function $V(\cdot) : D \times \mathbb{R} \to \mathbb{R}$ such that

  (i) $V(\bar{x},t) \geq v_T$, for $\bar{x} \notin T$, $t \in \mathbb{R}$;

  (ii) $V(\bar{x},t) < v_0^+$, for $\bar{x} \in CT$, $t \in \mathbb{R}$;

  (iii) for each $\bar{u} \in \bar{P}(\bar{x},t)$ there is $T_p(\bar{x}^0,t_0,\bar{\phi}(\cdot)) > 0$ yielding

$$\int_{t_0}^{t_0 + T_p} [\frac{\partial V(\bar{x},t)}{\partial t} + \nabla_{\bar{x}} V(\bar{x},t)^T \cdot \bar{f}(\bar{x},\bar{u},\bar{w},t)] dt \leq -(v_0^+ - v_T)$$

(5.3.9)

  for all $\bar{w} \in W$.

It follows by the same argument as for Conditions 5.3.2 but with (5.3.4) made a weak inequality and $V(\bar{x}^0,t_0) < v_0^+$, for the non-trivial case $\bar{x}^0 \notin$ int T . Corollaries 5.3.1, 5.3.2, 5.3.3 apply for penetration as well, and so do the stipulated time Conditions 5.3.3, 5.3.4.

In Section 3.3 we discussed stabilization below and above some E-levels and controlling the motions to a specific energy cup in Δ_L. However, it has been difficult to establish the means of reaching prescribed positions at a given energy level. In the case related to Fig. 3.5, paths C and D , we aimed at a target by switching off the dissipative controller (3.3.12) after a suitable E-level has been reached and then using the conservative controller calculated from (3.3.16) in order to travel along this E-level up to the desired destination. Looking for a periodic steady state, we may use (4.1.18). In order to get over a threshold, we had to use in turn an accumulative controller (3.3.17). The test function used for stabilization below and above some E-level is referred to as $E^{\pm}$ respectively. Assuming $v_0 = v_0^+$, $v_T = v_B$ the cited controllers are those obtainable from, and satisfying, the corollaries of our present section on collision and penetration, when the boundary of the target is determined by some E-level inside or outside a cup.

The conditions for collision and penetration are however more generally applicable, as they allow us to attain a target totally located on a single

E-level as well. The latter, together with reaching the prescribed levels, makes up for what we may call *maneuvering*, that is, sequential colliding and penetrating of a sequence of prescribed targets located on a desired route of the system. The time intervals between these collisions and penetrations may obviously be stipulated. This is for instance the case of a so called pick-and-place robotic manipulator working between a conveyer and, say, a milling machine.

When aiming at collision irrespective of energy levels or on a given energy level, the test function used may not necessarily be energy. It may for instance be a function related to a product of state variables, like $V = x_1 \cdot x_2 \cdot, \ldots, \cdot x_{N-2}$, which for even N will be positive within a certain region of attraction to T enclosing Δ_0, with simple partial derivatives $\partial V / \partial x_i = x_1 \cdot, \ldots, \cdot x_{i-1} \cdot x_{i+1} \cdot, \ldots, \cdot x_{N-2}$ likely to produce negative inner products with components of $\bar{f}$. In practical use of the conditions for various types of collision and several targets, it is convenient to cover the area concerned by several functions $V(\cdot)$, with overlapping domains rather than to search for a single function covering everything, see Stalford [2], Diligenski [1]. The latter may prove very difficult to find and certainly will always be difficult to adjust if any need for a change arises. We may for instance use a sequence of functions $V_\sigma(\cdot)$, $\sigma = 1, \ldots, \ell < \infty$ each defined on a subset D_σ of D, these subsets not necessarily disjoint. Then, however, we may have the problem of interface, particularly when smooth functions are needed and thus some "corner conditions" would have to be satisfied for continuity of the derivative. Take for instance the system

$$\dot{x}_1 = -ux_1 + wx_1$$
$$\dot{x}_2 = -ux_2 - wx_2 \qquad (5.3.10)$$

with a target about $(0,0)$. In order to cover an arbitrary Δ_0 in $\mathbb{R}^2$ it seems best to choose two test functions over two regions: for $x_1 x_2 > 0$, use $V_1 = x_1 x_2 > 0$ with $\dot{V}_1 = x_2(-ux_1 + wx_1) + x_1(-ux_2 - wx_2) = -2ux_1 x_2$, and for $x_1 x_2 < 0$, use $V_2 = -x_1 x_2 > 0$ with $\dot{V}_2 = 2ux_1 x_2$, without difficulty in interfacing the domains at $x_1 x_2 = 0$. In other cases, however, such an interface may prove difficult. On the other hand, the interface may not be needed if the union of D_σ's fills up Δ_0 and if we may use the relaxed Conditions 5.3.1 which do not require smooth $V(\cdot)$.

In our control for maneuvering (sequential collision and penetration) we may come across steady state limit sets, but they rarely are our present targets since, by definition, they mean capture rather than collision.

EXAMPLE 5.3.3. Let us continue Examples 1.1.1, 3.2.1, 3.3.1, with the perturbed system, cf. (3.2.19), (3.3.17),

$$\left. \begin{array}{l} \dot{x}_1 = x_2 \\ \dot{x}_2 = -ax_1 + bx_1^3 - cx_1^5 - d|x_2|x_2 + u + R(t) \end{array} \right\} \qquad (5.3.11)$$

and phase-plane pattern of Fig. 5.12, redrawn from Fig. 3.7. For simulation purposes, let $a = c = 1$, $b = 2.1$, $d \in [-1,1]$, unknown, yielding stable equilibria at $x_1^e = 0, \pm 1.17$ and unstable equilibria at $x_1^e = \pm 0.85$.

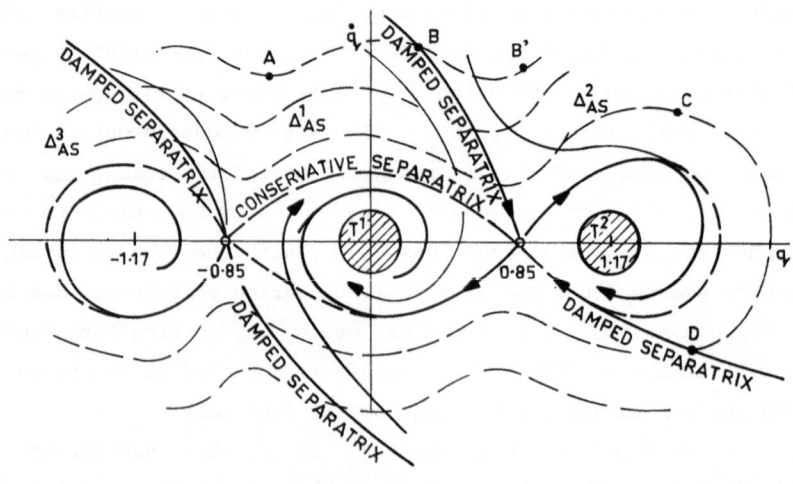

Fig. 5.12

Following Example 3.2.1 we have at our disposal the dissipative control program which is a modified (3.2.23) or similarly designed accumulative and conservative controllers. The dissipative controller is obtained from (3.2.21) and $\dot{E}^-(\bar{x}) \leq -(\delta h/T_c)$, where $\delta h > 0$ is the energy change from initial values to the target, the accumulative and conservative controllers are obtained from (3.2.21) with $\dot{E}^+(\bar{x}) \geq \delta h/T_c$, and $\dot{E}^-(\bar{x}) = 0$, respectively. Similarly to (3.2.22) we obtain for the dissipative control condition, see also (3.3.12):

$$\min_u (u \, \mathrm{sgn}\, x_2) \leq -R(t) \, \mathrm{sgn}\, x_2 + \min_d (dx_2^2) - \frac{\delta h}{|x_2|T_c} \qquad (5.3.12)$$

implied by the dissipative controller defined as

$$\min_u (u \, \mathrm{sgn}\, x_2) \leq \begin{cases} -R(t) \, \mathrm{sgn}\, x_2 - x_2^2 - \dfrac{\delta h}{|x_2|T_c} \, , & |x_2| \geq \beta \\[2ex] -R(t) \, \mathrm{sgn}\, x_2 - x_2^2 \, , & |x_2| < \beta \end{cases} \qquad (5.3.13)$$

where $\beta = \mathrm{const}$ is calculated as in (3.3.13). The accumulative control condition (3.3.17) gives

$$\max_{u}\ (u \operatorname{sgn} x_2)\ \geq\ -P(t) \operatorname{sgn} x_2 + \max_{d}\ (dx_2^2) + \frac{\delta h}{|x_2| T_c} \qquad (5.3.14)$$

implied by the accumulative controller

$$\max_{u}\ (u \operatorname{sgn} x_2)\ \geq\ \begin{cases} -R(t) \operatorname{sgn} x_2 + x_2^2 + \dfrac{\delta h}{|x_2| T_c} & |x_2| \geq \beta\ , \\[3mm] -R(t) \operatorname{sgn} x_2 + x_2^2\ , & |x_2| < \beta\ . \end{cases} \qquad (5.3.15)$$

By the same argument, the conservative control condition follows from
(3.3.15) or specifically (3.2.22) with c = 0 :

$$\min_{u}\ (u \operatorname{sgn} x_2)\ =\ -R(t) \operatorname{sgn} x_2 - x_2^2\ , \qquad (5.3.16)$$

and no need to introduce β . Alternatively the controller may be obtained
directly from (3.2.20):

$$u = d|x_2|x_2 - R(t)\ . \qquad (5.3.17)$$

Let us now specify two targets, T^1 about the basic equilibrium and
T^2 about $\bar{x}^e = (1.17,0)$, see Fig. 5.12. For the free system u ≡ 0 ,
R ≡ 0 with positive damping d = 1 , we would have three stable attractors
at (0,0) , (1.17,0) and (-1.17,0) with the corresponding regions of
asymptotic stability Δ_{AS}^1 , Δ_{AS}^2 , Δ_{AS}^3 indicated in Fig. 5.12, and separated
by the damped separatrices. For the perturbed system R(t) ≠ 0 and subject
to undertain d , possibly negative, we need the controllers introduced
above, their application depending upon initial conditions and the selection
of target.

Suppose we start from Δ_{AS}^1 as indicated in Fig. 5.12 by point A and
aim at collision not with T^1 but T^2 . The obvious choice is to use the
conservative controller (5.3.16) transferring our t_0 -family of motions to
the damped separatrix at point B and then switching to the dissipative
controller which was calculated from (5.3.12), but this is only to offset
the perturbation and negative damping, that is, for d ∈ [-1,0] . The
δh = h^0 - h_T^2 to be substituted is known, being obtained from $h^0 = E^-(\bar{x}^0)$
and ∂T^2 : $E^-(\bar{x}) = h_T^2$. Offsetting R(t) and d ∈ [-1,0] will let us
follow the damped separatrix to the saddle (0.85,0) and then along it to
T^2 . The same effect is achieved, perhaps more economically by using
d ∈ [-1,1] , if we do not switch the controller at B but use the conservative trajectory to some other position in Δ_{AS}^2 , say B' and then use the
dissipative controller. This applies as long as B' is still in Δ . The
same method of control may apply for passage from Δ_{AS}^3 to T^1 . Programming

for collision with T^1 from Δ^2_{AS}, say from point C, we would use the conservative controller until or past D, and then switch to the dissipative controller.

If we wanted sequential collision, say after hitting T^1 to collide with T^2, the obvious way of doing it is to rise the motion with the accumulative controller to the threshold $h_E = E^-(0.85,0)$ and then use the dissipative controller.   □

One of the obvious subcases of the sequential collision in the sense discussed above is the transfer from one stable equilibrium to another, partly indicated in the previous example. It applies to various physical scenarios, one of the typical being for instance the reorientation of a satellite from one stable periodic orbit into another, see Anchev-Melinkyan [1].

EXERCISES 5.3

5.3.1 Consider the system

$$\ddot{q} + d|q|\dot{q} + aq + bq^3 = u$$

with $a,b,d > 0$, $q \in \mathbb{R}$, $|u| \le \hat{u}$ and initial conditions $q^0 = 0$, $\dot{q}^0 = 0$. Find the control program which will rise the trajectory to the orbit $\frac{1}{2}\dot{q}^2 + \frac{1}{2}aq^2 + \frac{1}{4}bq^4 = h$ and hold it there for rotation π, then drop it down to the orbit at $\frac{1}{2}h$ for another π and then drop it down to $(0,0)$. Convert the angular distances into time intervals.

5.3.2 For the system

$$\ddot{q} + d\dot{q} + aq - bq^3 = u ,$$

with $q \in \mathbb{R}$, $d,a,b > 0$, $|u| \le \hat{u}$, find the equilibria and the separatrix. Then given the target

$$T : -\varepsilon \le q \le \varepsilon , \qquad\qquad -\varepsilon \le \dot{q} \le \varepsilon ,$$

specify the controller which generates the collision from all points within the energy cup. What would the saturation value $\hat{u}$ have to be, if the trajectories started outside the cup?

5.3.3 Consider the double pendulum of Example 1.1.3 with the motion equations (1.1.21) shown in Fig. 1.8. Calculate the total energy, power and equilibria of the system. Reducing the work region Δ

of the system to the range $-\pi \le q_i \le \pi$, $i = 1,2$, establish Δ_L and the thresholds. Then, assuming initial states outside Δ_L, find the controller and calculate the time of colliding with a sequence of targets surrounding the local minima of H. Find the controller allowing the transfer from one such minimum to another.

5.3.4   A homogenous slender rod of length ℓ is pinned at one end so that it is free to swing in a vertical plane from its initial horizontal (angular) position $q^0 = \pi/4$ to the vertical position $q^f = \pi/2$. The weight of the rod is located at its C.G. within the distance $\ell/2$ from either end. Find a control program for an actuator located at the pinned joint with the objective of colliding with q^f at zero velocity.

5.3.5   The system $\ddot{q} + 2d\dot{q} + 9q = u$ has a large damping $d = \sqrt{10}$ for small amplitudes $q \in [-2,2]$ and ignorable damping $d = 0$ outside this interval. Given initial conditions $q^0 = 0$, $\dot{q}^0 = 0$ design the controller such that the motion amplitudes:

    (i)   do not exceed some $\hat{q} = const$;

    (ii) cross some $\tilde{q} = const$ only once.

5.4   REGIONS OF CONTROLLABILITY

As mentioned at the opening of Section 5.3, determining the regions of controllability (that is, wherefrom the controllers may be used) is the second basic question in our control problem. It has been investigated, ever since the controllability problem appeared, along a considerable number of avenues. However, the Liapunov formalism seems to be the most successful method employed, in spite of the fact that the estimates of the regions must be qualified by the choice of $V(\cdot)$. In fact the search for "the best" Liapunov function, that is, the $V(\cdot)$ which offers the maximal region of controllability, is a frequently discussed topic, beginning with Hall [1], Tarnave [1], Willems [1], Shilds [1], Storey [1], Mansour [1], Noldus-Galle-Josson [1], later Lewandowska [1], Vanneli-Vidyasagar [1], and recently Blinov [1] and Olas [2].

The works on estimating the regions start with investigating regions of attraction cited already in Section 3.4, and reviewed recently by Chiang-Hirsch-Wu [1]. Referring to controllability estimates, we recently have two main lines of approaching the problem: *analytic* or geometric, and *numerical*, often combined. The analytic line started with the early work by Roxin-Spinadel [1] developing the concept of reachable sets, dual to controllability, followed by Storey [1], Noldus-Galle-Josson [1], Grantham-Vincent [1], Gayek-Vincent [1], and a few other works. It may be considered a sideline for the study on the degree of controllability developed in the recent decade, associated mainly with the problems of large flexible space structures (LSS), and reviewed by Schmitendorf [2] whose results we already applied. Generalizing this, Skowronski [38] and Skowronski-Stonier [2] suggested a method for determining an estimate of the maximal controllable Δ_0 using the necessary and sufficient conditions for controllability and the concept of semi-barriers between corresponding strong regions.

The numerical approach is based upon retro-integration (finding trajectories for $-t$) from the target boundary either for a specific duration, when we want to define $\Delta_c(T_c)$, or for $t \to -\infty$, aiming at $\partial\Delta_c$. The method has been widely used by Isaacs [1] and in terms of the Liapunov formalism, that is, taking the retro-trajectories across V-levels, it was developed by Grantham [1], Grantham-Vincent [1], Vincent-Skowronski [1] and Vincent [3]. It appears now that the Liapunov levels, generated by retro-integration, agree very conveniently with the cell-to-cell mapping method originated by Hsu, see Hsu [1], and developed with the successful use of V-levels, see Flashner-Hsu [1], Guttalu-Flashner [1], Flashner-Guttalu [1] and Guttalu-Skowronski [1].

First, given $V(\cdot)$ and $\bar{P}(\cdot)$ we look for the *strong recovery region* $\Delta_c(P)$, later specified by stipulated time. Assuming that ∂T and $\partial\Delta_c(P)$ are defined by V-levels, see (5.3.6), by Corollary 5.3.1, Condition (5.3.3) defines $\Delta_c(P)$ and given $\bar{u}^*, \bar{w}^*$ it yields the drop in $V(\bar{x})$ estimated by

$$\int_{t_0}^{t_c^+} \nabla V(\bar{x})^T \bar{f}(\bar{x}, \bar{u}^*, \bar{w}^*, t) dt \ .$$

Then by the nesting property of the continuous T_c^+-family of isochronal levels $\partial\Delta_c(P, T_c)$ for any two T_c', T_c'' such that $0 < T_c'' < T_c' < \infty$, we have $V(\bar{x}(t_c'')) < V(\bar{x}(t_c'))$. Note that upon approach to a threshold, we may use a negative test function but the inequality is still preserved. It follows that for each $\varepsilon > 0$ there is $T(\varepsilon)$ such that

$$\left| \int_{t_c''}^{t_c'} \nabla V(\bar{x})^T \bar{f}(\bar{x}, \bar{u}^*, \bar{w}^*, t) dt \right| < \varepsilon \tag{5.4.1}$$

for $T < t_c'' < t_c' < \infty$, which by the Cauchy criterion for convergence of improper integrals implies that there exists a finite limit for $t_c^+ = t_0 + T_c^+$, expressed by

$$v_T + \lim_{T_c^+ \to \infty} \int_{t_0}^{t_c^+} \nabla V(\bar{x})^T \bar{f}(\bar{x}, \bar{u}^*, \bar{w}^*, t) \, dt = v_c < \infty \, . \tag{5.4.2}$$

The corresponding level $V(\bar{x}) = v_c$, determines the boundary of the region $\Delta_c(P, T_c^+)$, $T_c^+ \to \infty$, thus of $\Delta_c(P)$. Obviously (5.4.2) specifies also the degree of strong controllability, namely

$$v_c = \lim_{T_c^+ \to \infty} \rho(T_c^+) \, . \tag{5.4.3}$$

Now that we know that the limit v_c specifying the boundary $\partial\Delta_c(P)$ does exist, we may find methods for determining it, both for $\Delta_c(P)$ and $\Delta_c(P, T_c)$. One of such methods is the mentioned retro-integration from ∂T. In classical terms it applied to unique trajectories of autonomous systems, but its usage may be augmented. By definition, $\Delta_c(P)$ forms the envelope of the $(\bar{x}^0, t_0)$ – family of $\Phi(\bar{x}^0, t_0, [t_0, t_c^+])$:

$$\text{int } \Delta_c(P, T_c) = \{ \cup \Phi(\bar{x}^0, t_0, [t_0, t_c]) \,|\, \bar{x}^0 \in \Delta_c(P) \,, \, t_0 \in \mathbb{R} \} \, . \tag{5.4.4}$$

Let us look closer at one of the motion cones of such a family in the process of approaching ∂T. From the quoted Filippov conditions, we know that given $\bar{u}^*, \bar{w}^*$ we obtain a motion from the selector equation (2.2.5)'. Since $\Delta_c(P)$ is covered by uniform strong controllability, at each $\bar{x}^0$ of this region we can identify a t_0-family of such motions. Note that all such families of $\Phi(\bar{x}^0, \mathbb{R}, [t_0, t_c])$ pass through the same $\bar{x}^0$. Hence this point may be determined by a retrograde trajectory $\bar{\phi}(\bar{x}^0, [t_c, t_0])$ of (2.2.5), generated by $\bar{u}^*, \bar{w}^*$, equivalent to one of the t_0-families of motions and starting from some $\bar{\phi}(\bar{x}^0, t_c) = \bar{x}^c \in \partial T$. Such a retrograde trajectory, represented time-discretely by a point-to-point mapping, is in turn equivalent to forward integration of the equation:

$$\dot{\bar{x}}(\tau) = -\bar{f}^*(\bar{x}(\tau)) \,, \quad \tau = -t \leq 0 \tag{5.4.5}$$

where $\bar{f}^*(\bar{x}) = \bar{f}(\bar{x}, \bar{u}^*, \bar{w}^*)$ of (2.2.5). By (5.4.4) and the above, $\Delta_c(P)$ is filled up by the retro-trajectories concerned, see Fig. 5.13.

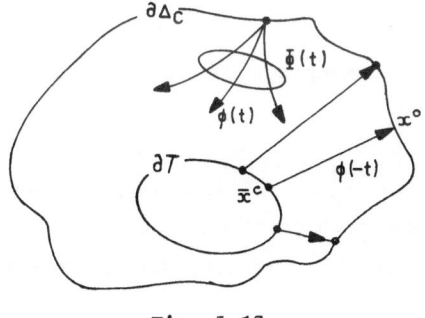

Fig. 5.13

Consequently, the boundary $\partial\Delta_C(P)$ is that of the family of trajectories of (5.4.5) for $\tau \to -t_c$. Now we ask: How can we determine the boundary of such a family of trajectories?

Let $C_0 : V(\bar{x}) = \text{const} = v_T$ and let $\{t_c + \tau_j\}$, $j = 0,1,2,\ldots,\tau_j \leq 0$ be a decreasing sequence of time instants. Denote C_j as the set of images of points on C_0 under the mapping along the trajectory $\tilde{\phi}(x(\tau_0),\tau)$, $\tau_0 = 0$ of (5.4.5) at the instant τ_j . Then upon departing from the target, each $C_j \subset C_{j+1}$, with the distances measured in terms of $V(\cdot)$.

There are two cases possible. Either the surface ∂T is disjoint from the hypothetical $\partial\Delta_C(P)$, or it is not. In the first case, the boundary $\partial\Delta_C(P)$ is formed only by an appropriate terminal C_j . Let $C_{-\infty}$ be such a terminal map of C_0 obtained for $\tau_j \to \infty$. By (5.4.2) we know that $C_{-\infty}$ exists, is finite and defined by $V(\bar{x}) = v_c$. The $\partial\Delta_C(P)$ so obtained is a candidate which must be confirmed by Conditions 5.3.2 with $v_0^+ = v_c$. The stipulated time candidate region $\Delta_C(P,T_c)$ is then formed by retracing points along $\tilde{\phi}(x(\tau),\mathbb{R}^-)$ to their values $\tilde{\phi}(x(\tau),\tau_c)$ where $\tau_c = -t_c$. The corresponding boundary $\partial\Delta_C(P,T_c)$ is thus formed by the cut-off V-level C_c .

In the second case, when $\partial T \cap \partial\Delta_C(P) \neq \phi$, the surface ∂T cuts into $\partial\Delta_C(P)$ at least in one point and $\partial\Delta_C(P)$ has two ends separated by ∂T . The retro-trajectories emanating from these ends form the boundary $\partial\Delta_C(P)$ together with an appropriate terminal V-level, C_j , that cuts off the whole family of retro-trajectories. This also applies to the stipulated time $\partial\Delta_C(P,T_c)$.

COMPUTATIONAL ALGORITHM:

Step 0: Choose a test function $V(\cdot)$. Select stopping criteria, a distance index between consecutive levels C_j and C_{j+1} , denoted $d_0(C_j,C_{j+1})$ and measured in terms of $V(\cdot)$, a convergence criterion $\varepsilon > 0$ and a maximum number of backward mappings K . Select the integration time step $\delta\tau$.

Step 1: Choose p points P_r^j , $r = 1,2,\ldots,p$ on the level curves $C_j : V(\bar{x}) = v_j > 0$. Map each point once forward. If all points map inside C_j , take C_j to be C_0 . If not, set j to $\frac{1}{2}j$ and repeat this step.

Step 2: Using equation (5.4.5) map the points of C_j from $\tau = \tau_j$ to $\tau_{j+1} = \tau_j + \delta\tau$, with initial conditions $\bar{x} = P_r^j$, $r = 1,2,\ldots,p$. The result of the integration are points P_r^{j+1}, $r = 1,2,\ldots,p$, which define C_{j+1}. Set $j = j+1$ and repeat.

Step 3: If $d_0(C_j, C_{j+1}) < \epsilon$, $j > K$ stop. The level C_j is the required boundary $\partial\Delta_c(P)$. If not, go to Step 1.

Step 4: Retrace the integration to the level obtained at $t = t_c$ which results in the boundary $\partial\Delta_c(P, T_c)$.

EXAMPLE 5.4.1. Let us consider the autonomous Duffing system in the following particular format

$$\left.\begin{array}{l} \dot{x}_1 = x_2 \\[2mm] \dot{x}_2 = -ux_2 - kx_1 + wx_1^3, \quad k > 0 \end{array}\right\} \tag{5.4.6}$$

with the range of uncertainty $w \in [0,2]$ and the control constraints $U : u \in [u^-, u^+]$. The system has a stable equilibrium at $(0,0)$ and two thresholds at the unstable equilibria $x_1 = \pm\sqrt{k/w}$, $x_2 = 0$, depending upon w and fixed for w^* which comes from extremizing the energy

$$E(x_1, x_2, w) = \tfrac{1}{2} x_2^2 + \tfrac{1}{2} kx_1^2 - \tfrac{1}{4} wx_1^4. \tag{5.4.7}$$

If we now locate the target T about $(0,0)$, then we shall let $V(\bar{x}) \overset{\Delta}{=} E^-(x_1, x_2) = \tfrac{1}{2} x_2^2 + \tfrac{1}{2} kx_1^2 - \tfrac{1}{2} x_1^4$ with $w^* = 2$ minimizing $E(x_1, x_2, w)$. Consequently the thresholds are at $x_1 = \pm\sqrt{k/2}$, $x_2 = 0$, and $v_T = V(\sqrt{k/2}, 0)$, that is, taking ∂T as the conservative separatrix:

$$\partial T : x_2^2 + kx_1^2 - x_1^4 = \tfrac{1}{2} k^2. \tag{5.4.8}$$

We choose the program $P : u = c|x_2|$, $c > 0$ satisfying Conditions 5.3.2 by yielding $\dot{V} = -ux_2^2 = -c|x_2|^3$. We may now investigate the retro-trajectories. First, let us write (5.4.5) for the present example in the form

$$\left.\begin{array}{l} \dot{x}_1 = x_2 \\[2mm] \dot{x}_2 = -c|x_2|x_2 - kx_1 + 2x_1^3 \end{array}\right\} \tag{5.4.9}$$

with retro-trajectories starting at (5.4.8). As no trajectory from $\bar{x}^0$ lying outside $[-\sqrt{k/2}, \sqrt{k/2}]$ may reach T (see Example 1.1.1, Fig. 1.4), the boundary ∂T is not disjoint from $\partial\Delta_c(P)$ and these two meet at $(-\sqrt{k/2}, 0)$ and $(+\sqrt{k/2}, 0)$. Thus the retro-trajectories from these two points form part of $\partial\Delta_c(P)$ which is completed (closed off) by a suitable $C_{-\infty}$ defined by $E^-(x_1, x_2) = v_c$. Compare here the damping separatrices in Example 1.1.1. For computing the above retro-trajectores, we use the Hsu method of point-to-point mapping letting $c = 1.0$, $k = 2.0$. Initially 5000 points were distributed uniformly on a circle of radius 0.1 about $(0,0)$. The first closed curve in Figure 5.14 is the retro-trajectory obtained by integrating from $\tau = 0$ to $\tau = 2.3$. The second curve was obtained by continuing the integration to $\tau = 2.7$. The pattern in Fig. 5.12 may be compared with the analytic solutions (equation (5.4.9) is integrable) obtained in Example 1.1.1.   □

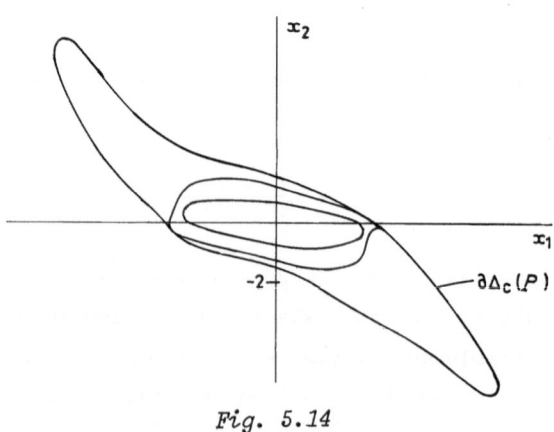

Fig. 5.14

Let us now see what can be done in estimating the region of strong controllability Δ_c, without qualifying it by given $\bar{P}(\cdot)$ or T_c, or both. Since (5.3.3) of Conditions 5.3.2 is necessary and sufficient subject to (5.3.6), it suffices to find uncertainty $\bar{w}$ contradicting the existence of robust (winning) $\bar{P}(\cdot)$ at some $(\bar{x},t)$, in order to ensure that such $(\bar{x},t)$ does not belong to Δ_c. The set of such points closest (in the sense of Section 3.1) to the target T specifies the boundary $\partial\Delta_c$. For instance, if the energy threshold about some cup bounds Δ_c, some $\bar{w}(\cdot)$ may pull a motion over the threshold to the next cup, see Example 5.4.1 below. We formalize the above in terms of the weak $\bar{P}(\cdot)$ - barrier of Section 3.1 with Q specified as collision, see Fig. 5.15.

Let $V(\cdot)$ be the test function and $\bar{P}(\cdot)$ the controller of Conditions 5.3.2, and let ∂T, $\partial\Delta_c$ be defined by the V-levels v_T, v_c respectively, see (5.3.6), the first known, the second to be determined.

Then let us take $V_B(\bar{x})$ of Conditions 3.1.2 as $V_B(\bar{x}) \overset{\Delta}{=} V(\bar{x})$ and the hypothetical $\mathcal{S}$ defined by a V_B-level bounding below its semi-neighborhood determined by the inequality

$$\nabla V(\bar{x})^T \bar{f}(\bar{x},\bar{u},\bar{w}^*,t) \geq 0 \, , \qquad\qquad (5.4.10)$$

for suitable $\bar{u} \in \bar{P}(\bar{x},t)$. The weak $\bar{P}(\cdot)$ - barrier is one of these surfaces, selected by the criterion that there is no other surface like that between it and $\partial \Delta_C : V(\bar{x}) = v_c$. Unfortunately, we do not know v_c . However, we know that there is no such $\mathcal{S}$ surface in Δ_C , so we may use the criterion that there will not be a weak $\bar{P}(\cdot)$ - nonpermeable surface between the pro-posed surface and the boundary $\partial T : V(\bar{x}) = v_T$, which we know. In practical terms, we search for the lower bound of the set defined by (5.4.10) and the surface $\partial \Delta_C = B_C^P$ is the level $V(\bar{x}) = v_c$ that passes through such a bound. Indeed, when the motion concerned starts from a point satisfying (5.4.10), the latter condition implies for each $t \geq t_0$, that

$$\int_{t_0}^{t} f_0(\bar{x},\bar{u},\bar{w}^*,t)dt > 0$$

contrary to (5.3.3). Thus the motion never crosses B_C^P again.

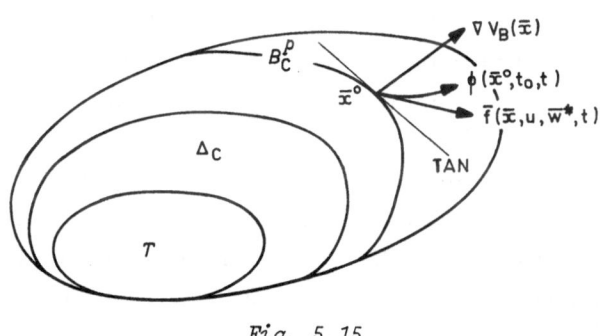

Fig. 5.15

When the target T is located about a Dirichlet stable equilibrium in some energy cup, as mentioned several times before, for strong controlla-bility for collision we use $V(\bar{x}) = E^-(\bar{x})$ and a dissipative controller $\bar{P}(\cdot)$ generated by Corollary 5.3.2 or $\min_{\bar{u}} \max_{\bar{w}} f_0(\bar{x},\bar{u},\bar{w},t) < 0$. By Conditions 5.3.2, such a controller secures the required controllability against all $\bar{w}(\cdot)$'s on the *unknown* Δ_C . As the E^--levels increase from $\partial T : E^-(\bar{x}) = v_T$ to the threshold, we come across the level $E^-(\bar{x}) = v_C$ at the first appearance of the maximizing $\bar{w}^*$ which actually breaks up the robustness of the dissipative strong controller, by generating $f_0(\bar{x},\bar{u},\bar{w}^*,t) = \dot{E}^-(\bar{x}) \geq 0$, that is, (5.4.10) along some motion. Such a level forms the weak $\bar{P}(\cdot)$ - barrier $B_C^P = \partial \Delta_C$. As the latter is the set defined by $\dot{E}^-(\bar{x}) = 0$ by Proposition 4.1.1 it may become a negative limit set Λ^- , or a positive limit set for the retro-trajectories from ∂T . In this sense both methods, retro-integration and the weak barrier, complement each other.

Obviously the breakup of the strong $P(\cdot)$ may not occur below the threshold, or even immediately above it. The weak $P(\cdot)$-barrier may in fact surround several energy cups. Where the program breaks up will very much depend on whether the t_0-family of motions crosses the unstable equilibrium (saddle position) or avoids it in the process of entering the cup concerned. In Example 5.3.3, the strong region for collision with T^1 from the point C through D and the saddle, that is, crossing over the threshold. In fact, by a similar procedure we can reach T^1, or sequentially T^1, T^2 from any point in Δ, as long as the control values remain unsaturated. So the strong controllability for collision with T^1 is global and complete.

When the target is located about an unstable equilibrium (the threshold, as above in sequential collision), we said we must lift the motion across the E-levels, so we apply the above discussion with the signs reversed. As mentioned several times already, we take $V(\bar{x}) = h_E^+ - E^+(\bar{x})$, if we want to lift the motion to the threshold or another suitable h_C^+ for other purposes. Then $\dot{V}(\bar{x}) = -\dot{E}^+(\bar{x}) \leq -\delta h/T_C$ or $f_0(\bar{x}, \bar{u}^*, \bar{w}^*, t) \geq \delta h/T_C$ for the max-min pair $\bar{u}^*, \bar{w}^*$, which gives the accumulative controller (3.3.17) required for the rise to the threshold level. After this, we use a conservative controller to reach equilibrium. The conservative controller this time is obtained by taking $\delta h = 0$ in the accumulative controller rather than in dissipative as done in Example 5.3.3, see (5.3.16).

EXERCISES 5.4

5.4.1 Return to the Van der Pol system of Exercise 4.1.4, consider the unperturbed case (i):

$$\ddot{q} + u(q^2 - 1)\dot{q} + q = 0 ,$$

and assume the control program such as to make u hold sign. Show that the trajectories for $u = \hat{u} > 0$ coincide with the retrograde trajectories ($\tau = -t$) of $u = -\hat{u} < 0$, and conversely, but the limit cycle is unchanged.

5.4.2 Consider the system

$$\dot{x}_1 = x_2 , \quad \dot{x}_2 = -(g/\ell) \sin x_1 - w|x_2|x_2 + u ,$$

with Δ : $-\pi \le x_1 \le \pi$, $-\infty < x_2 < \infty$. Find the equilibria and taking $w \in [1,2]$, design a control program securing collision with the target T : $x_1^2 + x_2^2 \le 0.2$ during 10 sec. Then, using retrograde trajectories, find the boundary of the corresponding retrieving region.

5.4.3 A conservative single DOF system consists of a solid magnet suspended from a ceiling on a linear spring with characteristic $2q$ over a steel covered bench exerting the magnetic force $u = -32/(12 - q)$. The magnet is also subject to the gravitational force $-4\ddot{q}$. Derive the motion equation and find the range of q where the magnet is effective (retrieving region). Answer $\hat{q} = 12$. If the magnet had to lift a mass M from the bench, how would it affect the range?

5.5 CONDITIONS FOR CAPTURE

Conditions for permanent holding of motions in a target must be as a rule somewhat more demanding than that for collision only. We must require the entry field H^- surrounding the target. The existing literature investigating conditions for such "permanent holding" for systems with untruncated nonlinearity is not vast, but extends considerably if we include the cases of controlled real-time attraction, which amounts to the same idea. The original work on ultimate boundedness belongs to Seibert [1], see Seibert-Auslander [1]. It was later developed by Yoshizawa [1], applied to nonlinear mechanical systems by Skowronski-Ziemba [4],[5] and Skowronski [4],[9], and to systems with uncertainty by Barmish-Leitmann [1] and Barmish-Petersen-Feuer [1].

For such systems, in reference to control for capture, the ultimate boundedness has been used by Bertsekas-Rhodes [2] and Delfour-Mitter [1], and later by Skowronski [27],[29]. Obviously control for asymptotic stability implies capture and thus the works on such stability are directly related to our present topic. Interested readers may recall the relevant literature from Sections 3.1 and 3.2.

We shall use the conditions introduced in Vincent-Skowronski [1], Skowronski-Vincent [1], Skowronski [29], and developed in Skowronski [32]. Similarly as for collision, we let Δ_0 be a candidate for the set in Δ which is strongly controllable for capture in T , but unlike collision, we also have some given $T_{\mathcal{C}}$ serving as a candidate for the capturing subtarget in T. The candidates may come from necessary conditions or an educated guess. Let $CT_{\mathfrak{C}} \overset{\Delta}{=} \Delta_0 - T_{\mathfrak{C}}$ and introduce a C^1-function $V(\cdot) : D \to \mathbb{R}$, $D(\text{open}) \supset \overline{CT_{\mathfrak{C}}}$, with

$$v_0^- = \inf V(\bar{x}) \mid \bar{x} \in \partial \Delta_0 , \quad v_{TC} \overset{\Delta}{=} \inf V(\bar{x}) \mid \bar{x} \in \partial T_{\mathfrak{C}} . \qquad (5.5.1)$$

CONDITIONS 5.5.1. The system (2.2.6)' is strongly controllable on Δ_0 for collision with capture in $T_{\mathfrak{C}}$, if there are $P(\cdot)$ and $V(\cdot)$ such that

(i) $V(\bar{x}) \geq v_{TC}$, for $\bar{x} \notin T_{\mathfrak{C}}$;

(ii) $V(\bar{x}) \leq v_0^-$, for $\bar{x} \in CT_{\mathfrak{C}}$;

(iii)   for all $\bar{u} \in \bar{P}(\bar{x},t)$, there is a continuous positive valued $c(\|\bar{x}\|)$ such that

$$\nabla V(\bar{x})^T \cdot \bar{f}(\bar{x},t,\bar{u},\bar{w}) \leq -c(\|\bar{x}\|) , \qquad (5.5.2)$$

for all $\bar{w}(t) \in W$.

To prove the Conditions, we show first that Δ_0 is positively strongly invariant under $P(\cdot)$. Suppose not, then $(\bar{x}^0,t_0) \in \Delta_0 \times \mathbb{R}$ generates $\bar{\phi}(\cdot) \in K(\bar{x}^0,t_0)$ such that for some $t_1 \geq t_0$, we have $\bar{\phi}(\bar{u},\bar{x}^0,t_0,t_1)$ $= \bar{x}^1 \in \partial\Delta_0$. Then by (ii), $V(\bar{x}^1) \geq V(\bar{x}^0)$ contradicting (5.5.2) and proving the invariance postulated.

Now take any $\bar{\phi}(\cdot) \in K(\bar{x}^0,t_0)$, $(\bar{x}^0,t_0) \in \overline{CT_{\mathfrak{C}}} \times \mathbb{R}$. From (5.5.2), integrating, one obtains the estimate

$$t - t_0 \leq (V(\bar{x}^0) - V(\bar{x})) / c(\|\bar{x}\|) \qquad (5.5.3)$$

for the interval of time spent in $\overline{CT_{\mathfrak{C}}}$. From (i), (ii), we have $V(\bar{x}) - v_{TC} \geq 0$, $V(\bar{x}^0) - v_0 \leq 0$ or $V(\bar{x}^0,t_0) - V(\bar{x}) \leq v_0 - v_{TC}$, whence there is $c^- \leq c(\|\bar{x}\|)$, $\forall \bar{x} \in CT_{\mathfrak{C}}$ such that

$$t \leq t_0 + \frac{1}{c^-}(v_0^- - v_{TC}) . \qquad (5.5.4)$$

Letting $T_{\mathfrak{C}} = \frac{1}{c^-}(v_0^- - v_{TC})$ we conclude that for $t \geq t_0 + T_{\mathfrak{C}}$ the motion must leave $CT_{\mathfrak{C}}$. As it cannot leave the strongly invariant Δ_0 , it must enter $T_{\mathfrak{C}}$. A return to $\overline{CT_{\mathfrak{C}}}$ is not possible as then (iii) and (i) contradict. Since we have the above for any motion from any point in $CT_{\mathfrak{C}} \times \mathbb{R}$ the theorem is proved.

REMARK 5.5.1. Note that if $\Delta_0, T_{\mathfrak{C}}$ are defined by some V-levels:

$$\partial\Delta_0 : V(\bar{x}) = v_0^- ; \quad \partial T_{\mathfrak{C}} : V(\bar{x}) = v_{TC} , \qquad (5.5.5)$$

then conditions (i), (ii) hold automatically.

252

CONDITIONS 5.5.2. Consider a C^1-function $V(\cdot) : D \to \mathbb{R}$ and constants $v_0^- > v_{TC} > 0$ such that (5.5.5) holds. Then in order for (2.2.6)' to be strongly controllable on Δ_0 for collision with capture in $T_{\mathbf{c}}$ it is *necessary and sufficient* that there is $\bar{P}(\cdot)$ and a continuous positive valued $c(\cdot)$ such that for all $\bar{u} \in \bar{P}(\bar{x},t)$,

$$\nabla v(\bar{x})^T \cdot \bar{f}(\bar{x},t,\bar{u},\bar{w}) \leq -c(\|\bar{x}\|) , \tag{5.5.2}$$

for all $\bar{w}(t) \in W$.

Proof of sufficiency follows from Conditions 5.5.1. To show the necessity, suppose along some $\bar{\phi}(\cdot) \in K(\bar{x}^0,t_0)$, $(\bar{x}^0,t_0) \in CT_{\mathbf{c}} \times \mathbb{R}$, for each $c(\|\bar{x}\|) > 0$, we have $\dot{V}(\bar{x}(t)) > -c(\|\bar{x}\|)$. Then no reaching is possible.

COROLLARY 5.5.1. *Given* $(\bar{x}^0,t_0) \in C_{T\mathbf{c}} \times \mathbb{R}$, *if there is a pair* $\bar{u}*,\bar{w}*$ *such that*

$$L(\bar{x},t,\bar{u}*,\bar{w}*) = \min_{\bar{u}} \max_{\bar{w}} L(\bar{x},t,\bar{u},\bar{w}) \leq -c(\|\bar{x}\|) , \tag{5.5.2'}$$

the condition (5.5.2) *is met with* $\bar{u} = \bar{P}(\bar{x},t)$, *and one can deduce the control program from* (5.5.2)' .

The proof follows from Conditions 5.5.1 by the same argument as for Corollary 3.2.2. Similarly as before for collision, we may either use (5.5.2)' directly to produce the control program $\bar{P}(\cdot)$, or indirectly using the L-G controller (5.3.12). The following short technical example illustrates the calculations.

EXAMPLE 5.5.1. Consider the system

$$\dot{x}_1 = -(x_1^{1/2} + x_2^{1/3})^2 ,$$

$$\dot{x}_2 = -x_2 + ax_1 + ux_2^{1/3} ,$$

where $a = \text{const} \neq 0$. We let $\Delta_0 : x_1^{2/3} + x_2^{2/3} \leq (2a)^{2/3}$,

$$T_{\mathbf{c}} : x_1^{2/3} + x_2^{2/3} \leq a^{2/3} \text{ with } V(x_1,x_2) = x_1^{2/3} + x_2^{2/3} - a^{2/3} ,$$

see Fig. 5.16. Then also $v_{TC} = 0$, $v_0^- = (2a)^{2/3} - a^{2/3}$ which makes (i), (ii) satisfied. Then

$$\dot{V} = \tfrac{2}{3} x^{-1/3} [-(x_1^{1/2} + x_2^{1/3})^2] + \tfrac{2}{3} x_2^{-1/3} (-x_2 + ax_1 + ux_2^{1/3})$$

$$= \tfrac{2}{3} (-x_1^{2/3} - x_2^{2/3} - 2x_1^{1/6} x_2^{1/3} - x_1^{-1/3} x_2^{2/3} + ax_1 x_2^{-1/3} + u) .$$

The controller $u = 2x_1^{1/6} x_2^{1/3} + (x_2^{2/3}/x_1^{1/3}) - (ax_1/x_2^{1/3})$, $x_1,x_2 \neq 0$ produces $\dot{V} \leq -c^- = -\tfrac{2}{3} (2a)^{2/3}$, as required. □

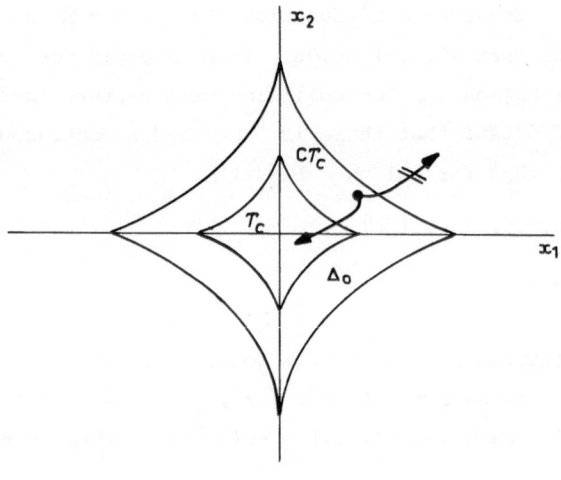

Fig. 5.16

REMARK 5.5.2. Note that given the bound c^- of the rate of change of $V(\cdot)$ in Conditions 5.5.2 we may find $T_{\mathfrak{C}} = (v_0^- - v_{TC})/c^-$ or alternatively, given $T_{\mathfrak{C}}$, the constant c^- may be found:

$$c^- = \frac{v_0^- - v_{TC}}{T_{\mathfrak{C}}} \ . \tag{5.5.6}$$

By definition, there may be no such thing as an objective composed of a sequence of captures, but one of the obvious applications of the study on collision with capture is the objective composed of *sequential collisions*, cf. Section 5.1, *terminating with capture*. Then, if the reachings must be in *stipulated time*, so must be the terminal capture, which in turn *requires stipulated* $T_{\mathfrak{C}}$, see Fig. 5.17. Also in the case of $\bar{x}$ represented by variables relative to the target, that is, in path tracking, both the path $T_{\mathfrak{C}}$ and $T_{\mathfrak{C}}$ are stipulated. Obviously there are other situations when arbitrary $T_{\mathfrak{C}}$, $T_{\mathfrak{C}}$ cannot be allowed.

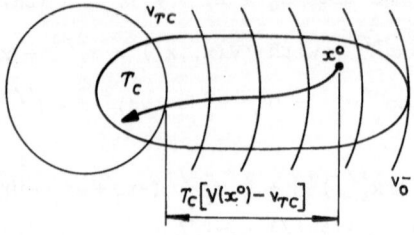

Fig. 5.17

CONDITIONS 5.5.3. The system (2.2.6)' is strongly controllable on Δ_0 for collision with capture in $T_{\mathfrak{C}}$ after $T_{\mathfrak{C}}$ if there is $\bar{P}(\cdot)$ defined on $CT_{\mathfrak{C}} \overset{\Delta}{=} \Delta_0 - T_{\mathfrak{C}}$ and a C^1-function $V(\cdot) : D \to \mathbb{R}$, $D(\text{open}) \supset \overline{CT_{\mathfrak{C}}}$, such that

    (i) $V(\bar{x}) \geq v_{TC}$ for $\bar{x} \notin T_{\mathfrak{C}}$;

    (ii) $V(\bar{x}) \leq v_0^-$ for $\bar{x} \in CT_{\mathfrak{C}}$;

    (iii) for all $\bar{u} \in \bar{P}(\bar{x},t)$,

$$\nabla V(\bar{x})^T \cdot \bar{f}(\bar{x},t,\bar{u},\bar{w}) \leq - \frac{v_0^- - v_{TC}}{T_C} \tag{5.5.7}$$

for all $\bar{w}(t) \in W$.

The proof follows from Conditions 5.5.1, if we specify c^- from (5.5.6). From Conditions 5.5.2, we have

COROLLARY 5.5.2. *The system (2.2.6)' is strongly controllable on* Δ_0 *defined by (5.5.5) for collision with capture of* $T_{\mathfrak{C}}$ *defined also by (5.5.5) after stipulated* $T_{\mathfrak{C}}$ *if and only if Conditions 5.5.2 hold with* c *in (5.5.2) specified by (5.5.7).*

Sufficiency follows from Conditions 5.5.3. To show necessity, observe that if not (iii) then for some $\tilde{u}, \tilde{w}$ we have

$$\dot{V}(t) > - \frac{1}{T_{\mathfrak{C}}} (v_0^- - v_{TC}) \ .$$

Integrating it along an arbitrary motion from anywhere in $CT_{\mathfrak{C}}$,

$$V(\bar{x}) - V(\bar{x}^0) > - \frac{v_0^- - v_{TC}}{T_{\mathfrak{C}}} (t - t_0) \ ,$$

or

$$V(\bar{x}^0) - V(\bar{x}) < \frac{v_0^- - v_{TC}}{T_{\mathfrak{C}}} (t - t_0) \ ,$$

which means that the drop in the value of $V(\cdot)$ is not sufficient to cross $\partial T_{\mathfrak{C}}$ for all $t - t_0 \geq T_{\mathfrak{C}}$, which contradicts collision and thus capture.

The constant $-c^-$ in (5.5.7) determines the upper bound of the dissipation rate along the motions concerned, and given $v_0^- - v_{TC} = h_0 - h_{TC} \overset{\Delta}{=} \delta h$ and the time interval $T_{\mathfrak{C}}$, c^- may be calculated. In turn, knowing this *speed of dissipation* and $T_{\mathfrak{C}}$, we may find δh for the objective involved. By the same argument as before, we have the following corollary.

COROLLARY 5.5.3. *Given* $(\bar{x}^0, t_0) \in \Delta_0 \times \mathbb{R}$, *if there is a pair* $(\bar{u}^*, \bar{w}^*)$ *such that*

$$L(\bar{x}, t, \bar{u}^*, \bar{w}^*) = \min_{\bar{u}} \max_{\bar{w}} L(\bar{x}, t, \bar{u}, \bar{w}) \leq -\frac{v_0^- - v_{TC}}{T_c}, \qquad (5.5.8)$$

then the condition (iii) of Conditions 5.5.3 is met with $\bar{u} \in \bar{P}(\bar{x}, t)$, *that is, it may be possible to deduce a winning control program from (5.5.8).*

In terms of the mechanical system, the control condition for dissipative controllers become

$$\min_{\bar{u}} \max_{\bar{w}} f_0(\bar{x}, \bar{u}, \bar{w}, t) \leq -\frac{h_0 - h_{TC}}{T_c} \qquad (5.5.9)$$

with our standard choice of the test function $V(\bar{x}) = E^-(\bar{x})$, and when we use the controller all the way to the capturing energy level $v_{TC} = h_{TC}$, from $\partial \Delta_0$. If not, the corresponding $\delta h = h^0 - h_{TC}$, $h^0 = E^-(\bar{x}^0)$, applies instead of the total drop $h_0 - h_{TC} \geq h^0$. For the accumulative controller, we take $v(\bar{x}) = h_0 - E^+(\bar{x})$ as before for collision, and thus the condition is

$$\max_{\bar{u}} \min_{\bar{w}} f_0(\bar{x}, \bar{u}, \bar{w}, t) \geq \delta h / T \qquad (5.5.10)$$

for generating the influx across a desired $\delta h > 0$ over the time interval T. Observe that (5.5.10) gives the measure of the rate of energy usage by the non-potential forces. Then the work of such forces is the corresponding energy flux discussed in Section 5.3, measuring the amount of the energy flow $h(h^0, t_0, t)$ along the motion concerned. For design of the conservative controller we use $\delta h \equiv 0$ and the equality in either (5.5.9) or (5.5.10) depending on whether the motion comes from H^- or H^+. The procedure of controlling for capture is the same as for collision discussed in detail in Section 5.3 and illustrated in Example 5.3.3. Our targets may however differ from those considered in the collision case, as capture is a terminal type of objective. It is thus natural to deal with the minimal invariant positive limit sets filling up the capturing subtargets T_c. The subtarget itself may be designed as a real time attractor below or above some E-level, as well as enclosed between two E-levels, the latter in the case of steady state periodic orbit being the limit set. In the first two cases, we use the technique of control described in Section 3.3 with the controllers designed so as to satisfy conditions (3.3.12)' and (3.3.17) for capture below and above some E-level, respectively. We shall give an applied example of this in Section 5.9. In the second case of sandwiching T_c between two levels we must obviously apply both conditions leading to

(4.1.11), but some comment may be required here.

Applying both conditions (3.3.12)' from above and (3.3.17) from
below, we actually use the capturability Conditions 5.5.2 twice, with two
different test functions $V = E^{\pm}(\bar{x})$ thus also formally aiming at two
different T_c's: T_{CA} above and T_{CB} below. Figure 5.18 shows the pattern
within the basic energy cup. When we do not need to win over the uncertainty, in some circumstances a single function $V(\cdot)$ may be used with
the Proposition 4.1.1 adjusted to capture: (4.1.7) replaces (5.5.2). The
proof is obvious as asymptotic stabilization implies capture.

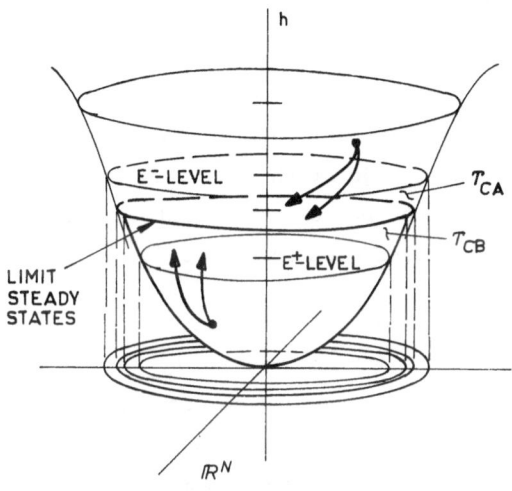

Fig. 5.18

EXAMPLE 5.5.2. Consider the system

$$\dot{x}_1 = -x_1^4 x_2 - x_1 x_2^2 - x_2^3 + u_1$$

$$\dot{x}_2 = 25x_2 + x_1^5 - x_2^3 + u_2$$

with $\Delta_0 = \Delta$ and $T_c : 19/4 \leq x_1^2 + x_2^2 \leq 21/4$, while the function
$V(x_1, x_2) = x_1^2 + x_2^2$, see Fig. 5.19. Letting the controller $u_1 = 25x_1 - x_1^3$,
$u_2 = -x_2 x_1^2 + x_1 x_2^2$, we obtain

$$\dot{V} = 2x_1(-x_1^4 x_2 - x_1 x_2^2 - x_2^3 + 25x_1 - x_1^3) + 2x_2(25x_2 + x_1^5 - x_2^3 - x_2 x_1^2 + x_1 x_2^2)$$

$$= 2(x_1^2 + x_2^2)[25 - (x_1^2 + x_2^2)] .$$

The above $V(\cdot)$ satisfies (i), (ii) of Conditions 5.5.1 automatically,
$\dot{V}(\bar{x})$ is negative for $(x_1^2 + x_2^2)^{\frac{1}{2}} > 5$, positive for $(x_1^2 + x_2^2)^{\frac{1}{2}} < 5$, which
satisfies condition (4.1.7) in the neighborhood of $x_1^2 + x_2^2 = 5$. $\Box$

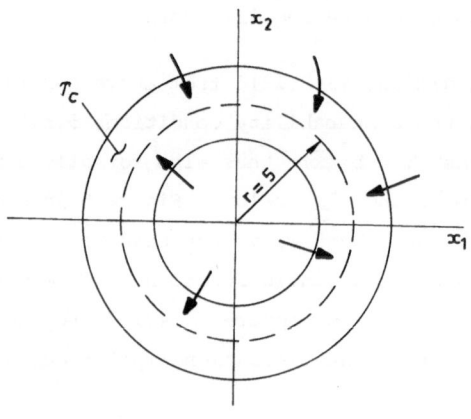

Fig. 5.19

Note here that the T_c between two levels may be located anywhere in Δ, not only inside a cup. Turning back to Example 5.3.3, we can for instance have it covering the joint threshold for all three cups, or the threshold of one cup, say the basic. The first case is marked by the dashed area, the other by the crossed area in Fig. 5.20.

Let us now return to the general discussion and observe that the design of the steady state limit set within T_c may be attained by using the condition (4.1.19) and a corresponding conservative version of the previous controller, either dissipative or accumulative, depending on from which side of the limit set the approach is made.

Consequently to the above, and to what has been said regarding the objectives of sequential collision, we may now state that the design of a controller for the global study, outside neighborhoods of equilibria, or as we shall briefly say a *global* controller, requires the ability to

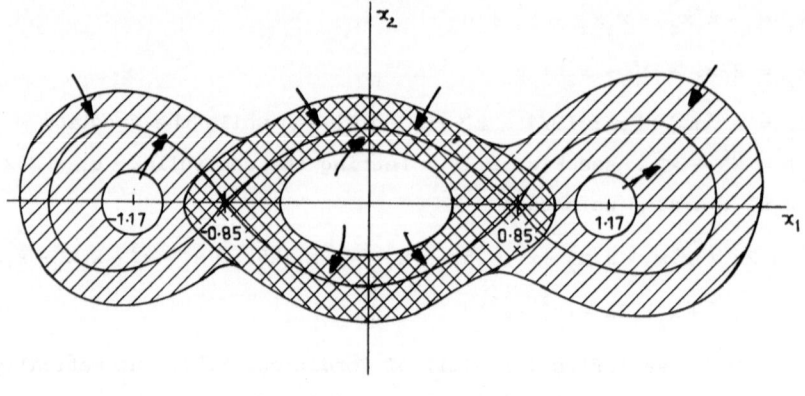

Fig. 5.20

maneuver for collision with several subsequent targets and then capture in some subtarget, in most cases enclosing steady state limit set. For such maneuvering we need the map of options in the state space, given on one hand by the conservative reference frame, that is, the E-level system, and on the other by the pattern of state space trajectories of the unperturbed and uncontrolled system without uncertainty, that is, (2.2.5). Such a map of options forms the basis for *planning the desired path of the motions* in the maneuver concerned, and thus for selecting the right succession of available types of controllers. The need for such a map has been quite visible from Example 5.3.3. It is particularly needed for the design of global controllers with composite objectives. The rules for introducing such a map have been discussed in Section 4.1 where the Properties of Disjointness and Identification together with separating sets were introduced. A more detailed map can only be made for case studies with specific scenarios and objectives. The switching states and instants between the controllers may be calculated from the condition of planning "the best" path of motions. The best is taken according to some cost criterion, possibly using necessary conditions for optimal strong controllability, see Section 5.2.

Whatever the outcome of the planning, the proposed path of motions must satisfy sufficient conditions for controllability or strong controllability, whatever applicable, see Definitions 5.2.1, 5.2.2. To this effect we may use conditions given in Section 5.3 for collision objectives and of this section for capture.

Then, if we want to adjoin some cost-optimal objective for additional qualification of the path, we may use the sufficient conditions for optimal strong controllability which follow. Such conditions may be obviously used also to confirm the candidates for the sequence of switching states and instants found from the necessary conditions for optimal path.

Referring to capture we shall specify the terminal instant in (5.2.14) as $t_f = t_0 + T_C$. The corresponding strong region will be denoted by Δ_C^* and we obviously have $\Delta_C^* \subset \Delta_C$. Then since the corresponding T_C^* must satisfy $T_C^* \subset \Delta_C^* \cap T$, it may differ from T_C. Using our usual argument, we stipulate T_C^* that satisfies the necessary conditions, but this time from the cost-non-permeability that is (5.2.11), passing ∂T^* through a cost-non-permeable surface closest to $\partial\Delta_C^*$. It also justifies a candidate Δ_0 for Δ_C^* itself. Then the pair T_C^*, Δ_0 must be confirmed by the sufficient conditions.

The sufficient conditions for optimization used below have been introduced by Stalford-Leitmann [1]. Their application to optimal capture was investigated in Skowronski [27], we use this work here. Let $CT_{\mathbb{C}}^* \triangleq \Delta_0 - T_{\mathbb{C}}^*$ and consider a function $V(\cdot) : D \to \mathbb{R}$, $D(\text{open}) \supset \overline{CT_{\mathbb{C}}^*}$ with v_0^*, v_{TC}^* defined as v_0^-, v_{TC} in (5.5.1).

CONDITIONS 5.5.4. The system (2.2.6)' is optimally strongly controllable on Δ_0 for collision with capture in $T_{\mathbb{C}}^*$, if there is $\bar{P}_*(\cdot)$ defined on D and a C^1-function $V(\cdot) : D \to \mathbb{R}$ such that

(i) $V(\bar{x}) \geq v_{TC}^*$, for $\bar{x} \notin T_{\mathbb{C}}^*$;

(ii) $V(\bar{x}) \leq v_0^*$, for $\bar{x} \in CT_{\mathbb{C}}^*$;

(iii) for all $\bar{u}^* \in \bar{P}_*(\bar{x},t)$,

$$V(\bar{x})^T \cdot \bar{f}(\bar{x},t,\bar{u}^*,\bar{w}) \leq -f_0^V(\bar{x},t,\bar{u}^*,\bar{w}) , \qquad (5.5.11)$$

for all $\bar{w} \in W$ and with

$$0 \leq v_0^* - v_{TC}^* \leq T_{\mathbb{C}} \cdot f_0^V(\bar{x},t,\bar{u}^*,\bar{w}) ; \qquad (5.5.12)$$

(iv) there is $\bar{w}^*(\cdot)$ such that

$$\nabla V(\bar{x})^T \cdot \bar{f}(\bar{x},t,\bar{u},\bar{w}^*) \geq -f_0^V(\bar{x},t,\bar{u},\bar{w}^*) \qquad (5.5.13)$$

for all $\bar{u} \in \bar{P}(\bar{x},t)$, and all $\bar{P}(\cdot)$ winning capture.

Since $0 < f_0^V(\bar{x},t,\bar{u},\bar{w}) < \infty$, the conditions (i) - (iii) satisfy Conditions 5.5.1, yielding strong controllability for capture. Thus we only have to prove the optimality. From (iii), (iv) one obtains respectively

$$\left. \begin{array}{l} f_0^V(\bar{x},t,\bar{u}^*,\bar{w}) + \nabla V(\bar{x})^T \cdot \bar{f}(\bar{x},t,\bar{u}^*,\bar{w}) = -\ell(t) , \\[2mm] f_0^V(\bar{x},t,\bar{u},\bar{w}^*) + \nabla V(\bar{x})^T \cdot \bar{f}(\bar{x},t,\bar{u},\bar{w}^*) = \ell(t) , \end{array} \right\} \qquad (5.5.14)$$

for some $\ell(t) > 0$ and all $t \geq t_0$. Integrating both equations (5.5.14) successively between t_0 and $t_{\mathbb{C}} = t_0 + T_{\mathbb{C}}$, we obtain

$$\left. \begin{array}{l} \displaystyle\int_{t_0}^{t_{\mathbb{C}}} f_0^V(\bar{x},t,\bar{u}^*,\bar{w})\,dt = V(\bar{x}^0) - V(\bar{x}^{\mathbb{C}}) - \int_{t_0}^{t_{\mathbb{C}}} \ell(t)\,dt , \\[5mm] \displaystyle\int_{t_0}^{t_{\mathbb{C}}} f_0^V(\bar{x},t,\bar{u},\bar{w}^*)\,dt = V(\bar{x}^0) - V(\bar{x}^{\mathbb{C}}) + \int_{t_0}^{t_{\mathbb{C}}} \ell(t)\,dt , \end{array} \right\} \qquad (5.5.15)$$

where $\bar{x}^{\mathbb{C}} = \bar{\phi}(\bar{x}^0,t_0,t_{\mathbb{C}})$. On the other hand, from both (iii) and (iv) together, we have

$$f_0^V(\bar{x},t,\bar{u}^*,\bar{w}^*) + \nabla V(\bar{x})^T \cdot \bar{f}(x,t,u^*,w^*) = 0 \qquad (5.5.16)$$

and integrating

$$\int_{t_0}^{t_{\mathcal{C}}} f_0^v(\bar{x},t,\bar{u}^*,\bar{w}^*)dt = V(\bar{x}^0) - V(\bar{x}^{\mathcal{C}}) . \tag{5.5.17}$$

Comparing (5.5.15), (5.5.16) with (5.5.17) gives the optimality defined by the saddle condition (5.2.4), thus completing the proof.

Observe that given the Liapunov function the condition (5.2.14) with $t_f = t_0 + T_{\mathcal{C}}$ estimates the size of $\Delta_{\mathcal{C}}^*$. Observe further that from Remark 5.5.2 and (5.5.7), we have

$$c^- \leq f_0(\bar{x},t,\bar{u}^*,\bar{w}) , \quad \forall \, \bar{w} \in W , \tag{5.5.18}$$

which yields

$$T_C = \frac{1}{f_0^v} (v_0^* - v_{TC}^*) . \tag{5.5.19}$$

Thus either $T_{\mathcal{C}}$ may be calculated from given $V(\cdot)$ or $\Delta_{\mathcal{C}}^*$ may be established with stipulated $T_{\mathcal{C}}$ which is our present case. On the other hand, the necessary conditions for optimality (5.2.11) produce the control program $\bar{P}*(\cdot)$ the same way as the Corollary 5.5.1 produced $\bar{P}(\cdot)$ winning capture. If such $\bar{P}(\cdot)$ satisfies Conditions 5.5.4, it must also satisfy Conditions 5.5.3.

Basically the same philosophy, but without formal checking against the sufficient conditions, may be found in Grantham-Chingcuanco [1], with more details than we can provide in this text. See also Section 5.6.

When the cost is time, we have $f_0^v(\bar{x},\bar{u},\bar{w},t) \equiv 1$ and the so called time optimization, a somewhat simpler problem. The latter may perhaps be the reason for its popular use, often beyond applicability, for instance minimizing work intervals for particular machines in assembly line where stipulated time is essential and minimization may destabilize the system, see Barmish-Feuer [1].

On the other hand the cost, which is most appropriate in our search for the best qualitative path of motions, is the energy flux minimization, obtained via the Pontriagin Principle confirmed by sufficient conditions. We may use here the routine of parametric optimization proposed by Klein-Briggs [1]: make the state variables dependent not only upon t, but also upon some other parameters such as initial and terminal positions, switching states and instants, etc. Analysing the energy flux $f_0^v(\bar{x},\bar{u},\bar{w},t) \overset{\Delta}{\equiv} f_0(\bar{x},\bar{u},\bar{w},t) \equiv \dot{E}(\bar{x},\bar{w})$ this way, we may specify the most energy-economic path of motions.

Turning now to the size of both $\Delta_{\mathfrak{c}}$ and the strong region $\Delta_{\mathfrak{c}}$, let us observe that $T_{\mathfrak{c}}$ if it exists must be positively invariant, whence collision with $T_{\mathfrak{c}}$ becomes capture. Consequently the regions of controllability Δ_c and strong controllability Δ_C for collision with $T_{\mathfrak{c}}$ found by the methods discussed in Section 5.4 are identical with the regions $\Delta_{\mathfrak{c}}, \Delta_{\mathfrak{c}}$ respectively. So once $T_{\mathfrak{c}}$ is either assumed or confirmed by sufficient conditions, we establish $\partial\Delta_{\mathfrak{c}}$ by the methods of Section 5.4. The same argument applies to $\Delta_{\mathfrak{c}}^*, \Delta_{\mathfrak{c}}^*$ given $T_{\mathfrak{c}}^*$.

EXERCISES 5.5

5.5.1 For the system

$$\dot{x}_1 = x_2, \quad \dot{x}_2 = -(g/\ell)\sin x_1 - w|x_2|x_2 + u ,$$

find a feedback control program securing strong capture in the target $T_{\mathfrak{c}}$: $0.45 \le x_1^2 + x_2^2 \le 0.5$ with $w \in [0,1]$. Calculate the time $T_{\mathfrak{c}}$ of such capture.

5.5.2 Prove that $u = x_3$ secures capture in $T_{\mathfrak{c}}$: $x_1^2 + x_2^2 + x_3^2 \le 1$ for trajectories of the system

$$\dot{x}_1 = -x_1 - 2u ,$$

$$\dot{x}_2 = -x_2 + 2u ,$$

$$\dot{x}_3 = -u .$$

Find the region of controllability.

5.5.3 Find the control program that generates controllability for capture in a target about $(0,0)$ for the system $\ddot{q} + 2u|\dot{q}| + 2q + q^2 = 0$. Find the region and the saturation control $\hat{u}$.

5.5.4 Find the retrieving region for capture in a target about $(1,0)$ of trajectories of the linear system $\ddot{q} + \dot{q} + q = u$, under the non-linear controller $u = \sqrt{|q|}$. Note that the equilibrium is shifted to $q^e = 1$ which may be shown asymptotically stable.

5.5.5 Show that the controller

$$u = \begin{cases} c/2x_2 , \ (x_1^2 - 1) , & |x_2| \ge \beta , \\ 0 , & |x_2| < \beta , \end{cases}$$

secures the strong controllability for capture in a target T : $x_1^2 + x_2^2 \le 1$ of trajectories of the Van der Pol system $\dot{x}_1 = x_2, \ \dot{x}_2 = -uw(x_1^2 - 1) - x$, with $w \in [1,2]$. Assuming $\hat{u}$, calculate β.

5.5.6 Let the origin of $\mathbb{R}^2$ be the source of a potential with intensity inversely proportional to the square distance $V(x_1, x_2) = k/(x_1^2 + x_2^2)$ and consider a point mass with a given initial kinetic energy $T = \frac{1}{2}[(x_1^0)^2 + (x_2^0)^2]$. Ignoring other forces, find a control program $P(\cdot)$ under which the mass is charged with the full potential V_{max}, that is, the point stays (rendezvous) for some time interval in a target located about the source.

5.5.7 Consider the system $\ddot{q} + wq = u$, $q \in \mathbb{R}$, with the controller $u = \pm k$ for $\dot{x} < 0$, $\dot{x} > 0$ respectively, and $w \in [1,2]$. By completing the square, show that the trajectories consist of ellipses centered at $q = \pm k/w$, $\dot{q} = 0$ for $\dot{x} > 0$, $\dot{x} < 0$, respectively. Then assuming T_c about $(0,0)$ verify the strong controllability for capture in finite time. Calculate this time.

5.5.8 Find the controller generating capture in $T_c : x_1^2 + x_2^2 \leq 1$ within the time $T_c = 10$ sec for trajectories of the system

$$\dot{x}_1 = -x_1 - ux_2^2,$$

$$\dot{x}_2 = x_2(u - x_1^2).$$

Establish the region of controllability for such a capture.

5.6 SPACECRAFT LARGE-ANGLE REORIENTATION

An illustrative example of controlled maneuvering that should end with capture is the case of controlling a spacecraft for stable reorientation/ slew maneuvers, which is the basic objective in any space mission. The conventional single-axis, small articulation-angle models are no longer adequate for present demands in such control. Consequently the system considered is nonlinear (involving trigonometric functions) with nonlinearity which may not be truncated more than the angular amplitude of rotation indicates - see our comments in Example 1.1.1 - and the study must be global.

The rigid body model of the craft may have its orientation represented by a direction cosine matrix, Euler angles or, more recently, in terms of *Euler parameters* sometimes also called *quaternions*, as explained below. Such a representation has been applied for the Space Shuttle and Galileo, see Wong-Breckenridge [1]. The controllers used are either of the standard feedback type discussed so far, see Wie-Barba [1], or adaptive (model reference, variable structure, etc.) to which we refer later in Chapter 7. These controllers are implemented by means of reaction wheels, usually for slow maneuvers, or using on-off thrusters (jet actuators), or both. The combination of both provides better fuel economy.

Ignoring gravity (potential) and damping forces, the motion equations about the principal axes of inertia take the Euler format, well known from Mechanics,

$$
\left.\begin{array}{l}
J_1\dot{\omega}_1 + (J_3 - J_2)\omega_2\omega_3 = u_1 \ , \\[2mm]
J_2\dot{\omega}_2 + (J_1 - J_3)\omega_1\omega_3 = u_2 \ , \\[2mm]
J_3\dot{\omega}_3 + (J_2 - J_1)\omega_1\omega_2 = u_3 \ .
\end{array}\right\}
\tag{5.6.1}
$$

In the above J_i , $i = 1,2,3$ are the principal moments of inertia, $\omega_i = \dot{\theta}_i$, $i = 1,2,3$ are the angular rates of the rotations θ_i , and u_i , $i = 1,2,3$ are the control torques supplied by the actuators.

First let us consider briefly the options offered by the Euler angle method. Choosing the state variables $x_i = \theta_i$, $x_{3+i} = \omega_i$, $i = 1,2,3$, we can rewrite (5.6.1) as

$$
\left.\begin{array}{l}
\dot{x}_1 = x_4 \ , \\[2mm]
\dot{x}_2 = x_5 \ , \\[2mm]
\dot{x}_3 = x_6 \ , \\[2mm]
\dot{x}_4 = -[(J_3 - J_2)/J_1]x_5x_6 + u_1/J_1 \ , \\[2mm]
\dot{x}_5 = -[(J_1 - J_3)/J_2]x_4x_6 + u_2/J_2 \ , \\[2mm]
\dot{x}_6 = -[(J_2 - J_1)/J_3]x_4x_5 + u_3/J_3 \ .
\end{array}\right\}
\tag{5.6.2}
$$

Given $x_1^0,\ldots,x_6^0$ we aim at capturing the state trajectories of (5.6.2) in a target about $(0,\ldots,0)$ with $\partial T_{\mathfrak{C}}$ determined by constant angular velocities $x_{3+i}^{\mathfrak{C}}$, which leads the system to the specific target orientation

$$
x_i^{\mathfrak{C}} = \int_0^{t_{\mathfrak{C}}} x_{3+i}^{\mathfrak{C}} \ dt
$$

in the *stipulated time* $t_{\mathfrak{C}}$.

In view of our assumptions, the potential energy of the system vanishes and thus the total energy reduces to kinetic:
$E(\bar{x}) = \frac{1}{2}(J_1\omega_1^2 + J_2\omega_2^2 + J_3\omega_3^2)$ which is the first integral of the free (uncontrolled) system: $u_1 = u_2 = u_3 = 0$ and is taken as our test function $V(\bar{x})$. Indeed, simple calculation shows that for the free system $\dot{V}(\bar{x}) = 0$,

which agrees with the fact that the free coasting spacecraft exhibits stability (but not asymptotic stability) and drifts along the conservative trajectories. Then, the control condition required by Corollary 5.5.3 is

$$\min_{u} \sum_{i=1}^{3} x_{3+i} u_i \leq - \frac{h^0 - h_{TC}}{t_{\mathcal{C}}} \qquad (5.6.3)$$

satisfied by the controller

$$u_i \, \text{sgn} \, (x_{3+i}) \begin{cases} \leq (h_{TC} - h^0)/3t_{\mathcal{C}} |x_{3+i}| \, , \quad \forall \, |x_{3+i}| \geq \beta_i \, , \\ = 0 \, , \quad \forall \, |x_{3+i}| < \beta_i \, , \quad i = 1,2,3 \, , \end{cases} \qquad (5.6.4)$$

with

$$h^0 = \sum_{i=1}^{3} J_i (x_{3+i}^0)^2/2 \, , \quad h_{TC} = \sum_{i=1}^{3} J_i (x_{3+i}^{\mathcal{C}})^2/2 \, ,$$

stipulated time $t_{\mathcal{C}}$ and β_i calculated as in (3.3.13) and (5.3.13).

We turn now to the Euler parameter or quaternion method. From the Euler theorem, cf. Hughes [3], we know that the rigid body attitude can be changed from one orientation to another by rotating the body by the angle ϕ about the so called Euler axis specified by the unit vector $\bar{a} = (a_1, a_2, a_3)^T$, fixed to the body and stationary in the inertial frame of reference. The above mentioned method of Euler parameters measures the change of attitude concerned in terms of such parameters which determine the rotation of the Euler axis. Let ϕ be the magnitude of this rotation. We define them as

$$\begin{aligned} \varepsilon_0 &\overset{\Delta}{=} \cos(\phi/2) \, , \\ \varepsilon_i &\overset{\Delta}{=} c_i \sin(\phi/2) \, , \quad i = 1,2,3 \, , \end{aligned} \right\} \qquad (5.6.5)$$

where the c_i are direction cosines of $\bar{a}$, which in turn form the rotation matrix for the axis, see Hughes [3]. Differentiating (5.6.5) with respect to time and inserting the formulae for $\dot{\phi}$ and $\dot{\bar{a}}$, we obtain the kinematics of rotation in the format of the following four linear equations for determining the parameters $\varepsilon_0, \dots, \varepsilon_3$:

$$\begin{aligned} \dot{\varepsilon}_0 &= \tfrac{1}{2} (-\omega_1 \varepsilon_1 - \omega_2 \varepsilon_2 - \omega_3 \varepsilon_3) \, , \\ \dot{\varepsilon}_1 &= \tfrac{1}{2} (\omega_1 \varepsilon_0 - \omega_2 \varepsilon_3 + \omega_3 \varepsilon_2) \, , \\ \dot{\varepsilon}_2 &= \tfrac{1}{2} (\omega_1 \varepsilon_3 + \omega_2 \varepsilon_0 - \omega_3 \varepsilon_1) \, , \\ \dot{\varepsilon}_3 &= \tfrac{1}{2} (-\omega_1 \varepsilon_2 + \omega_2 \varepsilon_1 + \omega_3 \varepsilon_0) \, ; \end{aligned} \right\} \qquad (5.6.6)$$

with variable coefficients $\omega_1, \omega_2, \omega_3$ given from (5.6.1), and with initial conditions $\varepsilon_0^0 = \varepsilon_0(0), \dots, \varepsilon_3^0 = \varepsilon_3(0)$. In particular (1,0,0,0) determines the initial alignment with the inertial reference frame.

Hughes [3] argues convincingly that the method of Euler parameters is computationally superior to all others available (Euler angles, direction cosines, Euler-Rodriguez, etc.), which is essential when dealing with the type of small computers available on-board the spacecraft. Unfortunately the method has its shortcomings too. It may be shown that the system matrix of (5.6.6), A(t) in our notation of Section 3.3, has vanishing real parts of some eigenvalues yielding the equilibrium $\varepsilon_i = 0$, $i = 0,...,3$ stable, but not asymptotically stable, while the customary methods of numerical integration have stability boundaries that may not cover the case, see Wie-Barba [1]. Nevertheless, Euler parameters have one decisive formal advantage when taken as state variables instead of the standard θ_i, ω_i, $i = 1,2,3$, namely that the transformation (5.6.5) hides the nonlinearity of the trigonometric representation. This will become clear below.

Wie-Barba [1] propose the following quaternional controller,

$$
\left.
\begin{aligned}
u_1 &= -\hat{u}(G\varepsilon_{e1} + G_1\omega_1) \ , \\
u_2 &= -\hat{u}(G\varepsilon_{e2} + G_2\omega_2) \ , \\
u_3 &= -\hat{u}(G\varepsilon_{e3} + G_3\omega_3) \ ,
\end{aligned}
\right\}
\tag{5.6.7}
$$

with alternative modification to

$$
\left.
\begin{aligned}
u_1 &= -\hat{u}(G\varepsilon_{e1}\ \mathrm{sgn}\ \varepsilon_{e0} + G_1\omega_1) \ , \\
u_2 &= -\hat{u}(G\varepsilon_{e2}\ \mathrm{sgn}\ \varepsilon_{e0} + G_2\omega_2) \ , \\
u_3 &= -\hat{u}(G\varepsilon_{e3}\ \mathrm{sgn}\ \varepsilon_{e0} + G_3\omega_3) \ ,
\end{aligned}
\right\}
\tag{5.6.8}
$$

where

$$
\begin{pmatrix} \varepsilon_{e1} \\ \varepsilon_{e2} \\ \varepsilon_{e3} \\ \varepsilon_{e0} \end{pmatrix} =
\begin{pmatrix}
\hat{\varepsilon}_0 & \hat{\varepsilon}_3 & -\hat{\varepsilon}_2 & -\hat{\varepsilon}_1 \\
-\hat{\varepsilon}_3 & \hat{\varepsilon}_0 & \hat{\varepsilon}_1 & -\hat{\varepsilon}_2 \\
\hat{\varepsilon}_2 & -\hat{\varepsilon}_1 & \hat{\varepsilon}_0 & -\hat{\varepsilon}_3 \\
\hat{\varepsilon}_1 & \hat{\varepsilon}_2 & \hat{\varepsilon}_3 & \hat{\varepsilon}_0
\end{pmatrix}
\begin{pmatrix} \varepsilon_1 \\ \varepsilon_2 \\ \varepsilon_3 \\ \varepsilon_0 \end{pmatrix}
\tag{5.6.9}
$$

with ε_{ei} being control *error Euler parameters*, $\hat{u}$ the saturation level of actuators and G, G_i linear feedback coefficients (positive control gains). Equation (5.6.9) is obtained via successive parameter rotation using the multiplication and inversion rules. For $\hat{\varepsilon}_0,...,\hat{\varepsilon}_3$ stipulated as the initial $(1,0,0,0)$, $\varepsilon_{e0},...,\varepsilon_{e3}$ become the current $\varepsilon_0(t),...,\varepsilon_3(t)$.

For the sake of illustration, we first simplify the model (5.6.1) to a single DOF, say rotation about axis 1, and represent it in the format

(1.5.19). Then introducing the attitude error $\phi_e(t) \overset{\Delta}{=} \phi(t) - \hat{\phi}$ with respect to the target attitude $\hat{\phi}$, and substituting the controller (5.6.7), we obtain the oscillator

$$J_1\ddot{\phi}_e + \hat{u}G_1\dot{\phi}_e + \hat{u}Gc_1 \sin(\phi_e/2) = 0 \qquad (5.6.10)$$

similar to that discussed in Examples 1.1.1, 3.2.1 and 5.3.3. Following our practice so far, it would be natural to choose the state variables as $x_1 \overset{\Delta}{=} \phi_e$, $x_2 \overset{\Delta}{=} \dot{\phi}_e = \dot{\phi} = \omega_1$ in the state-phase-plane $(x_1, x_2) = (\phi, \omega_1)$ in Figs. 1.1, 5.12 with the equilibria $\phi_e^e = 0, \pm 2\pi, \pm 4\pi, \ldots$, $\omega_1^e = 0$. The damped separatrices pass through the unstable equilibria $\phi_e^e = \pm 2\pi, \pm 6\pi, \ldots$, corresponding to $\varepsilon_0 = -1$. These separatrices act as weak $\bar{P}$-barriers between regions of strong recovery for collision or capture in targets about the stable equilibria, the latter selected from the sequence $\phi_e^e = 0, \pm 4\pi, \pm 8\pi, \ldots$, corresponding to $\varepsilon_0 = 1$. Before the controller was chosen, we could have aimed at capturing the trajectories in *any such target* T_c about a predetermined stable equilibrium, not necessarily the basic one. This could have been attained after stipulating a number of left or right rotations, while reorienting the craft from some initial position. This would lead to a specific amount of maneuvering in the state plane, exactly in the manner described in Example 5.3.3.

The reader may appreciate the fact that the nonlinearity in (5.6.10) which must be recognized depends precisely upon the number of rotations or parts of them which have to be made, and thus the number of equilibria involved. However, since our controller is already selected as (5.6.7), and substituted in (5.6.10), we can only talk about the recovery regions mentioned and the trajectories depend upon their initial states only. Obviously there is still a certain freedom of choice left by selecting the gains G, G_1. Observe that when $G_1 = 0$, (5.6.10) becomes conservative with trajectories that cross the damped separatrices along E-levels in a similar manner as was shown in Section 5.3 and illustrated in Example 5.3.3, see route AB' in Fig. 5.12.

Substituting the controller (5.6.8) instead of (5.6.7), one obtains a slightly different shape of H:

$$I_1\ddot{\phi}_e + \hat{u}G_1\dot{\phi}_e + \hat{u}Gc_1 \, \text{sgn}[\cos(\phi_e/2)] \sin(\phi_e/2) = 0 \qquad (5.6.11)$$

with unstable equilibria $\phi_e = \pm\pi, \pm 3\pi, \ldots$ and stable $\phi_e = 0, \pm 2\pi, \pm 4\pi, \ldots$.

The role of transformation (5.6.5) is seen directly from (5.6.10), (5.6.11). Taking the state variables in terms of the Euler parameters, we hide the trigonometric functions of (5.6.10), (5.6.11) within them, thus

simplifying the motion equations. The procedure is however purely formal
and does not change the physical nature of the system which still has to
have several equilibria and rotate when maneuvering.

We may turn now to the quaternion state representation, taking advantage of (5.6.5) and consider the full three DOF system (5.6.1). Let
$\hat{\varepsilon}_0,\ldots,\hat{\varepsilon}_3$ be the values of the parameters $\varepsilon_0,\ldots,\varepsilon_3$ specifying the
desired target attitude and define the relative variables
$\delta\varepsilon_i(t) \triangleq \varepsilon_i(t) - \hat{\varepsilon}_i$, $i = 0,1,2,3$. We shall choose our state variables
now as $x_i = \delta\varepsilon_i$, $x_{3+i} = \omega_i$, $i = 1,2,3$. The intended reorientation
must be performed with the aim of capturing the trajectories of the joint
system (5.6.1), (5.6.6) in a target about a stipulated equilibrium
attained after a desired number of rotations.

We may now use and verify the full 3D controller (5.6.7). To this aim
let us set up

$$V(\bar{x}) \triangleq \frac{1}{2} \sum_{i=1}^{3} J_i \omega_i^2 + \hat{u}G \sum_{i=0}^{3} \delta\varepsilon_i^2 \qquad (5.6.12)$$

which is positive definite with $V(\bar{x}) \to \infty$ as $|\bar{x}| \to \infty$. Substitute (5.6.7)
and (5.6.9) into (5.6.1) and multiply each resulting equation by corres-
ponding ω_i. Then multiply each equation (5.6.6) by the corresponding
$\hat{u} G \delta\varepsilon_i$, $i = 0,\ldots,3$ and add the resultant to the previously modified
(5.6.1). We obtain

$$
\begin{aligned}
\dot{V}(\bar{x}) &= \sum_{i=1}^{3} J_i \dot{\omega}_i \omega_i + 2\hat{u} G \sum_{i=0}^{3} \dot{\varepsilon}_i (\delta\varepsilon_i) \\
&= -\hat{u} \sum_{i=1}^{3} G_i \omega_i^2
\end{aligned}
\qquad (5.6.13)
$$

which secures the controller (5.6.7) as dissipative, and with a suitable
choice of $G_i > 0$ makes the Corollary 5.5.3 satisfied between the initial
h^0 and the desired h_{TC}.

EXAMPLE 5.6.1. Specifying now the target $T = (0\ 0\ 0\ 1)^T$ and assuming
the simulation values $J_1 = 10,000$ km/m^2, $J_2 = 9,000$ km/m^2,
$J_3 = 12,000$ km/m^2, $\hat{u} = 20$ N/m, $\omega_1^0 = 0.53$ deg/s, $\omega_2^0 = 0.55$ deg/set,
$\omega_3^0 = 0.053$ deg/sec, $G = 0.6$, $G_i = 177$ and $\varepsilon_1^0 = 0.95$, $\varepsilon_2^0 = 0.7$,
$\varepsilon_3^0 = 0.15$, $\varepsilon_0^0 = 0.15$, $\phi = 162°$ we obtain the simulation results seen
in Fig. 5.21.

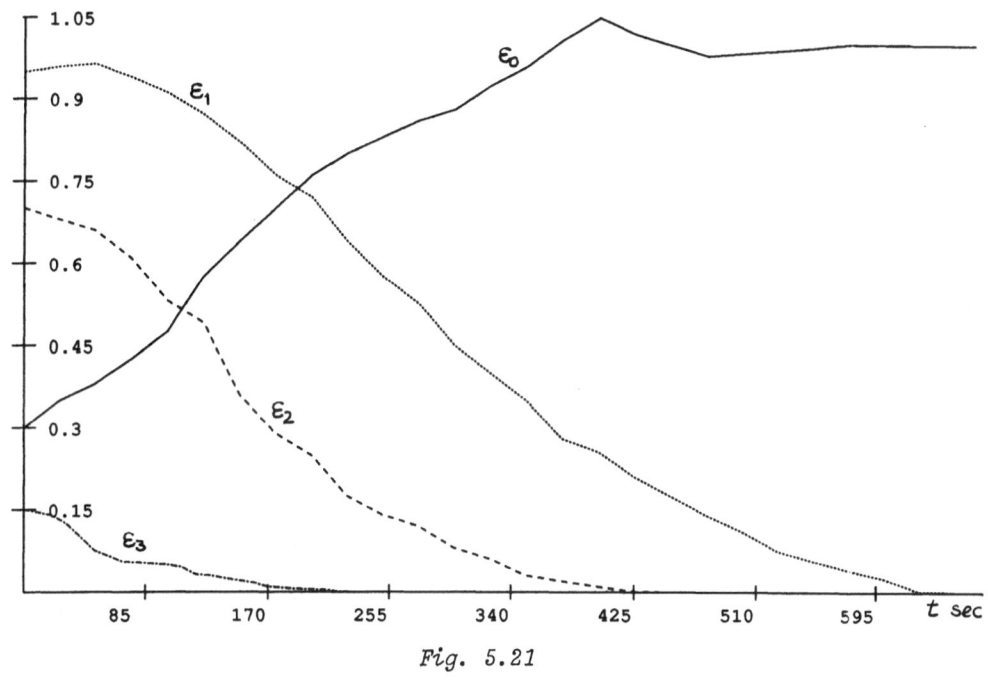

Fig. 5.21

5.7 CONTROL OF WHIRL-WHIP IN ROTORS

Amplitude control of a steady state limit trajectory in self-exciting
(accumulative) vibrations of a rotor in gas turbine, jet engine, pump or
compressor, induced by a flow of the fluid which generates the motion, is
also a good example to illustrate controllability for capture or stabilization between two stipulated energy levels. The reader is asked to recall
our comments at the opening of Section 4.3 on, among other cases, the case
of *whirl-whip self-exciting vibrations*, occurring in rotors supported by
fluid lubricated bearings. Similarly, like flutter described in Section
4.3, the whirl-whip phenomenon is fluid induced. However, now it is oil
or other lubricant at the bearing, instead of air, which provides the
energy source for accumulation. The latter is caused by coupling between
the rotation, which produces the fluid dynamic forces in bearings, and
lateral vibrations generated by such forces. The mechanism can be briefly
described as follows, see Muszynska-Bently [1]. The rotating shaft is
embedded in fluid at a bearing or a seal and pulls the fluid into a forward
rotary motion in the direction of the shaft rotation. The fluid flow is
generally three diminsional, but the circumferential component usually may
appear most significant. This component generates the effect of the fluid
rotary force which in turn causes the shaft lateral vibrations concerned.

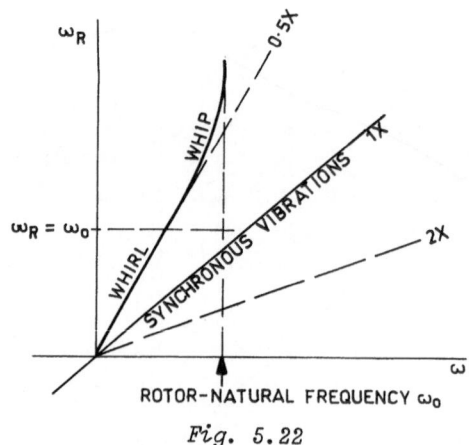

Fig. 5.22

A more detailed description follows from the graph of the angular speed of the rotor ω_R versus the frequency ω of the induced shaft lateral vibrations shown in Fig. 5.22.

The straight line through the origin with slope 1/1 marks the *synchronous* vibrations: ω_R the same as the vibration frequency ω (1× = one per one rotation). For small rotation speed ω_R the effect of the fluid action is minimal, and we have a small amplitude synchronous lateral vibration about the stable (basic) equilibrium caused by the imbalance of the rotor only - no negative damping. At higher ω_R, but still below the first balance resonance, that is, the first natural frequency of the rotor, the fluid dynamic forces begin to be effective. They destabilize the equilibrium and, along the synchronous vibrations, there appears a self-accumulative motion called *whirl*, produced by the mentioned coupling. It exists simultaneously with the synchronous component and establishes the vibrations at some subsynchronous frequency, usually around $\frac{1}{2}×$, generating a stable limit steady state Λ^+. With increasing rotor speed, the whirl frequencies maintain the constant ratio with it. The amplitudes of the whirl are usually higher than those of the synchronous vibrations forced by the imbalance alone, and they remain nearly constant. At the bearing the vibration amplitude may cover entire radial clearance. When ω_R grows further, and approaches the value of the natural frequency of the rotor, the whirl itself becomes unstable, disappears and the motion returns to the forced synchronous vibrations only, which in this range of speeds have high (resonant) amplitudes. With further growth of the rotative speed, the whirl returns. When the rotative speed reaches about twice the first balance resonant speed, the whirl becomes replaced by self-exciting vibrations called *whip*, which are characterized by the frequency equal to the rotor natural frequency.

Such a whip produces lateral forward precessional subharmonic vibrations of the rotor with constant frequency, independently upon the increase of ω_R, and corresponding to the first bending mode, see Muszynska [2]. The whip vibrations of the rotor form a stable limit steady state Λ^+ with amplitudes which may become very high (resonant frequency) and dangerous, even if they are limited at the end by the clearance of the bearing. In fact, in linear models with single equilibrium, the amplitude will appear growing indefinitely. Fortunately the reality is nonlinear and such growth is replaced by the above described limit steady states, see Fillod-Piranda-Bonnecase [1].

It may be interesting to note that whirl/whip does not appear when the rotor is under a heavy radial load, resulting in high shaft eccentricity at the bearing. Such a situation cancels self-vibrations due to changes in the flow pattern = no more predominance of the circumferential flow.

We obviously may want to control whirl/whip towards lower amplitudes, that is, lower energy levels. The presently used passive methods are based on redesigning the system in order to avoid all types of instability influencing the system (imbalance, misalignment, rotor-to-stator rub, internal friction, electro-dynamic forces in actuators, wear and tear effect making parts loose, etc.). Such methods help to reduce whirl and whip, but they work as long as one can identify the cause of vibrations. This is not always possible, off-line at least, and to satisfactory effect. A common remedy used is to increase the overall external damping which, obviously, produces considerable loss of power. Hence the recent trend to *active control*, designed to provide damping only while it is needed, that is, on-line, cf. Hagedorn-Kelkel-Weltin [1]. The presently popular electro-magnetic damping is one of the most significant representatives of the method, see Gondhalekar-Holmes [1], Nakai-Okada-Matsuda-Kibune [1], Salm-Schweitzer [1], Ulbrick-Anton [1], for an up-to-date review of the problem see Muszynska-Franklin-Bently [1]. In the latter work, another method of active control is proposed. The method introduced and developed in the Bently Rotor Dynamics Research Corporation seems to be more efficient than the electro-magnetic damping, as it affects directly the cause of the vibrations, the circumferential flow.

The basic features of the flow concerned are the fluid *average circumferential velocity ratio* λ and the *radial stiffness* $\tilde{K}_F$ of the fluid film. Both λ_F and $\tilde{K}_F$ are functions of many parameters, the principal being the

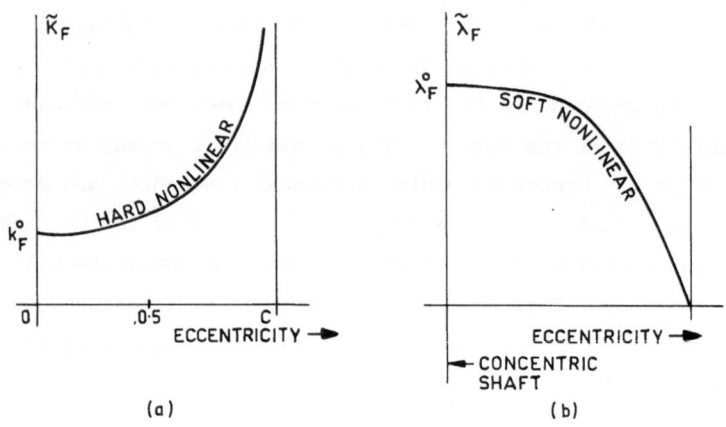

Fig. 5.23

shaft eccentricity: $\tilde{K}_F$ is "hard" nonlinear, that is, stiffly increases with the eccentricity, while λ_F decreases with it, with the decrease accelerating at high eccentricity values up to cutting off the circumferential flow completely at the so called "wall value", see Fig. 5.23(a), (b) where C means the radial clearance of the bearing.

The fluid rotates with angular circular frequency $\lambda\omega_R$, where ω_R is the angular speed. The Bently method is based on injecting into the bearing a controlling external flow directed tangentially and opposite to rotation. Such a controlling flow reduces λ and lowers the amplitude of the limit trajectory below a stipulated energy level.

Consider a slender shaft with the mid-span located disc supported rigidly without friction on one side and in an oil 360° lubricated cylindrical bearing on the other, as shown in Fig. 5.24. The following lumped mass model has been derived in Muszynska [2]. Let $z_i = x_i + jy_i$, $i = 1,2$ be the displacement vectors at the disc and at the bearing, respectively. The complex representation allows us to reduce the system

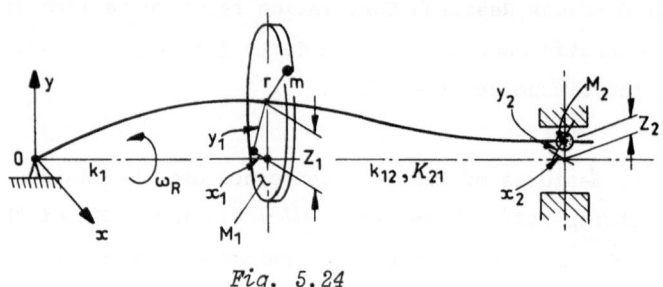

Fig. 5.24

to that with two complex DOF. Moreover, as mentioned in the description of the whip, we assume the rotor to be at its first (lowest) bending mode. The Cartesian motion equations represented in complex coordinates $z_i = x_i + jy_i$, $|z_i| = \sqrt{x_i^2 + y_i^2}$, $i = 1,2$, can be written as follows:

$$
\left.
\begin{aligned}
& M_1\ddot{z}_1 + d\dot{z}_1 + k_1 z_1 + k_{12}(z_1 - z_2) = mr\omega_R^2 e^{j\omega_R t} \\
& (M_2 + M_F)\ddot{z}_2 - 2jM_F\tilde{\lambda}_F\omega_R\dot{z}_2 - M_F\tilde{\lambda}_F^2\omega_R^2 z_2 \\
& \quad + [d_F + \tilde{\psi}_D(|z_2|)](\dot{z}_2 - j\tilde{\lambda}_F\omega_R z_2 + juz_2) \\
& \quad + [k_F + \tilde{\psi}_K(|z_2|)]z_2 + k_{21}(z_2 - z_1) + k_2 z_2 = 0
\end{aligned}
\right\}
\qquad (5.7.1)
$$

where M_1, M_2 are rotor modal masses, d is the external viscous damping coefficient, k_1, k_{21}, k_{12}, k_2 are the modal elasticity coefficients with k_2 covering also the external spring stiffness; m and r are the unbalanced mass and its radius in the rotor; $K_F(z_2) = k_F z_2 + \tilde{\psi}_K(|z_2|)z_2$ and $D_F(\dot{z}_2) = d_F\dot{z}_2 + \tilde{\psi}_D(|z|)\dot{z}_2$ are the fluid radial elasticity and damping characteristics, respectively, with $\tilde{\psi}_K(\cdot)$, $\tilde{\psi}_D(\cdot)$ hard-nonlinear positive functions and k_F, d_F positive coefficients; M_F is the fluid inertia coefficient and u is the air jet inlet external flow average tangential velocity representing the control variable.

We choose the state variables $x_1 \triangleq x_1$, $x_2 \triangleq x_2$, $x_3 \triangleq y_1$, $x_4 \triangleq y_2$, $x_5 \triangleq \dot{x}_1$, $x_6 \triangleq \dot{x}_2$, $x_7 \triangleq \dot{y}_1$, $x_8 \triangleq \dot{y}_2$ and denote $d \triangleq d/M_1$, $k_1 \triangleq k_1/M_1$, $k_{12} \triangleq k_{12}/M_1$, $m \triangleq m/M_1$, $\gamma \triangleq M_F/(M_2 + M_F)$, $d_F \triangleq d_F/(M_2 + M_F)$, $k_F \triangleq k_F/(M_2 + M_F)$, $\lambda_F \triangleq \tilde{\lambda}_F/(M_2 + M_F)$, $\psi_K \triangleq \tilde{\psi}_K/(M_2 + M_F)$, $\psi_D \triangleq \tilde{\psi}_D/(M_2 + M_F)$, $k_2 \triangleq k_2/(M_2 + M_F)$, $k_{21} = k_{21}/(M_2 + M_F)$. Then the state equations take the format

$$
\left.
\begin{aligned}
\dot{x}_1 &= x_5 , \\
\dot{x}_2 &= x_6 , \\
\dot{x}_3 &= x_7 , \\
\dot{x}_4 &= x_8 , \\
\dot{x}_5 &= -dx_5 - k_1 x_1 - k_{12}(x_1 - x_2) + mr\omega_R^2 \cos\omega_R t , \\
\dot{x}_6 &= -2\gamma\lambda_F\omega_R x_8 - k_2 x_2 - k_{21}(x_2 - x_1) \\
& \quad - [d_F + \psi_D(\sqrt{x_2^2 + x_4^2})](x_6 + \lambda_F\omega_R x_4 - ux_4) \\
& \quad - [k_F + \psi_K(\sqrt{x_2^2 + x_4^2}) - \gamma\lambda_F^2\omega_R^2]x_2 , \\
\dot{x}_7 &= -dx_7 - k_1 x_3 - k_{12}(x_3 - x_4) + mr\omega_R^2 \sin\omega_R t , \\
\dot{x}_8 &= 2\gamma\lambda_F\omega_R x_6 - k_2 x_4 - k_{21}(x_4 - x_3) - [d_F + \psi_D(\sqrt{x_2^2 + x_4^2})](x_8 - \\
& \quad - \lambda_F\omega_R x_2 + ux_2) - [k_F + \psi_K(\sqrt{x_2^2 + x_4^2}) - \gamma\lambda_F^2\omega_R^2]x_4 .
\end{aligned}
\right\}
\qquad (5.7.2)
$$

In the above $\lambda_F \overset{\Delta}{=} \lambda_F(\sqrt{x_2^2 + x_4^2})$ is a nonlinear decreasing function of shaft eccentricity with $\lambda_F(0) \overset{\Delta}{=} \lambda_F^0$, and the control variable is specified by a program $P(\cdot)$ which depends not only upon the vector $\bar{x} = (x_1, \ldots, x_8)^T$ but also upon the air pressure λ_p, which may be either a given constant, as assumed at the moment, or an adaptable parameter, as discussed later.

The equilibria are established by the equations

$$\left.\begin{aligned}
-k_1 x_1 - k_{12}(x_1 - x_2) &= 0 , \\
-k_2 x_2 - k_{21}(x_2 - x_1) - (k_F + \psi_K - \gamma\lambda_F^2\omega_R^2)x_2 &= 0 , \\
-k_1 x_3 - k_{12}(x_3 - x_4) &= 0 , \\
-k_2 x_4 - k_{21}(x_4 - x_3) - (k_F + \psi_K - \gamma\lambda_F^2\omega_R^2)x_4 &= 0 .
\end{aligned}\right\} \quad (5.7.3)$$

Note that $\psi_K(\sqrt{x_2^2 + x_4^2})$, $\lambda_F(\sqrt{x_2^2 + x_4^2})$ are "hard" and "soft" nonlinear functions of the eccentricity, respectively, thus with negative $-\gamma\lambda_F^2\omega_R^2$ we expect the bracket to be positive, see Fig. 5.23. We conclude that there is only one equilibrium at the origin of the state space, which means no energy thresholds and the cup about such equilibrium covering Δ. This is immediately confirmed if we calculate the potential and kinetic energies. Granted the natural assumption $k_{12} = k_{21}$, we have

$$\left.\begin{aligned}
V &= \tfrac{1}{2} k_1 (x_1^2 + x_2^2) + \tfrac{1}{2} (k_2 + k_F)(x_2^2 + x_4^2) + \tfrac{1}{2} k_{12}(x_1 - x_2)^2 \\
&\quad + \tfrac{1}{2} k_{12}(x_3 - x_4)^2 + \int \{\tilde{\psi}_K(\sqrt{x_2^2 + x_4^2}) - M_F[\tilde{\lambda}_F(\sqrt{x_2^2 + x_4^2})]^2\omega_R^2\}x_2\,dx_2 \\
&\quad + \int \{\tilde{\psi}_K(\sqrt{x_2^2 + x_4^2}) - M_F[\tilde{\lambda}_F(\sqrt{x_2^2 + x_4^2})]^2\omega_R^2\}x_4\,dx_4
\end{aligned}\right\}$$

$$(5.7.4)$$

$$T = \tfrac{1}{2} M_1 (x_5^2 + x_7^2) + \tfrac{1}{2} (M_2 + M_F)(x_6^2 + x_8^2) . \qquad (5.7.5)$$

The power characteristic $f_0(\cdot)$ is then determined as

$$\left.\begin{aligned}
f_0(\bar{x}, u, t, \omega_R, \lambda_F) &= mr\omega_R^2(x_5 \cos\omega_R t + x_7 \sin\omega_R t) - d(x_5^2 + x_7^2) \\
&\quad - (d_F + \psi_D)[(x_6^2 + x_8^2) + (\lambda_F\omega_R - u)(x_4 x_6 - x_2 x_8)] ,
\end{aligned}\right\}$$

$$(5.7.6)$$

with ω_R, λ_F appearing as system parameters.

If the system were linear: $\psi_D \equiv 0$, $\psi_F \equiv 0$, $\lambda_F \equiv 1.5$, then the stability of the basic equilibrium might have been qualified similarly as for the equation (3.1.4), by eigenvalues of the system matrix A having negative real parts - the matrix being thus stable, see Section 3.1. It may be shown that the real parts of eigenvalues remain negative up to a certain value of $\omega_R = \omega_R^{ST}$, which thus forms the ω_R-*based stability threshold*, see Muszynska [2]:

$$\omega_R^{ST} = \frac{1}{1.5} \left[\frac{k_1}{M_1} + \frac{k_{12}(k_F + k_2 - M_2 k_1/M_1)}{M_1[k_2 + (k_F - k_2 - M_2 k_1/M_1)]} \right]^{\frac{1}{2}} \tag{5.7.7}$$

such that for $\omega_R \geq \omega_R^{ST}$ we obtain exponentially increasing amplitudes of the linear systems, and the described whirl phenomenon should appear nearly above ω_R^{ST} for the nonlinear system. We may thus propose to control the system for capture below some E-level within the basic cup, the level being within the stipulated distance δh above that corresponding to ω_R^{ST}. To attain this objective, we use Proposition 4.1.1 and Corollary 4.1.1 with f_0 specified by (5.7.6).

Let us set up a candidate Δ_0 with h_0 defining $\partial\Delta_0$ reasonably high, close to $\partial\Delta$. Then, for control condition, we use the dissipative (5.5.9) with (5.7.6) substituted:

$$f(\bar{x}, u, t, \omega_R, \lambda_F) \leq \frac{h_0 - h_{TC}}{T_C} \tag{5.7.8}$$

with h_{TC} such as to make the ∂T_C level above that specified by ω_R^{ST} by a suitable δh. Deleting the negative terms from (5.7.6) the dissipative control condition reduces to the following:

$$\left. \begin{aligned} &m r \omega_R^2 (x_5 \cos\omega_R t + x_7 \sin\omega_R t) - (d_F + \psi_D)(\lambda_F \omega_R - u)(x_6 x_4 - x_2 x_8) \\ &\qquad = -(h_0 - h_{TC})/T_C \quad . \end{aligned} \right\} \tag{5.7.9}$$

Hence the control program implying (5.7.8) may be designed in the following format:

$$\left. \begin{aligned} u(t) &= [1/(d_F + \psi_D)(x_4 x_6 - x_2 x_8)]\{[-(h_0 - h_{TC})/T_C] \\ &+ (d_F + \psi_D)(x_6 x_4 - x_2 x_8)\lambda_F \omega_R - m r \omega_R^2 (x_5 \cos\omega_R t + x_7 \sin\omega_R t)\} \\ &\qquad\qquad\qquad\qquad \text{for} \quad |x_i| > \beta_i, \text{ and} \\ u(t) &= 0, \quad \text{for} \quad |x_i| < \beta_i, \quad i = 5,6,7,8 \; . \end{aligned} \right\} \tag{5.7.10}$$

In the above, β_i's have the role as described in Section 3.3 and can be calculated in the same way. Obviously we could have left the negative terms of (5.7.6) in (5.7.9) and thus in (5.7.10), then however the control effort would have to increase, and moreover (5.7.9) would have to be left as an inequality, and so would (5.7.10). For a small mass-imbalance m or when the effect of imbalance is minimal, say at high whip, we may delete the last trigonometric terms in (5.7.10), whose only role is to cancel the perturbation generated by imbalance. In the latter case, as well as in a number of other features appearing at different values of ω_R and λ_F we may see the parametric role of these variables and thus the possible need for adaptive control programs. We discuss this topic later in the text.

The control program (5.7.10), or in any other format of fluid dissipative algorithm, is implemented by air jets located in the stator of the bearing which introduce a backward circumferential (reverse) flow into the bearing clearance. As mentioned, u(t) is the tangential component of the average velocity of such a flow. The jets start acting on some pre-set value of the monitored amplitude of the lateral vibrations which is related to our pre-set energy level h_0. They switch themselves off at another amplitude corresponding to the level of $\partial T_{\mathcal{C}}$. There is no need to use the accumulative controller for forcing the motions to $T_{\mathcal{C}}$ from below as this is done automatically by the unbalanced rotor.

Simulation results implementing the controller (5.7.10) have been obtained using the following data $\omega_R = 209$ rad/sec , $\psi_K = 4000/(1-e^2)^3$, with $e = \sqrt{x_2^2 + x_4^2}$, $k_F = 0$, $\psi_D = 40/(1-e^2)^2$, $d_F = 0$, $\lambda_F = 0.48(1-e^2)^{1/5}$, $d = 4.0$ ℓb sec/in , $m = 6$ ℓb , $r = 1.2$ in , $\gamma = 0.5$ and the stiffness coefficients $k_1 = 2000$ ℓb/in , $k_2 = 10,000$ ℓb/in , $k_{12} = 1000$ ℓb/in . The corresponding motion components may be seen in Figs. 5.25, 5.26. Observe that the time taken for the motion between the initial energy level h^0 to h_T is approximately $T_{\mathcal{C}} = 8$ sec .

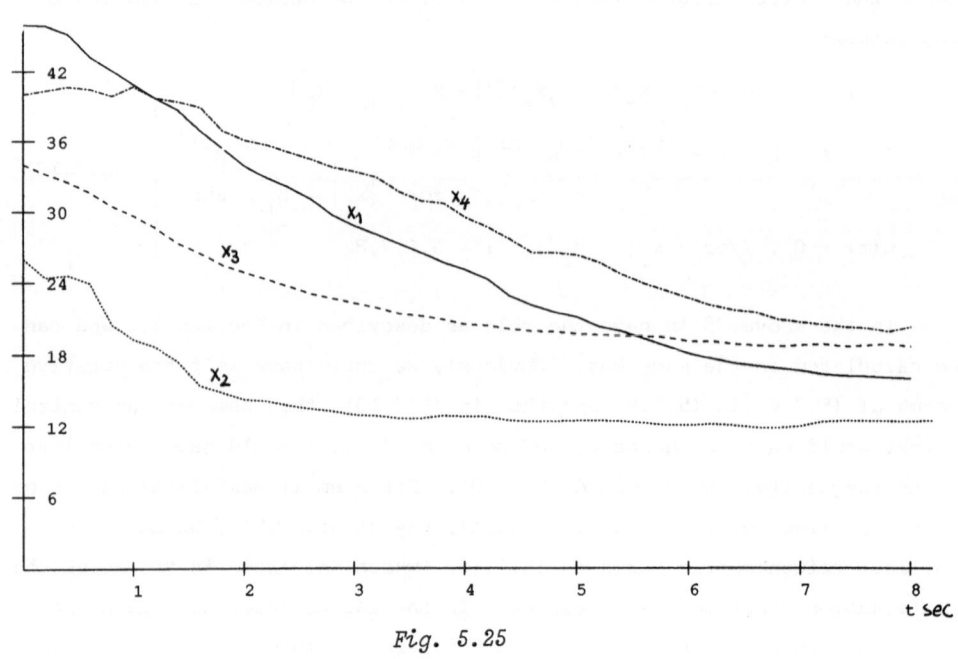

Fig. 5.25

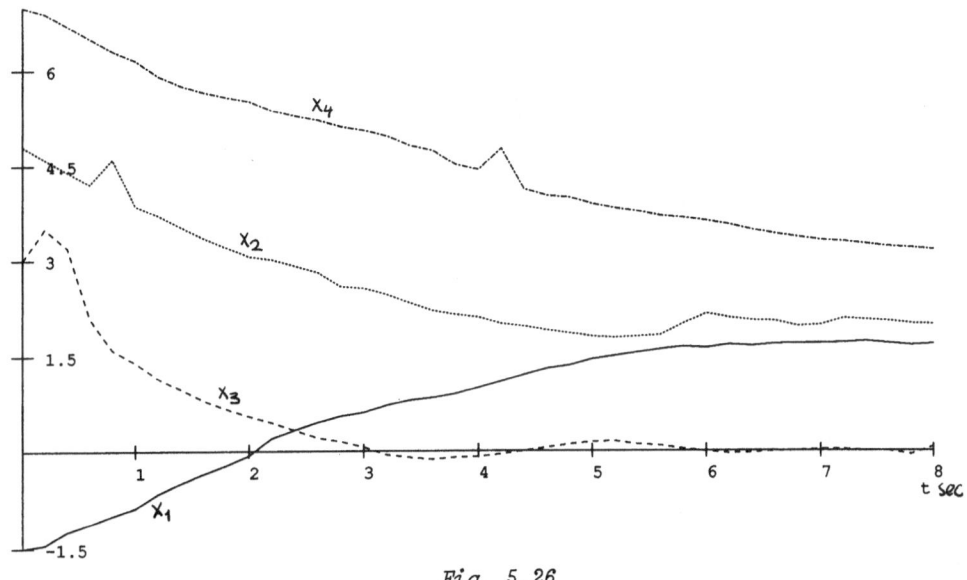

$$Fig.\ 5.26$$

5.8 CONDITIONS FOR RENDEZVOUS

We shall now look closer at the case when the system fails to attain capture in some target T. More precisely, we shall consider the systems which are strongly controllable at some $\bar{x}^0 \in \Delta$ for collision or penetration *without capture* in T, which occurs if and only if such a system is strongly controllable at $\bar{x}^0$ for collision or penetration, and for each $\bar{P}(\cdot)$ that wins this objective, there is $\bar{w}(\cdot)$ which denies capture, as specified by (5.1.13) of Definition 5.1.5.

The corresponding region of such strong controllability is denoted $\Delta_{\mathcal{C}}$. By the definitions of the objectives concerned, such a region is the complement of $\Delta_{\mathcal{C}}$ to either Δ_C or Δ_Z, see (5.1.16), and encloses the *noncapturing* part of T, namely $T_{\mathcal{C}} \triangleq T - \Delta_{\mathcal{C}}$.

Consequently to the above, conditions sufficient for strong controllability for collision or penetration without capture reduce to Conditions 5.3.2 for collision or Conditions 5.3.5 for penetration accompanied by contradicting (5.5.2) of Conditions 5.5.2 necessary for capture. Hence there is no need for separate formalization of conditions for collision or penetration without capture, unless we want a specified subcase of such an objective which is the rendezvous, see (5.1.13) in Definition 5.1.5, and Fig. 5.27.

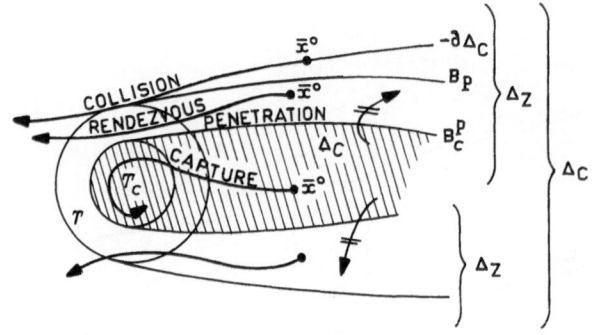

Fig. 5.27

To establish attention, let us consider the *collision with rendezvous*.
From Definition 5.1.5 it follows that there is a proper subset T_Z of T
consisting of points where the rendezvous takes place. Obviously $T_Z \subset T_{\mathcal{C}}$
as much as the strong region for rendezvous $\Delta_Z \subset \Delta_{\mathcal{C}} = \Delta_C - \Delta_{\mathcal{C}}$. It is
easily seen also that, given T_Z and taking it as a target, the region of
strong controllability for collision with T_Z is the region Δ_Z. This
makes the methods of Section 5.4 immediately applicable for determining
Δ_Z.

Clearly $\Delta_Z \cap T_Z \neq \phi$ and $T_Z \subset \Delta_Z \cap T$, see Fig. 5.26. If
$T_Z = \Delta_Z \cap T$, we may have $T_Z = \phi$ yielding collision without rendezvous,
that is, rejection, see Section 5.1.

Similarly as with capture, we attempt to force the motions of
$K(\bar{x}^0, t_0)$, $(\bar{x}^0, t_0) \in \Delta_0 \times \mathbb{R}$ into T_Z but now require that they stay in
T only for T_Z. This is secured by the following conditions introduced
in Skowronski [38].

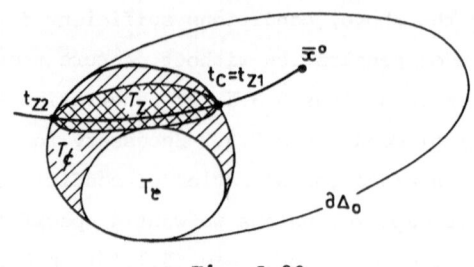

Fig. 5.28

As before, we stipulate candidates T_Z, Δ_0, with the options $\Delta_0 = \Delta_C$, $T_Z = T_{\mathscr{C}}$ if Δ_C, $\Delta_{\mathscr{C}}$ are known. Then we need sufficient conditions to verify the candidates. Denote $CT_Z \triangleq \Delta_0 - T_Z$, see Fig. 5.28, and consider a C^1-function $V(\cdot) : D \to \mathbb{R}$ with $D(\text{open}) \supset \Delta_0$ (or $D \supset \Delta_0 - \Delta_{\mathscr{C}}$, if $\Delta_{\mathscr{C}}$ known), and with

$$\left.\begin{array}{l} v_Z = \sup V(\bar{x}) \,\big|\, \bar{x} \in \partial\Delta_0 \ , \\[2mm] v_{TZ} = \inf V(\bar{x}) \,\big|\, \bar{x} \in \partial T_Z \ , \\[2mm] v_T = \inf V(\bar{x}) \,\big|\, \bar{x} \in \partial T \ . \end{array}\right\} \tag{5.8.1}$$

CONDITIONS 5.8.1. The system (2.2.6)' is strongly controllable on Δ_0 for collision with rendezvous in T; if given $T_Z \subset T$ there is $\bar{P}(\cdot)$ defined on $D \times \mathbb{R}$ and $V(\cdot)$ of above such that

(i) $V(\bar{x}) \le v_Z$,   for $\bar{x} \in CT_Z$;

(ii) $V(\bar{x}) > v_{TZ}$,   for $\bar{x} \notin T_Z$;

(iii) $V(\bar{x}) \le v_T$,   for $\bar{x} \in T_{\mathscr{C}}$;

(iv) for any $(\bar{x},t) \in CT_Z \times \mathbb{R}$, $\bar{\phi}(\cdot) \in K(\bar{x},t)$, $\bar{u} \in \bar{P}(\bar{x})$, there is a positive constant $T_c < \infty$ such that

$$\int_{t_0}^{t_0 + T_c} [\nabla V(x)^T \cdot \bar{f}(\bar{x},\bar{u},\bar{w},t)]dt \le -(v_Z - v_{TZ}) \tag{5.8.2}$$

for all $\bar{w} \in W$;

(v)   for all $(\bar{x},t) \in T_{\mathscr{C}} \times \mathbb{R}$, $\bar{\phi}(\cdot) \in K(\bar{x},t)$, $\bar{u} \in \bar{P}(\bar{x})$, there is a constant $c_Z > 0$ such that

$$0 < \nabla V(\bar{x})^T \bar{f}(\bar{x},\bar{u},\bar{w},t) < c_Z , \quad \forall\, \bar{w} \in W . \tag{5.8.3}$$

If $\Delta_{\mathscr{C}}$ is unknown, $T_{\mathscr{C}}$ in (iii), (v) must be replaced by T.

By Conditions 5.3.5, (i), (ii) and (iv) imply strong controllability on Δ_0 for collision with T_Z. It remains to show that no motion from T_Z may stay in T indefinitely or leave before T_Z.

The first assertion follows as an immediate conclusion from (iii) and the fact that by (v), $\dot{V}(\bar{\phi}(t)) > 0$. To show the second, consider an arbitrary motion $\bar{\phi}(\cdot)$ from T_Z leaving T at $t_2 \ge t_0$: $\bar{\phi}(x^0,t_0,t_2)$ $= \bar{x}^2 \in \partial T$. Due to collision it must have entered T_Z at some $t_c = t_1 \le t_2$: $\bar{\phi}(\bar{x}^0,t_0,t_1) = \bar{x}^1 \in \partial T_Z$. Integrating (5.8.3) we obtain the time estimate

$$t_2 - t_1 \geq \frac{1}{c_Z} \left[V(\bar{x}^2) - V(\bar{x}^1) \right] . \qquad (5.8.4)$$

Letting

$$T_Z(\bar{x}^0, t_0) = \frac{1}{c_Z} (v_T - v_{TZ}) ,$$

(5.8.1) implies $t_2 - t_1 \geq T_Z$ which proves the hypothesis.

Observe that as T_Z depends upon c_Z, which is the upper bound of the rate of change of $V(\bar{\phi}(t))$, it can be called the measure of slowing down in $T_{\mathscr{C}}$. We have

$$c_Z(\bar{x}^0, t_0, \bar{\phi}(\cdot)) = \frac{1}{T_Z} \left[v_{TZ} - V(\bar{x}^0) \right] \qquad (5.8.5)$$

with $v_{TZ} - V(\bar{x}^0) \geq 0$ by (ii). Given T_Z, we may design c_Z suitable for the required duration T_Z. Alternatively, given the rate bound c_Z the candidate T_Z may be adjusted to fit the time T_Z. Indeed

$$T_Z c_Z = v_{TZ} - V(\bar{x}^0) \geq 0 \qquad (5.8.6)$$

provides the condition for such adjustment. As mentioned in Section 5.1, the rendezvous becomes capture when $T_Z \to \infty$. Indeed then, in view of (5.8.5), we have $c_Z \to 0$, and the condition (5.8.3) must be replaced by $\dot{V} < 0$ implying capture. On the other hand, removing the slow-down restriction $c_Z < \infty$, that is, letting $c_Z \to \infty$ implies $T_Z \to 0$, that is, collision without rendezvous. In conclusion, similarly as before, we have the following corollary.

COROLLARY 5.8.1. *When* $\partial \Delta_0$, ∂T, ∂T_Z *are specified by* V-*levels, the conditions (i)-(iii) hold automatically, and (iv), (v) become* necessary *as well as sufficient* for collision with rendezvous with *T.*

Suppose now that T_Z is stipulated. The corresponding $T_Z(T_Z)$ will depend upon T_Z and obviously may differ from T_Z even for the same T_Z. Moreover, we have $T_Z(T_Z') \subset T_Z(T_Z'')$ for any two T_Z', T_Z'' such that $T_Z' \geq T_Z''$. The corresponding region is denoted by $\Delta_Z(T_Z)$. It is the same as the region of strong controllability for collision with $T_Z(T_Z)$ which gives the method for its determination. Obviously $\Delta_Z(T_Z)$ is related to $T_Z(T_Z)$ and thus also to T_Z : $\Delta_Z(T_Z') \subset \Delta_Z(T_Z'')$ for $T_Z' \geq T_Z''$. Hence, capture with its $T_Z \to \infty$ gives the lower estimate for all T_Z, Δ_Z : $T_Z(T_Z) \supset T_{\mathscr{C}}$, $\Delta_Z(T_Z) \supset \Delta_{\mathscr{C}}$, for any T_Z.

CONDITIONS 5.8.2. The system (2.2.6)' is strongly controllable on Δ_0 for collision with rendezvous in T during T_Z, if given $T_Z(T_Z)$ the Conditions 5.8.1 hold with (v) replaced by:

(v)' for all $(\bar{x},t) \in T_{\mathfrak{c}} \times \mathbb{R}$, $\bar{u} \in \bar{P}(\bar{x},t)$, we have

$$0 < \nabla V(\bar{x}) \cdot f(\bar{x},t,\bar{u},\bar{w}) \leq \frac{v_T - v_{TZ}}{T_Z} , \quad \forall \, \bar{w} \in W . \qquad (5.8.7)$$

The proof follows from Conditions 5.8.1 by setting up

$$c_Z = \frac{v_T - v_{TZ}}{T_Z} . \qquad (5.8.8)$$

The min-max and L-G controllers are obtainable by the same methods as for collision or capture, from the conditions (5.8.2) and (5.8.3) or (5.8.7), via the corresponding corollaries. In order to imply (5.8.2) for the collision subobjective, we can use the controller obtained from Corollary (5.3.2) during some or stipulated T_c, then switch the controller to that implying (5.8.3) or (5.8.7) for the rendezvous subobjective. For the latter purpose, we use the following corollary.

COROLLARY 5.8.2. *Given* $(\bar{x},t) \in \Delta_0 \times \mathbb{R}$, *if there is a pair* $(u^*,w^*) \in U \times W$ *such that*

$$\min_{\bar{u}} \max_{\bar{w}} L(\bar{x},\bar{u},\bar{w},t) \leq - \frac{v_T - v_{TZ}}{T_Z} , \qquad (5.8.9)$$

then (5.8.3) or (5.8.7) are met with $u^* \in \bar{P}^*(\bar{x},t)$.

The corollary allows us to design the controller for rendezvous. It is obviously highly desirable that it is the same controller as for the collision subobjective, thus eliminating the switching after T_c.

5.9 AEROASSISTED ORBITAL TRANSFER

To illustrate our theory on maneuvering between targets combined with rendezvous, let us consider a case of aeroassisted orbital transfer for the presently developed aerospace plane diving in the uncertain atmosphere. The transfer refers to the change from a high Earth orbit (HEO), which is exoatmospheric, to capture in a low Earth orbit (LEO). It allows for an unpowered rendezvous flight in the atmosphere with significant fuel saving, see Talay-While-Naftel [1] and Mease-Vinh [1]. The controller for such a rendezvous must be robust against large uncertain fluctuations in atmospheric density (potholes), reaching up to 40% according to the STS-6 Space

Shuttle data. The optimizing candidate $\bar{P}*(\cdot)$ for such a transfer has been designed by Grantham-Lee [1], based upon the minimum fuel consumption before reaching LEO. We shall rely heavily on the model studied in this work, augmenting the objective by some qualitative features.

The plane is modelled by a point-mass, and the transfer is investigated in the Cartesian plane $0xz$ with origin at the centre of the Earth and the two concentric orbits HEO and LEO, see Fig. 5.29. The scenario assumes an initial tangential retro-thrust at HEO with radius r_1. The first stage is to decrease the velocity by $-\delta v_1$ and to position the plane on an elliptical orbit with perigee at the boundary of the atmosphere with radius R. The atmosphere is a Cartesian ball-shape target $T : r \leq R$, the idealization good enough for our purpose.

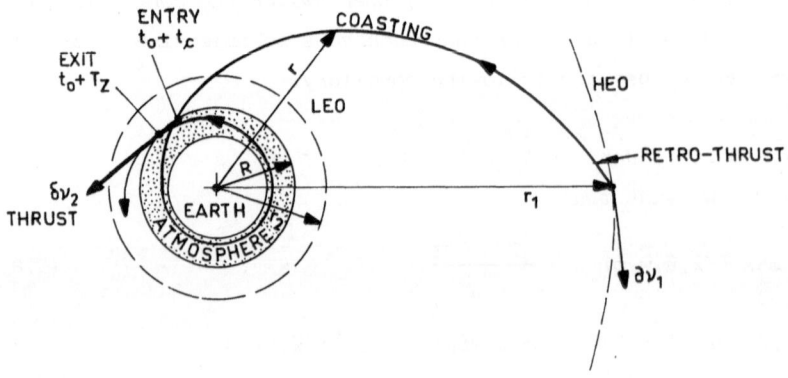

Fig. 5.29

It is expected that flying in the atmosphere a certain amount of kinetic energy will be converted to heat by drag and thus lost, producing outflux. If the plane were to fly along ∂T_z, that is, without penetration, it would obviously do it at minimum density with drag tending to zero, whence a suitable finite outflux would be produced in indefinite time. To avoid this, we need penetration and in fact the depth of it δh can be specified by the desired outflux

$$\delta r = \delta h = \int_{T_Z} f_0(\bar{x}, \bar{u}, \bar{w}) \, dt \qquad (5.9.1)$$

with the integral taken along the trajectory concerned over desired time interval T_Z. The latter is stipulated due to the constraint that the plane should make less than one revolution in T. After a suitable outflux of energy, the plane leaves T tangentially with zero lift and climbs in an elliptic orbit with apogee at LEO radius r_2 where the cap-

turing circular orbit is attained by a tangential circularizing burn
which increases the velocity by δv_2 .

The geometry of the atmospheric flight is shown in Fig. 5.30. Our
investigation refers to the radius $r(t)$, velocity $v(t)$ and the path
angle $\gamma(t)$ of the vehicle, independently of its angular position θ so
that, taking the lift and drag forces modelled successively by

$$L = \tfrac{1}{2}\,\rho SC_L v^2\ , \qquad D = \tfrac{1}{2}\,\rho SC_D v^2\ , \tag{5.9.2}$$

the equations of motion for the atmospheric flight reduce to the three-
dimensional system

$$\left.\begin{aligned}
\dot{r} &= v\sin\gamma\ , \\[4pt]
\dot{v} &= -(\rho SC_D v^2/2m) - (MG/r^2)\sin\gamma\ , \\[4pt]
v\dot\gamma &= (\rho SC_L v^2/2m) - [\,(MG/r^2) - v^2/r\,]\cos\gamma\ .
\end{aligned}\right\} \tag{5.9.3}$$

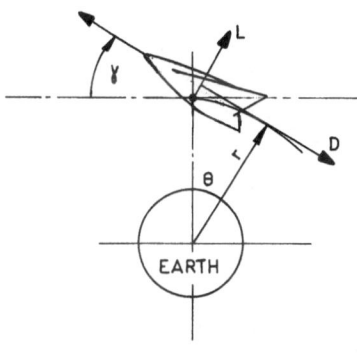

Fig. 5.30

In the above equations, C_L , C_D are the (angle of attack dependent) lift
and drag coefficients, see Section 4.3, with $C_D = C_{DO} + K\hat{C}_L^2$ where
$C_{DO} = C_D$ at $C_L = 0$ and $\hat{C}_L$ is the C_L measured at $[C_L/C_D]_{max} = \tfrac{1}{2}\sqrt{K C_{DO}}$.
Moreover S is the effective surface area normal to the velocity vector,
M and m are the masses of Earth and the vehicle, respectively, $G(r)$ is
the variable Newtonian gravity coefficient, while $\rho(t) \in [\hat\rho, \rho_0]$ is the
uncertain atmospheric density with $\rho = \hat\rho$ at $r = 40$ km and $\rho = \rho_0$ at
$r = R_E$. The entry v_{EN} , γ_{EN} and exit v_{EX} , γ_{EX} values of v,γ into
and out of the atmosphere must satisfy the conditions derived from the
conservation of energy and angular momentum outside the atmosphere, see
Grantham-Lee [1]:

$$(2 - v_{EN}^2)\hat{r}_1^2 - 2\hat{r}_1 + v_{EN}^2\cos^2\gamma_{EN} = 0\ , \tag{5.9.4}$$

$$(2 - v_{EX}^2)\hat{r}_2^2 - 2\hat{r}_2 + v_{EX}^2\cos^2\gamma_{EX} = 0\ , \tag{5.9.5}$$

283

where $\hat{r}_i = r_i/R$ are the nondimensional radii, with the entry and exit velocity changes δv_i expressed in dimensionless terms as

$$\delta v_1 = \sqrt{1/\hat{r}_1} - (v_{EN}/\hat{r}_1) \cos \gamma_{EN} , \qquad (5.9.6)$$

$$\delta v_2 = \sqrt{1/\hat{r}_2} - (v_{EX}/\hat{r}_2) \cos \gamma_{EX} . \qquad (5.9.7)$$

We shall now determine conditions for designing the controllers, and to do so, we need the state model. Let R_E be the idealized Earth radius and let $h = r - R_E$ be the altitude of the vehicle, with $h_E = R - R_E$ being the entry/exit altitude in and out of the atmospheric ball. We choose the dimensionless state variables

$$x_1 \overset{\Delta}{=} h/h_E , \qquad x_2 \overset{\Delta}{=} v/\sqrt{MG/R} , \qquad x_3 \overset{\Delta}{=} \gamma . \qquad (5.9.8)$$

The targets and controllable sets are seen in Fig. 5.31.

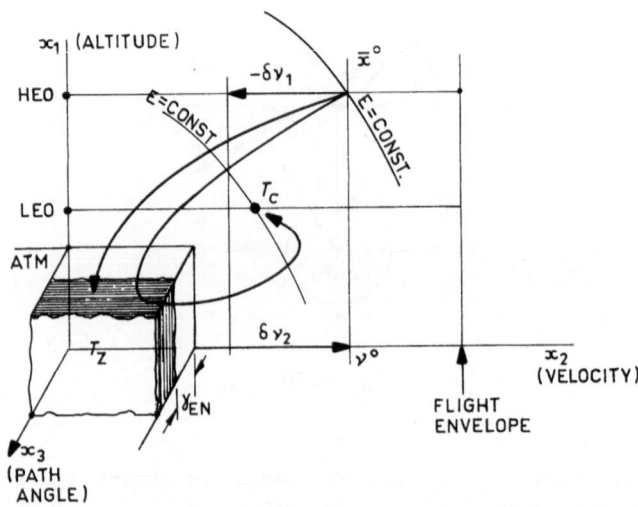

Fig. 5.31

The trajectory leaving HEO is an exoatmospheric descent with no air density, thus without lift and drag, as well as without uncertainty. Thus the energy outflux is produced totally with the power of the controlling retro-burn, the latter assumed to be an impulse force with constant value and short duration. Consequently to all the above, we may attempt to find a control condition implying (5.8.2) directly, instead of using a suitable corollary.

It is convenient to take $V(\bar{x}) \overset{\Delta}{=} E(x_1,x_2) + \frac{1}{2} x_3^2$, with the dimensionless $E(x_1,x_2) = \frac{1}{2}(x_1^2 + x_2^2) = (h^2/2h_E^2) + (v^2 R/2MG)$. We start at HEO :

$x_1 = (r_1 - R_E)/h_E$, but the initial V-level depends also upon the initial velocity $v^0 = v(t_0)$, which should be large enough to accommodate $-\delta v_1$. Given v^0 we can specify L_0 covering it, by the corresponding V-level:

$$\partial \Delta_0 \; : \; V(\bar{x}) = v_Z = \tfrac{1}{2}\{[(r_1 - R_E)^2/h_E^2] + [(v^0)^2 R/MG] + (\gamma^0)^2 . \qquad (5.9.9)$$

The collision stage ends when the trajectory reaches $T = T_Z = T_C$ at ∂T : $x_1 = 1$, $x_2 = v_{EN}/\sqrt{MG/R}$, $x_3 = \gamma_{EN}$. Horizontal entering of the atmosphere: $\gamma_{EN} = 0$ is fuel consumption optimal, but we must allow for some $\gamma_{EN} = \text{const} > 0$ in order to generate a lift. Such γ_{EN} substituted into (5.9.4) gives v_{EN} . With the latter two, we calculate

$$\left. \begin{aligned} v_T = v_{ZT} &= \{\inf[E(x_1,x_2) + \tfrac{1}{2}x_3^2] \,|\, \bar{x} \in \partial T\} \\ &= \tfrac{1}{2} + (v_{EN}^2 R/2MG) + \tfrac{1}{2}\gamma_{EN}^2 . \end{aligned} \right\} \qquad (5.9.10)$$

The collision time interval T_C is then estimated by

$$T_C \geq \frac{r_1 - R}{|-\delta v_1|} . \qquad (5.9.11)$$

Thus (5.8.2) becomes

$$u \int_{t_0}^{t_0 + T_C} x_2 dt \leq -(v_Z - v_{TZ}) \qquad (5.9.12)$$

with the constants specified by (5.9.9), (5.9.10), (5.9.11). Condition (5.9.12) forms the control condition for collision.

For the flight within the atmosphere, we use (5.9.3) in the state format. Substituting (5.9.8), introducing the dimensionless parameters

$$\tau \overset{\Delta}{=} (t\sqrt{MG/R})h_E ; \quad \delta \overset{\Delta}{=} \rho/\hat{\rho} ; \quad c \overset{\Delta}{=} R/h_E ; \quad \ell \overset{\Delta}{=} \hat{\rho}SR\hat{C}_L , \qquad (5.9.13)$$

and adding the lift control variable $u = C_L/\hat{C}_L$ to the uncontrolled (5.9.3), we obtain the state format, see Grantham-Lee [1]:

$$\left. \begin{aligned} \overset{\circ}{x}_1 &= x_2 \sin x_3 \\ \overset{\circ}{x}_2 &= -\frac{\ell\delta}{2[L/D]_{max}} (1 + u)x - \frac{c}{(c-1+x_1)^2} \sin x_3 \\ \overset{\circ}{x}_3 &= \ell\delta u x_2 + \frac{\cos x_3}{(c-1+x_1)} \left[x_2 - \frac{c}{(c-1+x_1)x_2} \right] \end{aligned} \right\} \qquad (5.9.14)$$

where $(\circ)$ denotes $d(\cdot)/d\tau$. The control values are bounded by the thrust constraint $|u| \leq \hat{u}$ and the uncertainty $\delta(t)$ is bounded by the bounds corresponding to $\rho_0 , \hat{\rho}$, namely $\delta \in [1,(\rho_0/\hat{\rho})]$. In order to use Corollary 5.8.2, we calculate

$$L(\bar{x}, \bar{u}, \bar{w}) = \sum_{i=1}^{3} \frac{\partial V}{\partial x_i} \dot{x}_i = x_1 x_2 \sin x_3 - x_2 \left[\frac{\ell \delta x_2^2 (1 + u^2)}{2[L/D]_{max}} + \frac{c \sin x_3}{(c - 1 + x_1)^2} \right]$$

$$+ x_3 \left\{ \ell \delta u x_2 + \frac{\cos x_3}{(c - 1 + x_1)} \left[x_2 - \frac{c}{(c - 1 + x_1) x_2} \right] \right\} .$$

Then the condition (5.8.9) becomes

$$\min_{u} \max_{\delta} \ell \delta x_2^2 \left[\frac{u x_3}{x_2} - \frac{x_2 u^2}{2[L/D]_{max}} \right] - \min_{\delta} x_2^2 \left(\frac{\ell \delta x_2}{2[L/D]_{max}} \right)$$

$$\le -c_Z + \left[\frac{c x_2}{(c - 1 + x_1)^2} - x_1 x_2 \right] \sin x_3 - \left[\frac{x_2 x_3}{(c - 1 + x_1)} - \frac{c x_3}{(c - 1 + x_1)^2} \right] \cos x_3 .$$

Substituting the maximizing $\delta*$ and minimizing δ_* we obtain the control condition for rendezvous in T:

$$\min_{u} \left[\frac{u x_3}{x_2} - \frac{u^2 x_2}{2[L/D]_{max}} \right] \le (-c_Z / \ell \delta* x_2^2) + \frac{\delta_* x_2}{2 \delta* [L/D]_{max}}$$

$$+ \left[\frac{c}{(c - 1 + x_1)^2} - x_1 \right] \frac{\sin x_3}{\ell \delta* x_2} - \left[\frac{x_2 x_3}{(c - 1 + x_1)} - \frac{c x_3}{(c - 1 + x_1)^2} \right] \frac{\cos x_3}{\ell \delta* x_2^2} .$$

$$(5.9.15)$$

In the above c_Z is estimated by $\delta h / T_Z$, where δh is the desired outflux of energy (5.9.1) between the E-level of entry and the E-level of exit into T. As mentioned already T_Z is determined by the demand of less than one revolution about Earth in T. After T_Z an extra burn produces the velocity increase δv_2 for reaching LEO. The vehicle again should leave horizontally, but small $\gamma_{EX} = \text{const} > 0$ is needed in order to climb out of the atmosphere T. Given γ_{EX}, we calculate v_{EX} from (5.9.5). Then the exit V-level is obtained as

$$V(\bar{x}) = v_{TEX} = \tfrac{1}{2} + v_{EX}^2 R / 2MG + \tfrac{1}{2} \gamma_{EX}^2 . \qquad (5.9.16)$$

The mentioned δh for c_Z may be expressed in terms of $v_T - v_{TEX}$ from (5.9.10) and (5.9.16).

The next target is the capturing T_C which is defined by the LEO's altitude and the kinetic energy corresponding to $v = v_{EX} + \delta v_2$, as well as the convenient limiting $\gamma = 0$:

$$T_C = \left\{ \bar{x} \;\middle|\; \left| x_1 - \frac{r_2 - R_E}{\sqrt{2}(R - R_E)} \right|, \; \left| x_2 - \frac{v_{EX} + \delta v_2}{\sqrt{2MG/R}} \right|, \; |x_3| \le \eta \right\} . \qquad (5.9.17)$$

In the above $\eta > 0$ is a secure bound for admissible deviations of the state variables from the trajectory desired at the LEO stage. The flight to LEO is again exoatmospheric with no air density and thus the same features as for the descent from HEO. This time, however, we are ascending in both the Cartesian space: $R \nearrow R_2$, and the state space with respect to E-levels (increasing velocity as well as altitude). Consequently, following the discussion in Section 5.5, when using the capture generating Corollary 5.5.3, we must adopt it to energy accumulation and apply the control condition (5.5.10). This requires adjustment in the test function: we take $V_{\mathcal{C}}(\bar{x}) = v_{TC} - V(\bar{x})$, where $V(\bar{x})$ is the function used so far, with

$$
v_{TC} = \frac{1}{2}\left(\frac{r_2 - R_E}{R - R_E}\right)^2 + (v_{EX} + \delta v_2)^2 R/2MG .
\tag{5.9.18}
$$

The capture time is then estimated by

$$
T_{\mathcal{C}} \leq \frac{r_2 - R}{\delta v_2}
\tag{5.9.19}
$$

and the control condition becomes

$$
\max_{u} (ux_2) \geq \frac{v_{TC} - v_{TEX}}{T_{\mathcal{C}}} .
\tag{5.9.20}
$$

After reaching $T_{\mathcal{C}}$ the controller is switched off and the vehicle follows the orbit under the conservative steady state regime. Observe that δv_2 enters (5.9.18)-(5.9.20) as a desired quantity to which the constant burn u is adjusted and from which the time of flight (5.9.19) is estimated. Consequently there is little chance for an overshot beyond the secure η of (5.9.17) and we do not have to consider another controller securing the avoidance of orbits higher than LEO, that is, stabilizing the trajectory from above.

Chapter 6
AVOIDANCE

6.1 AVOIDANCE RELATED OBJECTIVES

In applied control avoidance of obstacles, stationary or moving, is as
important as attaining collision or capture, and perhaps more so. When the
obstacles are controlled by the same agent, the avoidance control may be
part of coordination. When they are controlled by an independent agent,
the scenario may become an evasion part of a dynamic game, and when such an
agent does not have an objective on his own but simply acts to interfere
with ours, the game becomes a game "against nature" or against uncertainty.
The conditions for avoidance have been for a long time, and often still are,
obtained by contradicting the optimal collision, that is, in quantitative
terms, for instance by maximizing the minimal time or distance of collision
in an attempt to make it tending to infinity, cf. Isaacs [1]. Such a
technique simplifies the rather complex objective of avoidance to non-
collision, thus fails to satisfy many applied case goals such as avoidance in finite time, ultimate avoidance, collision with one target while
avoiding another, to name only a few from many possible scenarios which
include avoidance as a modular subobjective. Moreover, such a simplified
quantitative objective is studied via necessary conditions which, apart from
producing unconfirmed results, require numerical integration even for
systems of low dimensionality (≤ 3) so that only particular cases could be
discussed.

The problem of avoidance, independent from collision, was first posed
by Blaquière et al [1], Aggarwal-Leitmann [1] and later in terms of necessary conditions by Vincent [1],[2]. Sufficient conditions for strong
controllability for avoidance were introduced by Leitmann-Skowronski [1]

and Getz-Leitmann [1], followed by a number of works appearing soon afterwards, like Chikriy [1], Yeshmatatov [1], Litmanov [1], Ostapenko [1],
Schroeder-Schmitendorf [1], Barmish-Schmitendorf-Leitmann [1], Schmitendorf-
Barmish-Elenbogen [1], Leitmann [5], Leitmann-Liu [1], Leitmann-Skowronski
[2], Kaskosz [1],[2], Foley-Schmitendorf [1], Gutman [4], Shinar-Gutman
[1] and Krogh [1],[2], to quote only those referring to nonlinear systems.

In Section 1.6 we have discussed the models of antitargets T_A in the
Cartesian space and their transformation to configuration and velocity
antitargets T_{Aq} , $T_{A\dot{q}}$, respectively, in the state space, forming the state
space antitarget T_A as the Cartesian product $T_{Aq} \times T_{A\dot{q}} = T_A \subset \Delta$. Since
antitargets, both Cartesian and state, can have odd shapes, we often must
envelop their union in some *avoidance set* A in state space, equipped with
a smooth boundary for convenience of handling the avoidance problem.
Usually A is also a closed set in Δ, but not necessarily connected. It
would have the property that, given a control program $\bar{P}(\cdot)$, no motion of
(2.2.6)' may enter it, no matter what value of the bounded uncertainty is
applied. This obviously satisfies the requirement for complete avoidance
for all $t \in R^+$. This requirement may be scaled down as we shall see
below.

Let Δ_ε be the closure of an open subset of Δ such that $\Delta_\varepsilon \supset A$ and
$\partial \Delta_\varepsilon \cap \partial A = \phi$. We shall call $\Delta_A \overset{\Delta}{=} \Delta_\varepsilon - A$ the *safety zone*, see Fig. 6.1,
and define $CA \overset{\Delta}{=} \Delta - A$. The objective of avoidance is then specified as
follows.

DEFINITION 6.1.1. A motion of (2.2.6)' from outside A avoids this set if
and only if there is Δ_A such that $(\bar{x}^0, t_0) \in CA \times \mathbb{R}$ implies

$$\bar{\phi}(\bar{x}^0, t_0, \mathbb{R}) \cap A = \phi \ . \tag{6.1.1}$$

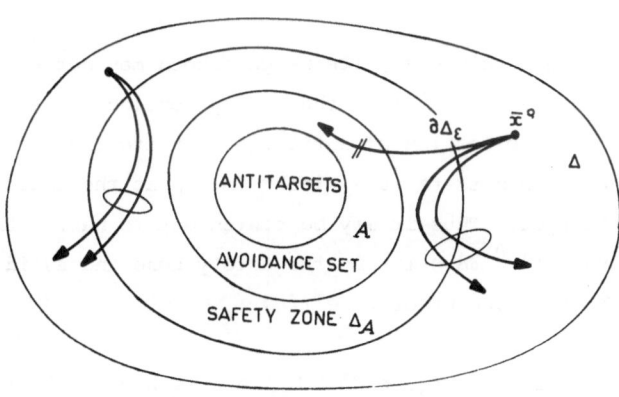

Fig. 6.1

289

The above also means that there is no finite interval of time during which $\bar{\phi}(\cdot)$ collides with A.

The definitions of *controllability* and *strong controllability for avoidance* follow from Definitions 3.1.4 and 3.1.5, respectively. Since Δ_A surrounds A by definition, if Δ_A is controllable or strongly controllable for avoidance of A, so is the whole Δ. Thus the region equals Δ and the controllabilities are automatically complete and global. We have

$$\Phi(C\!A \times \mathbb{R}, \mathbb{R}) \cap A = \phi \quad . \tag{6.1.2}$$

EXAMPLE 6.1.1. Consider the kinematic equation

$$\dot{q} = u\cos q + w\sin q \tag{6.1.3}$$

with $q \in \mathbb{R}$, control constraints $u \in [-1,1]$, $w \in [-1,1]$ and with the antitarget = avoidance set $A = \{0\}$ in Fig. 6.2. The control $u = 1$ wins

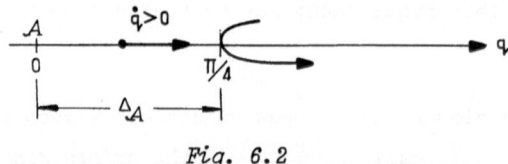

Fig. 6.2

strongly for all $q \in [0,\pi/4)$ generating $\dot{q} > 0$ without a chance of any effect by the action of w, even if it uses its best option $w = -1$. So $\Delta_A = [0,\pi/4)$ is the safety zone and it is easily seen that this way all $q^0 \geq \pi/4$ may not generate a trajectory entering A which makes the region of strong controllability equal to $[\pi/4,\infty)$ and the controllability complete and global. □

While A is given (desirable), its neighborhood may not be large enough or well defined enough to form the safety zone about it. For instance, envisage the collection of antitargets stretching to infinity while Δ is bounded. Then there is a "leak" in Δ_A at the intersection of $\partial\Delta$ with some antitarget. Thus it may be convenient to narrow the defence to $\bar{A}$ itself: make $\partial'A \stackrel{\Delta}{=} \partial A \cap \text{int } \Delta$ the safety zone and avoid $\text{int } A$. This way we come to the concept of *repelling*.

DEFINITION 6.1.2. Given $(\bar{x}^0, t_0) \in C\!A \times \mathbb{R}$, a motion of (2.2.6)' is *repelled* from A if and only if

$$\bar{\phi}(\bar{x}^0, t_0, \mathbb{R}) \cap \text{int } A = \phi \ . \qquad\qquad (6.1.3)$$

The motions of controllability and strong controllability are formed analogously to those of avoidance, that is, from Definitions 3.1.4 and 3.1.5, and by the same argument, these controllabilities are always complete and global.

Obviously the concept of repelling may refer equally well to CA, and hence to both int A and CA at the same time. Generalizing it, we let $A^1, \ldots, A^\ell$, $\ell < \infty$, be a sequence A of subsets of A such that $\cap_\sigma A^\sigma = \phi$ and $\cup_\sigma A^\sigma = A$, $\sigma = 1, \ldots, \ell$, representing an *avoidance archipelago*, see Fig. 6.3, and $B = \cap_\sigma \bar{A}^\sigma$ the *repelling mesh*, invariant under a given control.

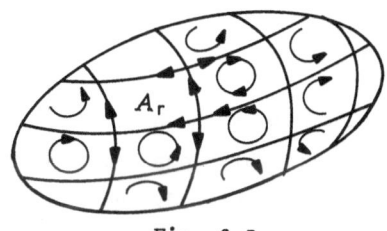

Fig. 6.3

In many situations of practical interest, the above requirement of avoidance for all time may be unduly restrictive and costly. In particular avoidance during or after a specified finite time interval may well suffice. It is called *real-time* and *ultimate avoidance*, respectively. We now let Δ_ϵ depend upon $\bar{x}^0$. As before $\Delta_\epsilon(\bar{x}^0) \supset A$ and $\partial\Delta_\epsilon(\bar{x}^0) \cap \partial A = \phi$, but also $\bar{x}^0 \in \partial\Delta_\epsilon(\bar{x}^0)$, that is, we consider motions starting from $\partial\Delta_\epsilon$. Then the safety zone also depends upon $\bar{x}^0$ which we write as $\Delta_A(\bar{x}^0) \triangleq \Delta_\epsilon(\bar{x}^0) - A$, and becomes a *slow-down zone* instead.

DEFINITION 6.1.3. Given $\bar{x}^0 \in CA$, a motion of (2.2.6)' *avoids A in real time* if and only if there is $T_A < \infty$ such that

$$\bar{\phi}(\bar{x}^0, t_0, t) \cap A = \phi, \quad \forall\, t \in [t_0, t_0 + T_A] \ . \qquad\qquad (6.1.4)$$

For stipulated T_A^+ we have *avoidance during* T_A^+.

Here again the motions of controllability and strong controllability are formed from Definitions 3.1.4 and 3.1.5, respectively.

A specific $T_A^+ < \infty$ marks the time passage between $\bar{x}^0$ on the corresponding $\partial\Delta_\epsilon(\bar{x}^0)$ and ∂A. It can be now called a *slow-down time*. Let

$$\rho(\bar{x}^0, A) = \min_{\bar{x} \in A} \|\bar{x} - \bar{x}^0\| \qquad (6.1.5)$$

be the distance between $\bar{x}^0$ and A while $\|\cdot\|$ is any norm in $\mathbb{R}^N$, and let $T_A^+ c_A \triangleq \rho(\bar{x}^0, A)$, $c_A > 0$ being the safe rate of approach to A during T_A^+. Given c_A, the corresponding safety zone $\Delta_A(T_A^+)$ depends upon T_A^+, and may be called a *slow-down zone*. The region of strong controllability, still being Δ, encloses such a zone, and for a different T_A^+ the complement to the zone $C\Delta_\varepsilon(\bar{x}^0) = \Delta - [\Delta_A(T_A^+) \cup A]$ is different: $T_A'' \geq T_A'$ implies

$$\Delta_A(T_A'') \cup A \subset \Delta_A(T_A') \cup A \qquad (6.1.6)$$

for any two T_A', T_A''. This agrees with the fact that the (infinite time) avoidance of Definition 6.1.1: $T_A^+ \to \infty$, implies any real time avoidance. Indeed $C\Delta_\varepsilon(\infty)$ is the smallest of all $C\Delta_\varepsilon(T_A^+)$ which complements $\Delta_\varepsilon(\bar{x}^0)$. This, on the other hand, means that the Δ_ε corresponding to such $C\Delta_\varepsilon(\infty)$ is the largest, and in the limit equals Δ, which has been specified as the region for indefinite time avoidance.

Finally we turn to the ultimate avoidance, that is, avoiding A after a certain finite interval of time, which again may be left unspecified and perhaps determined afterwards or stipulated, see Fig. 6.4.

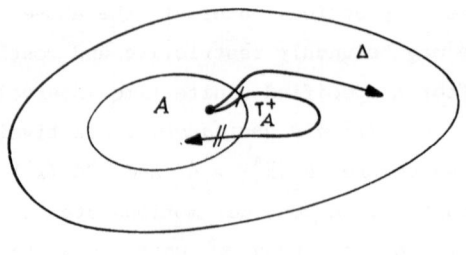

Fig. 6.4

DEFINITION 6.1.4. Given $(\bar{x}^0, t_0) \in \Delta \times \mathbb{R}$, a motion of $(2.2.6)'$ *ultimately avoids* A if and only if there is $T_A < \infty$ such that

$$\bar{\phi}(\bar{x}^0, t_0, t) \cap A = \phi , \quad \forall\, t \geq t_0 + T_A . \qquad (6.1.7)$$

For T_A stipulated, we qualify the objective as *ultimate avoidance after* T_A. It is easily seen that ultimate avoidance covers the case of a finite time *escape* from A, which may be specified by an additional requirement of $\bar{x}^0 \in A$.

Since the motions must either stay out of A all the time or leave it before the time $t_0 + T_A$, the complement $C\dot{A}$ must be strongly positively invariant under the $\bar{P}(\cdot)$ concerned. There is no need in this case for the safety zone, neither the need to narrow the controllabilities to any region smaller than Δ, so they are complete and global.

A few illustrative examples are provided in the next section.

6.2 CONDITIONS FOR AVOIDANCE

The essential features of avoidance control lie in designing a suitable feedback controller and defining the safety zone. Both can be found from the set of sufficient conditions we consider below, see Leitmann-Skowronski [1].

CONDITIONS 6.2.1. The system (2.2.6)' is completely strongly controllable for avoidance of A, if there is a safety zone Δ_A, $\bar{P}(\cdot)$ defined on Δ_A, and a C^1-function $V(\cdot) : D_A \times \mathbb{R} \to \mathbb{R}$, D_A(open) $\supset \bar{\Delta}_A$, such that for all $(\bar{x},t) \in \Delta_A \times \mathbb{R}$,

   (i) $V(\bar{x},t) > V(\bar{\xi},\tau)$, $\quad \bar{\xi} \in \partial A$, $\quad \tau \geq t$;

   (ii) for each $u \in \bar{P}(\bar{x},t)$,

$$\frac{\partial V(\bar{x},t)}{\partial t} + \nabla_x V(\bar{x},t)^T \cdot \bar{f}(\bar{x},\bar{u},\bar{w},t) \geq 0 \qquad (6.2.1)$$

      for all $\bar{w} \in W$.

Indeed, suppose some $\bar{\phi}(\cdot) \in K(\bar{x}^0,t_0)$, $\bar{x}^0 \notin A$, enters A. Then there is $t_1 > t_0$ such that $\bar{\phi}(\bar{x}^0,t_0,t_1) = x^1 \in \partial A$ and by (i), $V(\bar{x}^0,t_0) > V(x^1,t_1)$ contradicting (ii) and proving the conditions.

Similarly as in all the cases of collision and capture, Conditions 6.2.1 have an immediate corollary which may produce the controller. Recalling that

$$L(\bar{x},\bar{u},\bar{w},t) \triangleq \frac{\partial V(\bar{x},t)}{\partial t} + \nabla_x V(\bar{x},t)^T \cdot \bar{f}(\bar{x},\bar{u},\bar{w},t),$$

we have

COROLLARY 6.2.1. *Given* $(\bar{x},t) \in \Delta_A \times \mathbb{R}$, *if there is a pair* $(\bar{u}^*,\bar{w}^*)$ $\in U \times W$ *such that*

$$L(\bar{x},\bar{u}^*,\bar{w}^*,t) = \max_{\bar{u}} \min_{\bar{w}} L(\bar{x},\bar{u},\bar{w},t) \geq 0, \qquad (6.2.2)$$

then (ii) of Conditions 6.2.1 is met with $u* \in \bar{P}_*(\bar{x},t)$ *and it may be possible to deduce the winning control program from (6.2.2).*

The proof of the corollary follows immediately from the fact that

$$\max_{\bar{u}} \min_{\bar{w}} L(\bar{x},\bar{u},\bar{w},t) \leq \max_{\bar{u}} L(\bar{x},\bar{u},\bar{w},t) , \quad \forall \bar{w} \in W .$$

For the game case, the above result has been proved slightly later by Litmanov [1].

REMARK 6.2.1. Observe that the proving argument for Conditions 6.2.1 does not change if the inequality in (i) is made weak and that in (ii) made strong.

REMARK 6.2.2. Owing to the continuity of $\dot{V}(\cdot)$ and the continuity of $\bar{\phi}(\cdot)$ with respect to initial conditions, the test derivative $L(\bar{x},\bar{u},\bar{w},t)$ in (ii) could have been defined on ∂A only, with the same effect.

We may also note that the proving argument for Conditions 6.2.1 again does not change when the test function $V(\cdot)$ is assumed stationary.

CONDITIONS 6.2.2. The system (2.2.6)' is completely strongly controllable for avoidance of A, if there is Δ_A, $\bar{P}(\cdot)$ defined on Δ_A, and a C^1-function $V(\cdot) : D_A \rightarrow \mathbb{R}$, D_A(open) $\supset \bar{\Delta}_A$, such that for all $(\bar{x},t) \in \Delta_A \times \mathbb{R}$,

    (i) $V(\bar{x}) > v_A^+ = \sup V(\bar{x}) \mid \bar{x} \in \partial A$,

    (ii)   for each $\bar{u} \in \bar{P}(\bar{x})$,

$$\nabla V(\bar{x})^T \cdot \bar{f}(\bar{x},\bar{u},\bar{w},t) \geq 0 , \quad \forall \bar{w} \in W . \tag{6.2.3}$$

As for Conditions 6.2.1, upon entering A by some $\bar{\phi}(\cdot) \in K(\bar{x}^0,t_0)$, $\bar{x}^0 \not\in A$, there is $t_1 > t_0$ such that $\bar{\phi}(\bar{x}^0,t_0,t_1) = x^1 \in \partial A$ generating contradiction between (i) and (ii) of the above.

The Corollary 6.2.1 holds for the stationary $V(\cdot)$ without change, and we have another immediate conclusion.

COROLLARY 6.2.2. *When the avoidance set* A *in Conditions 6.2.2 is defined by the* V-*level:*

$$\partial A : V(\bar{x}) = v_A^+ \tag{6.2.4}$$

then (i) of Conditions 6.2.2 hold automatically and (ii) becomes also necessary.

Indeed, if not (ii), then $\dot{V} < 0$ generating collision.

Conditions 6.2.2 have been introduced in Skowronski [38]. They work better for our type of application, since the energy $E(\bar{x})$ is taken, as a rule, stationary. Moreover, with a choice of $V(\cdot)$ in terms of energy, the conditions are necessary and sufficient, thus defining the fields of avoidance and collision in terms of the map of H^+, H^- options for the controller.

Observe that Corollary 6.2.1 provides an accumulative controller. This, in our energy reference frame, means rising in E-levels away from A, which surrounds antitargets that have replaced the targets of Chapter 5. Such a controller works as long as we intend to avoid an E^+-level below us, that is, *avoid from above*. The situation is in fact the same as in stabilization above some E^+-level in Section 3.3, or collision/capture in a target located above the initial E-level, in $C\dot{A}$, with the methods of Sections 5.3, 5.5. In each case we satisfy (6.2.2) using the accumulative control condition (3.3.17). When the situation is reversed, that is, when we want to avoid A which is located above the initial E-level, that is, *avoid from below*, for instance avoid thresholds while in the energy cup, we use a dissipative controller to damp the motion energy below some E^--level, that is, again away from A with the same control condition as in Chapter 5: (5.3.7), (5.3.11) or (5.3.14) specified as in (3.3.12)'. Indeed, observe the following.

REMARK 6.2.3. The proving argument for Conditions 6.2.1 and 6.2.2 remains valid when the inequalities (i) and (ii) are simultaneously reversed.

This remark automatically applies to Corollary 6.2.1 as well as to the controller implementing it. However, two comments must be made. First, when inverting the inequality (6.2.2), $\bar{u}^*$ becomes the minimizer while $\bar{w}^*$ is the maximizer, as in collision. Second, inverting the inequalities while still applying the energy reference frame, we must let a new test function $V(\bar{x}) = \text{const} - E^-(\bar{x})$ or $V(\bar{x}) = 1/E^-(\bar{x})$, as both of them change the sign of the derivative. Alternatively, we may use the latter functions in Conditions 6.2.2 and return back to $V(\bar{x}) = E^-(\bar{x})$ for avoiding A from below.

The discussed specification of the H^+, H^- fields in service of avoidance is, for instance, used in robotic practice where the test functions $V(\bar{x}) \triangleq \text{const} - E(\bar{x})$, $1/E(\bar{x})$ are called the *breaking* or *push-off potentials* and located by design about obstacles. They work in lieu of vision, representing a less expensive alternative, see Khatib-LeMaitre [1]. Indeed, robotics raised considerably the interest in the artificial push-off potential fields. We may immediately quote the abovementioned and Khatib [1], Khatib-Burdick [1], Koditchek [1], Krogh [2], Newman-Hogan [1], Volpe-Koshla [1], to name only a few.

The standard way of using the push-off or *repulsive* potential is to insert it into the potential energy reference surface V of the system, usually within the energy cup, generating a local maximum on V, integrated smoothly into this surface. Spherical symmetry is a typical feature of the repulsive *hill*, see Koditchek [1]. The hill may increase cubically with radial distance inside of a circular threshold range. Some hills have gaussian shapes. In general, these hills should not create local minima about them, see Volpe-Khosla [1]. However, there are a few relatively popular repulsive functions which do so. One of them is the so called FIRAS function

$$V(r) = \frac{A}{2} \left(\frac{1}{r} - \frac{1}{r_0} \right)^2 , \qquad 0 < r < r_0$$

where r is the closest distance of the object surface, r_0 is the effective range and A is a scaling factor. To avoid the minima, the potential must be circular (in two dimensions), but the circular potential will not work properly on noncircular objects. A proposed alternative, see Khatib [1], is to use an n-ellipse:

$$(x/a)^{2n} + (y/b)^{2n} = 1 .$$

At the surface of the obstacle the potential levels should match the contour of the surface. This requires $n \to \infty$ at the surface, while away from it the contour must become spherical. Another potential, similar to FIRAS, is proposed by Kuntze-Schill [1]:

$$V(r) = \begin{cases} \dfrac{1}{r} - \dfrac{1}{\varepsilon} , & \text{for } r \leq \varepsilon \\ \\ 0 , & \text{for } r > \varepsilon \end{cases}$$

where $\varepsilon > 0$ is a small constant.

Since the artificial repulsive potential is a function of position, it requires some sensing of such position. The present algorithms are based on two methods of obstacle surveillance: *absolute* by fixed sensors

mounted on the ceiling, see Paul [2], or *relative* by sensors that are
mobile with the machine - say the arm of a manipulator - see Bejczy [1],
Calm-Phillips [1]. Obviously the second method lends itself better
to our purpose in that it requires a shorter sensing range and simpler
algorithms based on direct measuring of the distance between the obstacle
and the sensor. When position sensing is too expensive, it may be replaced
by force sensing.

EXAMPLE 6.2.1. Consider the linear system

$$\dot{\bar{x}} = A\bar{x} + B\bar{u} + C\bar{w} \tag{6.2.6}$$

with A,B,C constant matrices of appropriate dimensions. Here the reader
may like to recall equation (3.1.4) and the subsequent discussion. Suppose
-A is stable, that is, with non-positive real parts of the eigenvalues.
Let Q be a negative definite $N \times N$ matrix of the Liapunov Matrix Equation
(3.1.7) with solution being a positive definite symmetric matrix P . We
choose as in Section 3.1: $V(\bar{x}) = \bar{x}^T P \bar{x}$. If $A \overset{\Delta}{=} \{\bar{x} | \bar{x}^T P \bar{x} \leq const\}$, then
(i) of Conditions 6.2.2 holds. Furthermore, if there is a matrix D such
that $C = BD$, and if

$$U \overset{\Delta}{=} \{\bar{u} \mid \|\bar{u}\| < \rho_1 = const > 0\}$$

$$W \overset{\Delta}{=} \{\bar{w} \mid \|\bar{w}\| < \rho_2 = const > 0\}$$

with $\rho_1 \geq \|D\| \rho_2$, then (ii) of Conditions 6.2.2 is met and the avoidance
is generated by the corresponding L-G controller, cf. (5.3.12):

$$\bar{u} = \frac{B^T P \bar{x}}{\|B^T P \bar{x}\|} \rho_1 \tag{6.2.7}$$

for all $\bar{x} \notin N \overset{\Delta}{=} \{\bar{x} | B^T P \bar{x} = 0\}$. For $\bar{x} \in N$, $\bar{u}$ may take on any admissible
value, possibly the saturation value or zero. The set N replaces in our
first order system the set specified by $\bar{\beta} = const$ in (3.3.13), and plays
the same role.

The satisfaction of (i) in Conditions 6.2.2 follows at once, whereas
that of (ii) is readily seen by calculating

$$L(\bar{x},\bar{u},\bar{w}) = 2\bar{x}^T P (A\bar{x} + B\bar{u} + C\bar{w})$$

$$= \bar{x}^T (PA + A^T P) \bar{x} + 2\bar{x}^T PB\bar{u} + 2\bar{x}^T PBD\bar{w}$$

$$= -\bar{x}^T Q\bar{x} + 2(\bar{x}^T PB\bar{u} + \bar{x}^T PBD\bar{w})$$

$$\geq -\bar{x}^T Q\bar{x} + 2(\bar{x}^T PB\bar{u} - \|B^T P\bar{x}\| \cdot \|D\| \rho_2) .$$

REMARK 6.2.4. Suppose $-A$ is not stable but $(-A,-B)$ is stabilizable, that is, there exist a matrix N such that $(-A,-BN)$ is stable, then (6.2.7) is replaced by

$$\bar{u} = N\bar{x} + \frac{B^T P \bar{x}}{\|B^T P \bar{x}\|} \|D\| \rho_2 \qquad (6.2.7)'$$

and for this control program to be admissible, U must be such that $\bar{P}(\bar{x}) \subset U$ for all $\bar{x} \in \Delta_A$. In any event $\bar{P}(\cdot)$ must be piece-wise continuous.

To illustrate the case numerically, let us assume

$$A = \begin{bmatrix} 0 & 1 \\ 0 & 0 \end{bmatrix}, \quad B = C = \begin{bmatrix} 0 \\ 1 \end{bmatrix}, \quad W = \{\bar{w} \mid |\bar{w}| \le 1\} .$$

Since $-A$ is not stable, but $(-A,-B)$ is stabilizable, we determine first the linear part of (6.2.7)'. It is readily deduced to be $N\bar{x}$ with for instance $N = (-1,1)$. Then if Q is taken as the unit matrix, the solution of the Liapunov Matrix Equation (3.1.7) with A replaced by $A + BN$ is

$$P = \begin{bmatrix} 3/2 & -1/2 \\ -1/2 & 1 \end{bmatrix}$$

so that the nonlinear part of the control program is $\text{sgn}\,(-\frac{1}{2}x_1 + x_2)$. Thus provided A is a ball: $\bar{x}^T P \bar{x} \le a = \text{const} > 0$, the avoidance controller is

$$u = -x_1 + x_2 + \text{sgn}\,(-\frac{1}{2}x_1 + x_2) \qquad (6.2.8)$$

for all $\bar{x} \in N$. The control constraint set U depends on the choice of Δ_A. For instance, we may choose

$$\Delta_\varepsilon = \{\bar{x} \mid \bar{x}^T P \bar{x} \le a + \varepsilon, \ \varepsilon > 0\}$$

so that U must be such that

$$\bar{P}(\bar{x}) \subset U, \quad \forall \bar{x} \in \Delta_A - A, \quad t \in \mathbb{R} . \qquad \square$$

EXAMPLE 6.2.2. Following the case study by Bojadziev [1], we extend Example 1.1.3 to the n links physical pendulum, see Fig. 6.5, which may represent an open chain mechanical structure such as a robotic manipulator. We reduce the links to point masses m_i in the corresponding centers of gravity CG_i, $i = 1,\ldots,n$, at the distance a_i from the joints 0_{i-1}, while ℓ_i denotes the length of the link concerned and I_i is the corresponding moment of inertia. Then we abbreviate

$$. J_i = a_i m_i + \ell_i \sum_{k=1}^n m_k \qquad (6.2.9)$$

$$J_i^2 = a_i^2 m_i + \ell_i^2 \sum_{k=1}^{n} m_k + I_i \qquad (6.2.10)$$

and obtain the kinetic energy

$$T = \frac{1}{2} \sum_{i=1}^{n} \left[2J_i \dot{q}_i^2 \sum_{k=1}^{i-1} \ell_k \dot{q}_k \cos(q_i - q_k) + J_i^2 \dot{q}_i^2 \right] \qquad (6.2.11)$$

and the potential energy

$$V = g \sum_{i=1}^{n} J_i (1 - \cos q_i) \ . \qquad (6.2.12)$$

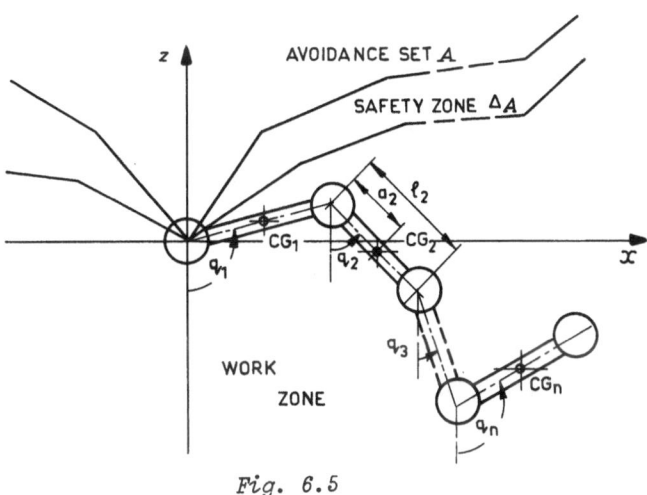

Fig. 6.5

This yields the Lagrange equations in the form

$$\left. \begin{array}{l} J_i \sum_{k=1}^{i-1} \ell_k [\cos(q_i - q_k)\ddot{q}_k + \sin(q_i - q_k)\dot{q}_k^2] \\[2mm] \quad + \ell_i \sum_{k=i+1}^{n} J_k [\cos(q_i - q_k)\ddot{q}_k + \sin(q_i - q_k)\dot{q}_k^2] \\[2mm] \quad + J_i^2 \ddot{q}_i + \lambda_i \dot{q}_i = u_i \end{array} \right\} \qquad (6.2.13)$$

where $\lambda_i \dot{q}_i = Q_i^{DD}$ represent linear damping forces and u_i is the control torque at each joint 0_i of the manipulator. From (6.2.11), (6.2.12) we obtain

$$\left. \begin{array}{l} E(\bar{q}, \dot{\bar{q}}) = \sum_{i=1}^{n} \left[J_i \dot{q}_i \sum_{k=1}^{i-1} \ell_k \dot{q}_k \cos(q_i - q_k) + J_i^2 \dot{q}_i^2 \right] \\[2mm] \quad + g \sum_{i=1}^{n} J_i (1 - \cos q_i) \end{array} \right\} \qquad (6.2.14)$$

with the power

$$\dot{E}(\bar{q}, \dot{\bar{q}}) = \sum_{i=1}^{n} (u_i \dot{q}_i - \lambda_i \dot{q}_i^2) = \bar{u}\,\dot{\bar{q}} - \bar{\lambda}\,\dot{\bar{q}}^2 \ . \qquad (6.2.15)$$

Note that (6.2.15) is the same as if (6.2.13) were inertially decoupled. The work region of the chain is described in the Cartesian frame Oxz by

$$W : x^2 + z^2 \le \left(\sum_{i=1}^{n} \ell_i \right)^2$$

and the chain is not allowed to cross its boundary ∂A of the avoidance set A infested by obstacles (antitargets), as marked in Fig. 6.5. A given shape-line of the chain is denoted $\text{Conf}(\bar{q})$ in Oxz, and the boundary ∂A is defined by

$$\partial A : \text{Conf}(\bar{q}^A) \cup \text{Conf}(-\bar{q}^A)$$

with $\bar{q}^A = (q_1^A, \ldots, q_n^A)^T$, $q_i^A > 0$, $i = 1, \ldots, n$ separating the antitargets from the work envelope in the state space corresponding to W, see Section 1.6. A safety zone Δ_A shown in Fig. 6.5 safeguards the chain from crossing ∂A. It is located between the latter and $\partial \Delta_\epsilon$:

$$\partial \Delta_A : \text{Conf}(\bar{q}^S) \cup \text{Conf}(-\bar{q}^S)$$

with $\bar{q}^S = (q_1^S, \ldots, q_n^S)^T$, $0 < q_i^S < q_i^A$, $i = 1, \ldots, n$. We may now use E-levels to specify A, Δ_ϵ:

$$A : E(\bar{q}, \dot{\bar{q}}) \ge h_A , \quad h_A \le h_{CE}$$

$$\Delta_\epsilon : E(\bar{q}, \dot{\bar{q}}) \ge h_\epsilon , \quad h_\epsilon < h_A$$

with $\Delta_A : h_\epsilon \le E(\bar{q}, \dot{\bar{q}}) \le h_A$ with $E(\bar{q}, \dot{\bar{q}})$ given by (6.2.14). Then the state work envelope corresponding to W is determined as the complement $\Delta_E - \Delta_\epsilon$, that is, by $E(\bar{q}, \dot{\bar{q}}) \le h_\epsilon$, within the basic energy cup. As we are avoiding A from below, the inequality in Corollary 6.2.1 is inverted and we are in search of a non-accumulative controller. By (6.2.15) we need

$$u_i \, \text{sgn} \, \dot{q}_i = \begin{cases} \lambda_i \dot{q}_i^2 / |\dot{q}_i| , & |\dot{q}_i| \ge \beta_i \\ 0 , & |\dot{q}_i| < \beta_i , \quad i = 1, \ldots, n . \end{cases} \tag{6.2.16}$$

EXAMPLE 6.2.3. The use of relative coordinates is demonstrated on avoidance of a tracking missile M which pursues the point-mass modelled craft C along the line-of-sight (LOS) automatic infrared guidance, see Fig. 6.6.

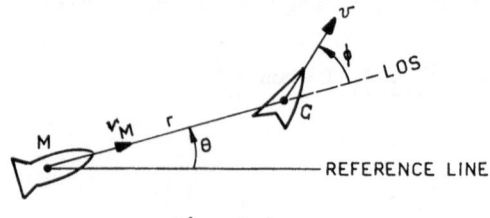

Fig. 6.6

We reduce the description of motion to kinematics. For constant pursuer
and evader speeds v_M and v respectively, the equations are

$$\left. \begin{aligned} \dot{r} &= v \cos \phi - v_M \ , \\ \dot{r}\theta &= v \sin \phi \ , \end{aligned} \right\} \tag{6.2.17}$$

while the evader's normal acceleration is $u = v(\dot\theta + \dot\phi)$, $|u| \le \hat{u}$. Denote
$(v_M/v) \stackrel{\Delta}{=} s \in (0,1)$ and choose $x_1 = r$, $x_2 = \phi$. Then the state equations
are

$$\left. \begin{aligned} \dot{x}_1 &= v \cos x_2 - v_M \ , \\ \dot{x}_2 &= (u/v) - (v/x_1) \sin x_2 \ . \end{aligned} \right\} \tag{6.2.18}$$

Note that the obstacle normal acceleration is a fixed constant. The objective is to avoid M, that is, the anti-target is defined as
$T_A = \{\bar{x} | x_1 \le a = \text{const} > 0\}$, in terms of the choice $V(\bar{x}) = x_1$. In choosing
the avoidance set $A \supset T_A$ we are guided by the following considerations.
When $\dot{x}_1 = \dot{x}_{1\max} = v - v_M > 0$, $(\cos x_2 = 1)$, M can be allowed to approach
to within a minimum distance a. However, when $\dot{x}_1 = \dot{x}_{1\min} = -v - v_M < 0$,
$(\cos x_2 = -1)$, M must be kept further away to give C sufficient time to
evade. Thus, if $\Delta \stackrel{\Delta}{=} \{\bar{x} | x_1 \in \mathbb{R}^+, |x_2| \le \pi\}$, we are led to design the avoid-
ance set as

$$A = \{\bar{x} | x_1 \le a + \bar{x} + \delta - \sqrt{(\pi - \delta)^2 - x_2^2} \ , \ \delta > 0\} \ ,$$

see Fig. 6.7. Using Corollary 6.2.1, we obtain

$$u = -\hat{u} \operatorname{sgn} x_2 \ . \tag{6.2.19}$$

Condition (6.2.2) is met for all $\bar{x} \in \Delta_A$ if

$$\hat{u} \ge v^2 (s - \cos x_2) \sqrt{(\pi + \delta)^2 - x_2^2} / |x_2| \ , \quad \forall \ \bar{x} \in \Delta_A \ .$$

Since $x_2 \in [-\pi, \pi]$, a conservative bound is given by

$$\hat{u} \ge v^2 (s + 1) \sqrt{(\pi + \delta)^2 - (\cos^{-1}s)^2} / \cos^{-1}s \ . \tag{6.2.20}$$

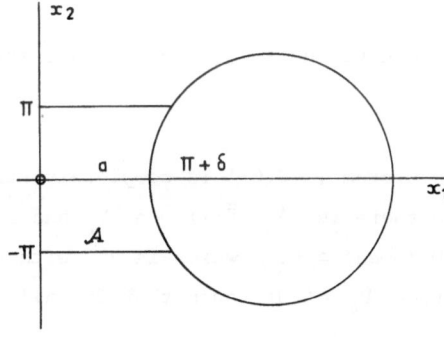

Fig. 6.7

Note that, the larger δ, the closer the missile M may approach the craft C when $\dot{x}_1 = \dot{x}_{1min}$. Hence the larger the required control $\hat{u}$. Also, as s increases, so does the lower bound of $\hat{u}$. $\qquad\qquad\square$

In case studies it may be difficult to find a C^1-test-function, see our comments in Section 3.4. Leitmann [5] introduces the sufficient conditions for avoidance with less demanding piece-wise C^1-function $V(\cdot)$.

DEFINITION 6.2.1. A denumerable decomposition $\mathcal{D}$ of a set $D \times \mathbb{R} \subset \mathbb{R}^{N+1}$ is a denumerable collection of disjoint sets whose union is D. We write $\mathcal{D} \triangleq \{D^j \mid j \in J\}$ where J is a denumerable index set of disjoint subsets.

DEFINITION 6.2.2. Let $D \times \mathbb{R}$ be a subset of $\mathbb{R}^{N+1}$ and $\mathcal{D}$ a denumerable decomposition of D. A real valued continuous function $V(\cdot)$ on $D \times \mathbb{R}$ is said to be of class C^1 with respect to $\mathcal{D}$ if and only if for each $j \in J$ there is a pair $W^j, V_j(\cdot)$ such that W^j is an open set containing D^j and $V_j(\cdot) : W^j \times \mathbb{R} \to \mathbb{R}$ is of class C^1 such that $V_j(\bar{x},t) = V(\bar{x},t)$, $\forall \bar{x} \in D^j$, $t \in \mathbb{R}$.

The following lemma was proved by Stalford [1].

LEMMA 6.2.1. *Let* $V(\cdot) : D \times \mathbb{R} \to \mathbb{R}$ *be* C^1 *with respect to decomposition* $\mathcal{D}$. *Let* $\{W^j, V_j(\cdot) \mid j \in J\}$ *be a collection of pairs associated with* $V(\cdot)$, *and let* $T_j \triangleq \{t \in [t_0, t_1] \mid \bar{x}(t) \in D^j\}$, $j \in J$. *Suppose that for each* $j \in J$, *we have*

$$\frac{d}{dt} V_j[x(t)] \geq 0, \quad a.e. \ T_j.$$

Then the function $g(\cdot) : [t_0, t_1] \to \mathbb{R}$ *defined by* $g(t) = V[x(t)]$ *is absolutely continuous and monotone nondecreasing.*

Using the above, we have the following generalization of Conditions 6.2.1, Leitmann [5]:

CONDITIONS 6.2.3. The system (2.2.6)' is completely strongly controllable for avoidance of A if there is $\Delta_A, \bar{P}(\cdot)$ on Δ_A and a function $V(\cdot) : D_A \times \mathbb{R} \to \mathbb{R}$, D_A(open) $\supset \bar{\Delta}_A$, which is C^1 with respect to a denumerable decomposition $\mathcal{D}_A$ of D, such that for all $(\bar{x},t) \in \Delta_A \times \mathbb{R}$,

(i) $V(\bar{x},t) > V(\bar{\xi},\tau)$, $\bar{\xi} \in \partial A$, $\tau \geq t$,

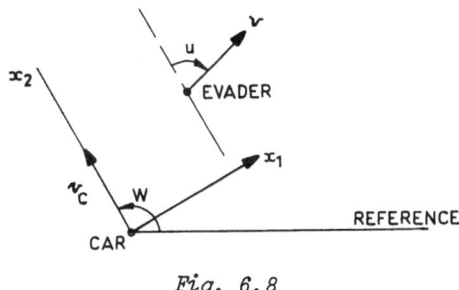

Fig. 6.8

(ii) for all $(\bar{x},t) \in D^j \times \mathbb{R}$, $\bar{u} \in \bar{P}(\bar{x},t)$,

$$\frac{\partial V(\bar{x},t)}{\partial t} + \nabla_x V(\bar{x},t)^T \cdot \bar{f}(\bar{x},\bar{u},\bar{w},t) \geq 0 \qquad\qquad (6.2.21)$$

for all $\bar{w} \in W$.

The proof is analogous to that for Conditions 6.2.1 differing only in that Lemma 6.2.1 is used. By the same argument as before, we also have the following corollary, Leitmann [5].

COROLLARY 6.2.3. *Given $(\bar{x},t) \in D^j \times \mathbb{R}$, $j \in J$, if there is $(\bar{u}^*,\bar{w}^*) \in U \times W$ such that Corollary 6.2.1 holds, then (ii) of Conditions 6.2.3 is met with $\bar{u}^* \in \bar{P}_*(\bar{x},t)$.*

EXAMPLE 6.2.4. We illustrate the use of Conditions 6.2.3 on the scenario of a pedestrian avoiding the homicidal driver on a bounded parking lot in the classical homicidal chauffeur game of Isaacs [1]. It is assumed that the strategy of the pursuing car w is unknown to the evader except for its limitations $w \in W$. With the geometry seen at Fig. 6.8, the kinematic equations of motion lead to the following state equations:

$$\left.\begin{aligned} \dot{x}_1 &= v \sin u + w x_2 , \\ \dot{x}_2 &= v \cos u - w x_1 - v_c , \end{aligned}\right\} \qquad\qquad (6.2.22)$$

where v,u are the evader's constant speed and control turning angle, respectively, while $v_c = \text{const}$ is the pursuing car's speed with its turning angle w bounded in $W : |w| \leq \hat{w}$.

The antitarget is a circular disc centered at the pursuer

$$T_A : x_1^2 + x_2^2 \leq \ell^2 , \quad \ell = \text{const} .$$

We design the avoidance set seen in Fig. 6.9 as

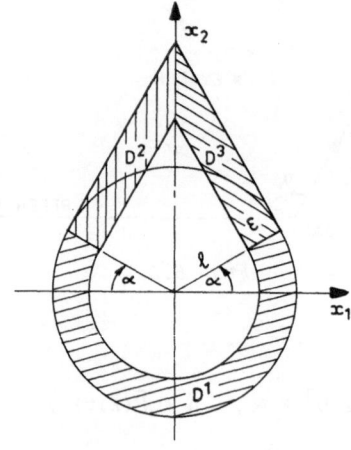

Fig. 6.9

$$A \triangleq \left\{ \begin{array}{l} (x_1,x_2) \left| x_1^2 + x_2^2 \leq \ell^2 , \quad x_2 \leq |x_1| \tan \alpha , \right. \\ (x_1,x_2) \left| |x_1| \tan \alpha \leq x_2 \leq -|x_1| \cot \alpha + \ell \csc \alpha . \right. \end{array} \right\}$$

We define Δ_ε as A but with ℓ replaced by $\ell + \varepsilon$, $\varepsilon = \text{const} > 0$ and let $D = D^1 \cup D^2 \cup D^3$ enclose Δ_A with

$$D^1 \triangleq \{(x_1,x_2) | \ell^2 < x_1^2 + x_2^2 \leq (\ell + \varepsilon)^2 , \quad x_2 \leq |x_1| \tan \alpha\} ,$$

$$D^2 \triangleq \{(x_1,x_2) | x_1 \leq 0 , x_2 > -x_1 \tan \alpha$$
$$x_1 \cot \alpha + \ell \csc \alpha < x_2 \leq x_1 \cot \alpha + (\ell + \varepsilon) \csc \alpha\} ,$$

$$D^3 \triangleq \{(x_1,x_2) | 0 < x_1 , x_2 > x_1 \tan \alpha ,$$
$$-x_1 \cot \alpha + \ell \csc \alpha < x_2 \leq -x_1 \cot \alpha + (\ell + \varepsilon) \csc \alpha\} .$$

To apply Conditions 6.2.3, assume

$$V_1(x,t) \triangleq x_1^2 + x_2^2 , \quad \text{for } \bar{x} \in D^1 , \quad t \in \mathbb{R} ,$$

$$V_2(x,t) \triangleq (x_2 - x_1 \cot \alpha) \ell \sin \alpha , \quad \text{for } \bar{x} \in D^2 , \quad t \in \mathbb{R} ,$$

$$V_3(x,t) \triangleq (x_2 + x_1 \cot \alpha) \ell \sin \alpha , \quad \text{for } \bar{x} \in D^3 , \quad t \in \mathbb{R} .$$

With this specification, the function $V(\cdot)$ is continuous and satisfies (i) of Conditions 6.2.3. On the use of Corollary 6.2.3, it is readily seen that (ii) of these conditions is met with the control condition

$$\sin \alpha \leq v/v_c \tag{6.2.23}$$

$$\hat{u} \leq (v - v_c \sin \alpha)/(\ell + \varepsilon) \cot \alpha \tag{6.2.24}$$

and the controller is defined by.

$$\left. \begin{array}{l} \sin u = x_1 / \sqrt{x_1^2 + x_2^2} , \\ \cos u = x_2 / \sqrt{x_1^2 + x_2^2} , \end{array} \right\} \quad \bar{x} \in D^1 \tag{6.2.25}$$

$$\left.\begin{array}{l} \sin u = -\cos \alpha \\ \cos u = \sin \alpha \end{array}\right\} \quad \bar{x} \in D^2 \tag{6.2.26}$$

$$\left.\begin{array}{l} \sin u = \cos \alpha \\ \cos u = \sin \alpha \end{array}\right\} \quad \bar{x} \in D^3 \ . \tag{6.2.27}$$

Note that (6.2.23) can be met also if $v < v_c$, that is, the evader is slower than the car. Moreover, if $v < v_c$ then there is no C^1-function $V(\cdot)$ for which Conditions 6.2.1 apply (T.L. Vincent, private communication) and we are forced to use Conditions 6.2.3. The control function is single valued at $\mathrm{int}\ D^j$, for all j , but the program $\bar{P}(\cdot)$ must be set valued for the intersections $\bar{D}^i \cap \bar{D}^j$, $i \neq j$. The latter is the common case when several test functions are in use, thus presenting another reason why we operate with set valued $\bar{P}(\cdot)$'s. □

As mentioned, it is sometimes inconvenient to design the safety zone Δ_A outside $\bar{A}$, we may then settle for repelling instead of avoidance, see Definition 6.1.2. The corresponding sufficient conditions follow.

CONDITIONS 6.2.4. The system (2.2.6)' is completely strongly controllable for repelling from A , if there is $\bar{P}(\cdot)$ defined on $\bar{A}$ and a C^1-function $V(\cdot) : \bar{A} \to \mathbb{R}$ such that for all $\bar{x} \in A$,

(i) $V(\bar{x}) < V(\bar{\xi})$, $\bar{\xi} \in \partial\bar{A}$,

(ii) for each $\bar{u} \in \bar{P}(\bar{x})$,

$$\nabla V(\bar{x})^T \bar{f}(\bar{x},\bar{u},\bar{w},t) \geq 0 , \quad \forall\ \bar{w} \in W . \tag{6.2.28}$$

The Conditions are implied by the following simple observation. Motions enter A only through $\partial'A \overset{\Delta}{=} \partial A \cap \Delta$, thus consider $\bar{\phi}(\bar{x}^0,t_0,\mathbb{R}^+)$ from $\bar{x}^0 \in \partial'A$. Suppose there is $t_1 > t_0$ such that $\bar{\phi}(\bar{x}^0,t_0,t_1) = \bar{x}^1 \in \mathrm{int}\,A$. Then by (i), $V(x^1) < V(x^0)$, contradicting (ii).

Note that here we may again invert both inequalities in (i) and (ii) simultaneously, without changing the proving argument thus the adjustment between the dissipative and accumulative controllers applies in the same way as before. We obviously may use Corollary 6.2.1 for repelling of the above, as much as we used it for implementing Conditions 6.2.2.

Recall now the repelling archipelago of Section 6.1 with the sequence $\mathbb{A}$, which is in fact a denumerable decomposition of A , and the mesh $\mathcal{B}$.

Following Conditions 6.2.4 and using Corollary 6.2.1, if we may find the sequence of C^1-functions $V_\sigma(\cdot) : \bar{A}^\sigma \to \mathbb{R}$ such that

$$A^\sigma = \{\bar{x} \in A \mid V_\sigma(\bar{x}) \leq v_A^\sigma\} \ , \quad \sigma = 1,\ldots,\ell \ ,$$

and satisfying (6.2.20), then B becomes the repelling mesh of Section 6.1. Obviously, since B encloses $A = \cup_\sigma A^\sigma$, then A is repelling as well. Then the controllers used for repelling from each A^σ serve two objectives: (1^0) keep the motions starting at A in B, (2^0) repel from A all motions starting outside this set.

EXAMPLE 6.2.5. We illustrate the repelling on the frequently used case of oscillator (1.1.6) followed in Examples 1.1.1, 3.2.1, 3.3.1 and 5.3.3. Let the system be described by (5.3.11), with the conservative separatrix (3.3.18):

$$\tfrac{1}{2}x_2^2 + \tfrac{1}{2}ax_1^2 - \tfrac{1}{4}bx_1^4 + \tfrac{1}{6}cx_1^6 = h_{CE} \ , \tag{6.2.29}$$

the antitargets inside the basic energy cup Z_E which thus asks for $A \overset{\Delta}{=} \Delta_E : E(\bar{x}) \leq h_{CE}$, with ∂A defined by (6.2.29) restricted to the cup, that is, constrained by condition (2.3.19)

$$(ax_1 + bx_1^3 - cx_1^5)x_1 \geq 0 \ ,$$

which makes ∂A defined by (6.2.29) and

$$x_1^2(b - cx_1^2) \geq -a \ . \tag{6.2.30}$$

It is obvious that no matter what the behavior of the motions outside A, upon collision with ∂A if at all, we must switch the controller to the conservative (5.3.17) which makes the motion follow (6.2.29), preventing its entry to int A, which is our objective.

Observe also that if such a controller is implemented before collision with ∂A, we obtain avoidance, as the motion follows a conservative path about all equilibria across damped separatrices, see Fig. 5.12. □

EXERCISES 6.2

6.2.1 Design controllers generating avoidance of the set A by trajectories of the following systems:

(i) $\dot{x}_1 = 2x_1^3 - 2x_1x_2^2$,

$\dot{x}_2 = -5ux_1^2 + 4x_2^3$;

$A : x_1^2 + x_2^2 \leq \varepsilon$, $\varepsilon > 0$ small constant .

(ii) $\dot{x}_1 = x_1 - ux_2^2$,

$\dot{x}_2 = x_2(u - x_2^2)$;

$A : x_1^2 + x_2^2 < 6$.

(iii) $\dot{x}_1 = x_1 + x_2$,

$\dot{x}_2 = x_2 + u$,

$\dot{x}_3 = -x_3 + x_2$,

$A : x_1^2 + x_2^2 + x_3^2 \leq (3.6)^2$.

(iv) $\dot{x}_1 = 3x_1 - x_1 x_2^2 + u_1$,

$\dot{x}_2 = x_2^3 + x_1 + u_2$,

$A : x_1^2 + x_2^2 \leq 1$.

(v) $\begin{pmatrix} \dot{x}_1 \\ \dot{x}_2 \end{pmatrix} = \begin{pmatrix} 0 & -1 \\ 3 & 6 \end{pmatrix} \begin{pmatrix} x_1 \\ x_2 \end{pmatrix} + \begin{pmatrix} u_1 \\ u_2 \end{pmatrix}$

$A \triangleq \{(0,0)\}$.

6.2.2 Find the controller securing complete controllability for avoidance
of the set $A : x_1^2 + x_2^2 + x_3^2 \leq a^2$ by trajectories of the system

$\dot{x}_1 = x_1 - x_2$,

$\dot{x}_2 = x_2 + u$,

$\dot{x}_3 = x_3 + x_2$.

Estimate a , for given $\hat{u}$.

6.2.3 What is the difference, if any, between the avoidance of Definition
6.1.1 and collision with stipulated $T_C \to \infty$?

6.2.4 Give an example of a manipulator arm travelling along a repelling
mesh B placed somewhere in H . Find the controller.

6.3 REAL TIME AVOIDANCE

Real time and ultimate avoidance are weaker objectives and thus may be implied by less demanding conditions. The reader may like to recall Definitions 6.1.3 and 6.1.4, and consider the C^1-function $V(\cdot)$: $D_A \times \mathbb{R} \to \mathbb{R}$, $D_A(\text{open}) \supset \overline{CA}$, with

$$
\left.
\begin{aligned}
v_A^+ &\triangleq \sup V(\bar{x},t) \,|\, (x,t) \in \partial A \times \mathbb{R} \ , \\
v_\epsilon^- &\triangleq \inf V(\bar{x},t) \,|\, (x,t) \in \partial \Delta_\epsilon \times \mathbb{R} \ .
\end{aligned}
\right\}
\tag{6.3.1}
$$

The following conditions are adopted from Leitmann-Skowronski [2].

CONDITIONS 6.3.1. The system (2.2.6)' is *strongly controllable on some* $\Delta_0 \subset \Delta$ *for real time avoidance of* A , if there are functions $\bar{P}(\cdot), V(\cdot)$ defined on $D_A \times \mathbb{R}$ such that

(i) $0 < v_A^+ < v_\epsilon^- < \infty$;

(ii) for all $u \in \bar{P}(\bar{x},t)$, there is $T_A < \infty$ such that

$$
\frac{\partial V(\bar{x},t)}{\partial t} + \nabla_x F(\bar{x},t)^T \bar{f}(\bar{x},\bar{u},\bar{w},t) \geq \frac{v_\epsilon^- - v_A^+}{T_A}
\tag{6.3.2}
$$

for all $\bar{w} \in W$. The conditions remain the same for stipulated $T_A = T_A^+$.

To verify the conditions, consider $\bar{\phi}(\bar{x}^0,t_0,\cdot) \in K(\bar{x}^0,t_0)$, $\bar{x}^0 \in \partial \Delta_\epsilon(\bar{x}^0)$ for all $\bar{x}^0 \notin A$, $t_0 \in \mathbb{R}$. If such a motion does not cross ∂A then it avoids A for indefinite time. Thus suppose $\bar{\phi}(\cdot)$ intersects ∂A . Let $t_1 \geq t_0$ be the time of the first intersection: $\bar{\phi}(\bar{x}^0,t_0,t_1) = \bar{x}^1 \in \partial A$. Integrating (6.3.2) along the motion we obtain

$$
t_1 - t_0 \geq \frac{V(\bar{x}^0,t_0) - V(\bar{x}^1,t_1)}{v_\epsilon^- - v_A^+} \, T_A \ ,
\tag{6.3.3}
$$

but in view of (i),

$$
\frac{V(\bar{x}^0,t_0) - V(\bar{x}^1,t_1)}{v_\epsilon^- - v_A^+} \geq 1 \ ,
$$

thus yielding $t_1 - t_0 \geq T_A$ which closes the proof. The proving argument remains the same for stipulated $T_A = T_A^+$.

Observe that in view of Definition 6.1.3 and (6.3.2) the avoidance in indefinite time of Section 6.2 is recovered for $T_A \to \infty$. Now let

$$v_A^- \stackrel{\Delta}{=} \inf V(\bar{x},t) \mid (\bar{x},t) \in \partial A \times \mathbb{R} \; ,$$
$$v_\varepsilon^+ \stackrel{\Delta}{=} \sup V(\bar{x},t) \mid (\bar{x},t) \in \partial \Delta_\varepsilon(\bar{x}^0) \times \mathbb{R} \; . \Bigg\} \tag{6.3.4}$$

Then

$$V(\bar{x}^1,t_1) - V(\bar{x}^0,t_0) \geq v_A^- - v_\varepsilon^+ \stackrel{\Delta}{=} -c_A T_A^+$$

so that $-c_A \leq 0$ is a lower bound for the time derivative of $V(\bar{x},t)$ along a motion from $\bar{x}^0 \in \partial \Delta_\varepsilon$ intersecting ∂A at $x^1 = \bar{\phi}(\bar{x}^0,t_0,t_1)$. Thus it is a safe rate of decrease of $\rho(\bar{x}^0,A)$ measured in terms of $V(\cdot)$. Conversely, as mentioned in Section 6.1, given c_A and T_A we can estimate $\Delta_A(T_A)$. Moreover, given c_A and $\bar{x}^0,t_0$ one may obtain $T_A = (v^+ - v_A^-)/c_A$. Finally, utilizing

$$c_A = \frac{(v_\varepsilon^+ - v_A^-)}{T_A^+} \geq \frac{(v_\varepsilon^- - v_A^+)}{T_A^+} \geq 0 \tag{6.3.5}$$

in (6.3.2), we see that $c_A = 0$ implies indefinite time avoidance, that is, $T_A^+ \to \infty$.

Obviously when $\partial \Delta_\varepsilon(\bar{x}^0)$ and ∂A are defined by V-levels, then $v_\varepsilon^- = v_\varepsilon^+ = v_\varepsilon$, $v_A^- = v_A^+ = v_A$. Often $V(\cdot)$ is selected from ∂A : $V(\bar{x},t) = \text{const} = v_A$.

COROLLARY 6.3.1. *Given* $\bar{x}^0 \in A$, $t_0 \in \mathbb{R}$, *if there is a pair* $(\bar{u}*,\bar{w}*)$ $\in U \times W$ *such that*

$$L(\bar{x},\bar{u}*,\bar{w}*,t) = \max_{\bar{u}} \min_{\bar{w}} L(\bar{x},\bar{u},\bar{w},t) \geq -\frac{v_\varepsilon^- - v_A^+}{T_A} \; , \tag{6.3.6}$$

then (ii) of Conditions 6.3.2 are met with $\bar{u}* \in \bar{P}_*(\bar{x},t)$.

The proving argument here is the same as for Corollary 6.2.1. We illustrate the use of the corollary in the following example.

EXAMPLE 6.3.1. A point-mass modelled craft C moving on a plane with a speed $v = \text{const}$ and with some controlled directional azimuth u tries to avoid during T_A a point-mass modelled body B moving with a greater speed $v_B > v$ in an unpredictable direction under azimuth w . The size of the body is such that its diameter forms a circular zone of radius r_B about its center. The geometry of the scenario is seen in Fig. 6.10 and the kinematic equations are

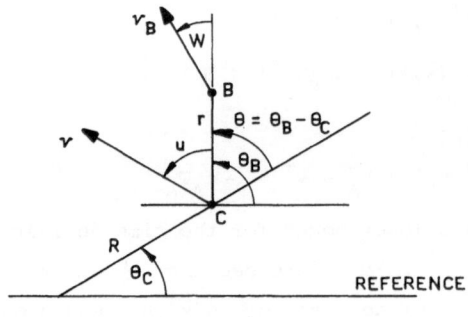

Fig. 6.10

$$\dot{R} = v \cos(\theta + u)$$

$$R\dot{\theta}_C = v \sin(\theta + u)$$

$$\dot{r} = v_B \cos w - v \cos u \qquad\qquad (6.3.7)$$

$$r\dot{\theta}_B = v_B \sin w - v \sin u$$

with the antitarget

$$T_A = \{(R, \theta_C, r, \theta_B) \in \mathbb{R}^4 \mid r \le r_B\} .$$

We let the avoidance set be $A : r \le r_B$ and take $V(R, \theta_C, r, \theta_B) = r$. Then in view of Corollary 6.3.1,

$$-v_B + v \ge \frac{r^0 - r_B}{T_A} \qquad\qquad (6.3.8)$$

where $r^0 = r(t_0)$ at the beginning of T_A. The control program $\bar{P}(\cdot)$ obtained from (6.3.8) is such that $u = u^* = -\pi$ for all $t \in [t_0, t_0 + T_A]$, assuming the avoidance of T_A during

$$T_A \le \frac{r_B - r^0}{v_B - v} . \qquad\qquad (6.3.9)$$

Conversely, of course, given T_A the evading craft can assure avoidance of T_A during T_A if

$$r_B - r^0 \ge (v_B - v) T_A . \qquad\qquad (6.3.10)$$

Rather a different result is obtained when T_A is not specified a-priori, but depends on an objective of the evading craft and hence on the controller. Suppose that, starting at $R^0 = R(t_0)$ the craft wishes to reach $R = R_C < R^0$ as rapidly as possible while assuring avoidance during this interval, that is, he wishes to attain $R = R_C$ before $r = r_B$. In this event one may consider $v \geq v_B$ as much as $v \leq v_B$. Now, condition (6.3.2) requires u to be chosen such that

$$\cos u \leq \frac{r^0 - r_B}{v T_A} - \frac{v_B}{v} \ . \tag{6.3.11}$$

However, since C desires to reach $R = R_C$ as rapidly as possible, it follows from the first equation of (6.3.7) that u must satisfy

$$\cos (\theta + u) = -1 \tag{6.3.12}$$

or

$$u = \pi - \theta \ . \tag{6.3.13}$$

Substitution of (6.3.13) into (6.3.11) gives

$$\cos \theta \geq \frac{v_B}{v} - \frac{r^0 - r_B}{v T_A} \ . \tag{6.3.14}$$

The interval T_A is obtained from the integration of the first equation of (6.3.7) with (6.3.12), that is,

$$T_A = \frac{R^0 - R_C}{v}$$

whence (6.3.14) becomes

$$\cos \theta \geq \frac{v_B}{v} - \frac{r^0 - r_B}{R^0 - R_C} \ . \tag{6.3.15}$$

This condition, which guarantees the satisfaction of (6.3.2), is assured if

$$\frac{r^0 - r_B}{R^0 - R_C} \geq 1 + (v_B / v) \ . \tag{6.3.16}$$

Thus, provided the initial values r^0, R^0 satisfy the above, the craft attains R_C before collision: $r = v_B$. $\qquad$ □

We turn now to ultimate avoidance and introduce conditions adapted from Leitmann-Skowronski [2].

CONDITIONS 6.3.2. The system (2.2.6)' is completely *strongly controllable for ultimate avoidance* if there is a function $\bar{P}(\cdot)$, a C^1-function $V(\cdot)$, both defined on $\Delta \times \mathbb{R}$, and two constants, $v_A > 0$ and $c_A > 0$, such that

    (i) $0 \leq V(\bar{x}, t) \leq v_A$, $\forall (\bar{x}, t) \in A \times \mathbb{R}$;

    (ii)   for all $(\bar{x}, t) \in \Delta \times \mathbb{R}$ and all $\bar{u} \in \bar{P}(\bar{x}, t)$,

$$\frac{\partial V(\bar{x},t)}{\partial t} + \nabla_{\bar{x}}V(\bar{x},t)^T\bar{f}(\bar{x},\bar{u},\bar{w},t) \geq c_A \qquad (6.3.17)$$

for all $\bar{w} \in W$.

In order to verify the conditions, we first consider the case of $\bar{\phi}(\cdot) \in K(\bar{x}^0,t_0)$ where $V(\bar{x}^0,t_0) \geq v_A$. By (ii), $V(\cdot)$ increases along $\bar{\phi}(\cdot)$. Thus by (i), there is no $t_1 > t_0$ such that $\bar{\phi}(\bar{x}^0,t_0,t_1) \in A$. Next we consider $(\bar{x}^0,t_0)$ such that $V(\bar{x}^0,t_0) < v_A$ so that for some $t > t_0$ it is possible that $\bar{\phi}(\bar{x}^0,t_0,t) \in A$, in view of (i). Again, by (ii), $V(\cdot)$ increases along the corresponding trajectory so that there is a $t_A > t_0$ and $\bar{x}^A = \bar{\phi}(\bar{x}^0,t_0,t_A)$ such that $V(\bar{x}^A,t_A) = v_A$ provided that $\bar{\phi}(\bar{x}^0,t_0,\cdot)$ is defined on $[t_0,t_1] \supset [t_0,t_A]$. Then, upon integrating (ii) along the trajectory and taking note of (i), we obtain

$$t_A - t_0 \leq \frac{v_A - V(\bar{x}^0,t_0)}{c_A} \leq v_A/c_A . \qquad (6.3.18)$$

The proviso concerning $[t_0,t_1] \supset [t_0,t_A]$ can always be met by adjusting the constant c_A. Of course, as in the first case, now there also is no $t_1 > t_A$ for which $\bar{\phi}(\bar{x}^0,t_0,t_1) \in A$. Finally, by (6.3.18),

$$T_A = v_A/c_A . \qquad (6.3.19)$$

The discussion of (6.3.19) is similar to that of (6.3.5) and is left to the reader.

Let us now briefly consider the case of ultimate avoidance with stipulated $T_A^- < \infty$.

CONDITIONS 6.3.3. The system (2.2.6)' is completely strongly controllable for ultimate avoidance after T_A^-, if there is a function $\bar{P}(\cdot)$ and a C^1-function $V(\cdot)$, both defined on $\Delta \times \mathbb{R}$, and a constant v_A, such that

(i) $0 \leq V(\bar{x},t) \leq v_A$, $\forall(\bar{x},t) \in A \times \mathbb{R}$;

(ii)   for all $(\bar{x},t) \in \Delta \times \mathbb{R}$ and $\bar{u} \in \bar{P}(\bar{x},t)$,

$$\frac{\partial V(\bar{x},t)}{\partial t} + \nabla_{\bar{x}}V(\bar{x},t)^T\bar{f}(\bar{x},\bar{u},\bar{w},t) \geq \frac{v_A}{T_A} , \qquad (6.3.20)$$

for all $\bar{w} \in W$;

(iii) $\bar{\phi}(\bar{x}^0,t_0,\cdot) \in K(\bar{x}^0,t_0)$ is defined on

$[t_0,t_1] \supset [t_0,t_0+T_A]$, $\forall(\bar{x}^0,t_0) \in \{(\bar{x},t) \in \Delta \times \mathbb{R} \mid 0 \leq V(\bar{x},t) < v_A\}$.

The proof is analogous to that of Conditions 6.3.2 with c_A replaced by (v_A/T_A).

COROLLARY 6.3.2. *Given* $\bar{x}, t \in \Delta \times \mathbb{R}$, *if there is a pair* $(u^*, w^*) \in U \times W$ *such that*

$$L(\bar{x}, \bar{u}^*, \bar{w}^*, t) = \max_{\bar{u}} \min_{\bar{w}} L(\bar{x}, \bar{u}, \bar{w}, t) \geq c_A , \qquad (6.3.21)$$

then (iii) of Conditions 6.3.3 is met with $\bar{u}^* \in \bar{P}_*(\bar{x}, t)$. *When* T_A *is stipulated* c_A *is replaced by* v_A^-/T_A .

EXAMPLE 6.3.2. Consider the same motion equations as in Example 6.3.1, and let the evading craft C have the same objective of reaching $R = R_C$ but this time having escaped from $A : r \leq r_A$, see Fig. 6.11, which is specified as a turbulence area B on route of C.

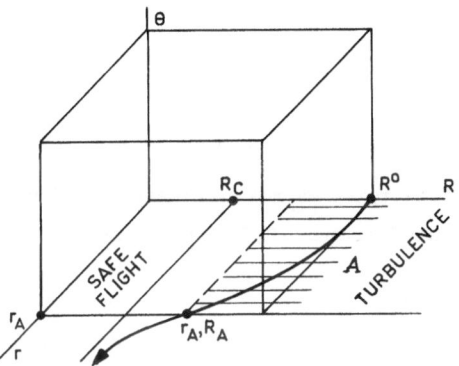

Fig. 6.11

There is no need for a safety zone and the time is

$$T_A \leq \frac{R_C - R^0}{\max \dot{R}} = \frac{R_C - R^0}{v} .$$

We choose again $V = r$, $v_A = r_A$, with (6.3.20):

$$v_B \cos w - v \cos u \geq \frac{r_A v}{R_C - R^0} , \quad \forall w \in W .$$

Since $|\cos w| < 1$ the above holds if

$$-v_B - v \cos u \geq \frac{r_A v}{R_C - R^0} ,$$

or

$$\cos (u + \pi) \geq \frac{r_A}{R_C - R^0} + (v_B/v) .$$

This condition allows us to choose a feasible controller provided

$$\frac{r_A}{R_C - R^0} + (v_B/v) \leq 1 \ ,$$

which means $r_A < R_C - R^0$, $v_B < v$, that is, the craft has to be faster than the turbulence. ▯

The reader may also recall here the escape from atmosphere in Section 5.9.

Closing the section, let us observe that the proving argument for the conditions does not change when stationary $V(\bar{x})$ is introduced, as shown in the examples.

EXERCISES 6.3

6.3.1 Consider the system

$$\dot{x}_1 = x_2 \ , \quad \dot{x}_2 = -3x_2 - x_1 + 2x_1^3 + u \ ,$$

and design a control program for ultimate avoidance after $T_A = 10\,\text{sec}$ of the set $\mathring{A}$ located about $(0,0)$ and bounded by the separatrices.

6.3.2 Form sufficient conditions for the real time avoidance of $\mathring{A}$ during T_A^+ and specify means of obtaining a corresponding control program.

6.3.3 Recall what was said that it happens to Definition 6.1.3 when $T_A \rightarrow \infty$. Discuss the alternative extreme case of $T_A \rightarrow 0$. How does it refer to collision?

6.3.4 Using necessary and sufficient conditions, show that ultimate avoidance after a stipulated time of the exterior set outside separatrices of the system $\ddot{q} + d\dot{q} + aq - bq^3 = u$, $a,b,d > 0$, is equivalent to finite time capture in the interior (inside separatrices) set surrounding $(0,0)$. Find the control program for both cases.

6.4 COLLISION OR CAPTURE FIRST

Let us now show how to combine two modular objectives in the relatively simple case of avoiding some antitarget (one or more) before colliding with another. Such a combination is obviously basic for many practical scenarios. In passing, it has been suggested already in Examples 6.3.1, 6.3.2. It means that, given a collection of bodies, we want to collide with a specific body *before* allowing collision with any other. Suggestion of the problem and sufficient conditions (for capture rather than collision) belongs to Getz-Leitmann [1], modified later by Stonier [2]. It had been used extensively for the game applications, see our references in Chapter 8, and for coordination control of robotic systems, see Skowronski [44].

DEFINITION 6.4.1. Given a target T and a set of antitargets enclosed in A in Δ, a motion of (2.2.6)' from $\bar{x}^0 \notin A$, $t_0 \in \mathbb{R}$, *collides* with T *first*, avoiding A before collision, if and only if there is $T_F < \infty$, such that

$$\bar{\phi}(\bar{x}^0, t_0, [t_0, t_0 + T_F]) \cap A = \phi \ , \tag{6.4.1}$$

$$\bar{\phi}(\bar{x}^0, t_0, t_0 + T_F) \in T \ . \tag{6.4.2}$$

When T_F is stipulated, we have *collision first during* T_F.

The above definition combines real time avoidance with real time collision but the emphasis is on the latter. It means that T_F is conditioned by collision during T_F with A subtracted: $\Delta_F \subset \Delta_C - A$, and the avoidance time adjusted. Here Δ_F denotes the region of strong controllability for collision first. The above means that for the avoidance part we should consider Definition 6.1.3, replacing T_A by T_F. Observe that there is no need to introduce the symmetric concept with emphasis on avoidance, say avoidance first, as the real time avoidance of Section 6.3 takes care of the case.

The fact that the avoidance time-interval must match (be equal or larger than) the interval before collision gives an additional condition from which the region of strong controllability Δ_F may be determined. It will be seen below following the calculation of these intervals.

The sufficient conditions for strong controllability for collision first are designed by combining the conditions for the particular subobjectives: collision and real time avoidance.

Given the sets T, A, $T \cap A = \phi$, and Δ_0 enclosed in $\Delta_C - A$, as well as $\Delta_\varepsilon(\bar{x}^0) \supset A$ with $\partial A \cap \partial \Delta_\varepsilon = \phi$, we adjust the complements: $C_F T \triangleq \Delta_F - T$, $CA \triangleq \Delta_F$ and introduce open sets $D_F \supset C_F$, $D_A \supset \Delta_F$. Moreover we introduce two C^1-test functions $V_F(\cdot) : D_F \to \mathbb{R}$, with v_0^+, v_{FT} defined by (5.3.1) and $V_A(\cdot) : D_A \to \mathbb{R}$, with v_ε^-, v_A^+ defined by (6.3.1).

CONDITIONS 6.4.1. The system (2.2.6)' is strongly controllable on Δ_0 for collision-first with T avoiding A, if there is a program $\bar{P}(\cdot)$ defined on $D_F \cup D_A$ and two C^1-functions $V_F(\cdot)$, $V_A(\cdot)$ such that

(i) $V_F(\bar{x}) > v_{FT}$, for $\bar{x} \notin T$;

(ii) $V_F(\bar{x}) \le v^+$, for $\bar{x} \in C_F T$;

(iii) $0 < v_A^+ < v_\varepsilon^- < \infty$;

(iv) for all $\bar{x} \in D_F$, $\bar{u} \in \bar{P}(\bar{x})\big|_{D_F}$ there is $T_F < \infty$ such that

$$\int_{t_0}^{t_0 + T_F} \nabla V_F(\bar{x})^T \bar{f}(\bar{x}, \bar{u}, \bar{w}, t)\, dt \le -(v_0^+ - v_{FT}) \qquad (6.4.3)$$

for all $\bar{w} \in W$; and

(v) for all $\bar{x} \in D_A$, $\bar{u} \in P(\bar{x})\big|_{D_A}$,

$$\nabla V_A(\bar{x})^T \bar{f}(\bar{x}, \bar{u}, \bar{w}, t) \ge -\frac{v_\varepsilon^- - v_A^+}{T_F} \qquad (6.4.4)$$

for all $\bar{w} \in W$.

The verification of the above follows immediately from Conditions 5.3.2 and 6.3.1 with $T_F = T_A^+$.

Obviously it may be possible to combine $V_F(\cdot)$, $V_A(\cdot)$ into a single function, very much along the lines that Stonier [2] did for the Getz-Leitmann's combined objective, but for reasons explained in Section 3.4, it is not necessarily convenient.

The corresponding corollary which allows us to determine $\bar{P}(\cdot)$ can be written immediately the same way as for Conditions 5.3.2, 6.3.1. From (6.4.3) we may obtain

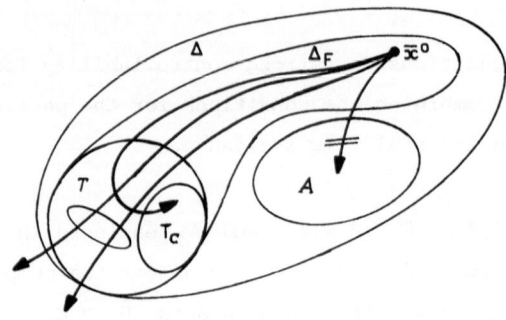

Fig. 6.12

$$T_F \leq - \frac{v_0^+ - v_{FT}}{v_0^+ - V_F(\bar{x}^0)} \tag{6.4.5}$$

while (6.4.4) yields

$$T_F \geq \frac{v_\epsilon^- - v_A^+}{c_A} \tag{6.4.6}$$

where $-c_A$ is the safe rate of decrease of $\rho(\bar{x}^0, A)$ measured in $V_A(\bar{x})$, see Section 6.3. From (6.4.5) and (6.4.6) we obtain

$$- \frac{v_0^+ - v_{FT}}{v_0^+ - V_F(\bar{x}^0)} \geq \frac{v_\epsilon^- - v_A^+}{c_A} \tag{6.4.7}$$

which defines Δ_F relative to c_A.

Since collision is the leading property, we may also have the objective of *collision first in stipulated time* T_F, definitions, conditions and corresponding controller yielding corollary are obtainable immediately from the above and left to the reader.

The mentioned alternative combination of capture in some target T after a time interval T_F during which we have avoided A, is a more demanding objective and needs stronger conditions.

DEFINITION 6.4.2. A motion of (2.2.6)' from $\bar{x}^0 \notin A$, $t_0 \in \mathbb{R}$ is *captured in T first, avoiding A* before such capture, if and only if there is $T_F < \infty$ such that

$$\bar{\phi}(\bar{x}^0, t_0, [t_0, t_0 + T_F]) \cap A = \phi , \tag{6.4.5}$$

$$\bar{\phi}(\bar{x}^0, t_0, t) \in T , \quad \forall \, t \geq t_0 + T_F . \tag{6.4.6}$$

For illustration, see trajectories enclosed in the positively invariant T_C in Fig. 6.12.

Sufficient conditions for the above objective have been introduced by Getz-Leitmann [1] and we quote them below adjusting the notation.

Given a candidate capturing subtarget T_C and the avoidance set A, we introduce a set Δ_0, enveloping but not equal to $T_C \cup A$, and two C^1-functions $V_F(\cdot), V_A(\cdot)$ defined on Δ_0 and such that

$$T_C \supset \Delta_1 \triangleq \{\bar{x} \in \Delta_0 \,|\, V_F(\bar{x}) \leq v_F\} \tag{6.4.6}$$

$$A \subset \Delta_2 \triangleq \{\bar{x} \in \Delta_0 \,|\, V_A(\bar{x}) \leq v_A\} \tag{6.4.7}$$

where $v_F, v_A > 0$ are suitable constants.

CONDITIONS 6.4.2. The system (2.2.6)' is *strongly controllable on* Δ_0 *for capture first* in T_C avoiding A, if there is a program $\bar{P}(\cdot)$ defined on Δ_0 and two C^1-functions $V_F(\cdot), V_A(\cdot)$ as defined above such that

(i) $V_F(\cdot)$ is radially unbounded: $V_F(\bar{x}) \to \infty$, $|\bar{x}| \to \infty$;

(ii) Δ_0 is strongly positively invariant;

(iii) for all $\bar{x} \in \Delta_0$, $\bar{u} \in \bar{P}(\bar{x})$ there are constants $c_F > 0$, and c_A such that

$$\sup_{\bar{w}} \nabla V_F(\bar{x})^T \bar{f}(\bar{x}, \bar{u}, \bar{w}, t) \leq -c_F , \qquad (6.4.8)$$

$$\inf_{\bar{w}} \nabla V_A(\bar{x})^T \bar{f}(\bar{x}, \bar{u}, \bar{w}, t) \geq -c_A . \qquad (6.4.9)$$

The proof of the above conditions is given in Getz-Leitmann [1], or may be derived by combining Conditions 5.5.1 and 6.3.1 with $T_A^+ = T_F$.

The reader might have observed already that the controller-implementing corollary is already included in the item (iii) of the conditions, by taking the inequalities with supremum and infimum, successively. While an assumed candidate Δ_0 may be confirmed by the conditions as a strongly controllable set, it is not yet the region of such controllability, that is, the largest strongly controllable set. To obtain the latter, we estimate T_F from both inequalities in (iii) and compare the results, similarly as for (6.4.7). This verifies the region of strong controllability for capture-first determined as the maximal

$$\Delta_0 = \left\{ \bar{x} \in \Delta - (T_C \cup \Delta_2) \;\middle|\; \frac{c_F}{V_F(\bar{x}) - v_F} > \frac{c_A}{V_A(\bar{x}) - v_A} \right\} , \qquad (6.4.10)$$

generated by a suitable maximizing choice of the constants $c_F > 0, c_A$, see Getz-Leitmann [1].

Following the latter work, we may now combine Example 6.3.1 with 6.3.2 to obtain an illustration of our present conditions.

EXAMPLE 6.4.1. The craft C of Example 6.3.1 attempts capture first: attain a shelter target T before being intercepted by a turbulence B moving faster $v_B > v$ and in uncertain direction given by $w \in W$. The geometry of the scenario is shown in Fig. 6.10, the motion equations are given by (6.3.7), with the target

$$T = \{ (R, \theta_C, r, \theta_B) \in \mathbb{R}^4 \mid R \leq R_C \} ,$$

see Example 6.3.2, and avoidance set $\Lambda : r \le r_B$, see Example 6.3.1, and Fig. 6.11. We select the functions $V_{\mathcal{F}} = R$, $V_A = r$ so that the sets Δ_1, Δ_2 are specified by V-levels: $v_{\mathcal{F}} = R_C$, $v_A = r_B$, that is,

$$T_{\mathcal{C}} = \Delta_1 : R < R_C \qquad (6.4.11)$$

$$A = \Delta_2 : r \le r_B \qquad (6.4.12)$$

and $\nabla V_{\mathcal{F}} = (1,0,0,0)$, $\nabla V_A = (0,0,1,0)$. Then (iii) of Conditions 6.4.2 give the control conditions

$$v \cos (\theta + u) \le -c_{\mathcal{F}} < 0 \qquad (6.4.13)$$

$$-v_B - v \cos u \ge -c_A \qquad (6.4.14)$$

and Δ_0 is defined by

$$r - r_B > \frac{c_A}{c_{\mathcal{F}}} (R - R_C) . \qquad (6.4.15)$$

The control condition (6.4.13) implies

$$-1 \le \cos (\theta + u) \le \frac{c_{\mathcal{F}}}{v} < 0$$

so that

$$c_{\mathcal{F}} = \frac{v}{1 + \delta} \quad \text{for} \quad \delta \in [0, \infty) .$$

Given $c_{\mathcal{F}}, \Delta_0$ is the largest when choosing the smallest c_A such that (6.4.14) is met for all possible θ. This results in

$$c_A = v_B + \frac{v}{1 + \delta}$$

whence

$$\frac{c_A}{c_{\mathcal{F}}} = 1 + (1 + \delta) \frac{v_B}{v} .$$

Thus $\Delta_{\mathcal{F}}$ is obtained when $\delta = 0$, yielding

$$\Delta_{\mathcal{F}} : r - r_B > \frac{v + v_B}{v} (R - R_C) . \qquad \square$$

EXAMPLE 6.4.2. As another example, let us investigate the "soft landing" problem of a craft which aims at avoiding the ground of a planet by suspending itself in a safe position at some distance h above such planet. Ignoring less influential forces, that is, all except gravity g of the planet and uncertain thrust uw, we obtain the motion equation $\ddot{q} = uw - g$ with the state equations

$$\left.\begin{array}{l} \dot{x}_1 = x_2 , \\[2mm] \dot{x}_2 = uw - g . \end{array}\right\} \tag{6.4.16}$$

We assume $\Delta : q \geq 0, \dot{q} \in \mathbb{R}, \ T \overset{\Delta}{=} \{h,0\}$, and $A \overset{\Delta}{=} \{(q,\dot{q}) \,|\, q = 0, \ \dot{q} > \hat{q}\}$ with $\hat{q}$ a safe value of velocity, see Fig. 6.13. The action of the thrust $uw = \text{const}$ balances the planet gravitation. The pilot of the craft may use $u \in [a_1, a_2]$, $a_2 > a_1 > 0$ subject to the uncertainty $w \in [b_1, b_2]$, $b_2 > b_1 > g > 0$. The equation (6.4.16) is integrable:

$$x_2^2 - 2(uw - g)x_1 = \text{const} \tag{6.4.17}$$

which would give the test function V if needed, but since we have trajectories in closed form, direct analysis applies. The pilot's best is to maximize the thrust: $u(t) \equiv a_2$ against all w, that is, against $w(t) \equiv b_1$ which generates the family of trajectories

$$x_2^2 - 2(a_2 b_1 - g)x_1 = \text{const} , \tag{6.4.18}$$

seen in Fig. 6.13. The trajectory that collides with the target T without switching the control $u \equiv a_2$ is thus given by

$$x_1 = \frac{x_2^2}{2(a_2 b_1 - g)} + h . \tag{6.4.19}$$

Motions on all other paths must switch control as shown in Fig. 6.13. On the other hand, crash-landing in spite of $u \equiv a_2$ occurs at $(0,0)$ along (6.4.19) with $h = 0$ substituted, which means that the strongly winning control of $u \equiv a_2$ ends before the line

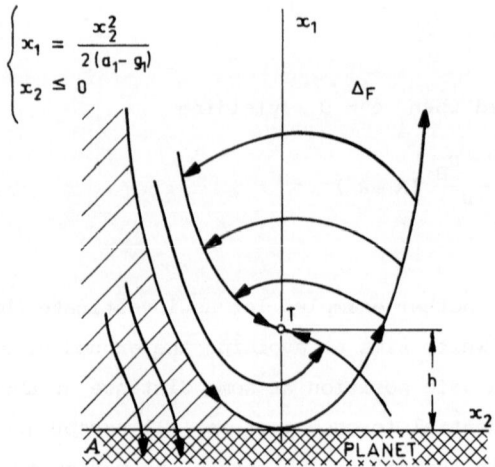

Fig. 6.13

$$
x_1 = \begin{cases} x_1^2/2\,(a_2 b_1 - g) \,, & x_2 < 0 \\ \\ 0 \,, & x_2 \geq 0 \end{cases}
$$

which determines the boundary $\partial\Delta_F$, making the region Δ_F a well determined open set. Note that so defined $\partial\Delta_F$ is weakly $\bar{P}(\cdot)$-nonpermeable, so that it also is a semi-barrier. Consequently $\Delta - \Delta_F$ is filled up with paths crashing into the ground A.

The scenario is slightly different if we allow for the effect of atmosphere which produces drag. Then the motion equations lead to

$$\dot{x}_1 = x_2$$

$$\dot{x}_2 = uw - g - dx_2$$

with $d > 0$ being some damping coefficient, possibly a function of x_1, x_2. The state equations are no longer integrable but we may use (6.4.18) to generate the test function

$$V = x_2^2 - 2\,(a_2 b_1 - g)\,x_1 \,.$$

Indeed, introducing the extremizing u, w,

$$\min_u \max_w \dot{V}(x_1, x_2) = -2d\,(x_1, x_2)\,x_2^2 < 0$$

generates our objective and confirms the same controls. □

EXERCISES 6.4

6.4.1 For the system

$$\dot{x}_1 = x_2 \,, \quad \dot{x}_1 = -5x_2 - x_1 + 2.1x_1^3 - x_1^5 + u$$

and avoidance set $A : x_1^2 + x_2^2 \leq 1$, find the control program which secures controllability for "capture first" about the equilibrium $x_1^e = 1.17$, $x_2^e = 0$, when starting from $x_1^0 = 1$, $x_2^0 > 1$.

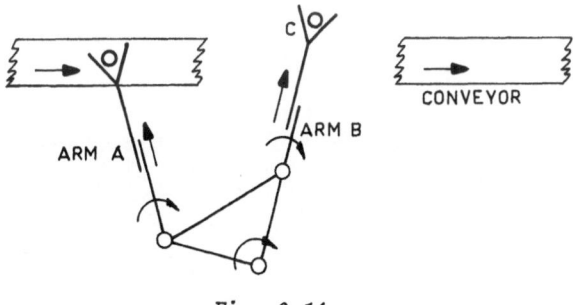

Fig. 6.14

6.4.2 Pick-and-place two-arm robot is shown in the figure below. Each
arm is of the RP-type (rotary-prismatic). We consider the scenario
from the viewpoint of arm A which aims at picking up an object
from the conveyor and depositing it at the spot C on a workbench
while avoiding collision with the arm B. The latter is considered
a moving obstacle in its attempt to pick up an object from C and
deposit it back on the conveyor. The conveyor moves with constant
speed. The system has 5 DOF indicated by arrows. Specify the
target and avoidance set and write the motion and state equation.
Find the feedback control program for arm A securing strong controllability for the pick and place objective described, robust
to the motion of arm B.

Chapter 7
ADAPTIVE TRACKING CONTROL

7.1 PATH AND MODEL FOLLOWING

The range of applications of path-tracking control is enormous for
any type of machines or, more generally, mechanical structures. It begins
from point-mass modelled craft dynamics in space, air, sea and on ground,
when controlling travel, pointing, station and formation keeping, or
generally traffic, etc., through multibody models of manufacturing
machinery, in particular robotic manipulators, to large flexible structures.
Consequently, the literature dealing with such applications is too vast to
be quoted here. For the same reason there is a large variety of methods
attacking the problem of such tracking, also with a broad literature. The
interested reader may look for a review in Luh [1],[2], Luh-Lin [1] and
Skowronski [38].

For the general nonlinear case of mechanical systems, it is usually
possible to attain the path-tracking objective by the methods discussed in
Chapter 5, as already indicated in a number of examples. We may proceed
in two different ways: either *directly*, taking the desired path as a
corresponding target, possibly moving, or in *relative coordinates*,
choosing the state variables to be the distance from the path in Cartesian
or Configuration spaces.

In most practical cases, the desired Cartesian path is pre-planned
and specified in terms of a small number of base positions and orientations
of the bodies concerned, chosen so that they allow a precise realization
of the desired path by interpolation. Then the path must be expressed in
closed (analytic) format. Usually such a path consists of straight lines

and curves of second or third degree which are easily identified after having stored a small number of *base points* in the controlling computer. For instance, for the welding robot, it is common to make the desired path of the gripper as an intersection of two planes combined with another intersection of a plane and a surface of second degree (cylindrical), see Markov-Zemanov-Nenchev [1]. Usually the base points are selected at corners of the path, and marked places where the controller may change its action: change of velocity, direction, pause, etc. What we have described seems like an alternative point-to-point sequential control. However, the difference is substantial and lies in a non-stopping motion at the base points. To achieve it, the whole path must be considered as a single target. The base points give the desired position and orientation of the structure at the desired time. For a full path planning, see Luh-Lin [1], we need also the desired corresponding velocities. Then, by inverse kinematics, see Section 1.6, the Configuration Space and lagrangian velocities space path can be designed: $\bar{q}_m(t)$, $\dot{\bar{q}}_m(t)$, $\forall\, t \geq t_0$. Now, the direct choice of state variables requires the vector $\bar{x}(t) = (\bar{q}(t), \dot{\bar{q}}(t))^T$ to be controlled to the target

$$T : (\bar{q}_m(t), \dot{\bar{q}}_m(t))^T = \bar{x}_m(t) = \bar{\gamma}(t) , \quad \forall\, t \geq t_0 , \qquad (7.1.1)$$

while the choice of relative state coordinates:

$$\bar{x}(t) \overset{\Delta}{=} (\bar{q}(t) - \bar{q}_m(t) , \dot{\bar{q}}(t) - \dot{\bar{q}}_m(t))^T$$

requires $\bar{x}(t)$ to be controlled to a target surrounding the origin of the corresponding state space. In either of these two cases, capture or at least a suitably prolonged rendezvous, rather than collision, is sought. It is worth noting that the target-path in the first case, and the target about the zero-deviation from such a path, that is, the origin in the second case are geometric constructions, in particular $\bar{\gamma}(t)$ is an arbitrarily designed curve. They have nothing to do with motions of the controlled system, or any other dynamical system for that matter. Allowing such independence is the advantage of using the methods discussed in Chapter 5. Past results in this direction on robust path tracking are quite limited.

More commonly $\bar{\gamma}(t)$ is considered some reference signal in terms of a motion or trajectory of the system concerned. We then pre-filter it through a suitable feedforward mechanism, perhaps adding an error compensating feedback terms, for design of the tracking controller. Such a motion or trajectory is then called *nominal*. Obviously the demand that $\bar{\gamma}(t)$ is a motion imposes extra constraints on both $\bar{\gamma}(t)$ and the system, narrowing the application of the study. However, for instance for the relative coordinates representation, the controller may act as a regulator to the nominal motion and when the initial $\bar{x}^0$ is not large (which commonly occurs) the system may be meaningfully linearized, see Paul [1], Bejczy [1], Raibert-Horn [1], Lee-Chung [1]. For large perturbations such an assumption may not be used and there are currently various techniques proposed. To quote a few based on the Liapunov method, we may quote Koditschek [1], Ha-Gilbert [1], the latter covering also the case of tracking an output signal instead of state, and Ryan-Leitmann-Corless [1] covering systems with uncertainty. In order to illustrate this avenue of study, we give below a brief outline of a technique presented in Skowronski [38] which seems to be the least restrictive in assumptions, but refers to systems without uncertainty, that is, (2.1.15) with (2.1.13).

Substituting $\bar{x}_m = \bar{\gamma}(t)$ into (2.1.15) we may evaluate the control $\bar{u} = \bar{u}_m$. In terms of the mechanical system (2.1.13), with $\bar{F}(\cdot)$ linear in $\bar{u}$: $F_i(\bar{q},\dot{\bar{q}},\bar{u},\bar{w}) \overset{\Delta}{=} B_i(\bar{q},\dot{\bar{q}},\bar{w})\bar{u}_i$, $B_i \neq 0$, we obtain

$$
\left.
\begin{aligned}
u_{mi} = \frac{1}{B_i(\bar{q}_m,\dot{\bar{q}}_m)} &\left[\ddot{q}_{mi} + \Gamma_i(\bar{q}_m,\dot{\bar{q}}_m) + D_i(\bar{q}_m,\dot{\bar{q}}_m) + \Pi_i(\bar{q}_m) \right. \\
&\left. - R_i(\bar{q}_m,\dot{\bar{q}}_m,t) \right] , \quad i = 1,\ldots,n
\end{aligned}
\right\}
\qquad (7.1.2)
$$

the open loop, feedforward only, controller. The obvious setback of such a controller is the on-line differentiation producing noise, in particular $\ddot{q}_{mi}$. There are various methods proposed for reducing this disadvantage, cf. Voroneckaya-Fomin [1], Gusev-Yakubovich [1], Gusev-Timofeev-Yakubovich [1], Aksenov-Fomin [1]. We leave this problem open as the case is well covered by our next discussion on the model following. Substituting the obtained $\bar{u}_m(t)$ into (2.1.15) we can expect to obtain the motions $\bar{\phi}_m(\bar{x}^0,t_0,\mathbb{R})$ which, provided the initial conditions are the same: $\bar{x}^0 = \bar{\gamma}(t_0)$, should be very close to the planned curve $\bar{\gamma}(t)$, $t \in \mathbb{R}$. It is then feasible to take $\bar{\phi}_m(\cdot)$ as the nominal motion. Note that $\bar{\phi}_m(t)$ is a given, well defined function of time and it does not depend upon any other solutions of (2.1.15). We now introduce the deviation from the nominal motion, $\bar{e}(t) \overset{\Delta}{=} \bar{x}(t) - \bar{\phi}_m(t)$, $t \geq t_0$, which transforms (2.1.15) into the relative state equation of the (2.1.6) type

$$
\dot{\bar{e}} = \bar{f}(\bar{e} + \bar{\phi}_m(t),\bar{u},t) - \bar{f}(\bar{\phi}_m(t),\bar{u}_m,t)
\qquad (7.1.3)
$$

with perturbation $\bar{f}(\bar{\phi}_m(t),\bar{u}_m,t)$ which is a given function of time and with the right hand side vanishing identically at $\bar{e} = 0$, see Section 2.1. Denote $\bar{f}_m(t) \overset{\Delta}{=} \bar{f}(\bar{\phi}_m(t),\bar{u}_m,t)$. By definition $\bar{f}_m(\cdot)$ is bounded the same way as $\bar{f}(\cdot)$ was. Obviously now the trivial solution $\bar{e}(t) \equiv 0$ represents the nominal motion and we use the controllers of Section 3.3 to

produce its asymptotic stability which in turn generates the desired
tracking objective. Alternatively we may use the controllers of Chapter
5 to produce capture or rendezvous in a desired target about $\bar{e}(t) \doteq 0$.

EXAMPLE 7.1.1. Consider the system

$$
\left.
\begin{aligned}
\dot{x}_1 &= x_2 \\
\dot{x}_2 &= -ax_1 + bx_1^3 - u
\end{aligned}
\right\} \tag{7.1.4}
$$

with $\gamma(t)$ to be followed defined by

$$
x_{m2}^2 + ax_{m1}^2 = 2h_C , \tag{7.1.5}
$$

where $h_C \le a^2/2b$ is the value marking an E-level of (7.1.4) within the
basic cup, but (7.1.5) is not such a level. Substituting (7.1.5) into
(7.1.4) we obtain $\dot{x}_{m1} = x_{m2}$, $\dot{x}_{m2} = -ax_{m1}$ provided

$$
u_m = bx_{m1}^2 \tag{7.1.6}
$$

which substituted to (7.1.4) generates the nominal trajectory coinciding
with the curve (7.1.5). Defining now $e_i \overset{\Delta}{=} x_i - x_{mi}$, $i = 1,2$, we
obtain

$$
\left.
\begin{aligned}
\dot{e}_1 &= e_2 \\
\dot{e}_2 &= -ae_1 + be_1^3 + 3be_1 x_{m1}(x_{m1} + e_1) + bx_{m1}^3 - u
\end{aligned}
\right\} . \tag{7.1.7}
$$

Letting $V(\bar{e}) = (1/2)e_2^2 + (a/2)e_1^2 - (b/4)e_1^4$, which is the obvious choice
via the first integral method, we obtain

$$
\dot{V} = 3be_1 x_{m1}(x_{m1} + e_1)e_2 - ax_{m1}e_2 + bx_{m1}^3 e_2 - ue_2
$$

or substituting the parametric representation of $\gamma(t) : x_{m1} = \sqrt{2h_C/a} \cos t$,
$x_{m2} = \sqrt{2h_C} \sin t$,

$$
\begin{aligned}
\dot{V} = 3be_1 e_2 (2ah_C \cos^2 t + e_1\sqrt{2h_C/a} \cos t) - ae_2\sqrt{2h_C/a} \cos t \\
+ be_2\sqrt{(2h_C/a)^3} \cos^3 t - ue_2 .
\end{aligned}
$$

Then let $\mu > 0$ be the required tracking estimate. We want to capture
$\bar{e}(t)$ in the subtarget $T_C : e_2^2 + ae_1^2 - (b/2)e_1^4 < 2\mu$, which by Conditions
5.5.1 is attained if $\dot{V}(\bar{e}) \le -c$. This is implemented by the control
condition

$$
\begin{aligned}
ue_2 \ge c - 3be_1 e_2 (2ah_C \cos^2 t + e_1\sqrt{2h_C/a} \cos t) \\
- ae_2\sqrt{2h_C/a} \cos t + be_2 (2h_C/a)^{3/2} \cos^3 t
\end{aligned} \tag{7.1.8}
$$

or using the minimizing equality, by

$$u = \begin{cases} (c/e_2) - 3be_1(2ah_C \cos^2 t + e_1\sqrt{2h_C/a} \cos t) \\ \quad - a\sqrt{2h_C/a} \cos t + b(2h_C/a)^{3/2} \cos^3 t, \quad |e_2| \geq \beta, \\ 0, \quad |e_2| < \beta, \end{cases} \qquad (7.1.9)$$

with β, c selected as always before. $\qquad\qquad\qquad\qquad$ $\square$

For systems with uncertainty, essential modifications must be made to the above technique. First, calculating $\bar{u}_m$ from $\bar{\gamma}(t)$ substituted into (2.2.5)', one obtains a function of time, but also of $\bar{w}$: $\bar{u}_m(t) \in \bar{P}_m(t,\bar{w})$ with $\bar{P}_m(\cdot)$ set valued. In particular (7.1.2) will now become

$$\begin{aligned} u_{m_i} &= \frac{1}{B_i(\bar{q}_m,\dot{\bar{q}}_m,\bar{w})} \, [\ddot{\bar{q}}_{m_i} + \Gamma_i(\bar{q}_m,\dot{\bar{q}}_m,\bar{w}) + D_i(\bar{q}_m,\dot{\bar{q}}_m,\bar{w}) \\ &\quad + \Pi_i(\bar{q}_m,\bar{w}) - R_i(\bar{q}_m,\dot{\bar{q}}_m,\bar{w},t)], \quad i = 1,\ldots,n. \end{aligned} \qquad (7.1.2)'$$

Substituting it back to (2.2.6)' one would not get a unique nominal trajectory $\bar{\phi}_m(\cdot)$, but a whole class $K_m(\bar{x}_m^0,t_0)$ of them. Subtracting them from $\bar{x}(t)$ to form $\bar{e}(t) \overset{\Delta}{=} \bar{x}(t) - \bar{\phi}_m(t)$ yields a $\bar{w}$-family of $\bar{e}(t)$ and (7.1.3) that becomes a selector in a contingent equation. We have

$$\begin{aligned} \dot{\bar{e}} &\in \{\bar{f}(\bar{e}+\bar{\phi}_m(t),\bar{u},\bar{w},t) - \bar{f}(\bar{\phi}_m(t),\bar{u}_m,\bar{w},t) \,|\, \bar{u} \in \bar{P}(\bar{x},t), \\ &\quad \bar{\phi}_m(\cdot) \in K_m(\bar{x}_m^0,t_0), \, \bar{w} \in W\} \end{aligned} \qquad (7.1.10)$$

with a $\bar{w}$-family of solutions $\bar{\phi}_e(\bar{e}^0,t_0,\cdot)$, which is to be either asymptotically stabilized about $\bar{e}(t) \equiv 0$ or captured in some T_c about this trivial solution. The method again does not differ from those discussed and illustrated in Chapter 5, so we do not elaborate on it any further.

In all our control problems so far, the control variables $\bar{u}(t)$ did change along the motion according to feedback programs, but the programs themselves have been designed and established off-line, even before the controlling process had begun. Such programs are sometimes called *memoryless programs*, see Corless-Leitmann [1],[2],[3]. They are less efficient and more costly in robustness against uncertainty. The effectiveness of control is certainly increased if we can adjust the program on-line, realizing the so called *signal adaptive control* usually in relation to changing parameters of the system, the latter possibly adjusted as well. The change is made according to a designed *adaptation law* which specifies some dynamics over the set of variable parameters. In general terms, (2.2.6)' is replaced by

$$\dot{\bar{x}} \in \{\bar{f}(\bar{x},\bar{u},\bar{w},\bar{\lambda},t) \,|\, \bar{u} \in \bar{P}(\bar{x},\bar{\lambda},t), \, \bar{w} \in W\} \qquad (7.1.11)$$

with the vector of adjustable parameters $\bar{\lambda}(t) \overset{\Delta}{=} (\lambda_1(t),\ldots,\lambda_l(t))^T \in \Lambda \subset \mathbb{R}^l$

where Λ is a given bounded set of values representing the parameter constraints. The adaptation law may be, again generally, expressed by the equation

$$\dot{\bar{\lambda}} = \bar{f}_a(\bar{x},t,\bar{\lambda}) \tag{7.1.12}$$

with solutions $\bar{\phi}_\lambda(\lambda^0,t_0,\cdot) : \mathbb{R} \to \Lambda$ for a suitably designed $\bar{f}_a(\cdot)$. The *adaptive controller*

$$\bar{u}(t) \in \bar{P}(\bar{x},\bar{\lambda},t) \tag{7.1.13}$$

may obviously be used in almost all of the so far discussed objectives, and in particular in the presently described path tracking. The topic has already a vast literature. The reader may like to see Le Borgue-Ibarra-Espiau [1] for review, and also Abele-Sturz [1], Aksenov-Formin [1], Gusev-Timofeev-Yakubovich [1], Corless-Leitmann [1]-[5], Corless-Leitmann-Ryan [1], Cvetkovic-Vukobratovic [1], Gusev-Yakubovich [1], Hanafi-Wright-Hewit [1], Horowitz-Tomizuka [1], Kulinich-Panev [1], Koivo-Paul [1], Luh-Lin [1], Ryan-Leitmann-Corless [1], Slotine-Sastry [1], Stoten [1], Takegaki-Arimoto [1],[2], Timofeev [1], Timofeev-Ekalo [1], Tkachenko-Brovinskaya-Kondratenko [1], to name a few. The technique leaves a choice in the type of the adaptive signals used, for example, local parametric optimization rules (see Asher-Matuszewski [1], Barnard [1], Kulinich-Panev [1]), Liapunov design (see Lindorf-Caroll [1]) or Popov hyperstability (see Jumarie [1]).

On the other hand, one may immediately observe that once $\bar{\gamma}(t)$ is made into a nominal trajectory of some dynamical system, we may as well not only track such a trajectory but also some set of *nominal parameters* $\bar{\lambda}_m = (\lambda_{m_1},\ldots,\lambda_{m\ell})^T \in \Lambda$ of the system, either constant or time varying, that is, we may require

$$\bar{e}(t) \to 0 , \quad \bar{e}_\lambda \overset{\Delta}{=} \bar{\lambda}(t) - \lambda_m \to 0 \tag{7.1.14}$$

as $t \to \infty$, or capturing $\bar{e}(t),\bar{e}_\lambda(t)$ in some $T_{\mathbf{c}}$ about zero. The parameter tracking is not necessary, but it makes the control more efficient. It amounts to tracking a desired dynamics rather than a path. We may then immediately ask the question whether the dynamical system to be followed must necessarily be the same type as the control system to follow. We answer the question negatively, that is, admitting an arbitrary system, perhaps much simpler, as the target, provided there is enough compatibility between the controlled and target systems for achieving the convergence of states and parameters, e.g. (7.1.14), cf. Shaked [1]. With this interpretation, our control objective changes to tracking of a *reference dynamical model*

$$\dot{\bar{x}}_m = \bar{f}_m(\bar{x}_m, \bar{u}_m, \bar{\lambda}_m) \qquad\qquad (7.1.15)$$

with trajectories $\bar{\phi}_m(\bar{x}_m^0, \cdot) : \mathbb{R} \to \Delta$, rather than a nominal trajectory. Obviously, in particular, $\bar{f}_m(\cdot)$ may simply be our tracking $\bar{f}(\cdot)$ with the *reference signal* or nominal control $\bar{u}_m$ substituted. Note that (7.1.15) allows for this option. The scheme of action is seen in Fig. 7.1.

Again in particular, the reference signal $\bar{u}_m$ may appear in (7.1.15) only as a factor reducing the misalignment between the two inputs and is then assumed as equal to the currently used control $\bar{u}_m(t) \equiv \bar{u}(t)$.

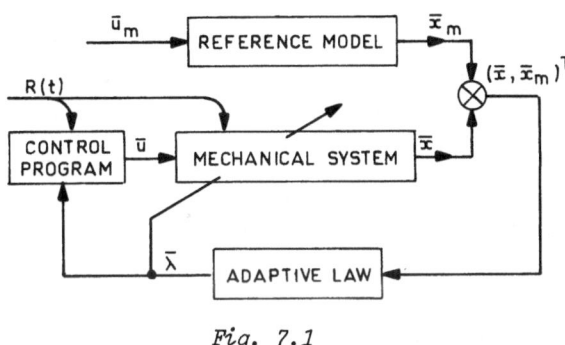

Fig. 7.1

In recent years, model tracking has become a very popular way of controlling mechanical structures. It began with autopilots, primarily in VTOL aircraft, see Narendra-Tripathi [1], and soon had been applied in all aircraft industries, see Landau-Curtiol [1], Kaufmann-Berry [1], Stein-Hartmann-Hedrick [1], Athans-Castanon-Dunn [1], Azab-Nouh [1], Vassar-Sherwood [1], Kanai-Uchikado [1], Ridgely-Banda [1], as well as other technologies, see for general reference Barnard [1], Winsor-Roy [1], Donaldson-Leondes [1], Hague-Monopoli [1], Van Amerongen-Nieuwenhuis [1], Laskin-Sirlin [1], just to name a few. Very good references to work at early stages is given in Asher-Ardasini-Dorato [1]. Recently the model following control dominates robotics, for references, see Skowronski [32], [38], and active control of space structures, see Nurre-Ryan-Scofield-Sims [1], Barnard [1], Klimentov-Prokopov [1], Kosut et al [1].

Model tracking is realized basically along two different methodologies:

    1. Self-Tuning Regulation (STR), and

    2. Model Reference Adaptive Control (MRAC).

The STR is stochastic in nature and usually reduces to three steps. First, we choose the parametric structure in such a way as to be able to adapt a time-discrete representation. Second, we estimate on-line the system

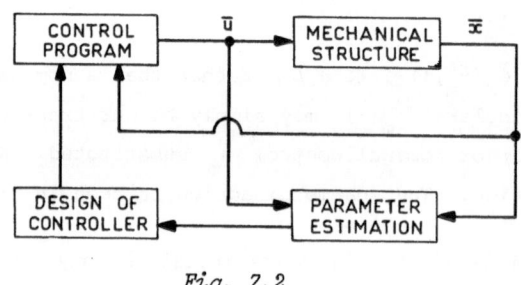

Fig. 7.2

parameters by using least squares or the maximum likelihood technique.
Third, we design the on-line controller based on the estimated parameters.
To do so, for linear systems it is possible to use the so called minimum
variance or pole-placement techniques. The block diagram of STR is shown
in Fig. 7.2. A very good review of the up-to-date development of STR may
be found in Tosunoglu-Tesar [1] or in Astrom [1]. The latter also compares
STR with MRAC.

MRAC is deterministic and allows parameter identification, if useful,
by an adjoint prediction technique which is also adaptive, and in practical
terms removes all influence of uncertainty in the system. MRAC started
long before STR but had been developed for a long time only for the class
of linear systems. It is described in Section 7.3.

Closing this section, we must comment on the option of being able to
extend our path and model following methods to the case of *path* and *model*
avoidance. The technique is entirely symmetric. This is another advantage
of using the control methods outlined in this section. Very little has
been developed in path and model avoidance otherwise, in spite of the
obvious fact that technological applications of control for such an objective are abundant - it is as much necessary to avoid an undesired path as
it is to follow a desired one.

The first group of methods in path avoidance relies, similarly as for
tracking, on considering the *anti-target path* $\bar{\gamma}_A(t)$ as T_A, envelop it
with a suitable A and use the conditions for avoidance described in
Chapter 6. The implementation of the above is immediate and it is left to
the reader. Still within this group of methods for avoiding T_A an adaptive controller may be used, interested readers are advised to read Corless-
Leitmann-Skowronski [1]. Model avoidance will require a different
approach and is discussed in Section 7.5.

EXERCISES 7.1

7.1.1 Find the control program that makes the system

$$\dot{x}_1 = x_2 , \quad \dot{x}_2 = -x_2 - ax_1 + bx_1^3 - cx_1^5 + u ,$$

with $a,b,c > 0$, $|u| \leq \hat{u}$, to follow the curve $\bar{\gamma} : ax_1^2 + x_2^2 = 2r^2$,
$r = $ const. Describe the state space pattern and determine the
region of controllability.

7.1.2 Design the controller which forces the system

$$\begin{cases} \ddot{q}_1 + d|\dot{q}_1|\dot{q}_1 + k_1 q_1 + k_{12}(q_1 - q_2) - b_{12}(q_1 - q_2)^3 = u_1 \\ \ddot{q}_2 + k_2 q_2 + k_{21}(q_2 - q_1) + b_{21}(q_2 - q_1)^3 = 0 , \end{cases}$$

with $k_{12} = -k_{21} < 0$, $b_{12} = -b_{21} < 0$, $k_1, k_2 > 0$, $d > 0$, to
track its own conservative subsystem on the global region Δ_L.

7.2 CONFIGURATION PARAMETERS

It is interesting to see which parameters of the system can be changed
to cause substantial and effective adjustment, that is, where we actually
locate our $\lambda_i(t)$, $i = 1,\ldots,l$ in the considered dynamics.

In terms of the specific model (2.2.1)', the selector equation of
(7.1.11):

$$\dot{\bar{x}} = \bar{f}(\bar{x},\bar{u},\bar{w},\bar{\lambda},t) \qquad\qquad (7.2.1)$$

will, in general, take the format

$$\left. \begin{array}{l} \ddot{\bar{q}} + \bar{\Gamma}(\bar{q},\dot{\bar{q}},\bar{w},\bar{\lambda}) + \bar{D}(\bar{q},\dot{\bar{q}},\bar{w},\bar{\lambda}) + \bar{\Pi}(\bar{q},\bar{w},\bar{\lambda}) \\ \quad = \bar{F}(\bar{q},\dot{\bar{q}},\bar{u},\bar{w},\bar{\lambda}) + \bar{R}(\bar{q},\dot{\bar{q}},\bar{w},\bar{\lambda},t) \end{array} \right\} \qquad (7.2.2)$$

and similarly for (2.2.2)'. In the above, all the characteristics include
$\bar{\lambda}$ not because the corresponding forces are adjustable, but because in
particular the inertia of the system could be made adjustable, and hence
subdividing by such inertia when decoupling the system dynamically, we make
all the characteristics related to $\bar{\lambda}$ in a purely formal way. Physically,
the option of adjusting parameters is limited to those of damping (viscosity,
general resistance, lubrication, etc.) and potential forces. In the latter
case, adjustment would mean a change in gravity via compensation, payload,
flight altitudes, etc., generating the change in inertia mentioned above.
For the case of elastic forces, we can expect adjustment in parameters
related to suspension of all kinds, joining of substructures, geometric

constraints, thus also vibration modes, and other types of what we call *configuration parameters*. Let us elaborate briefly on the case.

In general terms, the *configuration parameters describe relative positions of subsystems*, thus determining the configuration of the total system. In particular, it leads to specific constraints imposed upon $\bar{q}$, in turn resulting in corresponding reaction forces which, possibly, replace such constraints. To give examples, we may quote bearings transverse positions of a rotor or a crankshaft, shifts and tilts of bearings, cf. Parszewski-Krotkiewski [1], dimensions or positions of the bearing pedestals, interconnections between a turntable, tool supporting structure and tools in a lathe, cf. Parszewski-Chalko [1], positions of joints (lengths of rods or links) of a truss or a robotic arm, cf. our Section 1.6, Fig. 1.24, to name a few.

The reader may also recall our Section 5.7 and the eccentricity dependent radial stiffness $\tilde{K}_F$ of the suspending fluid film, see Fig. 5.23(a), which obviously is one of the force characteristics influencing the motions under study. In this sense the eccentricity is a configuration parameter acting on the joint between the bearing and the shaft. From our discussion in Section 5.7, it is clear that the eccentricity influences the amplitude of vibrations (whirl-whip) acting through $\tilde{K}_F$. In fact it may cause the same effect by producing reaction forces in the bearing. It is particularly significant in a multi-bearing shaft, see Parszewski [1]. We may thus think of controlling the vibrations by adjustment of the eccentricity.

Krotkiewski [1] verifies the above experimentally on a stand symbolically shown in Fig. 7.3 with the inner surface of the bearing replaced by an elastically suspended panel, whose deflections are measured by sensors S and which changes the eccentricity by changing its position. This change is adaptively controlled using two x,y-axial actuators F_x, F_y.

The stand is also a good illustration of the *electromagnetic suspension* with the electromagnetic field replacing the fluid, thus making the suspension and eccentricity more easily accessible for direct active control. We mentioned the case in Section 5.7 quoting the relevant literature. Such a direct control, without adaptation, simply ignores the influence of external forcing, assuming the system autonomous and thus acting through internal parameters. With the task of controlling the self-sustained part of vibrations, as discussed in Section 5.7, such posing of the problem suffices. However, for externally forced systems, the parameters should retain their parametric character while still influencing the controller.

This is done via some dynamics imposed on them and generated by the
adaptation mechanism (7.1.12), for our particular case shown in Fig. 7.3.
The controller generating an external forcing, a torque, acting upon the
shaft is also independently related to the state of the system through
another part of the control program which refers to other than eccentricity
influenced features of the motion concerned. In this setting, the resultant signal adaptive feedback control is more effective, as it may reflect
all aspects of the complex structural model concerned and of the composite
objective. Such a controller is shown in Fig. 7.3

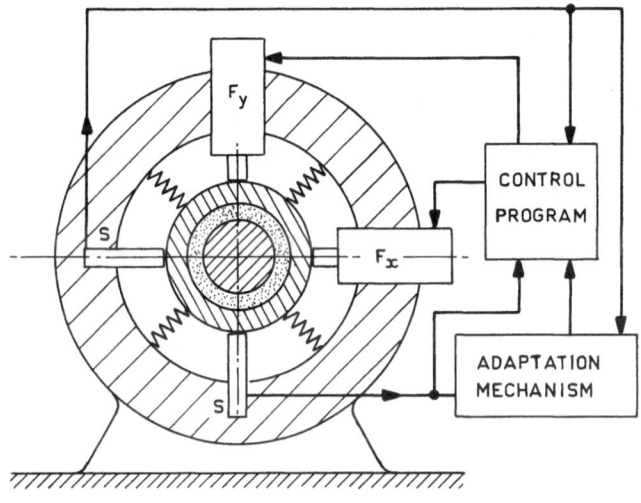

Fig. 7.3

The above is in fact applicable fairly generally. An adaptive, that
is, on-line adjustment of the configuration parameters of arbitrary nature
leads in general to on-line relative reconfiguration of subsystems, hence
also to *restructuring*. In control theoretic terms this means self-
organization of the system. It gives an additional *option* of *control via
selection and change of geometric constraints* specified by an extra designed
dynamics in terms of the adaptive laws (7.1.12).

The functional shape of potential forces, and thus potential characteristics, is determined by static measurements, that is, when the system
is assumed to be in equilibrium. This refers, in particular, to the
configuration parameters being part of such characteristics. When the
number of parameters does not exceed the number of the imposed equilibrium
conditions, the change of such parameters is of little effect on the system

dynamics, cf. Parszewski [1]. For instance, in the multi-bearing flexible
shaft system, a small variation in the length of links will change only
slightly the positions of the nodes, or small traverse displacement (shifts
and tilts) of bearings will have little effect on stiffness and damping,
provided the varying parameters are statically determinable, that is, may
be found from static equations of the system. In such circumstances adding
the adaptation mechanism to the external (time explicit dependent) feedback
control of the system is not really effective and thus not economic as well.

When, however, the number of parameters exceeds the number of static
conditions, that is, when the parameters become statically undeterminable
or *hyperstatic*, additional conditions can be imposed on the system
influencing the dynamics significantly. These conditions may be of three
kinds: Give *fixed values* of the parameters, give *algebraic conditions*
guaranteeing their move along a specific curve, or *give their dynamics*,
that is, establish them as solutions to some differential equation, e.g.
(7.1.12). The latter, obviously, makes them adaptive.

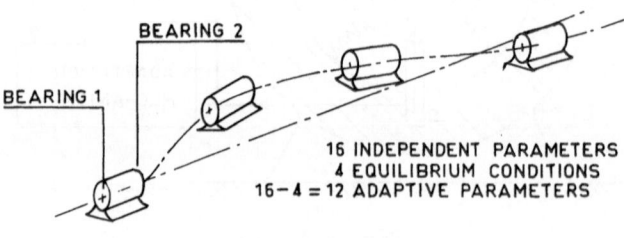

BEARING 2

BEARING 1

16 INDEPENDENT PARAMETERS
4 EQUILIBRIUM CONDITIONS
16 − 4 = 12 ADAPTIVE PARAMETERS

Fig. 7.4

EXAMPLE 7.2.1. In the multi-bearing rotor support system shown in Fig. 7.4,
see Parszewski [1], the shaft and each bearing may be considered a separate
subsystem with regard to shifts and tilts of the bearings. Then the number
of independent configuration parameters is four times the number of bearings, whereas the number of equilibrium conditions is four: two per axial
plane. The positions of centers of any two bearings are fixed and provide
a reference axis.

Moreover the conditions on the shaft and bearings can be expressed in
terms of the distribution of reactions in bearings. The reactions determine
bending moments, deflections of the shaft and the eccentricity parameters
concerned. It is shown, thus, that some reactions can be taken as controlling parameters, see Parszewski-Krotkiewski [1]. □

7.3 MODEL REFERENCE ADAPTIVE CONTROL

The method began in 1966 with the work by Butchard-Shackcloth [1] and Parks [1]. It then developed primarily along the avenue of applying the Liapunov formalism called *Liapunov Design*, see Hang-Parks [1], James [1], Monopoli [1], Narendra-Valavani [1] for exposition of basic facts in this direction. The other two avenues of research: Popov Hyperstability Theory and Variable Structure Systems (VSS) were developed later and on the margin of the Liapunov Design, cf. Taran [1], Emelianov [1], Drazenovic [1], Utkin [1]. The basic monograph on MRAC and the account of work in all directions was written by Landau [2], while a very good review of earlier works can be found in Landau [1]. Unfortunately most of the useable results, almost all until recently, are obtained for linear systems. This is due to the fact that the classical MRAC technique is based on the error equation resulting from subtracting the dynamics of the model from that of the controlled system. Such an equation is very difficult to obtain in terms of the error variable $\bar{e} \triangleq \dot{\bar{x}} - \bar{x}_m$ for nonlinear functions $\bar{f}(\bar{x},\bar{u},\bar{w},\bar{\lambda},t)$ and $\bar{f}_m(\bar{x}_m,\bar{u}_m,\bar{\lambda}_m)$. A nonlinear extension of MRAC was needed to cater for the real-life problems, see Choe-Nikiforuk [1]. The results toward such an extension are far less numerous than those on classical MRAC. Early suggestions belong to Lindorf [1], Krutova-Rutkovskii [1]-[5], Lowe-Rowland [1], Lal-Mehrorta [1], Klimentov-Prokopov [1], all including nonlinearities as partial characteristics.

A number of approximate techniques (local linearization, cancelling nonlinearity by the controller, decoupling) appeared later in MRAC applied to nonlinear robotic manipulators, see Tomizuka-Horowitz [1], Dubovsky-DesForges [1], Balestrino-DeMaria-Sciavicco [1], Stoten [1], Balestrino-DeMaria-Zinober [1], Erzberger [1], Liegeois-Fournier-Aldon [1]. The idea of abandoning the error equation altogether and replacing the asymptotic convergence of $\bar{e}(t)$ to $\bar{e}(t) \equiv 0$ by such convergence to a "diagonal" set in the product space of $(\bar{x},\bar{x}_m)^T$ belongs to Skowronski [26],[34], slightly later and in a different way suggested by Jayasuria-Rabins-Barnard [1]. It was later developed and applied to robotic manipulators and flexible structures by Skowronski [35]-[38],[40],[42],[44],[46]. Prior to this development, the nonlinear MRAC has been designed for the Leitmann system in Corless-Goodall-Leitmann-Ryan [1]. Finally Flashner-Skowronski [1],[2], Skowronski-Singh [1], returned to the error equation which appears to be easily handled for Hamiltonian models of mechanical systems.

We begin the description of MRAC with a brief outline of the linear technique, perhaps in a slightly simplified format. The block scheme is shown in Fig. 7.5. We consider the linear system (3.1.4) as

$$\dot{\bar{x}} = A(t)\bar{x} + B(t)\bar{u} \qquad (7.3.1)$$

where as before A, B are the system matrices of appropriate dimensions. We let the reference model be given as

$$\dot{\bar{x}}_m = A_m \bar{x}_m + B_m \bar{u}_m \qquad (7.3.2)$$

where $\bar{x}_m(t) \in \Delta$ is the model state vector, $\bar{u}_m(t) \in U$ model control vector while A_m, B_m are constant matrices of the same dimensions as $A(t), B(t)$. The control $\bar{u}_m$ is assumed to have been already selected to secure some desired behavior of the model, usually $\bar{u}_m \equiv \bar{u}$.

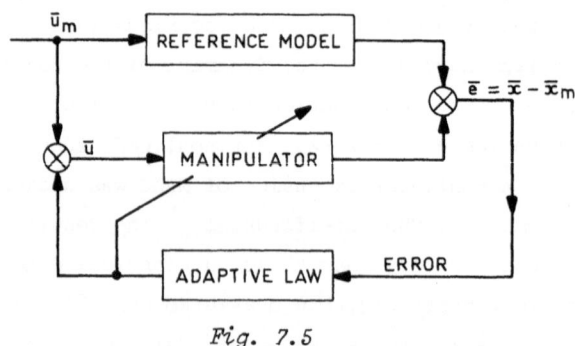

Fig. 7.5

We denote the state error again by $\bar{e}(t) = \bar{x}(t) - \bar{x}_m(t)$ for all $t \geq t_0$ and the parameter error by $A_e(t) \triangleq A(t) - A_m$, $B_e(t) \triangleq B(t) - B_m$, with $A(t), B(t)$ adjustable, $t \geq t_0$. Subtracting (7.3.2) from (7.3.1) gives the mentioned *error equation*

$$\dot{\bar{e}} = A_m \bar{e} + A_e(t)\bar{x} + B_e(t)\bar{u} . \qquad (7.3.3)$$

To concentrate on tracking rather than other fringe objectives, we assume that the control vector $\bar{u}$ has been already selected to secure some objective which is not incompatible with tracking. The plant system (7.3.1) follows the model, or more exactly in this linear version, the plant converges asymptotically to the model if

$$\bar{e}(t) \to 0, \quad A_e(t) \to 0, \quad B_e(t) \to 0 \quad \text{as} \quad t \to \infty \qquad (7.3.4)$$

which thus becomes our present tracking objective. The uniform asymptotic stability of the zero-error $\bar{e}(t) \equiv 0$ is attained when we have a Liapunov function that is a nesting square form and possesses a negative definite derivative. The obvious candidate is the traditional quadratic form

$$V(e, A_e, B_e) = \bar{e}^T P \bar{e} + \sum_{i=1}^{N} \sum_{j=1}^{N} a_{eij}^2 + \sum_{i=1}^{N} \sum_{j=1}^{n} b_{eij}^2 \qquad (7.3.5)$$

where $P = (p_{ij})$ is a positive definite symmetric matrix and a_{eij}, b_{eij} are components of $A_e(t), B_e(t)$, respectively. The only condition to check is the negative derivative, the other two being obviously satisfied. Differentiating (7.3.5) and substituting (7.3.3), we have

$$\dot{V}(t) = (A_m \bar{e} + A_m \bar{x} + B_e \bar{u})^T P \bar{e} + \bar{e}^T P (A_m \bar{e} + A_e \bar{x} + B_e \bar{u})$$
$$+ 2 \sum_i \sum_j a_{eij} \dot{a}_{eij} + 2 \sum_i \sum_j b_{eij} \dot{b}_{eij} \ ,$$

or

$$\left. \begin{array}{l} \dot{V}(t) = \bar{e}^T (A_m^T P + P A_m) \bar{e} + \bar{x}^T A_e^T (P + P^T) \bar{e} + \bar{u}^T B_e^T (P + P^T) \bar{e} \\[2mm] \qquad + 2 \sum_i \sum_j a_{eij} \dot{a}_{eij} + 2 \sum_i \sum_j b_{eij} \dot{b}_{eij} \ . \end{array} \right\} \qquad (7.3.6)$$

The negative definiteness of the first term on the right hand side of (7.3.6) follows immediately from the Liapunov Matrix Equation (3.1.5). Granted this, we need to reduce the other terms to zero. Simple calculation shows that the latter happens if the *adaptive laws* hold,

$$\left. \begin{array}{l} \dot{a}_{eij} = \frac{1}{2} x_j \sum_{k=1}^{N} (p_{ik} + p_{ki}) e_k \ , \\[3mm] \dot{b}_{eij} = \frac{1}{2} u_j \sum_{k=1}^{r} (p_{ik} + p_{ki}) e_k \ . \end{array} \right\} \qquad (7.3.7)$$

Note that since A_m, B_m are constant, we have $\dot{a}_{eij} = \dot{a}_{ij}$, $\dot{b}_{eij} = \dot{b}_{ij}$, where $\dot{a}_{ij}, \dot{b}_{ij}$ are coefficients of $A(t), B(t)$.

When (7.3.1), (7.3.2) are replaced by the nonlinear selector equation (7.2.1) and the reference model (7.1.15), respectively, the subtraction of (7.1.15) does not generate such "nice" functions of $\bar{e}(t)$ on the right hand side, like in the linear case. In fact such subtraction may not generate any function of $\bar{e}(t)$ at all. Consequently it may become feasible to use the *product-state-space technique* mentioned at the opening of this section.

Before describing the technique, let us make a few assumptions about the reference model (7.1.15). First, since it is our design, there is no uncertainty involved, and the model should be made as simple as possible to reduce computation time. We shall also tend to make it autonomous, and with well defined trajectories, but it will help considerably if it is *equilibria compatible* with the controlled system, that is, potential characteristics of the model are such that it has the same number of equilibria in similar locations or, at best, the same equilibria as the controlled system. If the equilibria of the model are for some reason prescribed, we can make the parameters of potential characteristics of the controller system adjusted so as to change its equilibria, see Section 2.1. Rewriting (7.1.15) in the mechanical format, we assume in general

$$\ddot{q}_{mi} + D_{mi}(\bar{q}_m, \dot{\bar{q}}_m, \bar{\lambda}_m) + \Pi_{mi}(\bar{q}_m, \bar{\lambda}_m) = F_{mi}(\bar{q}_m, \dot{\bar{q}}_{mi}, \bar{u}_m) , \quad i = 1, \ldots, n \quad (7.3.8)$$

where $\bar{q}_m(t) \in \Delta_q$, $\dot{\bar{q}}_m(t) \in \Delta_{\dot{q}}$ are the Lagrangian model displacement and velocity variables, and $D_{mi}(\cdot), \Pi_{mi}(\cdot)$, the non-potential (damping) and potential characteristics. Then for compatibility we assume the equilibria defined by the same relations as for the controlled system:

$$\left.\begin{aligned}
&\dot{q}_{mi} = 0 , \\
&\Pi_{mi}(\bar{q}_m^e, 0) = 0 , \\
&D_{mi}(\bar{q}_m^e, 0) = 0 , \quad \forall \bar{q}_m^e , \\
&F_{mi}(\bar{q}_m^e, 0, \bar{u}_m) = 0 , \quad \forall \bar{q}_m^e, \bar{u}_m ,
\end{aligned}\right\} \quad (7.3.9)$$

with $\bar{q}_m^e$ denoting the equilibrium positions of the model. If no adjustment of $\bar{q}^e$ is made by adaptation: $\bar{q}^e \neq \bar{q}_m^e$, then we better have the a-priori assertion that $\bar{q}^e = \bar{q}_m^e$.

Next, it is feasible to assume that we do not make the tracking all over $\mathbb{R}^N$, or even all over Δ. Let us take a subset Δ_0 of Δ, possibly equal to Δ, on which the tracking should occur and thus in which the model trajectories must be bounded by design. We assume Δ_0 strongly positively invariant under trajectories of (7.1.15), recall Section 3.1, or which is the same, positively Lagrange stable:

$$\bar{x}_m^0 \in \Delta_0 \implies \bar{\phi}_m(\bar{x}_m^0, t) \in \Delta_0 , \quad \forall t \geq t_0 , \quad t_0 \in \mathbb{R}, \quad (7.3.10)$$

inconsequent of whether this has been achieved by the choice of $\bar{u}_m, \bar{\lambda}_m$ or of $\bar{f}_m(\cdot)$ in (7.1.15). Trajectories do not leave Δ_0 unless passing through the part of the neighborhood $N(\partial\Delta_0)$ of the boundary $\partial\Delta_0$ which belongs to $\bar{\Delta}_0$, that is, the set $N(\partial\Delta_0) \cap \bar{\Delta}_0$. We shall stop them from doing so, if there is a C^1-function $V_m(\cdot) : N(\partial\Delta_0) \cap \bar{\Delta}_0 \rightarrow \mathbb{R}$, with

$$v_m = \inf V_m(\bar{x}_m) \,|\, \bar{x}_m \in \partial\Delta_0 \quad (7.3.11)$$

such that for all points $\bar{x}_m \in N(\partial\Delta_0) \cap \bar{\Delta}_0$,

$$\left.\begin{aligned}
&\text{(i)} \quad V_m(\bar{x}_m) \leq v_m , \\
&\text{(ii)} \quad \dot{V}_m(\bar{x}_m) < 0 .
\end{aligned}\right\} \quad (7.3.12)$$

Indeed, if any model trajectory $\bar{\phi}_m(\bar{x}_m^0, \mathbb{R}^+)$ from $\bar{x}_m^0 \in \Delta_0$, crosses $\partial\Delta_0$, then there is $t_1 > 0$ such that, by (i) we have $V_m(\bar{\phi}_m(t)) \geq V_m(\bar{x}_m^0)$, which contradicts (ii).

Consider now the total energy of the model $E_m(\bar{x}_m)$ together with the corresponding energy surface H_m. Since (7.3.8) has the same number of equilibria as (7.1.15), the extrema of H_m are close to those of H, with a correction to E^+ or E^- when our aim is to bound motions below (or above) some Z_C. Design of the regular regions of H_m is quite arbitrary, but we may like to follow the shape of H.

Choosing $V_m(\bar{x}_m) = E_m(\bar{x}_m)$, the conditions (i),(ii) hold if we design the model such that for all $\bar{x}_m \in N(\partial\Delta_0) \cap \bar{\Delta}_0$, we have

$$\nabla E_m(\bar{x}_m)^T \bar{x}_m \geq 0 , \qquad\qquad (7.3.13)$$

$$\nabla E_m(\bar{x}_m)^T f_m(\bar{x}_m,\bar{u}_m,\bar{\lambda}_m) < 0 . \qquad\qquad (7.3.14)$$

This corresponds to Δ_0 located in an energy-cup of H_m about some Dirichlet stable equilibrium (minimum). According to what we said about following the shape of H, such a cup would naturally belong to a local or in-the-large cup of H. If so, then we would like to invert the inequalities (7.3.13), (7.3.14), when our Δ_0 is located about a Dirichlet unstable equilibrium (threshold) of H. Indeed, observe that letting then $V_m(\bar{x}_m) = -E_m(\bar{x}_m)$ and inverting simultaneously both inequalities, the conditions (i),(ii) still hold, and the positive Lagrange stability is preserved.

The assumption (7.3.13) is slightly more demanding than the condition (2.4.21) which may be used to justify it, see Yoshizawa [1]. The properties of the energy flow which had led to (2.4.21) are valid here as well.

When $\partial\Delta_0$ is specified in terms of an E_m-level, it may be shown that (7.3.13), (7.3.14), or their inverted counterparts, are also necessary for the positive Lagrange stability on Δ_0 located in a cup or about a threshold respectively.

Our technique is based on the fact that $\bar{x} = \bar{x}_m$ is attained on a "generalized diagonal" set in the space of product vectors $(\bar{x},\bar{x}_m)^T$. Let us thus introduce the vector $\bar{X}_m(t) \triangleq (\bar{x}(t),\bar{x}_m(t))^T$ ranging in $\Delta^2 \triangleq \Delta \times \Delta \subset \mathbb{R}^{2N}$ and form the product system of (7.1.11), (7.1.12), (7.1.15) but with a well specified model (7.1.15) and with the adaptive laws (7.1.12) built-in, so that we obtain the vector $\bar{F} = (\bar{f},\bar{f}_m,\bar{f}_a)^T$. Then let $\bar{\alpha}(t) \triangleq \bar{\lambda}(t) - \bar{\lambda}_m$, $\bar{\lambda}_m = const$, generating $\dot{\bar{\alpha}}(t) = \dot{\bar{\lambda}}(t)$, $t \geq t_0$. The vector $\bar{\alpha}$ ranges in Λ as much as $\bar{\lambda}$ did. The above allows us to form the product selector equation

$$(\dot{\bar{x}}_m,\dot{\bar{\alpha}})^T = \bar{F}(\bar{x}_m,\bar{\alpha},\bar{u},\bar{w},t) \qquad (7.3.15)$$

of the contingent product system

$$(\dot{\bar{x}}_m,\dot{\bar{\alpha}})^T \in \{\bar{F}(\bar{x}_m,\bar{\alpha},\bar{u},\bar{w},t) \,|\, \bar{u} \in \bar{P}(\bar{x}_m,\bar{\alpha}), \bar{w} \in W\} \ , \qquad (7.3.16)$$

with the Filippov-type solutions $\bar{\phi}(\bar{x}_m^0,\bar{\alpha}^0,t_0,\cdot) : \mathbb{R} \to \Delta^2 \times \Lambda$, $\bar{x}_m^0 = \bar{x}_m(t_0)$, $\bar{\alpha}^0 = \bar{\alpha}(t_0)$, $t_0 \in \mathbb{R}$, generating at each $(\bar{x}_m^0,\bar{\alpha}^0) \in \Delta^2 \times \Lambda$ the class of motions $K(\bar{x}_m^0,\bar{\alpha}^0)$. Then we define the "diagonal" set in $\Delta^2 \times \Lambda$,

$$M = \{(\bar{x}_m,\bar{\alpha}) \in \Delta^2 \times \Lambda \,|\, \bar{x} = \bar{x}_m \,, \bar{\alpha} = 0\} \qquad (7.3.17)$$

and its μ-neighborhood:

$$M_\mu = \{(\bar{x}_m,\bar{\alpha}) \in \Delta^2 \times \Lambda \mid \|\bar{x} - \bar{x}_m\| < \mu,\ \|\alpha\| < \mu\} \qquad (7.3.18)$$

with $\|\cdot\|$ being a norm in $\mathbb{R}^{2N}$, $\mathbb{R}^l$ respectively, and $\mu > 0$ being a stipulated constant specifying the estimate of the precision in tracking. The following definition describes our present objective of stabilized tracking.

DEFINITION 7.3.1. The system (7.1.11) tracks the reference model (7.1.15) on Δ_0 with precision $\mu > 0$, if and only if there is a control program $\bar{P}(\cdot)$, an adaptive law $\bar{f}_a(\cdot)$ and a time interval $T_\mu < \infty$, such that Δ_0 is strongly positively invariant under $\bar{P}(\cdot)$ and $\bar{\phi}(\bar{x}_m^0,\bar{\alpha}^0,\cdot) \in K(\bar{x}_m^0,\bar{\alpha}^0)$, $(\bar{x}_m^0,\bar{\alpha}^0) \in \Delta_0^2 \times \Lambda$ implies

$$\bar{\phi}(\bar{x}_m^0,\bar{\alpha}^0,t) \in M_\mu \ , \quad \forall t \geq t_0 + T_\mu \ . \qquad (7.3.19)$$

In the above $\Delta_0^2 \overset{\Delta}{=} \Delta_0 \times \Delta_0$. When T_μ is stipulated, we refer to tracking *"after a given T_μ"* .

The block scheme of tracking is seen in Fig. 7.6. Given the set Δ_0 where we want the tracking to occur, let $N[\partial(\Delta_0^2 \times \Lambda)]$ be a neighborhood

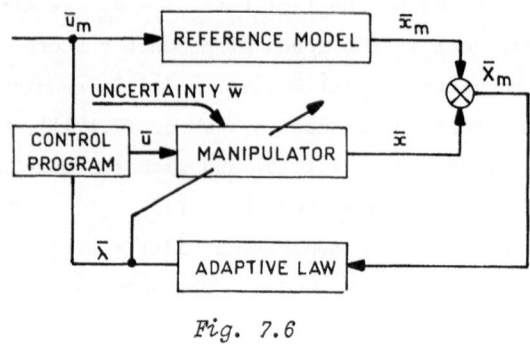

Fig. 7.6

of the boundary $\partial(\Delta_0^2 \times \Lambda)$ of the region $\Delta_0^2 \times \Lambda$ in $\mathbb{R}^{2N+l}$, and define the semi-neighborhood $N_s = N[\partial(\Delta_0^2 \times \Lambda)] \cap \overline{\Delta_0^2 \times \Lambda}$ and the relative complement $CM_\mu = (\Delta_0^2 \times \Lambda) - M_\mu$. Moreover, let $D(open) \supset \overline{CM}_\mu$ be such that $D \cap M = \phi$, and introduce two C^1-functions $V_s(\cdot) : N_s \to \mathbb{R}$ and $V_\mu(\cdot) : D \to \mathbb{R}$

$$
\left.
\begin{aligned}
v_s &= V_s(\bar{X}_m, \bar{\alpha}) \,|\, \forall (\bar{X}_m, \bar{\alpha}) \in \partial(\Delta_0^2 \times \Lambda) \ , \\
v_\mu^- &= \inf V_\mu(\bar{X}_m, \bar{\alpha}) \,|\, (\bar{X}_m, \bar{\alpha}) \in \partial M_\mu \cap \overline{CM}_\mu \ , \\
v_\mu^+ &= \sup(\bar{X}_m, \bar{\alpha}) \,|\, (\bar{X}_m, \bar{\alpha}) \in \partial(\Delta_0^2 \times \Lambda) \cap \overline{CM}_\mu \ .
\end{aligned}
\right\}
\tag{7.3.20}
$$

The first relation requires forming $V_s(\cdot)$ from a suitable $\partial(\Delta_0^2 \times \Lambda)$ taken as a level curve of this function, or forming the boundaries of Δ_0, Λ from level curves of suitable $V_s(\cdot)$. In the latter case, smaller Δ_0, Λ than those really desired will be the secure choice, see Fig. 7.7.

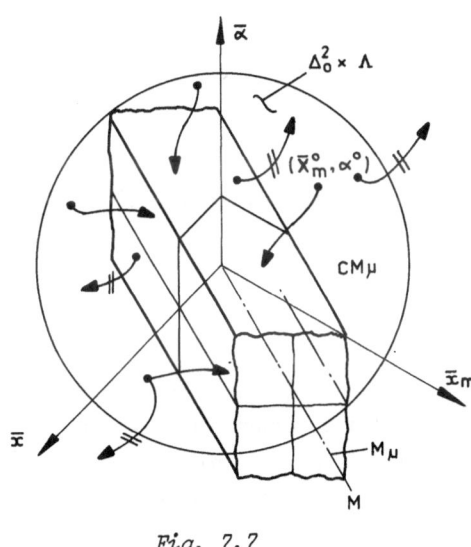

Fig. 7.7

The following conditions have been proved in Skowronski [34]-[38].

CONDITIONS 7.3.1. The system (7.1.11) is strongly controllable on Δ_0 for tracking the model (7.1.15) according to Definition 7.3.1, if given Δ_0, Λ, μ there is $\bar{P}(\cdot)$ and two C^1-functions $V_s(\cdot), V_\mu(\cdot)$ as defined above such that for all $(\bar{X}_m, \bar{\alpha}) \in \Delta_0^2 \times \Lambda$ we have

(i) $\quad V_s(\bar{X}_m, \bar{\alpha}) \le v_s$, $\quad \forall \ (\bar{X}_m, \bar{\alpha}) \in N_s$;

(ii) $\quad \forall \ \bar{u} \in \bar{P}(\bar{X}_m, \bar{\alpha})$,

$$
\nabla V_s(\bar{X}_m, \bar{\alpha})^T \cdot \bar{F}(\bar{X}_m, \bar{\alpha}, \bar{u}, \bar{w}, t) < 0 \ , \quad \forall \ \bar{w} \in W ;
\tag{7.3.21}
$$

(iii) $\quad 0 \le V_\mu(\bar{X}_m, \bar{\alpha}) < v_\mu^+$, $\quad \forall \ (\bar{X}_m, \bar{\alpha}) \in \overline{CM}_\mu$;

(iv) $\quad V_\mu(\bar{x}_m,\bar{\alpha}) \le v_\mu^-$, $\quad (\bar{x}_m,\bar{\alpha}) \in D \cap M_\mu$;

(v) $\quad \forall\ \bar{u} \in \bar{P}(\bar{x}_m,\bar{\alpha})$ there is $c(\|(\bar{x}_m,\bar{\alpha})\|) > 0$, continuously increasing, such that

$$\nabla V_\mu(\bar{x}_m,\bar{\alpha})^T \cdot \bar{F}(\bar{x}_m,\bar{\alpha},\bar{u},\bar{w},t) \le -c(\|(\bar{x}_m,\bar{\alpha})\|) \ , \ \forall\ \bar{w} \in W . \quad (7.3.22)$$

The first two conditions (i),(ii) imply the positive strong invariance of $\Delta_0^2 \times \Lambda$. Indeed, suppose that some product motion $\bar{\phi}(\bar{x}_m^0,\bar{\alpha}^0,\mathbb{R}^+)$ from N_s crosses $\partial(\Delta_0^2 \times \Lambda)$. Then by (i) there is $t_1 > t_0$ such that $V_s(\bar{\phi}(t_1))$ $= v_s \ge V_s(\bar{\phi}(t_0))$ which contradicts (ii), see Fig. 7.7. Granted the positive strong invariance of $\Delta_0^2 \times \Lambda$, let us consider an arbitrary $\bar{\phi}(\bar{x}_m^0,\bar{\alpha}^0,\mathbb{R}^+)$ from CM_μ . Integrating (7.3.22) along such a motion, we obtain the estimate of time spent in CM_μ :

$$t \le t_0 + \frac{[V_\mu(\bar{x}_m^0,\bar{\alpha}^0) - V_\mu(\bar{x}_m,\bar{\alpha})]}{c(\|\bar{x}_m,\bar{\alpha})\|)} \ . \quad (7.3.23)$$

By (iii), $\quad V_\mu(\bar{x}_m,\bar{\alpha}) \ge 0$, $\quad V_\mu(\bar{x}_m^0,\bar{\alpha}^0) - v_\mu^+ \le 0$ whence $V_\mu(\bar{x}_m^0,\bar{\alpha}^0) - V_\mu(\bar{x}_m,\bar{\alpha})$ $\le v_\mu^+$, and from (7.3.23), $\quad t \le t_0 + (v_\mu^+/c^-)$, $\quad c^- = \inf c(\|(\bar{x}_m,\bar{\alpha})\|) \,|\, (\bar{x}_m,\bar{\alpha})$ $\in \Delta_0^2 \times \Lambda$, which implies that there is

$$T_\mu = v_\mu^+ / c^- \ , \quad (7.3.24)$$

depending only upon the diameter of $\Delta_0^2 \times \Lambda$ and independent on the product motions, such that for any $t \ge t_0 + T_\mu$, the motions are in M_μ . There is no return to CM_μ as upon such return there would be $t_3 > t_2 \ge t_0 + T_\mu$, such that some $\bar{\phi}(\bar{x}_m^0,\bar{\alpha}^0,\mathbb{R}^+)$ from M_μ crosses ∂M_μ upon which by (iv), $V_\mu(\bar{\phi}(\bar{x}_m^0,\bar{\alpha}^0,t_3)) = v_\mu^- \ge V_\mu(\bar{x}_m^0,\bar{\alpha}^0)$ contradicting (v), and completing the proof.

REMARK 7.3.1. When $T_\mu > 0$ is stipulated, $c > 0$ is found from (7.3.24), and (7.3.22) becomes:

$$\nabla V_\mu((\bar{x}_m,\bar{\alpha})^T \cdot \bar{F}(\bar{x}_m,\bar{\alpha},\bar{u},\bar{w},t)) \le -(v_\mu^+ / T_\mu) \ , \quad (7.3.25)$$

for all $\bar{w} \in W$ and, together with other Conditions 7.3.1, implies the hypothesis.

Using the same argument as for Corollaries in Chapter 5, we obtain the following corollary which gives the program $\bar{P}(\bar{x}_m,\bar{\alpha})$. Denote

$$L_s(\bar{x}_m,\bar{\alpha},t,\bar{u},\bar{w}) \overset{\Delta}{=} \nabla V_s(\bar{x}_m,\bar{\alpha})^T \cdot \bar{F}(\bar{x}_m,\bar{\alpha},t,\bar{u},\bar{w}) \ ,$$

$$L_\mu(\bar{x}_m,\bar{\alpha},t,\bar{u},\bar{w}) \overset{\Delta}{=} \nabla V_\mu(\bar{x}_m,\bar{\alpha})^T \cdot \bar{F}(\bar{x}_m,\bar{\alpha},t,\bar{u},\bar{w}) \ .$$

COROLLARY 7.3.1. *Given* $(\bar{X}_m^0, \alpha^0) \in CM_\mu$, $t_0 \in \mathbb{R}$, *if there is a pair* $\bar{u}^*, \bar{w}^*$ *such that*

$$L_s(\bar{X}_m, \bar{\alpha}, t, \bar{u}^*, \bar{w}^*) = \min_{\bar{u}} \max_{\bar{w}} L_s(\bar{X}_m, \bar{\alpha}, t, \bar{u}, \bar{w}) < 0 , \qquad (7.3.26)$$

$$L_\mu(\bar{X}_m, \bar{\alpha}, t, \bar{u}^*, \bar{w}^*) = \max_{\bar{u}} \min_{\bar{w}} L_\mu(\bar{X}_m, \bar{\alpha}, t, \bar{u}, \bar{w}) \leq -c^- , \qquad (7.3.27)$$

then the conditions (ii), (v) are met with $\bar{u}^* \in \bar{P}_*(\bar{X}_m, \bar{\alpha})$ *and one can deduce* $\bar{P}_*(\cdot)$ *from (7.3.26), (7.3.27).*

Observe that the argument in the proof of Conditions 7.3.1 leading to contradiction of (i) and (ii), remains valid when we invert both inequalities. This implies the following.

COROLLARY 7.3.2. *Conditions 7.3.1 hold if the inequalities in (i) and (ii) are both simultaneously inverted.*

Let us now take Δ_0 located in an energy cup and propose the functions

$$V_s(\bar{X}_m, \bar{\alpha}) \stackrel{\Delta}{=} E_m(\bar{x}) + E_m(\bar{x}_m) + \bar{a}\bar{\alpha} , \qquad (7.3.28)$$

$$V_\mu(\bar{X}_m, \bar{\alpha}) \stackrel{\Delta}{=} \begin{cases} |E_m(\bar{x}) - E_m(\bar{x}_m)| + \bar{a}\bar{\alpha} , & \forall (\bar{X}_m, \bar{\alpha}) \in CM_\mu \\ \bar{a}\bar{\alpha} , & \forall (\bar{X}_m, \bar{\alpha}) \in M_\mu \cap D , \end{cases} \qquad (7.3.29)$$

where $\bar{a} \stackrel{\Delta}{=} (\text{sign } \alpha_1, \ldots, \text{sign } \alpha_l)$, with $\alpha_i(t) \neq 0$ for all $t \geq t_0$, and $E_m(\bar{x})$ is the energy function $E_m(\cdot)$ with $\bar{x}_m$ replaced by $\bar{x}$, that is, along the motions of the system (7.1.11). The condition $\alpha_i \neq 0$ follows from the adaptive laws introduced below, see (7.3.42), and is justified by the fact that adaptation is redundant for $\bar{\alpha} = 0$.

With a suitable choice of N_s covering $N(\partial\Delta_0) \cap \bar{\Delta}_0$ the assumption (7.3.13) implies (i) of Conditions 7.3.1. Also, by the character of the function of energy $E_m(\cdot)$, (iii), (iv) hold.

It remains to check (ii) and (v). To do so we differentiate (7.3.28), (7.3.29) with respect to time. Denoting $\delta E_m \stackrel{\Delta}{=} E_m(\bar{x}) - E_m(\bar{x}_m)$, one obtains

$$\dot{V}_s(\bar{X}_m, \bar{\alpha}) = \dot{E}_m(\bar{x}) + \dot{E}(\bar{x}_m) + \bar{a}\dot{\bar{\alpha}} , \qquad (7.3.30)$$

$$\dot{V}_\mu(\bar{X}_m, \bar{\alpha}) = \begin{cases} \dot{E}_m(\bar{x}) - \dot{E}_m(\bar{x}_m) + \bar{a}\dot{\bar{\alpha}} , & (\bar{X}_m, \bar{\alpha}) \in CM_\mu , \text{ for } \delta E_m \geq 0 , \\ \dot{E}_m(\bar{x}_m) - \dot{E}_m(\bar{x}) + \bar{a}\dot{\bar{\alpha}} , & (\bar{X}_m, \bar{\alpha}) \in CM_\mu , \text{ for } \delta E_m < 0 , \\ \bar{a}\dot{\bar{\alpha}} , & \text{ for } (\bar{X}_m, \bar{\alpha}) \in M_\mu \cap D , \end{cases} \qquad (7.3.31)$$

where

$$\dot{E}_m(\bar{x}_m) = \nabla E_m(\bar{x}_m)^T \bar{f}_m(\bar{x}_m, \bar{\lambda}_m, \bar{u}_m) \ , \tag{7.3.32}$$

$$\dot{E}_m(\bar{x}) = \nabla E_m(\bar{x})^T \bar{f}(\bar{x}, t, \bar{u}, \bar{\lambda}, \bar{w}) \ . \tag{7.3.33}$$

The following two conditions are based on Corollary 7.3.1 and secure (ii) and (v) of Conditions 7.3.1.

(I) Control Conditions.

$$\min_{\bar{u}} \max_{\bar{w}} \nabla E_m(\bar{x})^T \bar{f}(\bar{x}, t, \bar{u}, \bar{\lambda}, \bar{w}) \leq \dot{E}_m(\bar{x}_m) \ , \quad \text{for } \delta E_m \geq 0 \ , \tag{7.3.34}$$

$$\max_{\bar{u}} \min_{\bar{w}} \nabla E_m(\bar{x})^T \bar{f}(\bar{x}, t, \bar{u}, \bar{\lambda}, \bar{w}) \geq \dot{E}_m(\bar{x}_m) \ , \quad \text{for } \delta E_m < 0 \ , \tag{7.3.35}$$

(II) Adaptation Condition.

$$a\bar{\alpha}\dot{\bar{\alpha}} \leq -|\bar{\alpha}|\lceil c^- + |\dot{E}_m(\bar{x})| \rceil \ , \quad \bar{\alpha} \neq 0 \ . \tag{7.3.36}$$

Indeed, substituting (7.3.14), (7.3.36) into (7.3.30), we check (ii) to hold on N_s. Checking (v) is immediate upon substituting (I) and (II) into (7.3.31). In the above, $c^- > 0$ is a suitable constant for (v) in Conditions 7.3.1, but is calculated from (7.3.24) for the case of stipulated T_μ.

We must now design a control program $\bar{P}(\cdot)$ implying (I), and the adaptive law $\bar{f}_a(\cdot)$ of (7.1.12) to imply (II).

Substituting (7.3.8) into (7.3.32), we obtain

$$\dot{E}_m(\bar{x}_m) = \sum_{i=1}^n [F_{mi}(\bar{q}_m, \dot{\bar{q}}_m, \bar{u}_m) - D_{mi}(\bar{q}_m, \dot{\bar{q}}_m, \bar{\lambda}_m)]\dot{q}_{mi} \ . \tag{7.3.37}$$

Similarly, (7.3.9) substituted into (7.3.33) gives

$$\dot{E}_m(\bar{x}) = \sum_{i=1}^n [F_i(\bar{q}, \dot{\bar{q}}, \bar{u}) + R_i^r(\bar{q}, \dot{\bar{q}}, \bar{w}, t) - D_i(\bar{q}, \dot{\bar{q}}, \bar{w}, \lambda)]\dot{q}_i$$

$$+ \sum_{i=1}^n [\Pi_{mi}(\bar{q}, \bar{\lambda}_m) - \Pi_i(\bar{q}, \bar{\lambda}, \bar{w})]\dot{q}_i \ , \tag{7.3.38}$$

with the proviso of $\bar{\Gamma}(\bar{q}, \dot{\bar{q}}, \bar{w})\dot{\bar{q}} \equiv 0$, based on the earlier mentioned physical assumption that the Coriolis-gyro forces do not change the energy, thus that their power vanishes.

Now, let $\bar{w}^*$ be the corresponding extremizing value of $\bar{w}$ in (7.3.34), (7.3.35). The condition (7.3.34) is satisfied if

$$\min_{\bar{u}} F_i(\bar{q},\dot{\bar{q}},\bar{w}^*,\bar{u})\dot{q}_i \leq F_{mi}(\bar{q}_m,\dot{\bar{q}}_m,\bar{u}_m)\dot{q}_{mi} - R_i(\bar{q},\dot{\bar{q}},\bar{w}^*,t)\dot{q}_i$$
$$\left. + [D_i(\bar{q},\dot{\bar{q}},\bar{\lambda},\bar{w}^*)\dot{q}_i - D_{mi}(\bar{q}_m,\dot{\bar{q}}_m,\bar{\lambda}_m)\dot{q}_{mi}] \right\} \quad (7.3.39)$$
$$+ [\Pi_i(\bar{q},\bar{\lambda},\bar{w}^*)\dot{q}_i - \Pi_{mi}(\bar{q},\bar{\lambda}_m)\dot{q}_i] , \quad i = 1,\ldots,n ,$$

for $\delta E \geq 0$, and similarly (7.3.35) is satisfied if

$$\max_{\bar{u}} F_i(\bar{q},\dot{\bar{q}},\bar{w}^*,\bar{u})\dot{q}_i \geq F_{mi}(\bar{q}_m,\dot{\bar{q}}_m,\bar{u}_m)\dot{q}_{mi} - R_i(\bar{q},\dot{\bar{q}},\bar{w}^*,t)\dot{q}_i$$
$$\left. + [D_i(\bar{q},\dot{\bar{q}},\bar{\lambda},\bar{w}^*)\dot{q}_i - D_{mi}(\bar{q}_m,\dot{\bar{q}}_m,\bar{\lambda}_m)\dot{q}_{mi}] \right\} \quad (7.3.40)$$
$$+ [\Pi_i(\bar{q},\bar{\lambda},\bar{w}^*)\dot{q}_i - \Pi_{mi}(\bar{q},\bar{\lambda}_m)\dot{q}_i] , \quad i = 1,\ldots,n ,$$

for $\delta E < 0$. Specifying the gear function slightly further by collating the actuators with DOF, we may assume $F_i(\bar{q},\dot{\bar{q}},\bar{u},\bar{w}^*) \stackrel{\Delta}{=} B_i(\bar{q},\dot{\bar{q}},\bar{w}^*)u_i$, where $B_i(\cdot)$ are positive functions, with $r = n$, see Section 3.3. Then we may design $\bar{P}(\cdot)$ from (7.3.39) as

$$u_i \, \text{sgn} \, \dot{q}_i \begin{cases} \leq \dfrac{1}{B_i|\dot{q}_i|} \, [F_{mi}\dot{q}_{mi} - R_i\dot{q}_i + D_i\dot{q}_i - D_{mi}\dot{q}_{mi} + \Pi_i\dot{q}_i \\ \qquad\qquad - \Pi_{mi}(\bar{q})\dot{q}_i] , \; \forall \; |\dot{q}_i| \geq \beta_i ; \qquad (7.3.41) \\[2mm] = \text{suitable constant} , \; \forall \; |\dot{q}_i| < \beta_i , \; i = 1,\ldots,n , \end{cases}$$

for $\delta E \geq 0$. A similar controller is obviously obtained for $\delta E < 0$ from (7.3.40).

In the above β_i are calculated as many times before and the arguments of the functions dropped for clarity of expression whenever confusion may not occur. The inequality in (7.3.41) reflects the fact that the program is set valued.

When $r < n$, for some i we would have to design in (7.3.41) a sum over $n - r$ terms; apart from that, the procedure is the same. Obviously the above applies also to the $\delta E < 0$ counterpart of (7.3.41).

Turning now to the design of an adaptive law, generally written as (7.1.12), observe that (7.3.36) is implied by

$$\dot{\alpha}_i = -\alpha_i[c^- + |\dot{E}_m(\bar{x})|] , \quad \alpha_i^0 = \alpha_i(t_0) \neq 0, \; i = 1,\ldots,\ell .$$

Indeed, substituting $\alpha_i = |\alpha_i| / \text{sgn} \, \alpha_i$ into (7.3.42) and summing up over i , we obtain

$$\bar{a}\dot{\bar{\alpha}} = -[c^- + |\dot{E}_m(\bar{x})|] \sum_i |\alpha_i| .$$

Observe further that given $\bar{x}(t)$, (7.3.42) is a first order linear homogeneous equation in $\alpha_i(t)$ with exponential solutions in closed format, whose sign depends upon our choice of the initial α_i^0 : sgn α_i^0 = sgn $\alpha_i(t)$, $\forall t \geq t_0$.

Note here, that as long as the adaptive laws (7.3.42) imply the adaptation condition (II), (7.3.36), which in turn, together with (I) implies Conditions 7.3.1, we have the main goal of adaptation satisfied – and the fact that the solutions of (7.3.42) decay asymptotically to zero producing $\lambda(t) \rightarrow \lambda_m$ is secondary. In fact, in many applied scenarios, it may be irrelevant, $\bar{\lambda}_m$ may be ignored, technically put to zero, whence (7.3.42) would become a dynamics imposed upon $\overset{\leftarrow}{\lambda}(t)$ alone. In such a case it may be preferable to redesign (7.3.42) to, say,

$$\dot{\lambda}_i = c^- + \left| \dot{E}_m(\bar{x}) \right| , \quad i = 1,\ldots,l \tag{7.3.42}'$$

which gives the adaptation condition (II) as

$$\bar{a}\dot{\lambda} \leq -(c^- + \left| \dot{E}_m(\bar{x}) \right|)$$

which for suitable c^- also satisfies (ii),(v) when substituted jointly with (I). The preference follows the argument that we do not necessarily want to have $\lambda(t) \rightarrow 0$ during the tracking. In general, the suggested law depends on the designer, who must rely on physical requirements exercising his choice.

EXAMPLE 7.3.1.

A n-joint planar, open chain arm in a robotic system can be represented by a combination of two DOF, RP-units (Rotary-Prismatic) displayed in Fig. 7.8. Let us investigate such a unit. We take the Lagrange or joint

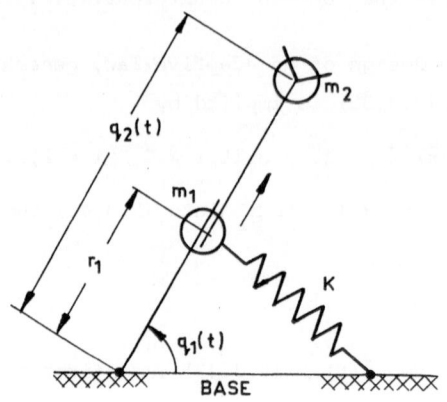

Fig. 7.8

coordinates $q_1 \overset{\Delta}{=} \theta$, $q_2 \overset{\Delta}{=} r$ as indicated, with $r = $ const at the rotary joint 1. The equations of motion (1.5.19) give

$$(m_1 r_1^2 + m_2 q_2^2)\ddot{q}_1 + 2m_2 q_2 \dot{q}_1 \dot{q}_2 + \lambda_1 |\dot{q}_1| \dot{q}_1 + g(m_1 r_1 + 2m_2 q_2) \cos q_1$$
$$- m_1 g r_1 + a q_1 + b q_1^3 = u_1 + Q_1^F(t) \qquad (7.3.43)$$

$$m_2 \ddot{q}_2 - m_2 q_2 \dot{q}_1^2 + \lambda_2 \dot{q}_2 + 2m_2 g \sin q_1 = u_2 + Q_2^R(t)$$

where now $Q_i^R(t)$, $i = 1,2$, represent external perturbations, and u_1, u_2 are actuator controls at joints 1,2 respectively. The equations (7.3.43) represent the plant.

We take the possible payload on the gripper as uncertain but within a known bound, which makes the mass m_2 and the perturbations Q_1^R, Q_2^R unknown but bounded in some known W, specified by

$$m^- \le m_2 \le m^+, \quad Q^- \le Q_i^R \le Q^+, \quad i = 1,2, \qquad (7.3.44)$$

with m^-, m^+, Q^-, Q^+ positive constants. Consequently inertia coefficients $(m_1 r_1^2 + m_2 q_2^2)$, m_2 and the gravity force $m_2 g \sin q_1$ are noise polluted. On the other hand, since the inertia coefficients are non-negative, we may subdivide the equations (7.3.43) by these coefficients respectively, thus decoupling the system inertially but noise polluting the other terms. Let us do this, and then aggregate all the potential forces (gravity and spring) into the potential characteristic functions:

$$\Pi_1(\bar{q}) = \frac{2g(m_1 r_1 + m_2 q_2) \cos q_1 - m_1 g r_1 + a q_1 + b q_1^3}{m_1 r_1^2 + m_2 q_2^2} \left.\vphantom{\frac{\frac{a}{a}}{\frac{a}{a}}}\right\}$$

$$\Pi_2(\bar{q}) = 2g \sin q_1 . \qquad (7.3.45)$$

Allowing just one rotation (three equilibria), we let $\sin q_1 = q_1 - q_1^3/6$, $\cos q_1 = 1 - q_1^2/2$. The characteristics (7.3.45) become

$$\Pi_1(\bar{q}) = \frac{a q_1 - \frac{1}{2} g m_1 r_1 q_1^2 + b q_1^3 - g m_2 q_2 q_1^2 + 2g m_2 q_2}{m_1 r_1^2 + m_2 q_2^2} \left.\vphantom{\frac{\frac{a}{a}}{\frac{a}{a}}}\right\}$$

$$\Pi_2(\bar{q}) = 2g q_1 - (g q_1^3/3) , \qquad (7.3.46)$$

which generates for (7.3.45) the three equilibria

$$q_1^e = 0, \pm \sqrt{6}, \left.\vphantom{\frac{\frac{a}{a}}{\frac{a}{a}}}\right\}$$

$$q^e = \frac{1}{4m_2 g} [\pm\sqrt{6}(a + 6b) - 3 g m_1 r_1] . \qquad (7.3.47)$$

Then the aggregate of all the nonpotential internal forces (centrifugal, damping) in terms of characteristic functions is:

$$\Phi_1 = \Gamma_1(\bar{q},\dot{\bar{q}}) + D_1(\bar{q},\dot{\bar{q}},\lambda_1) = \frac{2m_2 q_2 \dot{q}_1 \dot{q}_2 + \lambda_1 |\dot{q}_1| \dot{q}_1}{m_1 r_1^2 + m_2 q_2^2}$$

$$\Phi_2 = \Gamma_2(\bar{q},\dot{\bar{q}}) + D_2(\bar{q},\dot{\bar{q}},\lambda_2) = -q_2 \dot{q}_1^2 + \frac{1}{m_2} \lambda_2 \dot{q}_2 \; . \tag{7.3.48}$$

Finally, denoting $\bar{w}(t) \overset{\Delta}{=} (m_2, Q_1^R(t), Q_2^R(t))^T$ and

$$R_1(q_2,\bar{w}) = \frac{Q_1^R(t)}{m_1 r_1^2 + m_2 q_2^2} \; , \qquad \tilde{u}_1 = \frac{u_1}{m_1 r_1^2 + m_2 q_2^2}$$

$$R_2(\bar{w}) = Q_2^R(t)/m_2 \; , \qquad \tilde{u}_2 = u_2/m_2 \; , \tag{7.3.49}$$

we obtain for (7.3.43) the motion equations

$$\ddot{q}_i + \Phi_i(\bar{q},\dot{\bar{q}},\lambda_i,m_2) + \Pi_i(\bar{q},m_2) = R_i(q_2,\bar{w}) + \tilde{u}_i(q_2,\bar{w},t) \; , \tag{7.3.50}$$
$$i = 1,2 \; .$$

The reference model is now taken as the following:

$$\ddot{q}_{m1} + \lambda_{m1} \dot{q}_{m1} + aq_{m1} - \tfrac{1}{2} gr_1 q_{m1}^2 + bq_{m1}^3 - gq_{m2} q_{m1}^2 + 2gq_m = 0$$

$$\ddot{q}_{m2} + \lambda_{m2} \dot{q}_{m2} + 2gq_{m1} - (gq_{m1}^3/3) = 0 \; , \tag{7.3.51}$$

which has the same equilibria (7.3.47) for $\alpha_i \to 0$, $i = 1,\ldots,4$, provided $m_1 = 1$ and we approximate $m_2 = 1$. The total energy of the model is

$$E_m(\bar{q}_m,\dot{\bar{q}}_m) = \tfrac{1}{2}(\dot{q}_{m1}^2 + \dot{q}_{m2}^2) + \tfrac{1}{2} aq_{m1}^2 - \tfrac{1}{6} gr_1 q_{m1}^3 + \tfrac{1}{4} bq_{m1}^4$$
$$- \tfrac{1}{3} gq_{m2} q_{m1}^3 + 2gq_{m1} q_{m2} \; , \tag{7.3.52}$$

with the time derivative

$$\dot{E}_m(\bar{q}_m,\dot{\bar{q}}_m) = -(\lambda_{m1} \dot{q}_{m1}^2 + \lambda_{m2} \dot{q}_{m2}^2) \leq 0 \; , \tag{7.3.53}$$

securing the dissipativeness of the model, when $\lambda_{m1}, \lambda_{m2} > 0$. We also obtain

$$\dot{E}_m(\bar{q},\dot{\bar{q}}) = \sum_i [\tilde{u}_i + R_i - D_i + \Pi_{mi}(\bar{q},\bar{\lambda}_m) - \Pi_i] \dot{q}_i \tag{7.3.54}$$

where

$$\Pi_{m1}(\bar{q},\bar{\lambda}_m) = aq_1 - (gr_1 q_1^2/2) + bq_1^3 - gq_2 q_1^2 + 2gq_2 \; ,$$
$$\Pi_{m2}(\bar{q},\bar{\lambda}_m) = \Pi_2(\bar{q}) \; .$$

We want the manipulator to follow the model with the error less than $\mu > 0$ and not later than $t = t_\mu$. Two following conditions imply condition (I) of our case:

$$\min_{\bar{u}} \max_{\bar{w}} \left[\frac{-u_1\dot{q}_1 + Q_1^R\dot{q}_1 - \lambda_1|\dot{q}_1|\dot{q}_1^2}{m_1 r_1^2 + m_2 q_2^2} + (\Pi_{m_1} - \Pi_1)\dot{q}_1 \right] \le -\lambda_{m_1}\dot{q}_{m_1}^2$$

$$\min_{\bar{u}} \max_{\bar{w}} \left[\frac{-u_2\dot{q}_2 + Q_2^R\dot{q}_2 - \lambda_2\dot{q}_2^2}{m_2} \right] \le -\lambda_{m2}\dot{q}_{m2}^2 \qquad (7.3.55)$$

for $\delta E \ge 0$, and

$$\max_{\bar{u}} \min_{\bar{w}} \left[\frac{-u_1\dot{q}_1 + Q_1^R\dot{q}_1 - \lambda_1|\dot{q}_1|\dot{q}_1}{m_1 r_1^2 + m_2 q_2^2} + (\Pi_{m_1} - \Pi_1)\dot{q}_1 \right] \ge -\lambda_{m_1}\dot{q}_{m_1}^2$$

$$\max_{\bar{u}} \min_{\bar{w}} \left[\frac{-u_2\dot{q}_2 + Q_2^R\dot{q}_2 - \lambda_2\dot{q}_2^2}{m_2} \right] \ge -\lambda_{m2}\dot{q}_{m2}^2 \qquad (7.3.56)$$

for $\delta E < 0$. Substituting (7.3.44), conditions (7.3.55) are implied by

$$u_1^*\,\text{sgn}\,\dot{q}_1 \begin{cases} \le [(m_1 r_1^2 + m^- q_2^2)/|\dot{q}_1|]\cdot[(\Pi_1^* - \Pi_{m_1})\dot{q}_1 - \lambda_m\dot{q}_{m1}^2] \\ \qquad + \lambda_1\dot{q}_1^2 - (Q_1^*\dot{q}_1/|\dot{q}_1|)\,, \quad \forall\ |\dot{q}_1| \ge \beta_1\,, \\ = \text{suitable constant}\,, \quad \forall\ |\dot{q}_1| < \beta_1\,; \end{cases} \qquad (7.3.57)$$

$$u_2^*\,\text{sgn}\,\dot{q}_2 \begin{cases} \le -(\lambda_{m2}\dot{q}_{m2}^2 m^-/|\dot{q}_2|) + (\lambda_2\dot{q}_2 + Q_2^*)(\dot{q}_2/|\dot{q}_2|)\,, \\ \qquad\qquad\qquad \forall\ |\dot{q}_2| \ge \beta_2\,, \\ = \text{suitable constant}\,, \quad \forall\ |\dot{q}_2| < \beta_2\,; \end{cases} \qquad (7.3.58)$$

where $\Pi_1^* = \max_{m_2} \Pi_1(\bar{q}, m_2)$, $Q_i^*, \tilde{u}_i^*$ denote the extremizing values of $Q_i^R, \tilde{u}_i$, respectively and where the distances β_1, β_2 to the zero-velocity surface are calculated as always before.

The relations (7.3.57), (7.3.58) give the set valued control program $\bar{P}(\cdot)$ defined for $\delta E \ge 0$. An identical argument leads to relations implying (7.5.56) and thus giving the control program for the set defined by $\delta E < 0$ with the control functions u_i^* estimated by inverted inequalities (7.3.57), (7.3.58) with inverted estimates for m_2, Q^R. The adaptive laws become in our example

$$\dot{\alpha}_i = -\alpha_i[(v_\mu^+/t_\mu) + |\dot{E}_m(\bar{q}, \dot{\bar{q}})|]\,, \quad i = 1, 2\,, \qquad (7.3.59)$$

with $\dot{E}_m(\bar{q}, \dot{\bar{q}})$ defined by (7.3.54) and v_μ^+ found from $\partial\Delta_0 = \partial\Delta_L$ specified by

$$\dot{E}_m(\bar{q}, \dot{\bar{q}}) + \dot{E}_m(\bar{q}_m, \dot{\bar{q}}_m) + \lambda_1\,\text{sgn}\,\alpha_1 + \lambda_2\,\text{sgn}\,\alpha_2 = d$$

for a suitable constant $d > 0$, which yields $v_\mu^+ = v_s = d$. The numerical simulation was done for the following data:

$$g = 9.81 \text{ m/sec}^2 \ , \quad m_1 = 10.00 \text{ kg} \ , \quad m_2 \in [0.5,10] \ ,$$

$$\lambda_1 = 6.0 \text{ kg m}^2 \ , \quad \lambda_2 = 3.0 \text{ kg/sec} \ , \quad r_1 = 0.66 \text{ m} \ ,$$

$$a = 500.0 \text{ kg m}^2/\text{sec}^2 \ , \quad b = 3.0 \text{ kg m}^2/\text{sec}^2 \ ,$$

$$\lambda_{m_1} = 5.0 \text{ kg m}^2 \ , \quad \lambda_{m_2} = 2.0 \text{ kg/sec} \ , \quad d = 500 \ .$$

The integration method used was Runge-Kutta with step-size 0.02 sec.
Figure 7.9 shows the trajectories.

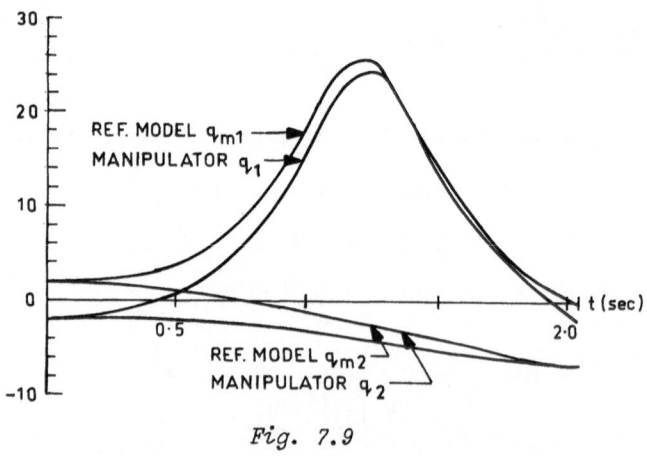

Fig. 7.9

7.4 MODEL TRACKING BY HAMILTONIAN SYSTEMS

The product-state-space method discussed in Section 7.2 has the disadvantage of doubling the dimension of the considered set of motion equations.
This is an unwelcome feature with regard to computation time. As we may
see in Skowronski [50], the problem can be solved approximately by using
less-dimensional observes to produce the feedback information needed for
controllers. On the other hand, using the error equation, we remain within
the dimensionality of the original system, and there have been at least a
few successful attempts to handle systems with untruncated nonlinearity, and
not necessarily decoupled, via the error equation method, see Johnson [1],
Krutova-Rutkovski [3]-[5], Lal-Mehrorta [1]. It also appeared recently, see
Flashner-Skowronski [1], that the error equation can be constructed for a
very large class of mechanical systems covered by the Hamiltonian format.
We describe the case briefly below.

Recall Section 1.7 with the motion equations (1.7.10), and recall its contingent augmentation (2.2.3)'. Introducing the parameters $\bar{\lambda}$ we obtain

$$
\left.
\begin{aligned}
\dot{q}_i &= \frac{\partial H(\bar{q},\bar{p},\bar{w},\bar{\lambda})}{\partial p_i} \quad , \\
\dot{p}_i &= -\frac{\partial H(\bar{q},\bar{p},\bar{w},\bar{\lambda})}{\partial q_i} + Q_i^D(\bar{q},\bar{p},\bar{w},^-) + Q_i^F(\bar{q},\bar{p},\bar{u},\bar{w}) + Q_i^R(\bar{q},\bar{p},\bar{w},t) \quad ,
\end{aligned}
\right\} \quad (7.4.1)
$$

$$
i = 1,\ldots,n \quad ,
$$

which serve now as the selector equations of the corresponding contingent subcase of (2.2.6)'. Note here, that since the control program is $\bar{\lambda}$-adjustable, $\bar{u} \in \bar{P}(\bar{x},t,\bar{\lambda})$, $\bar{x} = (\bar{q},\bar{p})^T$, then $Q_i^F(\cdot)$ is indirectly $\bar{\lambda}$-adjustable. The reference model (7.1.15) specified in terms of the Hamiltonian format becomes

$$
\left.
\begin{aligned}
\dot{q}_{mi} &= \frac{\partial H_m(\bar{q}_m,\bar{p}_m,\bar{\lambda}_m)}{\partial p_{mi}} \\
\dot{p}_{mi} &= -\frac{\partial H_m(\bar{q}_m,\bar{p}_m,\bar{\lambda}_m)}{\partial q_{mi}} + Q_{mi}^D(\bar{q}_m,\bar{p}_m,\bar{\lambda}_m)
\end{aligned}
\right\} \quad (7.4.2)
$$

assuming the reference input signal $\bar{u}_m \equiv 0$, and with $\bar{q}_m(t) \in \Delta_q$, $\bar{p}_m(t) \in \Delta_{\dot{q}}$ being the model displacements and momentum vectors. We uphold all the model assumptions made in Section 7.3. Then we define the misalignments in displacements and momenta as $\bar{e}_q(t) \overset{\Delta}{=} \bar{q}(t) - \bar{q}_m(t)$, and $\bar{e}_p(t) \overset{\Delta}{=} \bar{p}(t) - \bar{p}_m(t)$, respectively, let $\bar{\alpha}(t) \overset{\Delta}{=} \bar{\lambda}(t) - \bar{\lambda}_m$, as before, and form the sum of the Hamiltonians concerned

$$
H_e(\bar{e}_q,\bar{e}_p,\bar{q}_m,\bar{p}_m,\bar{w},\bar{\alpha}) \overset{\Delta}{=} H(\bar{q},\bar{p},\bar{w},\bar{\lambda}) + H_m(\bar{q}_m,\bar{p}_m,\bar{\lambda}_m) \quad . \qquad (7.4.3)
$$

Observe that $H_e(\cdot)$ is a function of $\bar{e}_q,\bar{e}_p$ through $\bar{q},\bar{p},\bar{q}_m,\bar{p}_m$ with parameters $\bar{\lambda}_m,\bar{q}_m,\bar{p}_m,\bar{w},\bar{\lambda}$. Of those, $\bar{\lambda}_m$ is given by design, $\bar{q}_m(t),\bar{p}_m(t)$ are known from (7.4.2) which is off-line calculable, if not directly designed. In particular, $\bar{q}_m(t),\bar{p}_m(t)$ may be given as the desired $\bar{\gamma}(t)$ filtered through the reference signal $u_m(t)$ as discussed before. Then $\bar{w}$ is unknown but bounded in its known range W and $\bar{\lambda}(t)$ is adjustable, thus for our present purposes, optional to the designer and varying in the fixed range Λ . The function $H_e(\cdot)$ is obviously obtained by substituting $\bar{q} = \bar{e}_q + \bar{q}_m$, $\bar{p} = \bar{e}_p + \bar{p}_m$, $\bar{\lambda} = \bar{\alpha} + \bar{\lambda}_m$ into $H(\bar{q},\bar{p},\bar{w},\bar{\lambda})$ of (7.4.3). Assuming, for all our practical purposes, that the Hamiltonians equal the corresponding energies, we follow our previous custom and let

$$
\left.
\begin{aligned}
H^-(\bar{q},\bar{p}) &\overset{\Delta}{=} \inf_{\bar{w},\bar{\lambda}} H(\bar{q},\bar{p},\bar{w},\bar{\lambda}) \quad , \\
H_e^-(\bar{e}_q,\bar{e}_p,\bar{q}_m,\bar{p}_m,\bar{\lambda}_m) &\overset{\Delta}{=} \inf_{\bar{w},\bar{\lambda}} H_e(\bar{e}_q,\bar{e}_p,\bar{q}_m,\bar{p}_m,\bar{\alpha},\bar{w}) \quad .
\end{aligned}
\right\} \quad (7.4.4)
$$

Subtracting (7.4.2) from (7.4.1) we have

$$\dot{e}_{q_i} = \frac{\partial H(\bar{q},\bar{p},\bar{\lambda},\bar{w})}{\partial p_i} - \frac{\partial H_m(\bar{q}_m,\bar{p}_m,\bar{\lambda}_m)}{\partial p_{mi}} \quad ,$$

$$\dot{e}_{p_i} = - \frac{\partial H(\bar{q},\bar{p},\bar{\lambda},\bar{w})}{\partial q_i} + Q_i^P(\bar{q},\bar{p},\bar{w},\bar{\lambda}) + Q_i^F(\bar{q},\bar{p},\bar{u},\bar{w}) + Q_i^R(\bar{q},\bar{p},\bar{w},t)$$

$$+ \frac{\partial H_m(\bar{q}_m,\bar{p}_m,\bar{\lambda}_m)}{\partial q_{mi}} - Q_{mi}^D(\bar{q}_m,\bar{p}_m,\bar{\lambda}_m) \quad .$$

Considering now that $q_i = e_{q_i} + q_{mi}$, $p_i = e_{p_i} + p_{mi}$, and $q_{mi} = q_i - e_{q_i}$, $p_{mi} = p_i - e_{p_i}$, we have

$$\left.\begin{aligned}
\frac{\partial H_e}{\partial e_{q_i}} &= \frac{\partial H(\bar{q},\bar{p},\bar{w},\bar{\lambda})}{\partial e_{q_i}} + \frac{\partial H_m(\bar{q}_m,\bar{p}_m,\bar{\lambda}_m)}{\partial e_{q_i}} \\[2mm]
&= \frac{\partial H}{\partial q_i}\frac{\partial q_i}{\partial e_{q_i}} + \frac{\partial H_m}{\partial q_{mi}}\frac{\partial q_{mi}}{\partial e_{q_i}} = \frac{\partial H}{\partial q_i} - \frac{\partial H_m}{\partial q_{mi}} \quad .
\end{aligned}\right\} \qquad (7.4.5)$$

Then

$$\dot{e}_{p_i} = - \frac{\partial H_e}{\partial e_{q_i}} + Q_i^D + Q_i^F + Q_i^R - Q_{mi}^D \quad .$$

Similarly

$$\left.\begin{aligned}
\frac{\partial H_e}{\partial e_{p_i}} &= \frac{\partial H(\bar{q},\bar{p},\bar{w},\bar{\lambda})}{\partial e_{p_i}} + \frac{\partial H_m(\bar{q}_m,\bar{p}_m,\bar{\lambda}_m)}{\partial e_{p_i}} \\[2mm]
&= \frac{\partial H}{\partial p_i}\frac{\partial p_i}{\partial e_{p_i}} + \frac{\partial H_m}{\partial p_{mi}}\frac{\partial p_{mi}}{\partial e_{p_i}} = \frac{\partial H}{\partial p_i} - \frac{\partial H_m}{\partial p_{mi}} \quad ,
\end{aligned}\right\} \qquad (7.4.6)$$

whence

$$\dot{e}_{q_i} = \frac{\partial H_e}{\partial e_{p_i}} \quad .$$

Thus the error equations are

$$\left.\begin{aligned}
\dot{e}_{q_i} &= \frac{\partial H_e}{\partial e_{p_i}} \\[3mm]
\dot{e}_{p_i} &= - \frac{\partial H_e}{\partial e_{q_i}} + (Q_i^D - Q_{mi}^D) + Q_i^F + Q_i^R \ , \quad i = 1,\ldots,n
\end{aligned}\right\} \qquad (7.4.7)$$

representing again a canonical form with the Hamiltonian H_e, displacements e_{q_i} and momenta e_{p_i}, $i = 1,\ldots,n$. Moreover, if the damping in both, the controlled system and the reference model, is representable in terms of the Rayleigh function:

$$Q_i^D \triangleq \frac{\partial \mathcal{R}(\bar{q},\bar{p},\bar{\lambda},\bar{w})}{\partial p_i} \quad , \qquad Q_{mi}^D \triangleq \frac{\partial \mathcal{R}_m(\bar{q}_m,\bar{p}_m,\bar{\lambda}_m)}{\partial p_{mi}} \quad ,$$

then, introducing $\mathcal{R}_e \triangleq \mathcal{R} + \mathcal{R}_m$ and

$$Q_{ei}^D(\bar{e}_q,\bar{q}_p,\bar{\alpha},\bar{q}_m,\bar{p}_m,\bar{w}) \triangleq Q_i^D(\bar{q},\bar{p},\bar{\lambda},\bar{w}) - Q_{mi}^D(\bar{q}_m,\bar{p}_m,\bar{\lambda}_m) \quad , \qquad (7.4.8)$$

we have by similar derivation as for H_e :

$$Q_{ei}^D = \partial \mathcal{R}_e(\bar{e}_q,\bar{e}_p,\bar{\alpha},\bar{q}_m,\bar{p}_m,\bar{w})/\partial e_{p_i}$$

wherefrom the error equations become

$$\left. \begin{array}{l} \dot{e}_{q_i} = \dfrac{\partial H_e}{\partial e_{p_i}} \quad , \\[20pt] \dot{e}_{p_i} = -\dfrac{\partial H_e}{\partial e_{q_i}} + Q_{ei}^D + Q_i^F + Q_i^R \, , \end{array} \right\} \qquad (7.4.9)$$

where all the functions have arguments $\bar{e}_q, \bar{e}_p$ and parameters $\bar{w},\bar{\alpha},\bar{q}_m$ and $\bar{p}_m$. In further discussion, we shall use either (7.4.7) or (7.4.9). Quite obviously both these canonical forms are selector equations of a subcase of (2.2.6)', with $\bar{x} \triangleq (\bar{e}_q,\bar{e}_p)^T$. Capture of the error-motions $\bar{\phi}(\bar{e}_q^0,\bar{e}_p^0,t_0,\cdot)$ of such a system in a target defined by

$$H_e^-(\bar{e}_q,\bar{e}_p) \le \mu \qquad (7.4.10)$$

secures our tracking objective, with an extra requirement that $\alpha(t) = \lambda(t) - \lambda_m$ satisfies $|\alpha(t)| < \mu$ for $t \ge t_0 + T_c$ of such capture, if such a requirement is feasible. To achieve this task, the controller and adaptive laws will be designed. We shall use Conditions 5.5.2, applied on a set $\Delta_0 \times \Lambda$ in the space $\mathbb{R}^{2n+l}$ of vectors $(\bar{e}_q,\bar{e}_p,\bar{\alpha})^T$ and refer to (7.4.9) as well as some adaptive law of the type (7.1.12).

Observe that we can adjust the origins $\bar{q} = 0$, $\bar{p} = 0$ and $\bar{q}_m = 0$, $\bar{p}_m = 0$ so that $H_e^-(0,0) = 0$. Moreover, with the same number of extrema of $H^-(\cdot)$ and $H_m(\cdot)$, there will be a H_e^--cup surrounding $H_e^-(0,0)$ with the threshold $H_e^-(\bar{e}_q,\bar{e}_p) \triangleq h_{eE}$. The latter constant is determined by the H_e^--level passing through a Dirichlet unstable equilibrium (saddle), which is the closest to the origin local maximum of H_e^- .

We let the test function for Conditions 5.5.2 be

$$v(\bar{e}_q,\bar{e}_p,\bar{\alpha}) \triangleq H_e^-(\bar{e}_q,\bar{e}_p,\bar{q}_m,\bar{p}_m,\bar{\lambda}_m) + \bar{a}\cdot\bar{\alpha} \, , \qquad (7.4.11)$$

where $\bar{a} \triangleq (\text{sgn } \alpha_1,\ldots,\text{sgn } \alpha_l)$, as in Section 7.3. Then let us define

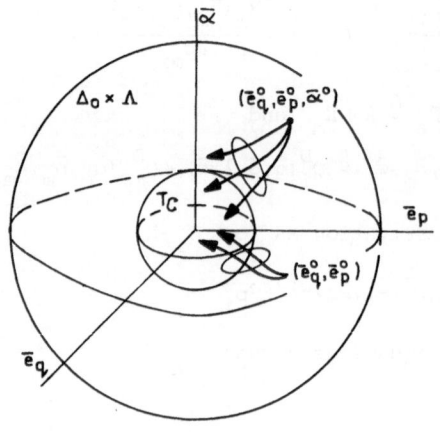

Fig. 7.10

$$\partial(\Delta_0 \times \Lambda) \ : \ V(\bar{e}_q, \bar{e}_p, \bar{\alpha}) = v_0^+ = h_{eE} + \alpha^+ \ , \qquad (7.4.12)$$

$$\partial T_{\mathfrak{C}} \ : \ V(\bar{e}_q, \bar{e}_p, \bar{\alpha}) = v_{TC} = \mu \ , \qquad (7.4.13)$$

with $\Lambda : 0 \leq \alpha_i \leq \alpha^+$, $i = 1, \ldots, l$, $\alpha^+ \in \mathbb{R}$, see Fig. 7.10. We use (5.5.2) implied by c^- defined from (5.5.6), for some $T_{\mathfrak{C}}$ required.

Since $\bar{q}_m(t), \bar{p}_m(t), \bar{\lambda}_m$ and the extremizing $\bar{\lambda}^*, \bar{w}^*$ in $H_E^-(\cdot)$ are known values, the formal differentiation of (7.4.11) gives

$$\dot{V}(\bar{e}_q, \bar{e}_p, \bar{\alpha}) = \sum_{i=1}^n [\, (Q_i^F + Q_i^R + Q_i^D - Q_{mi}^D)\dot{e}_{q_i}\,] + F + \bar{a}\dot{\bar{\alpha}}$$

where

$$F(\bar{e}_q, \bar{e}_p) \stackrel{\Delta}{=} \sum_{i=1}^n \left(\frac{\partial H^-}{\partial q_{mi}} \frac{\partial H_m}{\partial p_{mi}} - \frac{\partial H^-}{\partial p_{mi}} \frac{\partial H_m}{\partial q_{mi}} + \frac{\partial H_e^-}{\partial p_{mi}} \cdot Q_{mi}^D \right)$$

is the known perturbation due to the influence of known parameters $\bar{q}_m(t), \bar{p}_m(t)$.

Then, in order to satisfy (5.5.2)' of Corollary 5.5.1 or (5.5.8) of Corollary 5.5.3, we need

$$\left.\begin{array}{c} \min_{\bar{u}} \ \max_{\bar{w}} \ \sum_{i=1}^n [\, (Q_i^F + Q_i^R + Q_i^D - Q_{mi}^D)\dot{e}_{q_i}\,] + F + \bar{a}\dot{\bar{\alpha}} \\[3mm] \leq - \dfrac{h_{eE} + \alpha^+ - \mu}{T_{\mathfrak{C}}} \ . \end{array}\right\} \qquad (7.4.14)$$

This is obviously implied by the following two conditions.

(I)   Control Condition

$$\min_{\bar{u}} \max_{\bar{w}} \ (\bar{Q}^F + \bar{Q}^R + \bar{Q}^D) = \bar{Q}^D_m , \ \cdot \ \forall \ \dot{\bar{e}}_q \neq 0 , \qquad (7.4.15)$$

and

(II)   Adaptation Condition

$$\bar{a}\dot{\bar{\alpha}} \leq - \left(\frac{h_{eE} + \alpha^+ - \mu}{T_\mathfrak{C}} + |F| \right) |\bar{\alpha}| , \quad \forall \ \bar{\alpha} \neq 0 . \qquad (7.4.16)$$

The first is implied by the control program

$$\min_{\bar{u}} \max_{\bar{w}} \ \bar{Q}^F (\bar{q}, \bar{p}, \bar{u}, \bar{w})$$

$$= \begin{cases} \bar{Q}^D_m (\bar{q}_m, \bar{p}_m, \bar{\lambda}_m) - \max_{\bar{w}} [\bar{Q}^D (\bar{q}, \bar{p}, \bar{\lambda}, \bar{w}) + \bar{Q}^R (\bar{q}, \bar{p}, \bar{w}, t)] , \ \forall \ |\dot{\bar{e}}_q| \geq \beta, \\ \text{suitable constant} , \quad \forall \ |\dot{\bar{e}}_q| < \beta , \end{cases} \qquad (7.4.17)$$

with β obtainable as usually before.

Note that the controller (7.4.17) applies irrespective of whether or not we represent the misalignment in damping as $\bar{Q}^D - \bar{Q}^D_m$, or as the difference function $\bar{Q}^D_e(\cdot)$ derived from the Raynold's representation.

Unless $\bar{Q}^F(\cdot)$ is specified further, we cannot resolve (7.4.17) with respect to $\bar{u}$ any further than already done. Assuming, similarly as in the previous cases, $\bar{Q}^F \triangleq B(\bar{q}, \bar{p}, \bar{w})\bar{u}$ where B is a non-singular, positive definite, $N \times r$ matrix, we can implement (7.4.17) using

$$\bar{u}(t) = \begin{cases} -B^{-1}(\bar{q}, \bar{p}, \bar{w}^*) [\bar{Q}^D (\bar{q}, \bar{p}, \bar{\lambda}, \bar{w}^*) - \bar{Q}^D_m (\bar{q}_m, \bar{p}_m, \bar{\lambda}_m) \\ \qquad\qquad + \bar{Q}^R (\bar{q}, \bar{p}, \bar{w}^*, t)] , \ \forall \ |\dot{\bar{e}}_q| \geq \beta , \\ \text{suitable constant} , \quad \forall \ |\dot{\bar{e}}_q| < \beta , \end{cases} \qquad (7.4.18)$$

where $\bar{w}^*$ denotes the extremizing value of $\bar{w}(\cdot)$.

In turn, the adaptation condition (7.4.16) is implied by the following adaptive laws:

$$\dot{\alpha}_i = -\alpha_i \left[|F| + \frac{h_{eE} + \alpha^+ - \mu}{T_\mathfrak{C}} \right] , \quad \alpha_i^0 \neq 0 , \quad i = 1, \ldots, l . \qquad (7.4.19)$$

To justify the above, we use the same argument as in Section 7.3, when proving that (7.3.42) implies (7.3.36). Note that the type of equations and behavior of solutions of (7.4.19) is identical with those of (7.3.42).

355

EXAMPLE 7.4.1. Consider the scalar mechanical system

$$m\ddot{q} + w\dot{q} + \lambda(q - 2q^3 + q^5) = u \ , \tag{7.4.20}$$

with $\lambda, w \in [1,2]$ and let the reference model be

$$\ddot{q}_m + 5\dot{q}_m + 2(q_m - 2q_m^3 + q_m^5) = 0 \ . \tag{7.4.21}$$

In Hamiltonian format, these equations become

$$\left. \begin{aligned} \dot{q} &= p/m \ , \\ \dot{p} &= -\lambda(q - 2q^3 + q^5) + (wp/m) + u \ , \end{aligned} \right\} \tag{7.4.22}$$

with the Hamiltonian

$$H(q,p) = p^2/2m + \lambda \left[\tfrac{1}{2} q^2 - \tfrac{1}{2} q^4 + \tfrac{1}{6} q^6 \right] \ , \tag{7.4.23}$$

and

$$\begin{aligned} \dot{q}_m &= p_m \ , \\ \dot{p}_m &= -2(q_m - 2q_m^3 + q_m^5) + 5p_m \ , \end{aligned} \tag{7.4.24}$$

with the Hamiltonian

$$H_m(q_m, p_m) = \tfrac{1}{2} p_m^2 + 2 \left[\tfrac{1}{2} q_m^2 - \tfrac{1}{2} q_m^4 + \tfrac{1}{6} q_m^6 \right] \ . \tag{7.4.25}$$

Then considering that the potential energy in (7.4.23) is positive definite, we have the extremizing $\lambda^* = 2$ and

$$\left. \begin{aligned} \bar{H}_e(e_q, e_p, q_m, p_m) &= \frac{(e_p + p_m)^2}{2m} + (e_q + q_m)^2 - (e_q + p_m)^4 \\ &\quad + \tfrac{1}{3} (e_q + q_m)^6 + H_m(q_m, p_m) \ . \end{aligned} \right\} \tag{7.4.26}$$

Note that with $\lambda = \lambda^*$, $H(\cdot)$ and $H_m(\cdot)$ have the same extrema, that is, (7.4.20) and (7.4.21) have the same equilibria. Moreover, changes in $\lambda(t)$ do not produce incompatibility between our two systems regarding the equilibria. The control program (7.4.17) becomes

$$u = \begin{cases} 5p_m - (\max_w wp/m) \ , & \forall \ |p| \geq \beta \ ; \\ \hat{u} \ , & \forall \ |p| < \beta \ . \end{cases}$$

To calculate the adaptive law (7.4.19) we need F obtained from (7.4.25), (7.4.26):

$$\begin{aligned} F = \ & 2e_q p_m (1 + 2e_q^2 + e_q^4) + 2q_m p_m (1 + 6e_q^2 + 5e_q^4) + 4e_q p_m q_m^2 (3 + 20e_q^2) \\ & + 4q_m^3 p_m (1 + 5e_q^2) - 2(e_p + p_m)(q_m - 2q_m^3 + q_m^5)/m + 5p_m[e_p + (1 + m)p_m]/m \ . \end{aligned}$$

Substituting $m = 1.5\,\text{kg}$, we obtain trajectories shown in Fig. 7.11. $\square$

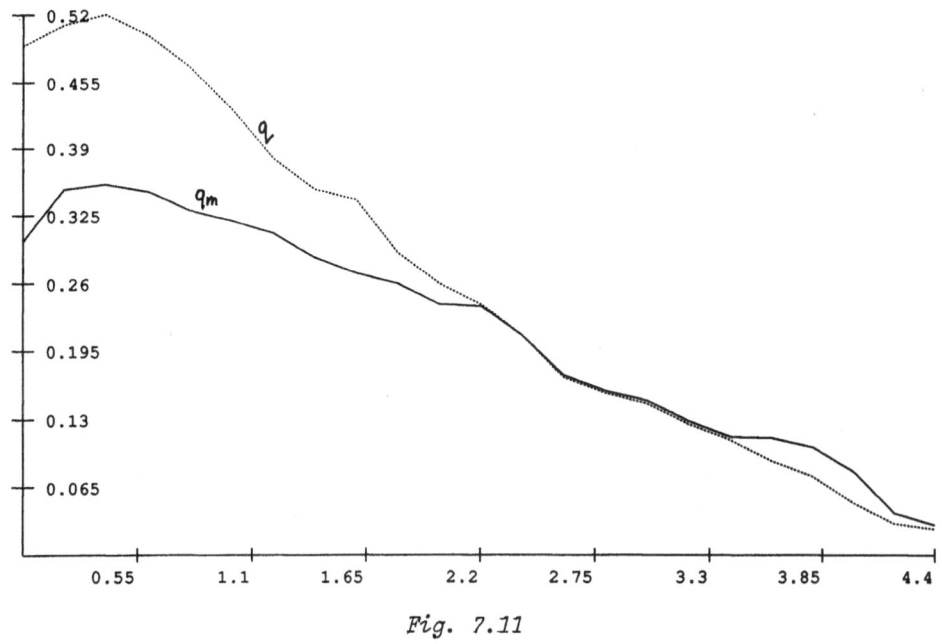

Fig. 7.11

7.5 ADAPTIVE COORDINATION CONTROL

Aircraft or spacecraft formation keeping, spacecraft rendezvous,
several robotic arms operating from the same conveyor and on the same
workspace, air traffic control over an airport, are but a few examples
where coordination control is essential. It may be considered separately,
as in our Chapter 8, or as an addition to tracking with avoidance. In
turn, the latter may be investigated in relative coordinates, utilizing
our conditions for capture-first described in Section 6.4, or as an
augmented version of mutual tracking with secured mutual avoidance of the
control systems concerned. For mechanical structures, such avoidance would
usually mean first an avoidance in Cartesian space, later transformed to
Configuration Space coordinates and finally discussed in State Space after
velocity objectives have been stipulated.

In mutual tracking we may have two approaches: either the objective
lies in *mutual tracking only*, that is, it does not matter where the con-
verging systems go, as long as they converge, or we aim at *mutual tracking
to a stipulated target set* in the state space. Obviously the second case
is more practical. The target may be fixed or moving. If moving, then
the motion may be given by a time parametrized curve or more generally by
a dynamical model. In the latter case, the problem may be made adaptive

and becomes a double MRAC with reference to a single model. The extension is called Mutual Reference Adaptive Control (MURAC) and was investigated in Skowronski [39],[41],[43],[44]. We shall follow the latter work here.

We require a target reference model, which is relatively simple, but compatible with both equations of the controlled systems. The axioms required are the same as those imposed upon the reference model in Section 7.3 and the design may include choosing an extra reference signal – control $\bar{u}_m$ to make the model trajectories coinciding with a given curve or curves in Δ, see Section 7.1.

Let us investigate in particular two mechanical structures in the form of interdependent chains, see Section 1.4, with the state equations (2.1.17) augmented by introduction of the uncertainty $\bar{w}$ and the adaptive parameters $\bar{\lambda}$ to the contingent format

$$\dot{\bar{x}}^j \in \{\bar{f}^j(\bar{x},\bar{u}^j,\bar{\lambda}^j,\bar{w}^j,t) \,|\, u^j \in P^j(\bar{x},\lambda^j,t), \bar{w}^j \in W^j\} , \quad j = 1,2 \quad (7.5.1)$$

with the vectors $\bar{w}^j,\bar{\lambda}^j$ defined identically as $\bar{w},\bar{\lambda}$ in Sections 2.2 and 7.1, respectively. The total system vector $\bar{x}(t)$ represents the state of all chains, while $\bar{x}^j(t)$ are chain state vectors, as in (2.1.17), and each chain is actuated by a separate control program

$$u^j(t) \in \bar{P}^j(\bar{x},\lambda^j,t) , \qquad\qquad\qquad\qquad (7.5.2)$$

with the option of feedback to all state vectors $\bar{x}^j$, $j = 1,2$. In the general format, the adaptive laws (7.1.12) to be designed are now

$$\dot{\bar{\lambda}}^j = \bar{f}_a^j(\bar{x},\lambda^j,t) , \quad j = 1,2 , \qquad\qquad (7.5.3)$$

and the reference model (7.1.15) remains the same with, as mentioned, the same assumptions.

In each chain j we have M^j bodies as specified in Section 1.6, with Cartesian body coordinate frame $0^j x_\nu^j, y_\nu^j, z_\nu^j$, $\nu = 1,\ldots,M^j$ which will in general have 6 DOF with respect to the base (3 translations and 3 rotations) provided the chain is not constrained. This gives a total number of DOF in the system $n = 6 \sum_j M^j$. With each of the controlled systems $j = 1,2$ referred to the model separately, augmenting our discussion to $j = 1,\ldots,m < \infty$, see Section 2.1, would not change any arguments or results. However $j = 1,2$ is more instructive.

The block scheme of the system is seen in Fig. 7.12. In the same way as for MRAC, we define the product vectors $\bar{x}_m^j(t) \triangleq (\bar{x}^j(t),\bar{x}_m(t))^T \in \Delta^2$, $\bar{a}^j(t) \triangleq \bar{\lambda}^j(t) - \bar{\lambda}_m$, $j = 1,2$ as well as the vectors $\bar{F}^j = (\bar{f}^j,\bar{f}_m,\bar{f}_a^j)^T$, $j = 1,2$ leading to the two product systems, cf. (7.3.16),

358

$$(\dot{\bar{x}}_m^j, \dot{\bar{\alpha}}^j)^T \in \{\bar{F}^j(\bar{x}_m^j, \bar{\alpha}^j, \bar{u}^j, \bar{w}^j, t) \mid \bar{u}^j \in \bar{P}^j(\bar{x}_m^j, \bar{\alpha}^j, t), \bar{w}^j \in W^j\} , \qquad (7.5.4)$$

$$j = 1,2 ,$$

with the Filippov type solutions $\phi^j(\bar{x}_m^{j0}, \bar{\alpha}^{j0}, t_0, t)$, $t \geq t_0$, within the class $K(\bar{x}_m^{j0}, \bar{\alpha}^{j0})$, $j = 1,2$. Then introduce the sets

$$M^j \triangleq \{ (\bar{x}_m^j, \bar{\alpha}^j) \in \Delta^2 \times \Lambda \mid \bar{x}^j = \bar{x}_m , \ \bar{\alpha}^j = 0 \} ,$$

$$M_\mu^j \triangleq \{ (\bar{x}_m^j, \bar{\alpha}^j) \in \Delta^2 \times \Lambda \mid \| \bar{x}^j - \bar{x}_m \| < \mu , \ \| \bar{\alpha}^j \| < \mu \} , \quad j = 1,2 ,$$

and again let $\Delta_0 \subset \Delta$ be the desired set for the mutual tracking to occur. We shall now have two objectives specified by the following two definitions.

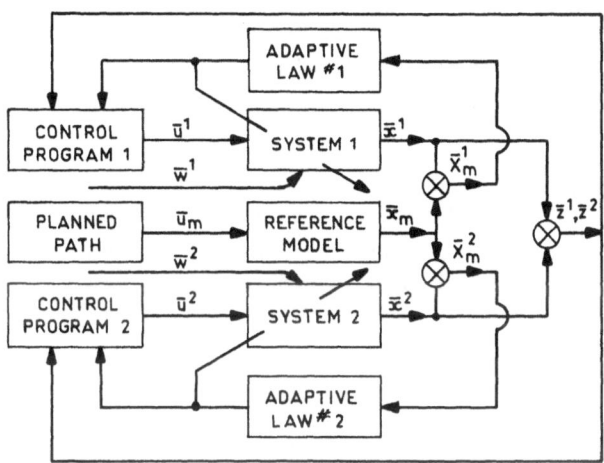

Fig. 7.12

DEFINITION 7.5.1. The systems (7.5.1) *mutually track* the reference model (7.1.15) on Δ_0 if and only if there is a pair of controllers $\bar{P}^j(\cdot)$ and a pair of adaptive law functions $\bar{F}_a^j(\cdot)$, $j = 1,2$ such that the set $\Delta_0^2 \times \Lambda$ is strongly positively invariant under the product motions of both classes $K(\bar{x}_m^{j0}, \bar{\alpha}^{j0})$, $j = 1,2$, and there are two time intervals $T_\mu^j < \infty$, such that for each $K(\bar{x}_m^{j0}, \bar{\alpha}^{j0})$, $j = 1,2$, $\phi(\bar{x}_m^{j0}, \bar{\alpha}^{j0}, t_0, \cdot) \in K(\bar{x}_m^{j0}, \bar{\alpha}^{j0})$ implies $\phi(\bar{x}_m^{j0}, \bar{\alpha}^{j0}, t_0, t) \in M_\mu^j$, $\forall t \geq t_0 + T_\mu$ where $T_\mu = \max(T_\mu^1, T_\mu^2)$. When T_μ is given a-priori we refer to tracking in stipulated time.

Let us now re-label the Cartesian coordinates in the general uncon-strained case and aggregate them in the vectors $\mathbf{z}^j = (\mathbf{z}_1^j, \ldots, \mathbf{z}_{3n}^j)^T$ $= (x_1, y_1, z_1, \ldots, x_n, y_n, z_n)^T$ and suppose that the transformation from Lagrangian to Cartesian coordinates (forward kinematics) is done off-line and given by

$$\mathbf{z}_i^j = \mathbf{z}_i^j(\bar{q}^j) , \quad i = 1, \ldots, 3n . \qquad (7.5.5)$$

Then let us denote $\bar{z}(t) \triangleq (\bar{x}^1(t), \bar{x}^2(t))^T$, $t \geq t_0$, generated along product trajectories $\bar{z}(t) = \bar{\phi}(\bar{z}^0, t_0, t)$, $t \geq t_0$, $\bar{z}^0 = \bar{z}(t_0)$, and define

$$M_A \triangleq \{\bar{z} \in \Delta_0^2 | \ |z_i^1 - z_j^2| \leq d, \ i,j = 1,\ldots,3n\},$$

the set of possible collision between the bodies to be avoided. Further, we design an avoidance set $A^2 \supset M_A$ with smooth boundary (best assumed as a Liapunov function level) and define $CA^2 = \Delta_0^2 - A^2$. Then also let $\Delta_\varepsilon^2 \supset A^2$, such that $\Delta_A^2 = \Delta_\varepsilon^2 - A^2$ is the required slow-down zone in CA^2 about A^2, cf. Section 6.1.

DEFINITION 7.5.2. The *mutually tracking systems* of Definition 7.5.1 *avoid collision* with precision $d > 0$ if there is Δ_A^2 in Δ_0^2 such that for any $\bar{z}^0 \in \Delta_A^2$ and any product motion $\bar{\phi}(\cdot) \in K(\bar{z}^0)$,

$$\bar{\phi}(\bar{z}^0, t_0, t) \in CA^2, \quad \forall t \geq t_0 .$$

Let $N_s \subset \Delta_0^2 \times \Lambda$ be defined as in Section 7.3, then also let $CM_\mu^j = (\Delta_0^2 \times \Lambda) - M_\mu^j$, $D^j \supset \overline{CM}$, while $D^j \cap M^j = \phi$, and introduce four C^1-functions $v_s^j(\cdot) : \bar{N}_s \to \mathbb{R}$, $v_\mu^j(\cdot) : D^j \to \mathbb{R}$, $j = 1,2$, with

$$\left.\begin{aligned}
v_s^j &= V_s(\bar{x}_m^j, \bar{\alpha}^j) | (\bar{x}_m^j, \bar{\alpha}^j) \in \partial(\Delta_0^2 \times \Lambda) ,\\
v_\mu^{j-} &= \inf v_\mu^j(\bar{x}_m^j, \bar{\alpha}^j) | (\bar{x}_m^j, \bar{\alpha}^j) \in \partial M_\mu^j \cap \overline{CM}_\mu^j ,\\
v_\mu^{j+} &= \sup v_\mu^j(\bar{x}_m^j, \bar{\alpha}^j) | (\bar{x}_m^j, \bar{\alpha}^j) \in \partial(\Delta_0^2 \times \Lambda) \cap \overline{CM}_\mu^j .
\end{aligned}\right\} \qquad (7.5.6)$$

Comments referring to the choice of $V_s(\cdot)$ of Section 7.3 apply here to $v_s^j(\cdot)$. We may now adjust the sufficient conditions for tracking introduced in Section 7.3, to our double MRAC-case. By the same argument as for Conditions 7.3.1, we have

CONDITIONS 7.5.1. Two systems (7.5.1) mutually track the reference model (7.1.15) according to Definition 7.5.1 if, given $\Delta_0, \Lambda, \mu > 0$ there are control programs $\bar{P}^j(\cdot)$ and C^1 functions $v_s^j(\cdot), v_\mu^j(\cdot)$ such that for all $(\bar{x}_m^j, \bar{\alpha}^j) \in \Delta_0^2 \times \Lambda$ we have

(i) $\quad v_s^j(\bar{x}_m^j, \bar{\alpha}^j) \leq v_s^j$, $\forall (\bar{x}_m^j, \bar{\alpha}^j) \in N_s$, $j = 1,2$;

(ii) $\quad$ for each $\bar{u}^j \in \bar{P}^j(\bar{x}_m^j, \bar{\alpha}^j)$,

$$\nabla v_s(\bar{x}_m^j, \bar{\alpha}^j)^T \cdot \bar{F}^j(\bar{x}_m^j, \bar{\lambda}^j, \bar{u}^j, \bar{w}^j, t) < 0, \quad \forall \bar{w}^j \in W^j; \qquad (7.5.7)$$

(iii) $\quad 0 \leq v_\mu^j(\bar{x}_m^j, \bar{\alpha}^j) < v_\mu^{j+}$, $\forall (\bar{x}_m^j, \bar{\alpha}^j) \in \overline{CM}_\mu^j$, $j = 1,2$;

(iv) $\quad v_\mu^j(\bar{x}_m^j, \bar{\alpha}^j) \leq v_\mu^{j-}$, $\forall (\bar{x}_m^j, \bar{\alpha}^j) \in D^j \cap M_\mu^j$, $j = 1,2$;

(v)   for each $\bar{u}^j \in \bar{P}^j(\bar{x}_m^j,\bar{\alpha}^j)$, there is $c_j(\|(\bar{x}_m^j,\bar{\alpha}^j)\|) > 0$
continuously increasing such that

$$\nabla v_\mu^j(\bar{x}_m^j,\bar{\alpha}^j)^T \cdot \bar{F}^j(\bar{x}_m^j,\bar{u}^j,\bar{\lambda}^j,\bar{w}^j,t) \le -c_j , \quad \forall \bar{w}^j \in W^j . \qquad (7.5.8)$$

REMARK 7.5.1. The tracking after stipulated time holds, if given $T_\mu^j < \infty$ the estimate c^- of the rate of change of $v_\mu^j(\cdot)$ is

$$c_j^- \le (v_\mu^{j+}/T_\mu^j) , \quad \forall (\bar{x}_m^j,\bar{\alpha}^j) \in \partial\Delta_0^2 \times \Lambda . \qquad (7.5.9)$$

Conditions 7.5.1 are proved by the same argument as Conditions 7.3.1, which is such that (i) and (ii) imply the strong positive invariance of $\Delta_0^2 \times \Lambda$ and are not used for anything else. Hence we can make the following

REMARK 7.5.2. When $\Delta_0^2 \times \Lambda$ is strongly positively invariant, conditions (i),(ii) are redundant for the hypothesis of Conditions 7.5.1.

We may now pass over to conditions sufficient for Definition 7.5.2.

CONDITIONS 7.5.2. The tracking systems (7.5.1) avoid collision according to Definition 7.5.2, if Conditions 7.5.1 hold and if there is a safety zone Δ_A with at least piecewise C^1 function $V_A(\cdot) : \Delta_A^2 \to \mathbb{R}$ such that for the tracking pair $\bar{P}^j(\cdot)$, $j = 1,2$, for all $\bar{z} \in \Delta_A^2$, we have

(vi) $V_A(\bar{z}) > V_A(\bar{z})$, $\forall \bar{z} \in \partial A^2$

(vii) for each $\bar{u}^j \in \bar{P}^j(\bar{z})$,

$$\nabla V_A(\bar{z})^T (\bar{f}^1(\bar{x}^1,\bar{u}^1,\bar{\lambda}^1,\bar{w}^1,t),\bar{f}^2(\bar{x}^2,\bar{u}^2,\bar{\lambda}^2,\bar{w}^2,t)) \ge 0 , \qquad (7.5.10)$$
for all $\bar{w}^j \in W^j$, $j = 1,2$.

The proof follows immediately from the fact that if some arbitrary product motion $\bar{\phi}(\bar{z}^0,t_0,t)$, $t \ge t_0$, $\bar{z}^0 \in \Delta_A^2$, crosses ∂A^2 at some $t_1 > t_0$, then by (vi), $V_A(\bar{z}(t_1)) < V_A(\bar{z}^0)$ which contradicts (vii).

The functions v_s^j and v_μ^j may be then defined by (7.3.28), (7.3.29):

$$v_s^j \overset{\Delta}{=} E_m(\bar{x}^j) + E_m(\bar{x}_m) + \bar{a}^j\bar{\alpha}^j , \quad \bar{\alpha}^j \neq 0 ; \qquad (7.5.11)$$

$$v_\mu^j = \begin{cases} |E_m(\bar{x}^j) - E_m(\bar{x}_m)| + \bar{a}^j\bar{\alpha}^j , & \forall (\bar{x}_m^j,\bar{\alpha}^j) \in CM_\mu^j , \\ \bar{a}^j\bar{\alpha}^j , & \forall (\bar{x}_m^j,\bar{\alpha}^j) \in M_\mu^j , \end{cases} \qquad (7.5.12)$$

with $\bar{\alpha}^j \neq 0$, $j = 1,2$, and $\bar{a}^j = (\text{sgn } \alpha_1^j,\ldots,\text{sgn } \alpha^j)^T$.

The function $V_A(\cdot)$ is defined by

$$V_A = \tfrac{1}{2} \lceil E_m(\bar{x}^1) - E_m(\bar{x}^2) \rceil^2 . \qquad (7.5.13)$$

Let us repeat the control conditions (I) and the adaptation condition (II) of Section 7.3.

(I) *Control Conditions*

$$\min_{\bar{u}^j} \max_{\bar{w}^j} \dot{E}_m(\bar{x}^j) \leq \dot{E}_m(\bar{x}_m) , \quad \text{for} \quad \delta E_m \geq 0 ,$$

$$\max_{\bar{u}^j} \min_{\bar{w}^j} \dot{E}_m(\bar{x}^j) \geq \dot{E}_m(\bar{x}_m) , \quad \text{for} \quad \delta E_m < 0 ;$$

where $\quad \delta E_m = E_m(\bar{x}^j) - E_m(\bar{x}_m)$.

(II) *Adaptation Condition*

$$\bar{a}^j \dot{\bar{\alpha}}^j \leq -|\bar{\alpha}^j| [c_j^- + |\dot{E}_m(\bar{x}^j)|] , \quad \bar{\alpha}^j \neq 0 ;$$

and add

(III) *Coordination Condition*

$$\max_{\bar{u}^1} \min_{\bar{w}^1} \dot{E}_m(\bar{x}^1) \geq \min_{\bar{u}^2} \max_{\bar{w}^2} \dot{E}_m(\bar{x}^2) , \quad \text{for} \quad E_m(\bar{x}^1) \geq E_m(\bar{x}^2) ,$$

$$\min_{\bar{u}^1} \max_{\bar{w}^1} \dot{E}_m(\bar{x}^1) < \max_{\bar{u}^2} \min_{\bar{w}^2} \dot{E}_m(\bar{x}^2) , \quad \text{for} \quad E_m(\bar{x}^1) < E_m(\bar{x}^2) .$$

By the same argument as in Section 7.3, the conditions (I) and (II) imply (i)-(v) of Conditions 7.5.1. To check upon (vi) of Conditions 7.5.2, note that from elementary mechanics, $V(\bar{q}^j) = V(\bar{z}^j)$ and since both potential energies are Taylor series developable, positive definite and increasing on $C\Delta_L$, then increasing some of the distances $|z_i^1 - z_j^2|$, $i,j = 1, \ldots, 3n$, one increases $V_A(\bar{z})$. We may thus define $\partial A^2 \colon V_A(\bar{z}) = v_A , v_A = \sup (\bar{z}) \, | \bar{z} \in M_A$, yielding (vi).

To check upon (vii) we differentiate (7.5.13) requiring

$$\dot{V}_A = [\dot{E}_m(\bar{x}^1) - \dot{E}(\bar{x}^2)] \cdot [E_m(\bar{x}^1) - E_m(\bar{x}^2)] \geq 0 \qquad (7.5.14)$$

and then we have to use conditions (I) and (III) combined. Let us investigate the relevant cases.

CASE $E_m(\bar{x}^j) \geq E_m(\bar{x}_m)$, $j = 1,2$:

We have two subcases:-

(a) $E_m(\bar{x}^1) \geq E_m(\bar{x}^2) \geq E_m(\bar{x}_m)$. Here we use (I): $\dot{E}_m(\bar{x}^1) \leq \dot{E}_m(\bar{x}_m)$ to calculate the control $\bar{u}^1$, and then substituting it, we use (III) to find $\bar{u}^2$ such that $\dot{E}_m(\bar{x}^2) \leq \dot{E}_m(\bar{x}^1)$. Since in turn $\dot{E}_m(\bar{x}^1) \leq \dot{E}_m(\bar{x}_m)$, then $\dot{E}_m(\bar{x}^2) \leq \dot{E}_m(\bar{x}_m)$, and $\bar{u}^2$ also satisfies (I).

(b) $E_m(\bar{x}^2) \geq E_m(\bar{x}^1) \geq E_m(\bar{x}_m)$. The roles of $E_m(\bar{x}^j)$ are reversed. Using (I) we find $\bar{u}^2$ and then using (III) we find $\bar{u}^1$ such that $\dot{E}_m(\bar{x}^1) \leq \dot{E}_m(\bar{x}^2) \leq \dot{E}_m(\bar{x}_m)$, thus also satisfying (I).

CASE $E_m(\bar{x}^j) < E_m(\bar{x}_m)$, $j = 1,2$:

Again, there are two subcases:-

(c) $E_m(\bar{x}^2) \leq E_m(\bar{x}^1) < E_m(\bar{x}_m)$. We use (I): $\dot{E}_m(\bar{x}^1) > \dot{E}_m(\bar{x}_m)$ to find $\bar{u}^1$ and substituting it, we use (III) to find $\bar{u}^2$ such that $\dot{E}_m(\bar{x}^2) \geq \dot{E}_m(\bar{x}^1)$ $> \dot{E}_m(\bar{x}_m)$ thus also satisfying (I).

(d) $E_m(\bar{x}^2) \leq E_m(\bar{x}^1) < E_m(\bar{x}_m)$. We use (I) to find $\bar{u}^2$ and (III) to find $\bar{u}^1$ such that $\dot{E}_m(\bar{x}^1) \geq \dot{E}_m(\bar{x}^2) > \dot{E}_m(\bar{x}_m)$ thus also satisfying (I), which closes our argument.

Substituting the j-chain correspondents of (7.3.37), (7.3.38) into the above analysis, the combined conditions, control (I) and coordination (III), obtain the following shape. For clarity of exposition, we substitute the extremizing values of $\bar{w}$ and drop the arguments of the functions involved.

$$\min_{\bar{u}^j} (\bar{F}^j)^T \dot{\bar{q}}^j \leq \bar{F}_m^T \dot{\bar{q}}_m - (\bar{R}^j)^T \dot{\bar{q}}^j + [(\bar{D}^j)^T \dot{\bar{q}}^j - \bar{D}_m^T \dot{\bar{q}}_m] + (\bar{\Pi}^j - \bar{\Pi}_m)^T \dot{\bar{q}}^j ,$$

$$(7.5.15)$$

$$\min_{\bar{u}^k} (\bar{F}^k)^T \dot{\bar{q}}^k + (\bar{R}^k)^T \dot{\bar{q}}^k - (\bar{D}^k)^T \dot{\bar{q}}^k - (\bar{\Pi}^k - \bar{\Pi}_m)^T \dot{\bar{q}}^k$$

$$\leq \max_{\bar{u}^j} (\bar{F}^j)^T \dot{\bar{q}}^j + (\bar{R}^j)^T \dot{\bar{q}}^j - (\bar{D}^j)^T \dot{\bar{q}}^j - (\bar{\Pi}^j - \bar{\Pi}_m)^T \dot{\bar{q}}^j$$

for all $(\bar{x}_m^{j,k}, \bar{\alpha}^{j,k}) \in CM_\mu^{j,k}$, such that $\delta E_m \geq 0$, $E_m(\bar{x}^j) \geq E_m(\bar{x}^k)$, where $j,k = 1,2$, $j \neq k$, and

$$\max_{\bar{u}^j} (\bar{F}^j)^T \dot{\bar{q}}^j \geq \bar{F}_m^T \dot{\bar{q}}_m - (\bar{R}^j)^T \dot{\bar{q}}^j + [(\bar{D}^j)^T \dot{\bar{q}}^j - \bar{D}_m^T \dot{\bar{q}}_m] + (\bar{\Pi}^j - \bar{\Pi}_m)^T \dot{\bar{q}}^j , \quad (7.5.16)a$$

$$\max_{\bar{u}^k} \; (\bar{F}^k)^T \dot{\bar{q}}^k + (\bar{R}^k)^T \dot{\bar{q}}^k - (\bar{D}^k)^T \dot{\bar{q}}^k - (\bar{\Pi}^k - \bar{\Pi}_m)^T \dot{\bar{q}}^k$$

$$\geq \min_{\bar{u}^j} \; (\bar{F}^j)^T \dot{\bar{q}}^j + (\bar{R}^j)^T \dot{\bar{q}}^j - (\bar{D}^j)^T \dot{\bar{q}}^j - (\bar{\Pi}^j - \bar{\Pi}_m)^T \dot{\bar{q}}^j \qquad (7.5.16)b$$

for all $(\bar{x}_m^{j,k}, \bar{a}^{j,k}) \in CM^{j,k}$, such that $\delta E_m < 0$, $E_m(\bar{x}^j) < E_m(\bar{x}^k)$ where $j,k = 1,2$, $j \neq k$.

If we now specify as before: $F_i^j(\bar{q}^j, \dot{\bar{q}}, \bar{w}^{j*}, \bar{u}^j) = B_i^j(\bar{q}^j, \dot{\bar{q}}^j, \bar{w}^{j*}) u_i^j$, $B_i^j > 0$, $i = 1, \ldots, n$, the above conditions are implied by the following control programs. For tracking,

$$u_i^j \operatorname{sgn} \dot{q}_i^j \begin{cases} \leq \dfrac{1}{|\dot{q}_i^j| B_i^j} [F_{mi} \dot{q}_{mi} - D_{mi} \dot{q}_{mi} - R_i^j \dot{q}_i^j + D_i^j \dot{q}_i^j \\ \qquad\qquad + (\Pi_i^j - \Pi_{mi}) \dot{q}_i^j], \; \forall \; |\dot{q}_i^j| \geq \beta_i^j, \qquad (7.5.17) \\ = \text{suitable constant}, \; \forall \; |\dot{q}_i^j| < \beta_i^j, \; i = 1, \ldots, n, \end{cases}$$

and for coordination

$$u_i^k \operatorname{sgn} \dot{q}_i^k \begin{cases} \leq \dfrac{1}{|\dot{q}_i^k| B_i^k} [\max_{u_i^j} \; (B_i^j u_i^j) \dot{q}_i^j + (R_i^j - D_i^j - \Pi_i^j + \Pi_{mi}) \dot{q}_i^j \\ \qquad\qquad - (R_i^k - D_i^k - \Pi_i^k + \Pi_{mi}) \dot{q}_i^k], \; \forall \; |\dot{q}_i^k| \geq \beta_i^k, \qquad (7.5.18) \\ = \text{suitable constant}, \; \forall \; |\dot{q}_i^k| < \beta_i^k, \; i = 1, \ldots, n, \end{cases}$$

with the control u_i^j obtained from (7.5.16), and $\beta_i^{j,k}$, $B_i^{j,k}$ playing identical roles as in (7.3.41).

The physical sense of the conditions (I) is that the power balance of the chains should track the power of the model, and (III) secures the fact that powers of the two chains diverge while their energy levels are different, thus counteracting possible collision.

The adaptation condition (II) is then implied by the adaptive law, cf. (7.5.13):

$$\dot{\alpha}_i^j = -\alpha_i^j [c_j^- + |\dot{E}_m(\bar{x}^j)|], \quad \alpha_i^{j0} \neq 0, \quad i = 1, \ldots, \ell, \quad j = 1,2 \qquad (7.5.19)$$

by the same argument as for (7.3.42).

EXAMPLE 7.5.1. We double the single arm system of Example 7.3.1 and form the structure shown in Fig. 7.13 with the double set of motion equations (7.3.43)-(7.3.49), the joint coordinates and velocities as well as forces with the superscript $j = 1,2$ added. The double reference model (7.3.51)

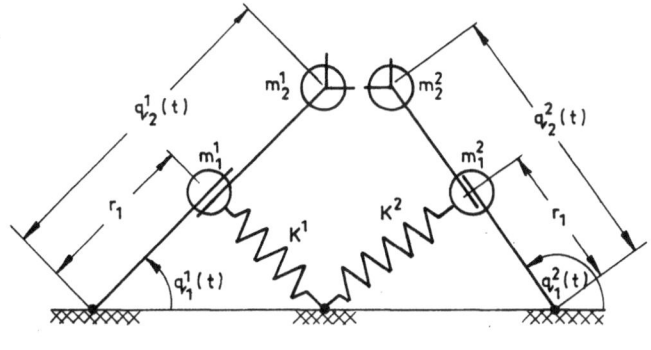

Fig. 7.13

remains unchanged, together with the corresponding total energy function $E_m(\cdot)$.

The tracking controller (7.5.17) in our present example is identical with (7.3.57),(7.3.58) of Example 7.3.1:

$$u_1^1 \operatorname{sgn} \dot{q}_1^1 \begin{cases} \leq \dfrac{m_1^1 (r_1^1)^2 + m^- (q_2^1)^2}{|\dot{q}_1^1|} \, [D_1^1 \dot{q}_1^1 - R_1^1 \dot{q}_1^1 - \lambda_{m1} \dot{q}_{m1}^2 \\ \qquad\qquad + (\Pi_1^{1*} - \Pi_{m1}) \dot{q}_1^1] \,, \quad \forall \, |\dot{q}_1^1| \geq \beta_1^1 \,, \qquad (7.5.20) \\ = \text{suitable constant}, \quad \forall \, |\dot{q}_1^1| < \beta_1^1 \,, \end{cases}$$

$$u_2^1 \operatorname{sgn} \dot{q}_2^1 \begin{cases} \leq \dfrac{m^-}{|\dot{q}_2^1|} \, [D_2^1 \dot{q}_2^1 - R_2^1 \dot{q}_2^1 - \lambda_{m2} \dot{q}_{m2}^2 + (\Pi_2^1 - \Pi_{m2}) \dot{q}_2^1] \,, \\ \qquad\qquad\qquad\qquad\qquad \forall \, |\dot{q}_2^1| \geq \beta_2^1 \,, \qquad (7.5.21) \\ = \text{suitable constant}, \quad \forall \, |\dot{q}_2^1| < \beta_2^1 \,, \end{cases}$$

while the coordination controller (7.5.18) becomes

$$u_1^2 \operatorname{sgn} \dot{q}_1^2 \begin{cases} \leq \dfrac{m_1^2 (r_1^2)^2 + m^- (q_2^2)^2}{|\dot{q}_1^2|} \, \Big[[B_1^{1*} u_1^{1*} + R_1^1 - D_1^1 - \Pi_1^{1*} + \Pi_{m1}] \dot{q}_1^1 \\ \qquad\qquad + [D_1^2 - R_1^2 + \Pi_1^{2*} - \Pi_{m1}] \dot{q}_1^2 \Big] \,, \quad \forall \, |\dot{q}_1^2| \geq \beta_1^2 \,, \quad (7.5.22) \\ = \text{suitable constant}, \quad \forall \, |\dot{q}_1^2| < \beta_1^2 \,; \end{cases}$$

$$u_2^2 \operatorname{sgn} \dot{q}_2^2 \leq \begin{cases} \dfrac{m^-}{|\dot{q}_2^2|} \, \Big[[B_2^{1*} u_2^{1*} + R_2^1 - D_2^1 - \Pi_2^1 + \Pi_{m2}] \dot{q}_2^1 \\ \qquad\qquad + [D_2^2 - R_2^2 + \Pi_2^2 - \Pi_{m2}] \dot{q}_2^2 \Big] \,, \quad \forall \, |\dot{q}_2^2| \geq \beta_2^2 \,, \quad (7.5.23) \\ \text{suitable constant}, \quad \forall \, |\dot{q}_2^2| < \beta_2^2 \,. \end{cases}$$

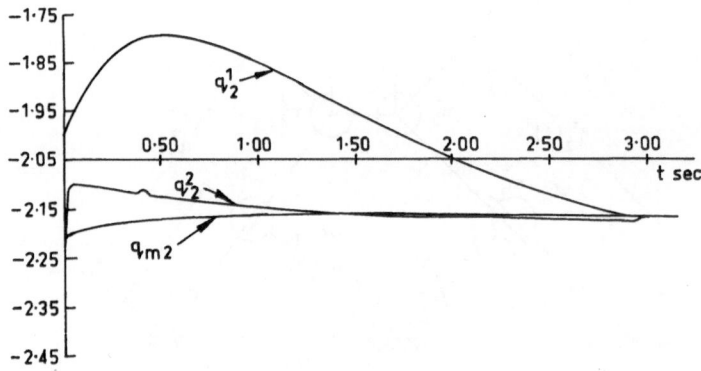

Fig. 7.14

In the above, $B_1^{1*} \triangleq 1/[m_1^1 (r_1^1)^2 + m^+ (q_2^1)^2]$, $B_2^{1*} \triangleq 1/m^+$. The arguments are omitted upon assuming that the extremizing values of m_2, ℓ_i have been substituted - confusion is unlikely, but the reader may want to compare (7.3.57), (7.3.58). The role of β_i^j is the same as that in (7.3.57), (7.3.58). Note that the program (7.5.22) for the control u_i^2 uses u_i^1 from (7.5.20), and (7.5.23) uses (7.5.21).

The adaptive laws (7.5.19) become now

$$\dot{\alpha}_1^j = -\alpha_1^j [c_j^- + \left| \frac{u_1^j \dot{q}_2^j}{m_1^j (r_1^j)^2 + m_2^j (q_2^j)^2} + R_1^j \dot{q}_1^j - D_1^j \dot{q}_1^j + (\Pi_{m_1} - \Pi_1^j) \dot{q}_1^j \right|] \quad (7.5.24)$$

$$\dot{\alpha}_2^j = -\alpha_2^j [c_j^- + | (u_2^j \dot{q}_2^j / m_2^j) + R_2^j \dot{q}_2^j - D_2^j \dot{q}_2^j + (\Pi_{m_2} - \Pi_2^j) \dot{q}_2^j |] , \quad j = 1,2 .$$

All the laws are integrable with exponential non-zero solutions. Here $c_j^- = v_\mu^{j+} / t_\mu^j$, $j = 1,2$ as in (7.3.56) when t_μ^j are stipulated, otherwise they can be any suitable positive constants. Assuming the arms symmetric in masses and length and taking the following data, the convergence between the arms and the model is achieved after 2 sec - the curves are shown in

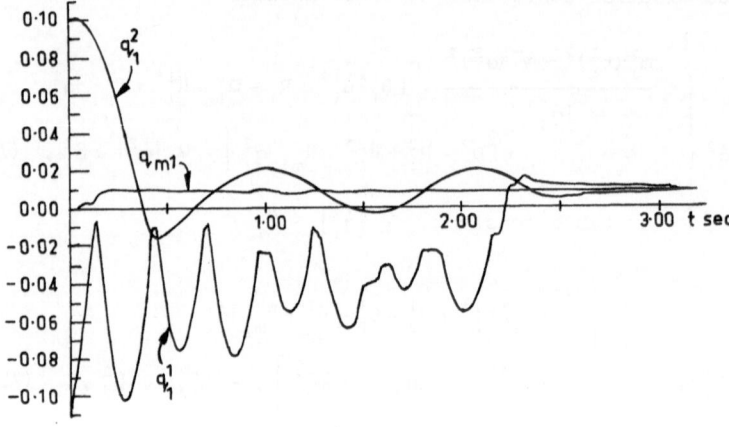

Fig. 7.15

Fig. 7.14 and Fig. 7.15. The data are $r = 0.66$, $a = 500$, $b = 3$, $m_1^1 = 1\,kg$, $m_1^2 = 1\,kg$, $m_2^1 \in [0.5, 10\,kg]$, $m_2^2 \in [0.5, 10\,kg]$, $c_1^- = 100$, $c_2^- = 200$, the time step 0.02 sec.

Closing this section. let us note briefly that Definition 7.5.2 of mutual avoidance of systems and Conditions 7.5.2 for such avoidance may stand independently of tracking. Only slight reformalization, which is left to the reader, suffices for that purpose. Details may be found in Skowronski [44]. In this text, for space reasons, we must leave them out.

Chapter 8

DYNAMIC GAMES

8.1 THE DYNAMICS OF CONFLICT

The description of a game by differential equations was called by
Isaacs [1] the *differential game*, and Blaquière-Gerard-Leitmann [1] called
the dynamical system subject to conflicting control a *dynamic game*. Both
names refer to, roughly speaking, the same set of problems, but the second
better illustrates the nature of our investigation, which is devoted
primarily to the state space pattern of behavior of dynamical, in particular
mechanical, systems under conflict. Introducing into the system (2.2.6)'
a second, competitive controlling agent with a different objective intended
on the same part of Δ, we produce the basic dynamical carrier for the
conflict that may arise. The applications seem to be everywhere within the
realm of mechanical systems, beginning from the classical, such as air
combat (see Ardema-Heymann [1]), air traffic, ship collision avoidance (see
Merz [1]), up to more recent applications like coordination control of
robotic arms on the manufacturing floor or in space (see Ardema-Skowronski
[1]), control of multi-legged vehicles (see McGhee [1], McGhee-Klein-Chas
[1], etc.) In either of these problems, as much as in many other real life
applications, the objectives attempted by the agents in conflict are
complex, qualitative and quantitative, subdivided into several phases of
achievement in time. For instance, in the aircraft duel with air-to-air
missiles, the composite objective for each of the agents consists of three
distinctive phases, each with a specific subobjective. First, it is to get
into a winning position for the launching of the missile, second, to get
the missile lockup onto the enemy in the shortest time, third, to escape
from the enemy's missile range. The end conditions of each phase must

overlap with the next. Much in the same way, the robot manipulator working on a bench must compose its successive operation subobjectives: pick up a screw from the conveyor, turn it to the proper position, place it into a specific hole, etc. The opposing agent (the enemy pilot or the other robot which may collide) will have a different goal for each of the phases or even quite different phases. Still, both players attempting to realize their objectives may be doing so against uncertainty, which is always present.

There is a large number of works, monographs, textbooks and even journals on differential games - far too vast to quote them here. In the direction of the present study, the reader may do well with the fundamental book by Isaacs [1], qualitative and quantitative games described by Blaquière-Gerard-Leitmann [1], the geometric approach by Krassovskii-Subbotin [1], Hajek [1], Basar-Olsder [1], and non-antagonistic games by Petrosian-Danilov [1]. A good source of up-to-date information is also given in Leondes (ed.), [1].

The large majority of past works refers to simple, single quantitative objective (optimal collision), takes a one sided view of a fixed player (pursuit-evasion, no role reversing) and uses necessary conditions as the operating technique. We already argued the need for recognizing the complexity of objectives met in case studies, see Section 3.1. In terms of the true scenario of conflict, the one sided view mentioned is also a definite shortcoming, see Merz [4], Olsder-Breakwell [1]. As mentioned, there is a number of works on the one sided pursuit-evasion: homicidal chauffeur game, Isaacs [1], Merz [3]; the game of two cars, Merz [2], Getz-Pachter [1]; and its applications to air combat, Prasad-Rajan-Rao [1], Jarmark-Merz-Breakwell [1], Peng-Vincent [1], Hagedorn-Breakwell [1],[2], and even on two-target games, but not two objectives: Pachter-Getz [1], Getz-Leitmann [1], Davidovitz-Shinar [1], Stonier [3]. It is only recently that the invertible roles of players are discussed, see Merz [4] and Olsder-Breakwell [1], and that the interface of such inverted roles is studied, see Skowronski-Stonier [1],[2].

Moreover, the first cited papers approach the games via the traditional Isaacs' technique of optimization and necessary conditions. Optimization may or may not be the basic feature in the complex objectives concerned. Quite often the fact of winning as such is more important, as we want to achieve a qualitative objective in favor of the player $i = 1,2$, call it the objective property $Q(i)$, see Section 3.1. The Isaacs' method of attaining the qualitative objective via optimization, for example, obtaining avoidance via maximizing to infinity the time of collision, was possible for

very simple objectives only. Furthermore, the analysis via necessary
conditions is usually very complicated and it is even made more difficult
by moving to realistic fan-shaped targets, firing zones. Moreover, it does
not yield a quick determination of the winning strategies and winning
regions. Since the necessary conditions produce only candidates for the
latter, they may lead to a long time digital search for the corresponding
trajectories without total security of success and on often small, PC-size
on-board computers. Major innovative ideas have been used by the authors
to overcome the problem. One of them is the reduction in dimensionality
and/or finding analytic solutions to the motion equations; see our comments
in Chapter 9 and also Davidovitz-Shinar [1] and Skowronski [41],[49].

The algorithms in the form of analytic solutions may often be obtained
from sufficient conditions for controllability for both qualitative and
quantitative objectives within the Liapunov Design, in the same way as was
shown in our previous chapters for the control case. The Liapunov differential game has been formalized in Skowronski [22]-[25]. Later it has been
shown in Krassovski [1]-[4], Krassovski-Subbotin [1], Leitmann-Skowronski
[1],[2], Pachter-Getz [1], Getz-Leitmann [1], Leitmann [5], Leitmann-Liu
[1], Sticht-Vincent-Schultz [1], Skowronski-Vincent [1], Skowronski [29],
Galperin-Skowronski [2],[3], Skowronski-Stonier [1],[2], Stonier [3], that
the Liapunov formalism serves the purpose very well providing both sufficient conditions for qualitative objectives and subsequent optimization.

It appears that the Liapunov Design means as much for the qualitative
objectives in the game as it did for control and stability. The Liapunov
type test function becomes a *qualitative performance "index"* of the game.
Conversely, the optimal pay-off for a player may frequently be shown to be
the Liapunov-type test function required qualitatively. Then the optimal
behavior implies the corresponding qualitative performance, which covers
Isaacs' games, but in a much broader sense of composite objectives with the
reversible role of the players. The latter is attained by interfacing two
semi-games, each for a player, and estimating within each such a semi-game,
the strong controllability for the objective of the player concerned. Each
of the players may change his role instantaneously. The typical example
is again the one already mentioned several times of air combat where the
same player may switch from evading to pursuing at any time of the game.
To accommodate this, we need a state space display of the said interface
of the two semi-games: regions of strong controllability separated by
neutral sets, draw sets and a barrier. Such a display represents a *map of
the game* revealing options for design of controllers and state constraints
for winning of the attempted objective.

8.2 CONTROLLABILITY IN THE GAME

Let us examine now the basic logical structure of the two person competitive dynamic game which consists of what may be called *moods of play* and corresponding underlying dynamics. We can write (2.2.6)' in the game format by introducing the second agent

$$\dot{\bar{x}} \in \{\bar{f}(\bar{x}, \bar{u}^1, \bar{u}^2, \bar{w}, t) \,|\, \bar{u}^i \in \bar{P}^i(\bar{x}, u^j, t), \bar{w} \in W\}, \quad i,j = 1,2, \quad i \neq j,$$

$$(8.2.1)$$

with the same conditions for the existence of solutions $\bar{\phi}(\bar{x}^0, t_0, \cdot) : \mathbb{R}^+ \to \Delta$ specified within the class $K(\bar{x}^0, t_0)$ by the triple $\bar{P}^1(\cdot), \bar{P}^2(\cdot), W$, and forming the attainable sets $\Phi(\bar{x}^0, t_0, t)$ as well as reachable sets $\Phi(\bar{x}^0, t_0, [t_0, t))$ at t. Such solutions will be called *game motions* (or *game trajectories* in autonomous case).

The reader must have noticed that we have made each program $\bar{P}^i(\cdot)$ dependent not only upon the state as required by feedback, but upon the opponent control. Formally, when there is no uncertainty in the system ($\bar{w}$ is assumed known) and the control $u^j(t)$ is known to the player i (u^j given) the program $\bar{P}^i(\cdot)$ becomes single valued and the system (8.2.1) generates unique motions on Δ. In the competitive game it may happen only in special circumstances, but it happens as a rule when the game becomes a coordination game for systems controlled by a single agency, for example, robotic arms. We shall briefly consider such a game in some examples, each time stating that it is the case. Generally, u^j is not known to the player i and must be treated in $P^i(\cdot)$ as bounded uncertainty thus adding up to the vector $\bar{w}$.

The *weak mood of play* answers the preliminary, but basic question whether the player i's objective $Q(i)$ is attainable at all. Considering therefore all possible selections of control programs for *both* players, this question is actually one of controllability for each of the player's defined objective. We shall term it briefly the i-controllability, meaning controllability in favor of the player i, given the favorable control program of the opposition j, $j \neq i$ and favorable uncertainty $\bar{w}(\cdot)$.

DEFINITION 8.2.1. The system (8.2.1) is i-controllable on some $\Delta_0 \subset \mathbb{R}^N$ for $Q(i)$, if and only if there is a triple $\bar{P}^1(\cdot), \bar{P}^2(\cdot), \bar{w}(\cdot)$ such that all game motions of $K(\bar{x}^0, t_0)$ generated on $x^0 \in \Delta_0$, $t_0 \in \mathbb{R}$, by this triple exhibit $Q(i)$.

The state $\bar{x}^0$ is called i-controllable for $Q(i)$ and the set Δ_q^i

being the maximal such Δ_0 for which Definition 8.2.1 holds is called the
region of i-controllability for $Q(i)$. There obviously may be pairs
$(\bar{P}^1(\cdot), \bar{P}^2(\cdot))$ generating independently $Q(1)$ and $Q(2)$ on the same set
in Δ, so that in general $\Delta_q^1 \cap \Delta_q^2$ is nonempty. We shall call the set
$\Delta_n \triangleq \Delta - (\Delta_q^1 \cup \Delta_q^2)$ if it is nonempty, the *dead zone*, a region of no controllability. If $\Delta_q^i = \Delta$ which we call a playing region (i = 1 or 2), the
i-controllability is called *complete*. It indicates that there is a possibility of a player achieving his objective at each $\bar{x}_0 \in \Delta$ depending, of
course, if an appropriate strategy is undertaken by the other player, and
subject to suitable $\bar{w}(\cdot)$. If Δ_q^i is larger than Δ, then we are interested only in $\Delta_q^i \cap \text{int } \Delta$ which is then called the *usable part* of Δ_q^i.
Any subset of such a usable part of Δ_q^i is i-controllable for $Q(i)$.

On the other hand the player i may have his own candidate set (say
coming from necessary conditions) which he would like to cover with
i-controllable states. Then we may not search for Δ_q^i, but only to secure
the *i-controllability on a stipulated set* $\Delta_0^i \subset \Delta_0 \cap \Delta_q^i \neq \phi$. Such
i-controllability is $\bar{x}^0, t_0$ - uniform, if the quantities implementing $Q(i)$
do not depend upon $\bar{x}^0, t_0$, respectively. For complex objectives it may
occur that some objective property $\tilde{Q}(i)$ implies $Q(i)$; for example,
capture implies collision. Then the i-controllability for $\tilde{Q}(i)$ implies
that for $Q(i)$ and we have

$$\tilde{\Delta}_q^i \subset \Delta_q^i . \tag{8.2.2}$$

EXAMPLE 8.2.1. In order to illustrate the case of i-controllability, let
us examine the turret game of Ardema-Heyman [1] applied to the combat
between player 1, a bomber, and player 2, a fighter plane. Player 1 moves
in a plane with arbitrary velocity relative to an inertial reference frame
Ox_0y_0 and can turn a ray weapon relative to a fixed direction at a bounded
angular rate $\dot{\alpha}$, see Fig. 8.1. Player 2 moves such that he is always at
a distance R from player 1, and he can traverse this circle at an angular
speed relative to a fixed direction at a bounded rate $\dot{\beta}$. Player 2 also
has a ray weapon that he can turn relative to the line of sight between
the two players at a bounded rate $\dot{\phi}$. The relative Cartesian reference
frame Oxy is such that the origin is located at player 1's position and
the y-axis lies along player 1's weapon.

Choosing the state variables in terms of the relative coordinates
$x_1 \triangleq \beta - \alpha$, $x_2 \triangleq \phi$, and the control variables $u^1 \triangleq \dot{\alpha}$, $u_1^2 \triangleq \dot{\beta}$,
$u_2^2 \triangleq -\dot{\phi}$, the kinematic state equations become

$$\dot{x}_1 = u_1^2 - u^1 \; , \qquad \left.\right\}$$
$$\dot{x}_2 = -u_2^2 \; , \qquad \left.\right\} \qquad (8.2.3)$$

with x_1, x_2 computed module 2π and the playing space

$$\Delta = \{ (x_1, x_2) \mid x_1, x_2 \in [0, \pi] \} \; .$$

The control constraints are $0 \le u^1 \le \hat{u}^1 \; , \quad u_1^2, u_2^2 \ge 0$ with

$$u_1^2/\hat{u}_1^2 + u_2^2/\hat{u}_2^2 \le 1 \; . \qquad (8.2.4)$$

The firing range of the weaponry is R and the target sets for player 1 are

$$T^1 \overset{\Delta}{=} \{ (x_1, x_2) \mid x_1 \le \varepsilon_1 \} = T_A^2 = A^2 \; ,$$
$$A^1 = T_A^1 \overset{\Delta}{=} \{ (x_1, x_2) \mid x_2 \le \varepsilon_2 \} = T^2 \; .$$

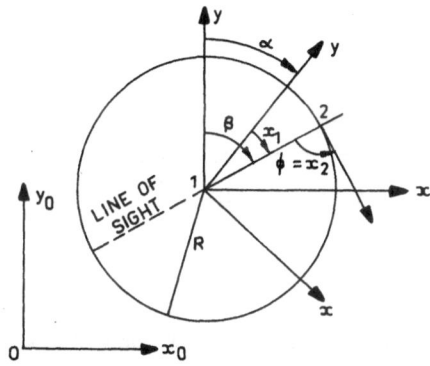

Fig. 8.1

His objective $Q(1)$ is collision-first, that is, to drive the state $\bar{x} = (x_1, x_2)^T$ by selection of control variable u^1 to hit the target set T^1 whilst avoiding T_A^1, that is, the strike zone T^2 of player 2, see Section 6.4; that is, there is a time $t_c > 0$ for which, cf. (6.4.1), (6.4.2),

$$x(\bar{t}) \in T^1 \; ,$$

and

$$x(t) \notin T_A^1 \quad \text{for} \quad t \in [0, t_c] \; .$$

Likewise, player 2's objective $Q(2)$ is also collision-first, but this time it is to drive the state by selection of control variable $(u_1^2, u_2^2)^T$ to hit target set T^2 whilst avoiding T_A^2, that is, the strike zone T^2 of player 1, see Fig. 8.2.

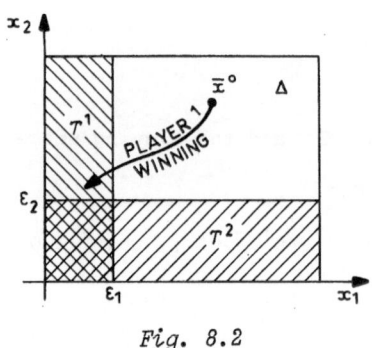

Fig. 8.2

Consider the strategy pair defined by the selection

$$u^{1*} = \hat{u}^1 \ , \quad u_1^{2*} = 0 \quad \text{and} \quad u_2^{2*} = 0 \tag{8.2.5}$$

(player 2 executes the null strategy). The state evolves according to

$$\dot{x}_1 = -\hat{u} \ , \quad \dot{x}_2 = 0 \ .$$

For $\bar{x}^0 \in \Delta$, $x_1(t)$ will decrease from $\bar{x}_1^0$ to ε_1 whilst $x_2(t)$ remains constant at $x_2^0 > \varepsilon_2$. This shows that for each $\bar{x}^0 \in \Delta$, the resulting trajectory obtained for this strategy pair hits T^1 before hitting T_A^1. Therefore $\Delta_q^1 = \Delta$. Similarly, we obtain $\Delta_q^2 = \Delta$ with the selection of the strategy pair

$$u^{1*} = 0 \ , \quad u_1^{2*} = 0 \ , \quad u_2^{2*} = \hat{u}_2^2 \ . \tag{8.2.6}$$

This shows that i-controllability for $i = 1$ and 2 is complete.

Note that in defining the player's qualitative objective property, we have specified that the state $\bar{x}(t)$ cannot lie in the opponent's firing zone at time t_c. This means that even though $\Delta_q^1 \cap \Delta_q^2 \neq \phi$ there is no pair $(\bar{P}^1, \bar{P}^2)$ for which properties $Q(1)$ and $Q(2)$ hold together on the system trajectories of (1). □

EXAMPLE 8.2.2. The turret game is particularly illustrative for the weak mood of play in another application, namely the *coordination game* for the two-arm pick-and-place robot, see Ardema-Skowronski [1]. The rearranged scenario is shown in Fig. 8.3. The two arms are considered players. Arm 1 has a rigid link of length r and a gripper both reduced to a point mass m_1 rotating about the base B_1 fixed at $(0,0)$ of the world coordinates (inertial) reference plane Oxy. Similarly arm 2 has a link of length r and a gripper m_2 rotating about the base B_2, which itself is fixed to a conveyor turntable rotating about B_1 with angular speed $\dot{\beta}(t)$. The radius of the table is r. The rotation angles of the arms are $\theta_i(t)$, $t \geq 0$, $i = 1,2$.

374

The gripper m_2 is supposed to pick up an object at some point 0_2 in the world coordinate plane outside the conveyor and deliver it to location B_1 by controlling the rotation of the turntable and the rotation of gripper m_2 relative to it. Simultaneously and independently, gripper m_1 is supposed to pick up an object at some point 0_1 in the world plane and deliver it to the conveyor at location B_2. To prevent collision with the conveyor, both grippers must deposit their objects with zero relative velocity. The turntable actuator acts as coordination controller.

Our goal is to seek control programs that guarantee successful task completion of each arm, despite the action of the other. Specifically we seek for each arm i collision-first: reaching the target point B_j while avoiding the arm j, $i, j = 1, 2$, $i \neq j$. Controllability for both objectives requires coordinated programs for both arms.

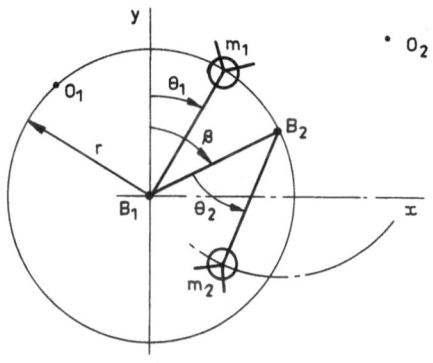

Fig. 8.3

Choosing the state variables as in Example 8.2.1, we have $x_1 \stackrel{\Delta}{=} \beta - \theta_1$, $x_2 \stackrel{\Delta}{=} \theta_2$. The control variables are similarly $u_1^1 = \dot{\theta}_1$, $u^2 = -\dot{\theta}_2$ and the coordination control u_1^2 with the same state equations (8.2.3), the same state constraints Δ and the same control constraints. Constraint (8.3.4) is a statement that arm 2 must allocate a fixed amount of control power between the components u_1^2 and u_2^2, balancing the coordination. The targets remain the same as in Example 8.2.1, with $\varepsilon_i r > 0$ determining a small neighborhood of B_j, where arm i must deposit each object. We still have $T^i = A_j$, $i, j = 1, 2$, $i \neq j$, for obvious reasons, see Fig. 8.2. We have to make a comment here. Since T^i are closed, so is their intersection. Hence the complements $\hat{T}^i = T^i - (T^1 \cap T^2)$ are open. This means that a convergent sequence of guaranteed safe trajectories for arm i terminating in $\hat{T}^i$ need not converge to a trajectory that terminates to a point in $\hat{T}^i$. To circumvent this difficulty, it is necessary to place an

375

open neighborhood around $T^1 \cap T^2$ and consider the complement of this open neighborhood embedded in both targets, Ardema-Heymann [1]. We assume such a neighborhood to be negligibly small.

With all the above, by the same argument as in Example 8.2.1, controllers (8.3.5),(8.3.6) generate complete 1,2-controllabilities for collision-first specified by the objectives of the arms. □

The latter comment as well as the discussion of i-controllability in Example 8.2.1 is a part of the more general problem of sufficient conditions for i-controllability. By definition, for the case of unknown $\bar{w}(\cdot)$, such conditions are identical with conditions for strong controllability for Q introduced in Chapters 3-7, while for the case of known $\bar{w}(\cdot)$, they coincide with conditions for controllability for Q. In turn those conditions are directly obtainable from the conditions for strong controllability for Q by fixing $\bar{w}(\cdot)$. The same applies now. We may use conditions for strong controllability for Q as conditions for i-controllability, fixing $\bar{w}(\cdot)$ when applicable.

The regions Δ_q^i are then found by the same methods as discussed in Section 5.4.

The competitive nature of our game appears in the *strong mood of play*, where a winning control program $\bar{P}^i(\cdot)$ must be found to secure $Q(i)$ against all the options of the opposition $P^j(\cdot)$, $j \neq i$, *and* all options of uncertainty $\bar{w}$. The corresponding dynamical carrier for the case is quite naturally the contingent equation

$$\overset{\circ}{\bar{x}} \in \{\bar{f}(\bar{x},\bar{u}^1,\bar{u}^2,\bar{w},t) \,|\, \bar{u}^i \in \bar{P}^i(\bar{x},\bar{u}^j,t)\,,\, u^j \in U_j\,,\, \bar{w} \in W\}\,, \qquad (8.2.7)$$
$$i,j = 1,2\,,\, i \neq j\,,$$

still with the game motions $\Phi(\bar{x}^0,t_0,\cdot) : \mathbb{R}^+ \to \Delta$, but within the class denoted $K^i(x^0,t_0)$, and with attainability sets $\Phi^i(\bar{x}^0,t_0,t)$ at $t \geq t_0$, and reachability sets $\Phi^i(\bar{x}^0,t_0,\lceil t_0,t))$ at $t \geq t_0$. The reachable cone $\Phi^i(x^0,t_0,\mathbb{R}^+)$ is called the *strong semi-game for player* i, briefly i-*game*.

DEFINITION 8.2.2. The game (8.2.1) is called strongly i-controllable at $\bar{x}^0 \in \Delta$ for $Q(i)$ if and only if there is $\bar{P}^i(\cdot)$ such that all game motions from $K^i(x^0,t_0)$, $t_0 \in \mathbb{R}^+$ of (8.2.7) exhibit $Q(i)$.

As before, $\bar{x}^0$ is strongly i-controllable for $Q(i)$, the set of all such $\bar{x}^0$ forms the *region of strong i-controllability for* $Q(i)$, denoted

Δ_Q^i. It is also called i-*strong region* or i-*winning region*. Any subset of Δ_Q^i is strongly i-controllable for $Q(i)$. Observe also that, as we defined it, the strong i-controllability is t_0-uniform. The corresponding program $\bar{P}^i(\cdot)$ is called a *winning program*.

In general terms, the winning program $\bar{P}^i(\cdot)$ for $Q(i)$ may be accompanied by a winning program $\bar{P}^j(\cdot)$ for $Q(j)$, $j \neq i$, on the same subset of Δ. Consequently we may have the *joint winning region*

$$W_Q^{12} \triangleq \Delta_Q^1 \cap \Delta_Q^2 \neq \phi . \tag{8.2.8}$$

This may be either undesired or convenient. For instance, in aircraft combat, it would mean mutual kill, in the case of the two robotic arms, operation on the same workpiece. Then, it may be of interest to define the *region of guaranteed* i-*winning* $\mathbb{W}_Q^i \triangleq \Delta_Q^i - \Delta_Q^j$ from which the winning $\bar{P}^j(\cdot)$ is excluded, and we always have

$$\mathbb{W}_Q^1 \cap \mathbb{W}_Q^2 = \phi . \tag{8.2.9}$$

On the other hand, the property (8.2.8) is contradicted in the *strictly competitive game* in which the opposition has no objective on its own but only attempts to contradict $Q(i)$, say $Q(j) = $ no $Q(i)$. In such a game we have

$$\Delta_Q^1 \cap \Delta_Q^2 = \phi \tag{8.2.10}$$

whence all i-winning regions are guaranteed, that is, $\mathbb{W}_Q^i = \Delta_Q^i$.

The reader might have observed that technically, when forming the strong semi-game, we adjoined the uncertainty vector $\bar{w}$ to the action of the opposing program $\bar{P}^j(\cdot)$. Such a game is then formally played against $U_j \times W$, see (8.2.7), although the control variables $\bar{u}^j$ are governed by $\bar{P}^j(\cdot)$. The reverse case is also possible. Suppose that the game is strictly competitive and that player j is a disorganized nature, having no control program. Then he becomes a "passive opponent" with undefined but bounded $\bar{u}^j(t) \in U_j$ which may be adjoined to the uncertainty vector $\bar{w}(t)$ augmenting it to $\bar{w}' \in U_j \times W$. We call such a scenario *game against nature* or *worst case design*. It obviously coincides with the control under certainty discussed since Section 2.2.

We conclude immediately that sufficient conditions for strong i-controllability for $Q(i)$ are the same as for strong controllability for Q, but with $\bar{w}' \in U_2 \times W$ replacing $\bar{w} \in W$. Hence the corollaries generating the controllers are the same and the regions Δ_Q^i are determined by the techniques of determining Δ_Q's, see Section 5.4.

The above applies to all objectives Q introduced in Chapters 3-7 which we would like now to turn into $Q(i)$'s of the strong semi-games concerned. In particular, it applies to collision-first or capture-first of Section 6.4 which seem to appear in the majority of game scenarios. We may as well state the conditions for one of the two, namely capture-first.

Given the sets T^i, T^i_C, A^i, $T^i \cap A^i = \phi, \Delta^i_0$ as well as functions $v^i_F(\cdot), v^i_A(\cdot) : \Delta^i_0 \rightarrow \mathbb{R}$ with $v^i_F, v^i_A > 0$ and sets Δ^i_1, Δ^i_2 defined in Section 6.4 without the superscripts, we have the following sufficient conditions.

CONDITIONS 8.2.1. The game (8.2.1) is strongly i-controllable on Δ^i_0 for capture-first in T^i_C avoiding A^i, if there is a program $\bar{P}^i(\cdot)$ and functions $v^i_F(\cdot), v^i_A(\cdot)$ defined on Δ^i_0 such that

(i) $v^i(\bar{x}) \rightarrow \infty$, as $|\bar{x}| \rightarrow \infty$;

(ii) Δ^i_0 is strongly positively invariant;

(iii) for all $\bar{x} \in \Delta_0$, $\bar{u}^i \in \bar{P}^i(\bar{x}, \bar{u}^j)$, there are constants $c^i_F \geq 0$, c^i_A such that

$$sup_{\bar{u}^j, \bar{w}} \ \nabla v^i_F(\bar{x})^T \cdot \bar{f}(\bar{x}, \bar{u}^i, \bar{u}^j, \bar{w}, t) \leq -c^i_F < 0 ; \qquad (8.2.11)$$

$$inf_{\bar{u}^j, \bar{w}} \ \nabla v^i_A(\bar{x})^T \cdot \bar{f}(\bar{x}, \bar{u}^i, \bar{u}^j, \bar{w}, t) \geq -c^i_A . \qquad (8.2.12)$$

The proof is the same as for Conditions 6.4.2. We also have the same estimate for the i-winning region which is the maximal

$$\Delta^i_0 = \left\{ \bar{x} \in \Delta - (T_C \cup \Delta_2) \ \middle| \ \frac{c^i_F}{v_F(\bar{x}) - v_F} > \frac{c^i_A}{v_A(\bar{x}) - v_A} \right\} , \qquad (8.2.13)$$

obtainable under suitably selected c^i_F, c^i_A.

We have mentioned already that the winning controllers may be found from corollaries designed in the same way as in Chapters 3-7. Here is one for our case, with the aim to satisfy (8.2.11), (8.2.12).

COROLLARY 8.2.1. *Given* $(\bar{x}^0, t_0) \in \Delta_0 \times \mathbb{R}$, *if there is a triple* $\bar{u}^1_*, \bar{u}^2_*, \bar{w}^* \in U_1 \times U_2 \times W$ *such that*

$$inf_{\bar{u}^i} \ sup_{\bar{u}^j, \bar{w}} \ \nabla v^i_F(\bar{x})^T \cdot \bar{f}(\bar{x}, \bar{u}^1, \bar{u}^2, \bar{w}, t) \leq -c^i_F ; \qquad (8.2.14)$$

$$sup_{\bar{u}^i} \ inf_{\bar{u}^j, \bar{w}} \ \nabla v^i_A(\bar{x})^T \cdot \bar{f}(\bar{x}, \bar{u}^1, \bar{u}^2, \bar{w}, t) \geq -c_A ; \qquad (8.2.15)$$

then (8.2.9), (8.2.10) are met with $\bar{u}^i_* \in \bar{P}^i_*(\bar{x}, \bar{u}^j, t)$.

EXAMPLE 8.2.3. Let us return to the turret game of Example 8.2.1 with the described air combat.

(A) We consider first the *1-game*, searching for the strong 1-controllability. Take $V_F^1 = x_1$, $V_A^1 = x_2$ and

$$T^1 \supset \Delta_1^1 = \{(x_1, x_2) \,|\, V_F^1 \le \varepsilon_1\},$$

$$A^1 \subset \Delta_2^1 = \{(x_1, x_2) \,|\, V_A^1 \le \varepsilon_2\}.$$

We need to find constants $c_A^1, c_F^1 > 0$, generating the maximal

$$\Delta_0^1 = \left\{ (x_1, x_2) \in \Delta \ \left| \ \frac{c_F^1}{V_F^1(\bar{x}) - \varepsilon_1} > \frac{c_A^1}{V_A^1(\bar{x}) - \varepsilon_2} \right. \right\}$$

such that for all $\bar{x}$ in this set there exists a control program $P_*^1(\cdot)$ generating $u_*^1(\cdot)$ for which

$$\sup_{u_1^2, u_2^2} \nabla V_F^1 \cdot \bar{f}(\bar{x}, u_*^1, u_1^2, u_2^2) \le -c_F^1 < 0 \tag{8.2.16}$$

$$\inf_{u_1^2, u_2^2} \nabla V_A^1 \cdot \bar{f}(\bar{x}, u^*, u_1^2, u_2^2) \ge -c_A^1. \tag{8.2.17}$$

Then such Δ_0^1 constitutes a winning region for player 1. From (8.2.16),

$$\left. \begin{aligned} \sup_{u_1^2, u_2^2} (u_1^2 - u^1) &\le -c_F^1 < 0, \\[2mm] \hat{u}_1^2 - u^1 &\le -c_A < 0. \end{aligned} \right\} \tag{8.2.18}$$

For a positive c_F^1 to exist, this necessarily implies that $\hat{u}_1^2 - \hat{u}^1 < 0$; that is, $\gamma_1 = \hat{u}_1^2 / \hat{u}^1 < 1$. Select the control program for player 1 to be $u_*^1 = \hat{u}^1$ for all time. Then the largest c_F^1 compatible with (8.2.18), independent of the state, is $c_F^1 = \hat{u}^1 - \hat{u}^2$. From (8.2.17),

$$\left. \begin{aligned} \inf_{u_1^2, u_2^2} (-u_2^2) &\ge -c_A^1, \\[2mm] c_A^1 &\ge \hat{u}_2^2. \end{aligned} \right\} \tag{8.2.19}$$

The smallest c_A^1 compatible with (8.2.19), independent of state, is $c_A^1 = \hat{u}_2^2$. Consequently for $\gamma_1 < 1$,

$$\begin{aligned} \Delta_0^1 &= \left\{ (x_1, x_2) \in \Delta \ \left| \ \frac{\hat{u}^1 - \hat{u}_1^2}{x_1 - \varepsilon_1} > \frac{\hat{u}_2^2}{x_2 - \varepsilon_2} \right. \right\} \\[3mm] &= \left\{ (x_1, x_2) \in \Delta \ \left| \ x_2 - \varepsilon_2 > \frac{\gamma_2}{1 - \gamma_1}(x_1 - \varepsilon_1) \right. \right\} \end{aligned} \tag{8.2.20}$$

where $\gamma_2 \overset{\Delta}{=} \hat{u}_2^2 / \hat{u}^1$. So the defined Δ_0^1 is easily shown to be a subset of Δ_F^1. Namely, integrating the state equations, we see that if player 1 always selects u_*^1 he will win from all initial states satisfying

$$(x_2 - \varepsilon_2) > \gamma_2 (x_1 - \varepsilon_1) \ . \tag{8.2.21}$$

Indeed, the greatest effect player 2 can have on the outcome is when his control selection is $u_1^2 = 0$ and $u_2^2 = \hat{u}_2^2$ to give the greatest rate of decrease of x_2 towards ε_2. So our Δ_0^1 is an underestimate of $\Delta_{\mathcal{F}}^1$ with the latter defined by (8.2.21).

For $\gamma_1 \geq 1$, we clearly have $\Delta_{\mathcal{F}}^1 = \phi$.

(B) Consider now the strong game for player 2, the *2-game*. Take $v_{\mathcal{F}}^2 = x_2$, $v_A^2 = x_1$ with reversed $v_{\mathcal{F}} = \varepsilon_2$ and $v_A = \varepsilon_1$ where

$$T^2 \supset \Delta_1^2 \quad \text{and} \quad A^2 \subset \Delta_2^2 \ .$$

Then the corresponding conditions to (8.2.16) and (8.2.17) are

$$\sup_{u^1} \nabla v_{\mathcal{F}}^2 \cdot \bar{f}(\bar{x}, u^1, u_1^{2*}, u_2^{2*}) \leq -c_{\mathcal{F}}^2 < 0 \tag{8.2.20}$$

$$\inf_{u^1} \nabla v_A^2 \cdot \bar{f}(\bar{x}, u^1, u_1^{2*}, u_2^{2*}) \geq -c_A^2 \tag{8.2.21}$$

wherefrom the program of selection of u_1^2, u_2^2 for player 2 is to be found. From (8.2.20), $\sup\limits_{u^1} (-u_2^2) \leq -c_{\mathcal{F}}^2 < 0$, or

$$-u_2^2 \leq -c_{\mathcal{F}}^2 < 0 \ . \tag{8.2.22}$$

This necessarily requires the selection of u_2^2 to be nonzero. From (8.2.21), $\inf\limits_{u^1} (u_1^2 - u^1) \geq -c_A^2$, or $u_1^2 - \hat{u}^1 \geq -c_A^2$ or

$$c_A^2 \geq \hat{u}^1 - u_1^2 \ . \tag{8.2.23}$$

Equations (8.2.22) and (8.2.23) are independent of the state variables. From (8.2.22), a maximum value of c_F^2 is given by $c_A^2 = \hat{u}_2^2/(1+\delta)$ for a selection of $u_2^2 = \hat{u}_2^2/(1+\delta)$ with $\delta \in [0,\infty)$. This means that acceptable values for u_1^2 satisfy

$$0 \leq u_1^2 \leq \left[1 - \frac{1}{1+\delta} \right] \hat{u}_1^2 = \frac{\delta}{1+\delta} \, \hat{u}_1^2 \ .$$

Hence we can write, selecting the smallest c_A^2 in (8.2.23), given a value of u_1^2,

$$\frac{c_A^2}{c_{\mathcal{F}}^2} = \frac{\hat{u}^1 - \beta\delta\hat{u}_1^2/(1+\delta)}{\hat{u}_2^2/(1+\delta)} = \frac{\hat{u}^1 + \delta\,(\hat{u}^1 - \beta\hat{u}_1^2)}{\hat{u}_2^2} = \frac{1 + \delta\,(1 - \beta\gamma_1)}{\gamma_2}$$

with $\beta \in [0,1]$.

Now *provided* $\gamma_1 \leq 1$, that is, $\hat{u}_1^2 \leq \hat{u}^1$, the smallest value of

$c_A^2 / c_{\mathbf{F}}^2$ is obtained when $\delta = 0$. So $c_A^2 / c_{\mathbf{F}}^2 = 1/\gamma_2$ when $u_2^{2*} = \hat{u}_2^2$ and $u_1^{2*} = 0$. In this case, the maximum winning region for player 2 is

$$\Delta_{\mathbf{F}}^2 = \{ (x_1, x_2) \in \Delta \mid x_2 - \varepsilon_2 < \gamma_2 (x_1 - \varepsilon_1) \} .$$

For $\gamma_1 \leq 1$, it is possible with $\beta = 1$ to select δ sufficiently large, this defining the selection of u_1^{2*} and u_2^{2*} to make the above expression for $c_A^2 / c_{\mathbf{F}}^2$ negative. Equivalently, for $\gamma_1 > 1$ we can select a $\beta \in (0,1)$ such that $\beta \hat{u}_1^2 = \hat{u}^1$. Then we require that player 2 play $u_1^{2*} = \beta \hat{u}_1^2$, $u_2^{2*} = \hat{u}_2^2 (1 - \beta)$. In this case then, $\Delta_{\mathbf{F}}^2$ must be all of Δ.

From the above, it may be seen that $\Delta_{\mathbf{F}}^1 \cap \Delta_{\mathbf{F}}^2 = \phi$ whence immediately $\Delta_{\mathbf{F}}^i = \mathbb{w}_{\mathbf{F}}^i$. The reader may as well take note of this fact which would be useful later. The above does not mean that the game is strictly competitive, as (8.2.10) was only necessary for such a game, that is, followed from it. $\qquad\square$

EXAMPLE 8.2.4. Recall Examples 6.3.1, 6.4.1 and rearrange the scenario into the game between the craft being an evader or player 2, with the speed $v_E = \text{const}$ and the turbulence being now a programmed pursuer, or player 1, with the speed $v_p = \text{const} > v_E$. The kinematic equations (6.3.7) become now

$$\left. \begin{aligned} \dot{R} &= v_E \cos (\theta + u^2) , \\ R\dot{\theta}_E &= v_E \sin (\theta + u^2) , \\ \dot{r} &= v_p \cos u^1 - v_E \cos u^2, \\ r\dot{\theta}_p &= v_p \sin u^1 - v_E \sin u^2 , \end{aligned} \right\} \qquad (8.2.24)$$

with the change of notation seen in Fig. 8.4.

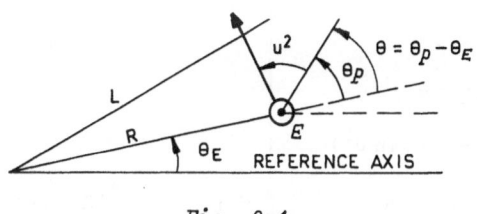

Fig. 8.4

The targets are now specified as

$$T^1 \triangleq \{ (R,\theta_E,r,\theta_P) \in \mathbb{R}^4 \mid r \leq r_P \} = A^2 \ ,$$
$$T^2 \triangleq \{ (R,\theta_E,r,\theta_P) \in \mathbb{R}^4 \mid R \leq r_E \} = A^1 \ ,$$

(8.2.25)

and the two strong semi-games are as follows.

2-game: The evader wishes to escape to his shelter before being intercepted by the pursuer, that is, capture T^2 avoiding $A^2 = T^1$.

1-game: The pursuer wants to intercept the evader before the evader reaches his shelter, that is, capture T^1 avoiding $A^1 = T^2$.

Each of these objectives is to be attained against all options of the opposition. The 2-game has been played under different notation in Example 6.4.1. It was found that the winning region was

$$\Delta_{\mathcal{F}}^2 : r - r_P > (v_E + v_P)(R - r_E)/v_E$$

(8.2.26)

and the control program was determined by $\cos(\theta + u^2) = -1$. Let us now complete the study by investigating the 1-game. The investigation is made easier by redefining the coordinate system as shown in Fig. 8.5. Observe that interchanging the positions of evader and pursuer does not change the results as long as the relevant connection between coordinates is realized.

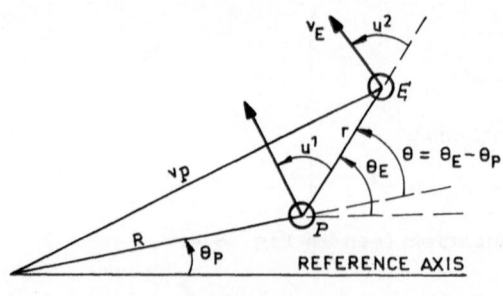

Fig. 8.5

Then the kinematic equations of motion become

$$\dot{R} = v_P \cos(\theta + u^1) \ ,$$
$$R\dot{\theta}_P = v_P \sin(\theta + u^1) \ ,$$
$$\dot{r} = v_E \cos u^2 - v_P \cos u^1 \ ,$$
$$r\dot{\theta}_E = (v_E \sin u^2 - v_P \sin u^1) \ .$$

(8.2.27)

To apply Conditions 8.2.1, we let $V_{\mathcal{F}}^1 \triangleq r$, $V_A^1 \triangleq R$ and $\Delta_1^1 : V_{\mathcal{F}}^1 \leq r_P$,

$\Delta_2^1 : V_A^1 \leq r_E$.   Then condition (iii) becomes

$$sup_{u^2} \ (v_E \cos u^2 - v_P \cos u^1) \leq -c_{\mathcal{F}}^1$$

or

$$v_E - v_P \cos u^1 \leq -c_{\mathcal{F}}^1 \ , \tag{8.2.28}$$

$$inf_{u^2} \ v_P \cos (\theta + u^1) \geq -c_A^1$$

or

$$v_P \cos (\theta + u^1) \geq -c_A^1 \ . \tag{8.2.29}$$

Then Δ_0 is defined by

$$R - r_E > \frac{c_A^1}{c_{\mathcal{F}}^1} \ (r - r_P) \ . \tag{8.2.30}$$

Now condition (8.2.28) implies

$$\frac{v_E}{v_P} - 1 \leq \frac{v_E}{v_P} - \cos u^1 \leq - \frac{c_{\mathcal{F}}^1}{v_P} < 0 \ .$$

This inequality imposes the restriction $v_P > v_E$.   Furthermore it is satisfied if

$$c_{\mathcal{F}}^1 = v_P (1 - \sigma)/(1 + \delta) \quad \text{for} \quad \delta \in [0, \infty) \ ,$$

where $\sigma = (v_E/v_P) < 1$.   We also observe that $\cos u^1 > (v_E/v_P)$.   For a given $c_{\mathcal{F}}^1$ the region Δ_0 is maximized by choosing the smallest c_A^1 such that condition (8.2.29) is met for all possible θ .   Now $c_A^1 \geq -v_P \cos (\theta + u^1)$.   The largest value of $-v_P \cos (\theta + u^1)$ is v_P , so we select $c_A^1 = v_P$.   Therefore

$$c_A^1/c_{\mathcal{F}}^1 = v_P (1 + \delta)/(v_P - v_E) \ .$$

The set Δ_0 is maximized into the winning region by taking $\delta = 0$ which gives

$$\Delta_{\mathcal{F}}^1 : R - r_E > v_P (r - r_P)/(v_P - v_E) \tag{8.2.31}$$

provided $v_P - v_E > 0$ as stated before. With $\delta = 0$, $c_{\mathcal{F}}^1 = v_P - v_E$ which is obtained when $\cos u^1 = 1$. This, thus, is the winning program for the pursuer. He heads always toward the evader along the line of sight. As this inequality is not dependent on θ_E, θ_P , returning to the original coordinate system, we obtain

$$\Delta_{\mathcal{F}}^1 : L - r_E > v_P (r - r_P)/(v_P - v_E) \ . \tag{8.2.32}$$

$\square$

8.2.1 Given the selector equations

$$\dot{x}_1 = x_2 , \qquad \dot{x}_2 = -u^2 x_2 - ax_1 - bx_1^3 + u^1 ,$$

with $a,b > 0$, $u^2 \in [0,1]$, write the 1-game in terms of the corresponding contingent equations and specifying suitable $\bar{P}^1(\cdot)$ define the attainability sets. Assuming the objectives as capture in respective targets, design the winning $\bar{P}^1(\cdot)$ and discuss the region of strong controllability.

8.2.2 Consider the system

$$\dot{x} = u^1 e^x + u^2 e^{-x}$$

with $x \in [-1,1] \subset \mathbb{R}$, $|u^i| \le 1$, $i = 1,2$, $x(0) = x^0$.

(i) Suppose $Q(1), Q(2)$ mean collisions with the targets $T_1 : x = 1$, $T_2 = x = -1$, respectively. Find the winning strategies and strong regions.

(ii) Suppose $Q(1), Q(2)$ mean avoidance of the sets $A_1 : x = 1$, $A_2 : -1$. Find the winning strategies and strong regions. Discuss the relation to the case (i).

8.2.3 Consider the system $\ddot{q} + u^1 + \Pi(q) = u^2$ with $q \in \Delta \subset \mathbb{R}$, $\Delta : \Pi(q)q \ge 0$; u^2 constrained by the limitation of power $|u^2 \dot{q}| \le N$, for some $N > 0$ and with the target $T_1 = \{(q,\dot{q}) | E(q,\dot{q}) \le 1\}$. Find the winning strategy and strong region for capture in T_1 subject to the restriction that $|u^1| < M(N)$. Specify the relationship between the bounds N and M.

8.2.4 In the game described by the dynamics

$$\begin{cases} \dot{x}_1 = 2x_2 - (1/2x_1)(x_1^2 + x_2^2 - u^1 u^2) , \\ \dot{x}_2 = -2x_1 - (1/2x_2)(x_1 + x_2^2 - u^1 u^2) , \end{cases}$$

with $u^1 \in [0,1]$, $u^2 \in [1,2]$ the player 1 wants to collide with the target $T_1 : x_1^2 + x_2^2 \le 1$ while the player 2 wants to avoid the interior of it, that is, $A_2 : x_1^2 + x_2^2 < 1$. Let $r^2 = x_1^2 + x_2^2$. Taking $V = r^2$, show that $\dot{V} < 0$ if $r > u^1 u^2$, $\dot{V} = 0$ if $r = u^1 u^2$ and $\dot{V} > 0$ if $r < u^1 u^2$ and consequently that both players achieve their objectives by setting $u^1 u^2 = 1$, that is, playing $u^1 = (1/u^2)$.

8.2.5 A unit mass located at the position $x(t), y(t)$ in the horizontal
plane Oxy is connected by two springs with coefficients k_1, k_2 to
the points $(a,0)$ and (a,b), respectively, see Fig. 8.6. Players
1 and 2 have non-stretchable ropes which are connected from (x,y)
to $(0,0)$ and $(0,b)$ respectively. Suppose the mass is initially at
rest at (x^0, y^0). Player 1 wishes to transfer (x,y) from (x^0, y^0)
to $(0,0)$, player 2 wants to transfer the same (x^0, y^0) to $(0,b)$.

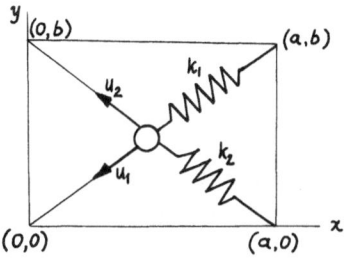

Fig. 8.6

The force at their disposal is $|u^i| \leq \hat{u}^i$, $i = 1,2$, respectively.
The motion equations are

$$\ddot{x} = \frac{k_1 (a - x)^2}{\sqrt{(a - x)^2 + y^2}} + \frac{k_2 (a - x)^2}{\sqrt{(a - x)^2 + (b - y)^2}}$$

$$- \frac{u_1 x}{\sqrt{x^2 + y^2}} - \frac{u_2 x}{\sqrt{x^2 + (b - y)^2}} \quad ;$$

$$\ddot{y} = \frac{-k_1 y^2}{\sqrt{(a - x)^2 + y^2}} + \frac{k_2 (b - y)^2}{\sqrt{(a - x)^2 + (b - y)^2}}$$

$$- \frac{u_1 y}{\sqrt{x^2 + y^2}} + \frac{u_2 (b - y)}{\sqrt{x^2 + (b - y)^2}} \quad .$$

Choose state variables, write the state equations of the game,
design controllers and specify strong regions of controllability for
attaining the objectives mentioned.

8.3 MAP OF A GAME

The selection of control programs and the determination of the regions of i-controllability and i-winning regions must be based on the interface of the two strong semi-games concerned. Such an interface can, in turn, be established by examining the decomposition of the playing region Δ into sets of options which we have called a map of the game.

Fig. 8.7 shows what may be concluded directly from the definitions of Section 8.2. At the outset, unless otherwise stated, we exclude the dead zone Δ_n from our study assuming $\Delta = \Delta_q^1 \cup \Delta_q^2$, which yields $\Delta_n = \phi$. Since strong i-controllability implies i-controllability for the same $Q(i)$, we have

$$\Delta_Q^i \subset \Delta_q^i , \tag{8.3.1}$$

as shown in Fig. 8.7. Let us now introduce the semi-neutral sets $\Delta_N^i \triangleq \Delta - \Delta_Q^i$, called i-*neutral*. By definition of Δ_Q^i, the strong i-controllability for $Q(i)$ is contradicted on Δ_N^i, namely for each $\bar{P}^i(\cdot)$ we have $\bar{P}_*^j(\cdot)$ such that there is at least one motion of $K^i(\bar{x}^0, t_0)$, $\bar{x}^0 \in \Delta_N^i$, which does not exhibit $Q(i)$. This is the *defining property* of Δ_N^i.

Let us now introduce the *neutral set*:

$$\Delta_N \triangleq \Delta_N^1 \cap \Delta_N^2 = \Delta - (\Delta_Q^1 \cup \Delta_Q^2) \tag{8.3.2}$$

on which there is no strong winning for either of the two players. We may now investigate the alternatives to such winning.

For $\bar{x}^0 \in \Delta_N \cap \Delta_q^i$ we know that there exist a pair $\hat{\bar{P}}^1(\cdot), \hat{\bar{P}}^2(\cdot)$ such that every resulting game motion has the property $Q(i)$. Suppose player i persists in playing $\hat{\bar{P}}^i(\cdot)$. If player j plays $\hat{\bar{P}}^j$, he will certainly let

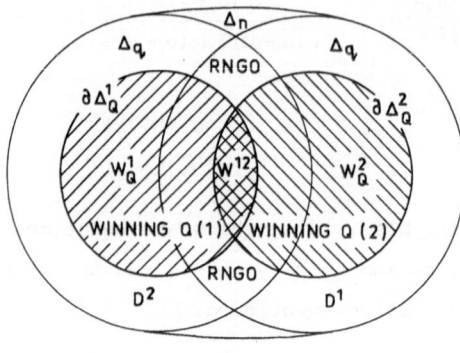

Fig. 8.7

player i achieve his objective. There is no certainty in his achieving $Q(j)$ unless $\bar{x}^0$ also belongs to Δ_q^j and $\hat{\bar{P}}^1(\cdot), \hat{\bar{P}}^2(\cdot)$ is a pair of programs for which every resulting motion of $K(\bar{x}^0, t)$ exhibits $Q(j)$. However, player j may choose for each $\hat{\bar{P}}^i(\cdot)$ to play $\bar{P}_*^j(\cdot)$, as above in the defining property for Δ_N^i. He then assures that player i cannot achieve his objective $Q(i)$. But again, if $\bar{x}^0 \in \Delta_q^j$ and the pair $\bar{P}^i(\cdot), \bar{P}_*^j(\cdot)$ is sufficient for j-controllability at $\bar{x}^0$, player j may still achieve his objective $Q(j)$.

If such $\bar{x}^0 \in \Delta_q^i - \Delta_q^j$, player j can force what we may term a *draw* by plying $\bar{P}_*^j(\cdot)$ for each program $\bar{P}^i(\cdot)$ of player i. (Neither of the properties $Q(1)$ and $Q(2)$ are attainable on trajectories under this strategy selection.) Similarly, if such $\bar{x}^0 \in \Delta_q^j - \Delta_q^i$, player i can force a draw.

We shall therefore call the set of these $\bar{x}^0$,

$$D \triangleq [\Delta_q^1 - (\Delta_q^2 \cup \Delta_Q^1)] \cup [\Delta_q^2 - (\Delta_q^1 \cup \Delta_Q^2)] \equiv D^2 \cup D^1,$$

the *guaranteed draw region* of the game, where D^i is the guaranteed draw region for player i, see Fig. 8.6. If $\bar{x}^0 \in \Delta_q^1 - \Delta_q^2$ *player 2's guaranteed draw strategy* will be to play $\bar{P}_*^2(\bar{P}^1)$ for each $\bar{P}^1$ of player 1. If $\bar{x}^0 \in \Delta_q^2 - \Delta_q^1$, *player 1's draw strategy* will be to play $\bar{P}_*^1(\bar{P}^2)$ for each $\bar{P}^2$ of player 2.

In the remaining set $(\Delta_q^1 \cap \Delta_q^2) \cap \Delta_N$, we may have all four options: (a) player 1 achieves $Q(1)$, player 2 does not achieve $Q(2)$; (b) player 1 does not achieve $Q(1)$, player 2 achieves $Q(2)$; (c) both players achieve their objective; (d) both players do not achieve their objective. For this reason, we can call this region the *region of no guaranteed outcome* (RNGO), see Fig. 8.6.

In practice, the regions W^i may become very small and players may be forced to use RNGO where they still have a chance to win, in preference to the draw regions. It will be seen that, in cases (a),(b) the set concerned becomes a barrier separating W^i's.

In those games in which a terminal time t_f is specified, there may be $\bar{x}^0$ in Δ_q^1 for which player 1 cannot achieve his objective $Q(1)$ within this specified time, but would do so if time t_f was extended. A similar case would exist for $\bar{x}^0 \in \Delta_q^2$. Under the former analysis, these $\bar{x}^0$ would belong to Δ_n. It would be realistic in some games, for example, the turret game, to define the union of D and Δ_n as the draw region.

We have found that in the turret game, see Fig. 8.8,

$$\Delta_q^1 = \Delta_q^2 = \Delta \; ; \qquad \Delta_Q^1 \cap \Delta_Q^2 = \phi \; ;$$

$$W^1 = \{ (x_1, x_2) \in \Delta \,|\, x_2 - \varepsilon_2 > \frac{\gamma_2}{1 - \gamma_1} (x_1 - \varepsilon_1) \} \quad \text{for} \quad \gamma_1 < 1 \; ;$$

$$W^1 = \phi \quad \text{for} \quad \gamma_1 \geq 1 \; ;$$

$$W^2 = \{ (x_1, x_2) \in \Delta \,|\, x_2 - \varepsilon_2 < \gamma_2 (x_1 - \varepsilon_1) \} \quad \text{for} \quad \gamma_1 \leq 1 \; ;$$

$$W^2 = \Delta \quad \text{for} \quad \gamma_1 > 1 \; .$$

Each of the diagrams in Fig. 8.8 displays the guaranteed winning regions for the players in each of the three cases $\gamma_1 < 1$, $\gamma_1 = 1$ and $\gamma_1 > 1$.

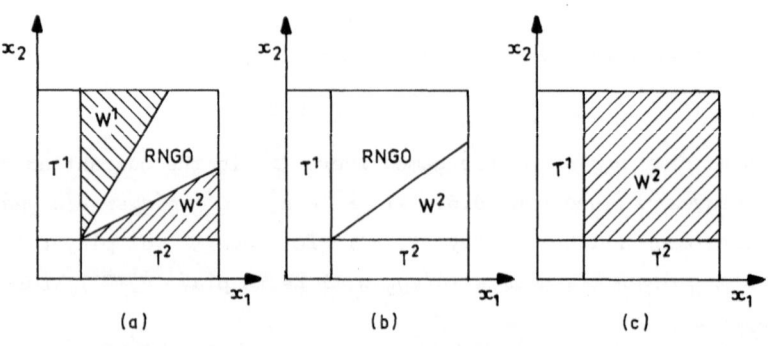

Fig. 8.8

In the case $\gamma_1 > 1$, Fig. 8.8(c), player 2 is guaranteed of a win in all of Δ with objective $Q(2)$; player 1 cannot hope to win unless player 2 does not play one of his guaranteed winning strategies. For the two remaining cases, there exist points in Δ not in W_1^1 and W_2^2, indeed they belong to $\Delta_q^1 \cap \Delta_q^2$, that is to the region of no guaranteed outcome. As mentioned above, both players cannot achieve their objective, thus case (c) is eliminated from our list of cases. Simple examination of programs and system equations shows that each of cases (a), (b) and (d) may occur.

These results may be compared with those given for minimum time pursuit-evasion analysis for the turret game in Ardema-Heymann [1]. However, it must be noted that the qualitative objectives defined here for the players *does not allow* for both target sets to be reached simultaneously.

Some of the following examples have been constructed to illustrate the reasoning developed in the previous subsections.

EXAMPLE 8.3.1. Consider the turret game of Examples 8.2.1, 8.2.3 with the added requirement for it to be completed in the interval $[0, t_f)$, $t_f = \pi/4$. Assume $\varepsilon_1 = \varepsilon_2 = \pi/4$, $\hat{u}^1 = 2$, $\hat{u}_1^2 = 1$ and $\hat{u}_2^2 = 2$. We determined previously that $\Delta_q^1 = \Delta$ with $u_1^2 = 0$, $u_2^2 = 0$ and u^1 non-zero. It is now restricted to $\Delta_q^1 = \{(x_1, x_2) \in \Delta \,|\, x_1 < 3\pi/4\}$, this being the maximum region for which the trajectory enters T^1, being specified by maximum $u^1 = 2$. Similar reasoning yields $\Delta_q^2 = \{(x_1, x_2) \in \Delta \,|\, x_1 < 3\pi/4\}$. Consequently we have

$$W^1 = \{(x_1, x_2) \in \Delta \,|\, x_2 > 2x_1 - \pi/4\} \,,$$

$$W^2 = \{(x_1, x_2) \in \Delta \,|\, x_2 < x_1 \text{ and } x_2 < 3\pi/4\} \,.$$

A dead zone now exists. It is $\Delta_n = \{(x_1, x_2) \in \Delta \,|\, x_1, x_2 \geq 3\pi/4\}$. The state space decomposition is shown in Fig. 8.9.

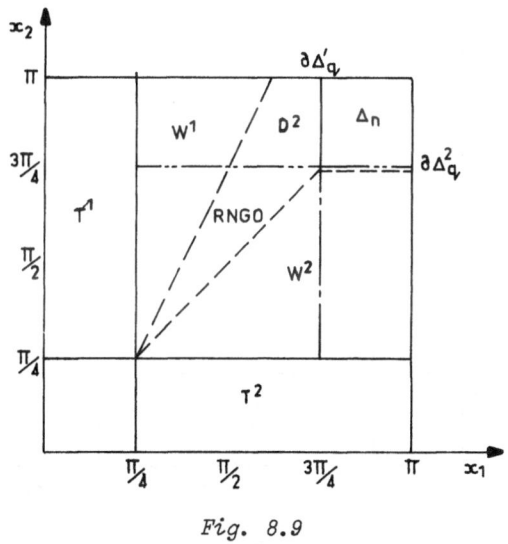

Fig. 8.9

The interesting feature of this example is that a draw region D_2 for player 2 exists, but none for player 1. Even though he cannot achieve his objective for any strategy $\bar{P}^2(\cdot)$, he can play for each $\bar{P}^1(\cdot)$ a strategy that will prevent player 1 achieving his objective in this region. The region of no guaranteed outcome is shown as the hatched region in the figure. □

EXAMPLE 8.3.2. We take the game dynamics

$$\dot{x}_1 = 1 - \hat{u} + u_1^2 \,,$$

$$\dot{x}_2 = -u_2^2 \,,$$

with $x_1, x_2 \in [0, \pi]$. Player 1 selects his control variable u^1 from $[2,3]$, and player 2 selects his control variables u_1^2 and u_2^2 that are constrained by $u_1^2 + u_2^2 = 1$ with $u_1^2, u_2^2 \geq 0$.

Again, take the same qualitative objective properties with time restriction in Example 8.3.1 and $\varepsilon_1 = \varepsilon_2 = \pi/4$. We find by straightforward analysis that

$$W^1 = \{(x_1, x_2) \in \Delta \mid x_2 > x_1 \text{ and } x_1 < \pi/2\},$$

$$W^2 = \{(x_1, x_2) \in \Delta \mid x_2 < x_1/2 + \pi/8 \text{ and } x_2 < \pi/2\},$$

$$\Delta_q^1 = \{(x_1, x_2) \in \Delta \mid x_1 < \pi/2\},$$

$$\Delta_q^2 = \{(x_1, x_2) \in \Delta \mid x_2 < x_1 \text{ and } x_2 < \pi/2\},$$

$$\Delta_n = \{(x_1, x_2) \mid x_1, x_2 \geq \pi/2\}.$$

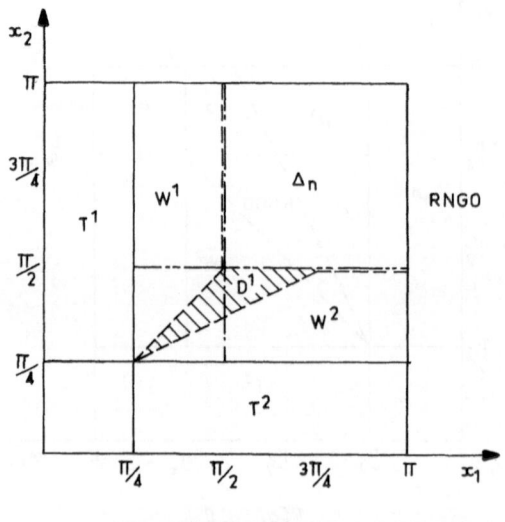

Fig. 8.10

□

In this example, there is again a dead zone Δ_n, a region of no guaranteed outcome, and this time a guaranteed draw region D^1 for player 1, in which he can prevent an outcome in favour of player 2 although he cannot hope to achieve his objective, see Fig. 8.10.

EXAMPLE 8.3.3. Consider the turret game with the same system dynamics and objectives (no time restriction) but with $\varepsilon_1 = \pi/4$, $\varepsilon_2 = \pi/4$ and with control selection given by $u^1 \in [1,2]$, $u_1^2 \in [0,1]$ and $u_2^2 \in [1,2]$. The following state decomposition is easily determined and shown in Fig. 8.11:

$$W^1 = \{(x_1, x_2) \in \Delta \mid x_2 > 2x_1 - \pi/4\},$$

$$W^2 = \{(x_1, x_2) \in \Delta \mid x_2 < 2x_1 - \pi/4\},$$

$$\Delta_q^1 = \{(x_1, x_2) \in \Delta \mid x_2 > x_1/2 + \pi/8\},$$

$$\Delta_q^2 = \Delta, \qquad \Delta_n = \phi.$$

We can also observe, what will be useful in the next section, that the set Δ in this example is separated into the two strong winning regions by the set $B = \{(x_1, x_2) \in \Delta \mid x_2 = 2x_1 - \pi/4\}$. This set is in fact the region of no guaranteed outcome. []

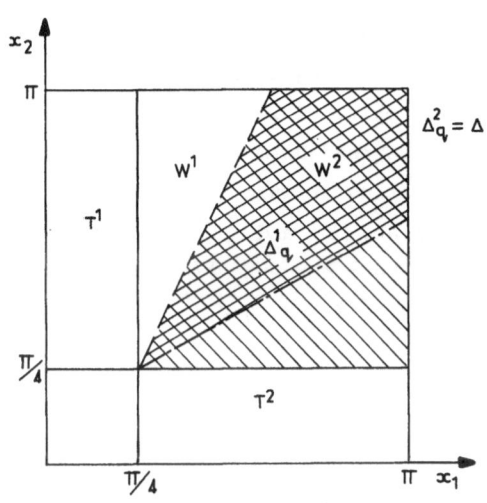

Fig. 8.11

EXAMPLE 8.3.4. Consider the game described in Example 8.3.3 with a slight change in the qualitative objectives of the players. Namely that for $Q(1)$ it is allowed for $x(\bar{t}) \in T^2$ when $x(\bar{t}) \in T^1$, similarly for $Q(2)$ (mutual kill). It is easy to show that

$$\Delta_Q^1 = \{(x_1, x_2) \in \Delta \mid x_2 \geq 2x_1 - \pi/4\},$$

$$\Delta_Q^2 = \{(x_1, x_2) \in \Delta \mid x_2 \leq 2x_1 - \pi/4\},$$

$$\Delta_q^1 = \{(x_1, x_2) \in \Delta \mid x_2 \geq x_1/2 + \pi/8\},$$

$$\Delta_q^2 = \Delta.$$

We now have a region of joint win

$$W^{12} = \Delta_Q^1 \cap \Delta_Q^2 = \{(x_1, x_2) \in \Delta \mid x_2 = 2x_1 - \pi/4\},$$

and the joint winning strategies are determined by $u^1 = 2$, $u_1^2 = 1$ and $u_2^2 = 2$. □

8.4 THE BARRIER

The last two examples indicate that the winning sets Δ_Q^i, $i = 1, 2$, may be separated (Example 8.3.3) or unseparable (Example 8.3.4). This fact is in each case decisive in characterizing the map of the game. The separation is established if we can find a barrier between them.

The reader is now asked to recall the concept of $\bar{P}(\cdot)$-nonpermeable surface (Definition 3.1.6), with necessary condition (3.1.21) and sufficient Conditions 3.1.2, as well as Corollary 3.1.2, and the concept of the weak $\bar{P}(\cdot)$-barrier B_q^P for the objective Q, see also Section 5.4, Fig. 5.15, and condition (5.4.10). We replace the uncertainty $\bar{w}(\cdot)$ in the above by the *active player in opposition* to our control and such uncertainty acting together, at the same time requiring of our control to be robust against all options of the opposing pair. This relates the nonpermeability to a strong semi-game and requires readjustment of the concepts concerned. We take Σ^i as a surface separating Δ into two disjoint open sets $\Delta^i \supset \Delta_Q^i$ and $C\Delta^i \stackrel{\Delta}{=} \Delta - \Delta^i$ which replace Δ^P and $C\Delta^P$, respectively, see Fig. 3.2. The surface is then said to be i-*nonpermeable*, if and only if for all $\bar{x}^0 \in \Sigma^i$, $t_0 \in \mathbb{R}$, there is $\bar{P}^j(\cdot)$, $j \neq i$, such that $\bar{\phi}(\bar{x}^0, t_0, \cdot) \in K^j(\bar{x}^0, t_0)$ implies

$$\bar{\phi}(\bar{x}^0, t_0, t) \notin \Delta^i, \quad \forall\, t \geq t_0 . \tag{8.4.1}$$

Similarly as in Section 3.1, we conclude that such Σ^i must belong to Δ_N^i and when it is smooth with non-zero gradient $\bar{n}^i$ directed away from Δ^i we have

$$(\bar{n}^i)^T \cdot \bar{f}(\bar{x}, \bar{u}^i, \bar{u}_*^j, \bar{w}, t) \geq 0 \tag{8.4.2}$$

for all $\bar{u}^i \in \bar{P}^i(\bar{x}, t)$, $\bar{w} \in W$, $t \geq t_0$. Sufficient conditions for i-nonpermeability may be formalized similarly to Conditions 3.1.2.

Define the sets $D^i \stackrel{\Delta}{=} \Delta_N^i \cap \bar{\Delta}^i$, $i = 1, 2$ and note that for some scenarios they may be empty.

CONDITIONS 8.4.1. A smooth surface Σ^i separating Δ into two disjoint open sets $\Delta^i, C\Delta^i$ is i-nonpermeable, if for each $\bar{P}^i(\cdot)$ on non-empty D^i there is $\bar{P}^j(\cdot)$ and a C^1-function $V_B^i(\cdot) : D^i \to \mathbb{R}$ such that for all $(\bar{x}, t) \in D^i \times \mathbb{R}^+$ we have

    (i) $V_B^i(\bar{x}) < V_B^i(\bar{\xi})$, $\quad \forall\, \bar{\xi} \in \Sigma^i$;

    (ii)   for each $\bar{u}^j \in \bar{P}^j(\bar{x}, t)$,

$$\nabla v_B^i (\bar{x})^T \cdot \bar{f}(\bar{x}, \bar{u}^i, \bar{u}^j, \bar{w}, t) \geq 0 \; , \qquad\qquad (8.4.3)$$

$$\text{for all} \quad \bar{u}^i \in \bar{P}^i(\bar{x}, t) \; , \quad \bar{w} \in W \; .$$

The proof follows by the same argument as for Conditions 3.1.2, that is, contradiction between (i) and (ii) if a motion from Σ^i were to enter Δ^i .

The obvious choice of $v_B^i(\cdot)$ is to make it an extension over Δ_N^i of the test function which generated the strong controllability in Δ_Q^j , say $v_{Q(j)}(\cdot)$, so that we have $v_B^i(\bar{x}) = v_{Q(j)}(\bar{x})$, and we can identify the levels of both functions. The case will be illustrated by examples later in this section. Identically as in Section 3.1, if Σ^i is specified by a v^i-level with $\nabla v^i(\bar{x}) = \bar{n}^i \neq 0$, condition (8.4.3) becomes necessary and sufficient.

Obviously $\Sigma^i \subset \Delta_N^i$, and there are many such surfaces. The surface whose Δ_i is minimal, that is, such that there is no other Σ^i between it and $\partial \Delta_Q^i$ is considered the semi-barrier for the player i , briefly $i\text{-}barrier$, denoted B_Q^i .

As we have seen in Section 5.4, B_Q^i estimates Δ_Q^i from outside with the largest size obtained when $B_Q^i = \partial' \Delta_Q^j \triangleq \Delta_Q^j \cap \Delta$, $j \neq i$. For all practical purposes, our success in separating Δ_Q^i's is assessed, if we can establish the mutual behavior of B_Q^i's. First let us consider the special case when $\Delta_N = \phi$. Then the pair Δ_Q^1, Δ_Q^2 partitions Δ and we may conclude that $\partial \Delta_Q^1 = \partial' \Delta_Q^2$ belongs to one of the two strong regions Δ_Q^i , the other left open. Such a joint boundary defines both strong regions, separating them, and determines completely the map of the game.

The alternative case of $\Delta_N \neq \phi$ is more complicated. When Δ_Q^1, Δ_Q^2 are disjoint: $\Delta_Q^i = W_Q^i$, $i = 1,2$, then such nonempty Δ_N separates them. To emphasise this separating role, let us introduce the *nonpermeable surfaces*

$$\Sigma \triangleq \Sigma^1 \cap \Sigma^2 \; , \qquad\qquad (8.4.4)$$

obviously located in Δ_N , each being both 1-nonpermeable and 2-nonpermeable. The family of such surfaces may be void or it may fill Δ_N , in the two extreme cases.

From the definition of Σ^i and (8.4.4) it follows that Σ is positively invariant under the pair $\bar{P}_*^1(\cdot), \bar{P}_*^2(\cdot)$ generating the 2,1-nonpermeabilities, respectively, and also that both necessary conditions (8.4.2) apply. In

particular, considering that at the points of Σ , $\bar{n}^1 = -\bar{n}^2$, it follows
that there is $\bar{P}_*^j(\cdot)$ for which

$$(\bar{n}^i)^T \cdot \bar{f}(\bar{x}, \bar{u}^i, \bar{u}_*^j, \bar{w}, t) \geq 0 \qquad (8.4.6)$$

for all $\bar{u}_*^j \in \bar{P}_*^j(\bar{x}, t)$, $\bar{u}^i \in U_i$, $\bar{w} \in W$, $t \geq t_0$, and that there is
$\bar{P}_*^i(\cdot)$ for which

$$(\bar{n}^i)^T \cdot \bar{f}(\bar{x}, \bar{u}_*^i, \bar{u}^j, \bar{w}, t) \leq 0 \qquad (8.4.7)$$

for all $\bar{u}_*^i \in \bar{P}^i(\bar{x}, t)$, $\bar{u}^j \in U_j$, $\bar{w} \in W$, $t \geq t_0$. We conclude

$$\min_{\bar{u}^i} \max_{\bar{u}^j} \; (\bar{n}^i)^T \cdot \bar{f}(\bar{x}, \bar{u}^i, \bar{u}^j, \bar{w}, t) = 0 \qquad (8.4.8)$$

for $i,j = 1,2$ and for all $\bar{w} \in W$.

The reader may like to recall our derivation of (5.2.11) and (5.2.16)
with $\bar{w}$ replaced now by $\bar{u}^j$. The arguments are the same. The sufficient
conditions for Σ are obtained by applying Conditions 8.4.1 twice, that is,
for $i,j = 1,2$ and $i,j = 2,1$. It is usually practical to take
$v_B^j(\bar{x}) \triangleq C - v_B^i(\bar{x})$, with $C = \text{const}$ calculated as a value of $v_B^i(\cdot)$ at
the boundary of its domain of definition. If the constant is difficult to
obtain, we may use the alternative $v_B^j(\bar{x}) = 1/v_B^i(\bar{x})$, with the same result.
The technique is illustrated in examples below. With the mentioned choice
of $v_B^j(\cdot)$, we in fact reduce the two functions to a single function $v_B^i(\cdot)$
and may state the sufficient conditions jointly as follows.

CONDITIONS 8.4.2. A surface Σ defined by (8.4.4) is nonpermeable if for
each $\bar{P}^i(\cdot)$ there is $\bar{P}^j(\cdot)$, $i,j = 1,2$, $i \neq j$, and a C^1-function
$V_B(\cdot) : \Delta_N \to \mathbb{R}$, with non-empty $D = D^1 \cup D^2$, such that for all
$(\bar{x}, t) \in D \times \mathbb{R}^+$ we have

> (i) $v_\Sigma^- < V_B(\bar{x}) < v_\Sigma^+$,
>
>> with $v_\Sigma^- = \sup V_B(\bar{x}) \mid \bar{x} \in \partial D^i$, $i = 1,2$,
>>
>> $v_\Sigma^+ = \inf V_B(\bar{x}) \mid \bar{x} \in \partial D^j$, $j = 2,1$;
>
> (ii) for each $\bar{u}_*^j \in \bar{P}_*^j(\bar{x}, t)$, $j = 2,1$, we have
>
>> $$\nabla V_B(\bar{x})^T \cdot \bar{f}(\bar{x}, \bar{u}^i, \bar{u}_*^j, \bar{w}, t) \geq 0 , \qquad (8.4.9)$$
>>
>> for all $\bar{u}^i \in U_i$, $\bar{w} \in W$, $(\bar{x}, t) \in D^i \times \mathbb{R}^+$, $i = 1,2$;
>
> (iii) for each $\bar{u}_*^i \in \bar{P}_*^i(\bar{x}, t)$, $i = 1,2$, we have
>
>> $$\nabla V_B(\bar{x})^T \cdot \bar{f}(\bar{x}, \bar{u}_*^i, \bar{u}^j, \bar{w}, t) \leq 0 , \qquad (8.4.10)$$
>>
>> for all $\bar{u}^j \in U_j$, $\bar{w} \in W$, $(\bar{x}, t) \in D^j \times \mathbb{R}^+$, $j = 2,1$.

The proof follows by the same argument as for Conditions 8.4.1 used twice, with $v_B^j(\bar{x}) \overset{\Delta}{=} c - v_B^i(\bar{x})$.

when Σ is represented by a V_B-level, similarly as for Conditions 8.4.1, (8.4.9) *and* (8.4.10) *become also necessary*, and by the same argument as for deriving (8.4.8) we have

$$\min_{\bar{u}^i} \max_{\bar{u}^j} \nabla V_B(\bar{x})^T \cdot \bar{f}(\bar{x}, \bar{u}^i, \bar{u}^j, \bar{w}, t) = 0 \tag{8.4.11}$$

for $i, j = 1, 2$, $i \neq j$, and for all $\bar{w} \in W$.

If there is *only one* Σ *separating* Δ_Q^1, Δ_Q^2 we conclude that $\Sigma^i = B_Q^i$. Conversely, if two semi-barriers B_Q^i intersect, we obtain Σ which is unique. We shall call it the *barrier*:

$$B \overset{\Delta}{=} B_Q^1 \cap B_Q^2 . \tag{8.4.12}$$

By such design, B is the minimal Δ_N , that is, it is the lower estimate of the neutral set. In fact any Δ_N larger than B may be considered a *dispersed barrier*, which is particularly useful when B is empty. Let us suppose that it is nonempty. In general B does not subdivide Δ into two disjoint sets. In fact, the barrier itself may be disjoint. For instance, as shown in Fig. 8.7, the barrier reduces to the two isolated points which constitute the intersection of the boundaries $\partial'\Delta_Q^1 \cap \partial'\Delta_Q^2$. On the other hand, when Δ_Q^1, Δ_Q^2 are arbitrary but open and disjoint, the nonempty B is shown in Fig. 8.12. Obviously the above cases occur only when $B_Q^i = \partial'\Delta_Q^j$, $i, j = 1, 2$, while $\partial'\Delta_Q^1 \cap \partial'\Delta_Q^2 \neq \phi$, which is a special case.

By definition the barrier is nonpermeable, thus positively invariant under the corresponding pair $\bar{P}_*^1(\cdot), \bar{P}_*^2(\cdot)$ and such that (8.4.8) holds, or if it is defined by some V_B-level, (8.4.11) holds. Thus in order to find the barrier, in the general case when the boundaries $\partial'\Delta_Q^i$ may not intersect, we execute the following three steps:

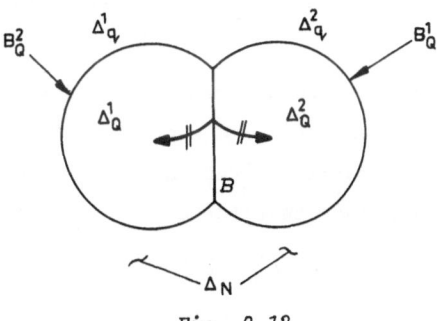

Fig. 8.12

(a) find a surface defined by (8.4.8);

(b) check whether it is nonpermeable;

(c) if yes, check whether it is single in Δ_N; if not single, go back to (a).

For item (b) we use twice Conditions 8.4.1, or Conditions 8.4.2. When we may find a smooth function $V_B(\cdot)$ satisfying (8.4.11) we may proceed directly to step (c).

The situation becomes simpler in the case when $\Delta_N^i \cap \Delta^i = \phi$, which makes Conditions 8.4.1 or 8.4.2 not only useless, but also redundant. Indeed, then $\Delta^i = \Delta_Q^i$, $i = 1,2$, implying that $\partial'\Delta_Q^1 \cap \partial'\Delta_Q^2 = \Delta_N \neq \phi$. Hence, by the definition of Δ_Q^i, both boundaries are j-nonpermeable and their intersection is the only nonpermeable surface between Δ_Q^1 and Δ_Q^2 implying

$$B = \partial\Delta_Q^1 \cap \partial\Delta_Q^2 . \qquad (8.4.13)$$

The latter is obviously the necessary condition, but if we can find the nonempty intersection of boundaries, we may consider it a reasonable candidate for the barrier. Such a candidate may be confirmed in three equivalent ways:

(a) Showing that $\partial\Delta_Q^j$ are i-nonpermeable, for $i,j = 1,2$, that is, that $\partial\Delta_Q^1 \cap \partial\Delta_Q^2 = \Sigma$. The fact that such Σ is unique: $\Sigma = B$ follows from the definitions of Δ_Q^i.

(b) Showing that

$$\Delta_N^i = \overline{C\Delta}^i = \Delta_Q^j , \quad i,j = 1,2 . \qquad (8.4.14)$$

In fact, since by definition $\Delta_Q^j \subset \Delta_N^i$, the condition (8.4.14) follows if only $\Delta_N^i = \overline{C\Delta}^i \subset \Delta_Q^j$, $i,j = 1,2$, see Example 8.4.3.

(c) Checking the candidate against the following.

CONDITIONS 8.4.3. Given two strong regions Δ_Q^i, $i = 1,2$ together with corresponding winning programs $\bar{P}^i(\cdot)$, such that $\partial\Delta_Q^1 \cap \partial\Delta_Q^2 \neq \phi$, the latter intersection is the barrier B if there are two C^1-functions $v_B^i(\cdot) : \Delta \rightarrow \mathbb{R}$, $i = 1,2$, such that for all $(\bar{x},t) \in \bar{\Delta}_N^i \times \mathbb{R}$ we have

(i) $v_B^i(\bar{x}) < v_B^i(\bar{\xi})$, $\forall \bar{\xi} \in \partial\Delta_Q^1 \cap \partial\Delta_Q^2$;

(ii) for each $\bar{u}^j \in P^j(\bar{x},t)$, $j = 2,1$,

$$\nabla v_B^i(\bar{x})^T \cdot \bar{f}(\bar{x},\bar{u}^i,\bar{u}^j,\bar{w},t) \leq 0 \qquad (8.4.14)$$

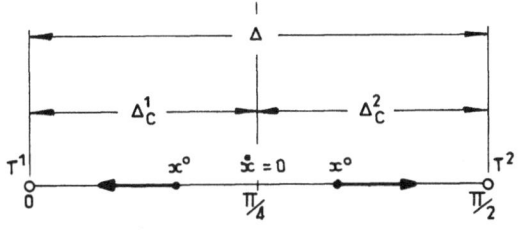

Fig. 8.13

for all $\bar{u}^i \in U^i$, $\bar{w} \in W$, $i = 1,2$.

The proof follows immediately from the fact that conditions (i),(ii) make Δ_N^i, $i = 1,2$ strong regions for $Q(j)$, $j = 1,2$, thus implying nonpermeability as in method (a). It is practical to take $v_B^i(\cdot)$ coinciding with $v_Q^j(\cdot)$ on Δ_Q^j:

$$v_B^i(\bar{x}) \mid \bar{x} \in \Delta_Q^j \equiv v_Q^j(\bar{x}) . \qquad (8.4.15)$$

EXAMPLE 8.4.1. We return to Example 5.1.2 but replace the pair $\bar{u},\bar{w}$ in (5.1.17) by our present $\bar{u}^1,\bar{u}^2$ with the same constraints, the same objectives including targets, and the same scenario. Then the state equation becomes

$$\dot{x} = u^1 \cos x + u^2 \sin x \qquad (8.4.16)$$

with $|u^i| \le 1$, $\Delta \triangleq \{x \in \mathbb{R} \mid 0 \le x \le \pi/2\}$ and the objectives are $Q(1)$: capture in $T_C^1 = \{0\}$, and $Q(2)$: capture in $T_C^2 = \{\pi/2\}$, see Fig. 8.13. We use Conditions 5.5.1 and let $v_C^1 \triangleq x$ and $v_C^2 \triangleq (\pi/2) - x$, satisfying conditions (i),(ii), see Fig. 8.14.

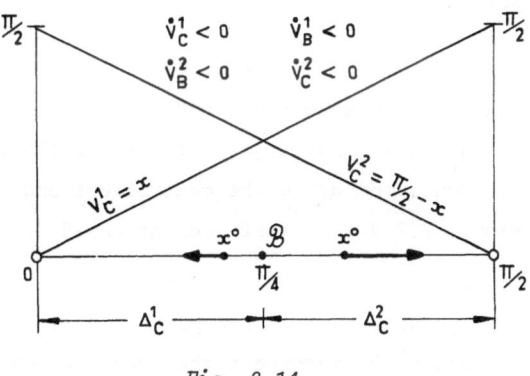

Fig. 8.14

Then we calculate along the trajectories:

$$\dot{V}_C^1 = \dot{x} = u^1 \cos x + u^2 \sin x \ , \left. \vphantom{\begin{matrix}1\\1\end{matrix}} \right\} \qquad (8.4.17)$$
$$\dot{V}_C^2 = -\dot{x} = -u^1 \cos x - u^2 \sin x \ ,$$

in order to check condition (iii). As seen in Example 5.1.2, Δ_C^1 : $x \in [0,\pi/4)$, and Δ_C^2 : $x \in (\pi/4 , \pi/2]$. Indeed, taking $V_C^1 = x$, we observe that for $x \in [0,\pi/2)$ there is $u^1 = -1$ winning $\dot{V}_C^1 = \dot{x} < 0$ against all $-1 \leq u^2 \leq 1$, which implies $x(t) \to 0$. On the other hand, taking $V_C^2 = (\pi/2) - x$, we observe that for $x \in (\pi/4 , \pi/2]$ there is $u^2 = 1$ winning $\dot{V}_C^2 = -\dot{x} < 0$ for all $-1 \leq u^1 \leq 1$, which results in $x(t) \to \pi/2$. It follows that $\partial\Delta_C^1 \cap \partial\Delta_C^2 = \{\pi/4\}$. Turning now to the barrier, observe that all points in Δ are V_C^1, V_C^2 - levels, and that at $\partial\Delta_C^2 = \{\pi/4\}$ we have $\dot{x} = (u^1 + u^2)(\sqrt{2}/2)$, so if player 2 plays $u^2 = 1$, then even for $u^1 = -1$ we still have $\dot{x} = 0$. Conversely if player 1 plays $u^1 = -1$, we also have $\dot{x} = 0$. So $x = \{\pi/4\}$ is a rest point, wherefrom no motion can leave, that is, nonpermeable, positively invariant Σ . By the properties of the surrounding Δ_C^i it is also a unique such point, thus we conclude that $B = B_C^1 \cap B_C^2 = \{\pi/4\}$, without having to use Conditions 8.4.3. For the sake of illustration, let us however consider $V_B^1 = (\pi/2) - x$, for $x \in [\pi/4 , \pi/2]$; $V_B^2 = x$, for $x \in [0,\pi/4]$, see Fig. 8.13. Satisfaction of Conditions 8.4.3 is seen immediately from our discussion of V_C^1 , V_C^2 . $\qquad \square$

EXAMPLE 8.4.2. We return now to the turret game of Examples 8.2.1, 8.2.3 with the scenario in Fig. 8.1. Recall the targets in Fig. 8.2 and the summing up discussion in Section 8.3 with Fig. 8.7. We continue here assuming the case $\gamma_1 < 1$, which gave the two winning regions

$$\Delta_F^1 = W_F^1 = \{(x_1,x_2) \in \Delta \mid x_2 - \varepsilon_2 > \gamma_2(x_1 - \varepsilon_1)\} \ , \left. \vphantom{\begin{matrix}1\\1\end{matrix}} \right\}$$
$$\Delta_F^2 = W_F^2 = \{(x_1,x_2) \in \Delta \mid x_2 - \varepsilon_2 < \gamma_2(x_1 - \varepsilon_1)\} \ . \qquad (8.4.18)$$

Then the nonempty set

$$B = \partial\Delta_F^1 \cap \partial\Delta_F^2 = \{(x_1,x_2) \in \Delta \mid x_2 - \varepsilon_2 = \gamma_2(x_1 - \varepsilon_1)\} \qquad (8.4.19)$$

is the obvious candidate for the barrier, and since by (8.4.18) we have $\Delta_N^i = \overline{C}\Delta^i = \Delta_F^j$ for $i,j = 1,2$ and $i,j = 2,1$, it is the confirmed candidate. It can be also confirmed using the controllers and functions $V_F^i(\cdot) , V_A^i(\cdot)$ of Example 8.2.3 to satisfy Conditions 8.4.3.

EXAMPLE 8.4.3. In the pursuit-evasion Example 8.2.4, for two strong semi-games: pursuer 1 attempting to intercept the evader 2 before he escapes to shelter, and evader 2 attempting to escape to the shelter before being intercepted by the pursuer, we obtained the two winning sets, (8.2.31) and (8.2.26),

$$
\left.
\begin{aligned}
\Delta_F^1 = W_F^1 : R - r_E &> \frac{v_P(r - r_P)}{v_P - v_E} \quad , \\[2mm]
\Delta_F^2 = W_F^2 : r - r_P &> (v_E + v_P)(R - r_E)/v_E \quad .
\end{aligned}
\right\}
\qquad (8.4.20)
$$

We shall show that the boundaries intercept: $\partial\Delta_F^1 \cap \partial\Delta_F^2 \neq \phi$, meaning that there is a set

$$
B = \left\{ (R,\theta_E,r,\theta_P) \in \Delta \,\Big|\, R - r_E = \frac{v_P(r - r_P)}{v_P - v_E} \,,\text{ and } r - r_P = \frac{v_E + v_P}{v_E}(R - r_E) \right\}
$$

$$(8.4.21)$$

which is nonempty. In the above L is such that

$$
L^2 = R^2 + r^2 + 2r R \cos\theta \quad . \qquad (8.4.22)
$$

Writing $L = ar + b$, $R = Ar + B$ where

$$
a = \frac{v_P}{v_P - v_E} \quad , \qquad b = r_E - \frac{v_P r_P}{v_P - v_E} \quad ,
$$

$$
A = \frac{v_E}{v_P + v_E} \quad , \qquad B = r_E - \frac{v_E r_P}{v_P + v_E} \quad ,
$$

and substituting into (8.4.22) we find

$$
(a^2 - 1 - A^2 - 2A\cos\theta)r^2 + (2ab - 2AB - 2B\cos\theta)r + b^2 - B^2 = 0 \quad . \qquad (8.4.23)
$$

A value for r in equation (8.4.23) is obtainable if

$$
N = B^2\cos^2\theta + 2b(Ab - aB)\cos\theta + (Ab - aB)^2 + (b^2 - B^2) \geq 0 \quad .
$$

Examining N we see that when $\cos\theta = 1$ $(\theta = 0)$,

$$
N = (Ab - aB + b)^2 \geq 0 \qquad (8.4.24)
$$

and when $\cos\theta = -1$, $(\theta = \pi)$,

$$
N = (Ab - ab - b)^2 \geq 0 \quad . \qquad (8.4.25)
$$

From (8.4.24) and (8.4.25) we deduce that, even if for values of $\cos\theta$ between -1 and 1, N becomes negative, there must still remain a nonempty set of θ for which $N \geq 0$ and hence solution for r from (8.4.23). This shows that the intersection of the two boundaries concerned is nonempty.

In general we obviously have $\Delta_N^i \supset \Delta_F^j$, for $i = 1$, $j = 2$ and $i = 2$, $j = 1$. In terms of (8.4.20) it means that the set defined by, see Section 8.2,

$$\Delta_F^2 : v_E(r - r_P) > (v_E + v_P)(R - r_E) \tag{8.4.26}$$

should be covered by the set

$$\Delta_N^1 : R - r_E \leq \frac{v_P(r - r_P)}{v_P - v_E} , \qquad v_P > v_E . \tag{8.4.27}$$

The condition (8.4.26) may be rewritten as

$$R - r_E < \frac{v_E(r - r_P)}{v_P + v_E} , \tag{8.4.28}$$

which by virtue of

$$\frac{v_E(r - r_P)}{v_P + v_E} < \frac{v_P(r - r_P)}{v_P - v_E} \tag{8.4.29}$$

proves our point, see Fig. 8.15.

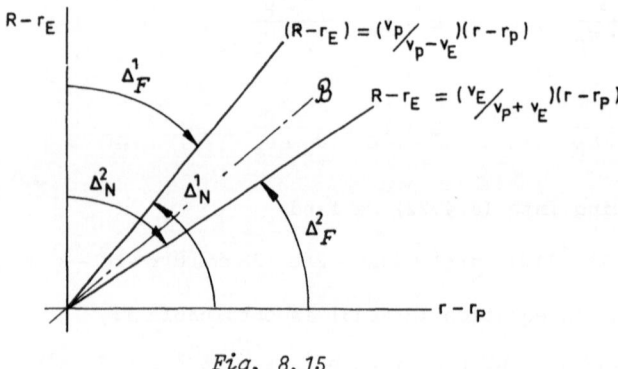

Fig. 8.15

Similarly,

$$\Delta_F^1 : R - r_E > \frac{v_P(r - r_P)}{v_P - v_E} , \tag{8.4.30}$$

should be covered by

$$\Delta_N^2 : v_E(r - r_P) \leq (v_E + v_P)(R - r_E) , \tag{8.4.31}$$

which may be rewritten as

$$R - r_E \geq \frac{v_E}{v_E + v_P}(r - r_P) . \tag{8.4.32}$$

In view of (8.4.29), comparing (8.4.30) and (8.4.32), we prove our second point, see Fig. 8.14. To check (8.4.14) we need $\Delta_N^i = \Delta_F^j$, $i,j = 1,2$ which occurs when, see (8.4.21),

$$\frac{v_P}{v_P - v_E} = \frac{v_E}{v_P + v_E} \quad ,$$

thus defining the barrier $\mathcal{B}$. $\qquad\qquad\qquad\qquad\qquad\qquad$ □

Closing this section, we may illustrate our comments made in Section 8.1 on how the present concept of the barrier relates to the Isaacs' type barrier obtained for a single objective (collision with a target) game via necessary conditions optimizing the cost, in particular the time of collision, see our Section 5.2 on optimal controllability. The following Example 8.4.4 establishes the barrier in the widely known *homicidal chauffeur game* discussed in detail by Isaacs [1]. He did not use the Liapunov formalism.

EXAMPLE 8.4.4. The action takes place in the plane. The pursuer 1 moves at a fixed speed v_1 but with his radius of curvature controlled up to a given quantity R. The evader 2 moves with fixed speed $v_2 < v_1$ and controls his direction of travel time instantaneously without the restriction of following along a smooth path. Collision occurs when the distance between the players is below or equal to a given quantity ℓ. Clearly, if R is large enough, ℓ small and v_1 not greatly exceeding v_2 the evader can always escape by persistent sidestepping. The problem is to find precise conditions, values of $R, \ell, (v_1/v_2)$ which demarcate this possibility.

As always before, we take the state variables relative to the pursuer 1, thus located at the origin $(0,0)$, with the position of the evader $E = (x_1, x_2)$ reflected anywhere outside the target $T : x_1^2 + x_2^2 \leq \ell^2$. The x_2-axis is always to be in the direction of v_1, see Fig. 8.16.

Denote the current center of curvature for the pursuer by $C = (R/u^1, 0)$, and its distance from the evader by d. Then the pursuer rotation about C is equivalent to a rotation of (x_1, x_2) about C in the opposite direction with the same angular speed. Thus (x_1, x_2) moves with speed $v_1(du^1/R)$ in a direction perpendicular to CE. Its velocity components are obtained by multiplying the speed by $-x_2/d$ and $(x_1 - R/u^1)/d$. Thus the dynamics are

$$\left.\begin{array}{l} \dot{x}_1 = -(v_1/R)x_2 u^1 + v_2 \sin u^2 , \\[2mm] \dot{x}_2 = (v_1/R)x_1 u^1 - v_1 + v_2 \cos u^2 ; \end{array}\right\} \qquad (8.4.33)$$

with $|u^2| \le 1$. Let $\alpha = v_1/v_2$, the ratio of speeds. Then (8.4.33) becomes

$$\left.\begin{aligned}\dot{x}_1 &= v_2(\sin u^1 - \alpha x_2 u^1/R) , \\ \dot{x}_2 &= v_2(\cos u^2 - \alpha + \alpha x_1 u^1/R) .\end{aligned}\right\} \qquad (8.4.34)$$

To construct a suitable test function $V(x_1, x_2)$ according to the method of the first integral, we take

$$u^1 = \begin{cases} 1 , & x_1 \le 0 , \\ -1 , & x_1 > 0 ; \end{cases}$$

$$u^2 = 0 .$$

This yields the auxiliary integrable system:

$$\dot{x} = v_2 \psi \qquad (8.4.35)$$

where $\psi = (\psi_1, \psi_2)^T$, with

$$\psi_1 = (\alpha x_2/R) \operatorname{sgn} x_1 , \qquad \psi_2 = (1-\alpha) - (\alpha x_1/R) \operatorname{sgn} x_1$$

which upon integration provides us with the test function

$$V(x_1, x_2) = v_2 \left[\frac{\alpha}{2R} (x_1^2 + x_2^2) + (\alpha-1)|x_1| \right] . \qquad (8.4.36)$$

The first term in the square bracket represents the distance between players, the second the relative surplus of speed. The function (8.4.36) consists of two C^1-branches

$$\left.\begin{aligned}V_1(x_1, x_2) &= v_2 \left[\frac{\alpha}{2R} (x_1^2 + x_2^2) - (\alpha-1)x_1 \right] , & x_1 < 0 , \\ V_2(x_1, x_2) &= v_2 \left[\frac{\alpha}{2R} (x_1^2 + x_2^2) + (\alpha-1)x_1 \right] , & x_1 > 0 .\end{aligned}\right\} \qquad (8.4.37)$$

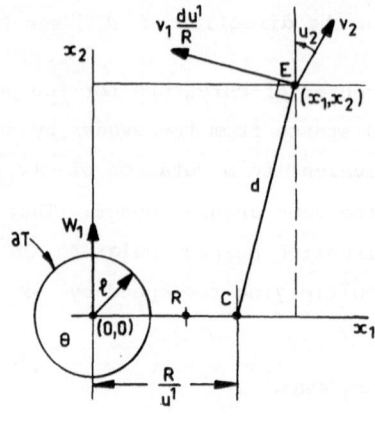

Fig. 8.16

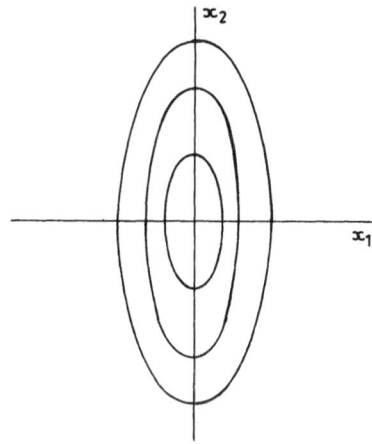

Fig. 8.17

It gives rise to a topographical reference system $\tilde{V}(x_1, x_2) = \text{const}$ on $\mathbb{R}^2$ displayed in Fig. 8.17. Also, V is positive definite on $\mathbb{R}^2$. Our first interest lies in the half plane $\{(x_1, x_2) \mid x_2 \geq 0\}$, since the velocity vector of the car is always directed along the positive x_2-axis. As the situation is totally symmetric about the x_2-axis, we need consider only the quadrant $\{(x_1, x_2) \mid x_1 > 0, \ x_2 > 0\}$. Now,

$$\frac{\partial V}{\partial x_1} = v_2 \left[\frac{\alpha x_1}{R} + (\alpha - 1) \right], \qquad \frac{\partial V}{\partial x_2} = w_2 \frac{\alpha x_2}{R} .$$

Substitution of these into $L(x, t, \tilde{u}^1, \tilde{u}^2)$ gives the Liapunov derivative

$$L = (v_2^2/R) [(\alpha x_1 + R(\alpha - 1)) \sin u^2 + \alpha x_2 \cos u^2 - \alpha(\alpha - 1) x_2 u^1 - \alpha^2 x_2] .$$

Player 1 selects $u_*^1 = 1$, which gives him the greatest minimizing effect. Player 2 wishes to maximize L:

$$\frac{\partial L}{\partial u^2} = [\alpha x_1 + R(\alpha - 1)] \cos u^2 - \alpha x_2 \sin u^2 ,$$

so the maximizing u^2 satisfies

$$\tan u_*^2 = \frac{\alpha x_1 + R(\alpha - 1)}{\alpha x_2} .$$

Setting $X = \alpha x_1 + R(\alpha - 1)$, $Y = \alpha x_2$ we find

$$L(x_1, x_2, u_*^1, u_*^2) = (v_2^2/2) [(X^2 + Y^2)^{\frac{1}{2}} - (2\alpha - 1) Y] . \tag{8.4.38}$$

Call this quantity L^*. In order that we can deduce the motion of solutions of (8.4.34) relative to the V-levels, we must know which regions satisfy $L^* > 0$ and $L^* < 0$. Accordingly we locate the curve $L^* = 0$. From (8.4.38),

$$(X^2 + Y^2)^{\frac{1}{2}} - (2\alpha - 1)Y = 0$$

or

$$Y^2(4\alpha^2 - 4\alpha) = X^2 \ .$$

That is,

$$Y = X/(2\sqrt{\alpha(\alpha - 1)}) \ . \qquad (8.4.39)$$

Returning to the original formulation, we have

$$x_2 = \frac{x_1}{2\sqrt{\alpha(\alpha - 1)}} + \frac{R\sqrt{\alpha - 1}}{2\alpha\sqrt{\alpha}} \qquad . \qquad (8.4.40)$$

We now locate the curve $\dot{x}_1 = 0$. This is equivalent to

$$\sin u_*^2 - \alpha u_*^1 x_2/R = 0$$

or

$$X/(X^2 + Y^2)^{\frac{1}{2}} = Y/R$$

in the abbreviated notation, so we have

$$X^2 R^2 = Y^2(X^2 + Y^2)$$

or

$$X^4 + X^2 Y^2 - X^2 R^2 = 0 \ .$$

This quadratic in Y^2 has the solution

$$Y = \frac{1}{\sqrt{2}} (X\sqrt{4R^2 + X^2} - X^2)^{\frac{1}{2}} \qquad (8.4.41)$$

valid in the domain of our interest.

At this point, we examine the asymptotic behavior of (8.4.41) as X becomes large: Consider $f(t) = (a + t)^{\frac{1}{2}}$. Expanding this formally into a Taylor Series about $t = 0$ we have

$$f(t) = a^{1/2} + \frac{1}{2} t a^{-1/2} - \frac{1}{4} \frac{t^2}{2!} a^{-3/2} + \frac{3}{2} \cdot \frac{1}{2} \cdot \frac{1}{2} \cdot \frac{t^3}{3!} a^{-5/2}$$

$$+ \text{ (terms involving higher powers of } a^{-1}) \ .$$

Let $X^2 = a$ and $t = 4R^2$. Then we have

$$\sqrt{4R^2 + X^2} \approx X + 2R^2/X - 2R^4/X^3 \ ,$$

for suitably large values of X . Using this in (2.3.38) gives

$$Y \approx \frac{1}{\sqrt{2}} (2R^2 - 2R^4/X^2)^{\frac{1}{2}}$$

or as $X \to \infty$, $Y \to R$. So in terms of (x_1, x_2) , $x_1 \to \infty$ implies $x_2 \to R/\alpha$.

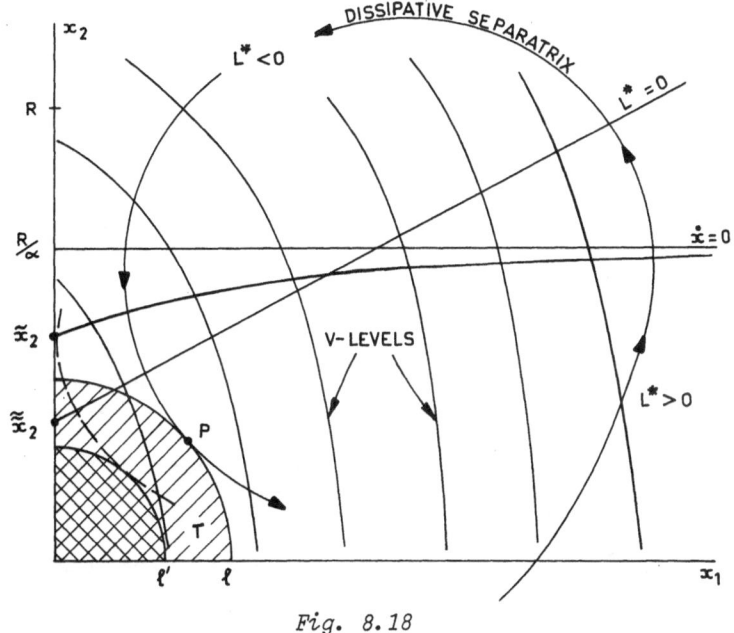

Fig. 8.18

Of further interest is the point at which the curve $\dot{x}_1 = 0$ intersects the x_2-axis. That is, $X = R(\alpha - 1)$, so (8.4.41) becomes

$$\tilde{x}_2 = \frac{1}{\alpha\sqrt{2}} \left[R(\alpha-1)\sqrt{4R^2 + R^2(\alpha-1)^2} - R^2(\alpha-1)^2 \right]^{\frac{1}{2}}$$

or

$$\tilde{x}_2 = \frac{R\sqrt{\alpha-1}}{\alpha\sqrt{2}} \left[\sqrt{4+(\alpha-1)^2} - (\alpha-1) \right]^{\frac{1}{2}} .$$

The corresponding intersection point of the curve $L^* = 0$ with the x_2-axis is

$$\tilde{\tilde{x}}_2 = (R\sqrt{\alpha-1})/2\alpha\sqrt{\alpha} .$$

As the curve $\dot{x}_1 = 0$ is bounded at $x_1 = \infty$ (by R), this means there is an intersection of these two curves in the first quadrant. This is displayed in Fig. 8.18.

REFERENCES AND SELECTED LITERATURE

Abele, E., and Sturz, W.
[1] Sensors for adaptive control of settling tasks with industrial robots, *Proc. 2nd Internat. Conf. Robot Vision and Sensory Controls*, Stuttgart, 1982.

Abraham, R.H.
[1] Dynamism; exploratory research in bifurcation using interactive computer graphics, in Gurel-Roessler (eds), *Bifurcation Theory and Applications*, New York Academy of Sciences Publ., 1979.
[2] *Complex Dynamical Systems*. Aerial Press, Santa Cruz, 1986.

Aeyels, D.
[1] Local and global stabilizability for nonlinear systems, in C.I. Byrnes, A. Linguist (eds), *Theory and Applic. of Nonlinear Control Systems*, Elsevier Sci. Publ., 1986, pp. 93-105.

Aggarwal, R., and Leitmann, G.
[1] Avoidance control, Trans. ASME J. Dynamic Systems Measurement Control and Vol. 94, 1972, pp. 152-154.

Ahmed, N.U., and Lim, S.S.
[1] Modelling and stabilization of flexible spacecraft under the influence of orbital perturbation, *Proc. 26th CDC*, Los Angeles, 1987, #FP11.

Aksenov, G.S., and Fomin, V.H.
[1] On the problem of adaptive control of a manipulator, in *Kybernetic Problems, Adaptive Systems*, pp. 164-168, Nauka, Moscow, 1976.

Alford, C.O., and Belyeu, S.M.
[1] Coordinated control of two robot arms, *Int. Conf. Robotics*, Atlanta, GA, March 1984, pp. 468-473.

Alimov, Y.I.
[1] On application of Liapunov Direct Method to differential equations with ambiguous right sides, Automation and Remote Control, Vol. 22, 1961, pp. 713-725.

Almahamedov, M.I.
[1] Design of differential equations with given curves as limit cycles (in Russian), Izv. Vys. Uch. Zav. Mat., 1965, pp. 12-16.

Anchev, A.A., and Melinkyan, A.A.
[1] On the optimal reorientation of satellite in a circular orbit, Izv. Acad. Nauk SSSR, MTT, 1980, #6.

Andronov, A.A., and Leontovich, E.A.
[1] Some cases of limit cycle depending upon parameters (in Russian), Uchennye Zapiski GG Univ., 1939, pp. 3-9.

Andronov, A.A., Vitt, A., and Chaikin, S.E.
[1] *Theory of Oscillators* (in Russian), Fiz. Nat. Giz., Moscow, 1959 (second edition), English edition, Pergamon, 1966.

Andronov, A.A., and Voznesensky, N.N.
[1] About works by Maxwell, Vyshegradsky and Stodola on Control Theory of machines, in *Theory of Automatic Control*, Publ. Ac. Sci. USSR, 1949.

Antonelli, P., and Skowronski, J.M.
[1] Adaptive identification of environmental stress for the management of plant growth, *Math. Comput. Modelling*, Vol. 10, 1988, pp. 27-35.
[2] Differential offensive-defensive games, IMA J. of Math. Applied in Med. & Biol., Vol. 3, 1986, pp. 319-340.

Antosiewicz, H.A.
[1] A survey of Liapunov second method, *Contributions to Theory of Nonlinear Oscillations*, Vol. 4, pp. 141-166, Princeton Univ. Press, Princeton, New Jersey, 1958.

Ardema, M.D., and Heymann, M.
  [1] A formulation and analysis of combat games, J. Opt. Th. Appl.,
       Vol. 46, 1985, #4.
Ardema, M.D., and Skowronski, J.M.
  [1] Coordination controllers for multi-arm manipulators - a case
       study, *Advances in Control and Dynamical Systems*, Vol. 34,
       Academic Press, 1990.
Aronsson, G.
  [1] Global controllability and bang-bang steering of certain nonlinear
       systems, SIAM J. Control, Vol. 11, 1973, pp. 607-619.
Asada, H., and Slotine, J.E.
  [1] *Robot Analysis and Control*. Wiley-Interscience, New York, 1986.
Asher, R.B., Ardasini, D., and Dorato, P.
  [1] Bibliography on Adaptive Control Systems, *Proc. IEEE*, Vol. 64,
       1976, pp. 1226-1240.
Asher, R.B., and Matuszewski, J.P.
  [1] Optimal guidance with maneuvering targets, J. Spacecraft Rockets,
       Vol. 11, 1974, pp. 204-206.
Astrom, K.J.
  [1] Theory and applications of adaptive control - a Survey, Automatica,
       Vol. 19, 1983, pp. 471-486.
Astrom, K.J., and Eykhoff, P.
  [1] System identification - a Survey, Automatica, Vol. 7, 1971,
       pp. 123-162.
Athans, M., Castanon, D., Dunn, K,P., Greene, C.S., Lee, W.H., Sandell,
       N.R., and Willsky, A.S.,
  [1] The stochastic control of F-8C aircraft using a multiple model
       adaptive control method, IEEE Trans. Autom. Control, Vol. AC-22,
       1977, pp. 768-780.
Atluri, S.N., and Amos, A.K. (eds)
  [1] *Large Space Structures: Dynamics and Control*. Springer, 1988.
Azab, A.A., and Nouh, A.
  [1] Adaptive model reference parameter tracking technique for aircraft,
       IFAC, *Adaptive Systems in Control and Signal Processing*, San
       Francisco, 1983, pp. 69-73.
Bakan, G.M.
  [1] Analytic design of dynamic system with a preassigned limiting set,
       Soviet Automatic Control, Vol. 7, 1974, pp. 12-21.
Balestrino, A., DeMaria, G., and Sciavicco, L.,
  [1] An adaptive model following control for robotic manipulators,
       Trans. ASME Dynamic Systems Measurement Control, Vol. 105, 1983,
       pp. 143-151.
Balestrino, A., DeMaria, G., and Zinober, A.S.I.
  [1] Nonlinear adaptive model following control, Automatica, Vol. 20,
       1984, pp. 559-568.
Banach, S.
  [1] *Mechanika*, Math. Monographs, Vol. 8, Lwow, Poland, 1938.
Banarjee, A.K., and Kane, T.R.
  [1] Tethered satellite retrieval with thruster augmented control,
       J. Guidance, Vol. 7, 1984, pp. 45-50.
Barbashin, E.A.
  [1] Construction of Liapunov functions for nonlinear systems, in
       Automatic and Remote Control, Butterworths, London, 1961, pp.
       943-947.
  [2] Construction of Liapunov functions, Differentialnyie Uravnienia,
       Vol. 4, 1968, pp. 2127-2158 (English translation).
  [3] *Introduction to Theory of Stability*. Walters-Noordhoff, Groningen,
       1970.
Barker, L.K., and Soloway, D.
  [1] Coordination of multiple robot arms, JPL Publ. 87-13, *Workshop on
       Space Telerobotics*, Vol. 2, pp. 301-306.

Barmish, B.R., Corless, M., and Leitmann, G.
[1] A new class of stabilizing controllers for uncertain dynamical systems, SIAM J. Control Optim., Vol. 21, 1983, pp. 246-252.

Barmish, B.R., and Feuer, A.
[1] Instability in optimal aim control, IEEE Trans. Automat. Control, Vol. AC-25, 1980, pp. 1250-1252.

Barmish, B.R., and Leitmann, G.
[1] On ultimate boundedness control of uncertain systems in the absence of matching conditions, IEEE Trans. Automat. Control, Vol. AC-27, 1982, pp. 153-155.

Barmish, B.R., Petersen, I.R., and Feuer, A.
[1] Linear ultimate boundedness control of uncertain dynamical systems, Automatica, Vol. 19, 1983, pp. 523-528.

Barmish, B.R., Schmitendorf, W.E., and Leitmann, G.
[1] A note on avoidance control, Trans. ASME, J. Dyn. Sys. Meas. Control, Vol. 103, 1981, pp. 69-70.

Barmish, B.R., Thomas, R.J., and Lin, Y.H.
[1] Convergence properties of a class of point-wise strategies, IEEE Trans. Automat. Control, Vol. AC-23, 1978, pp. 954-956.

Barnard, R.D.
[1] Optimal aim-control strategies applied to large scale nonlinear regulation and tracking system, IEEE Trans. Circuits and Systems, Vol. CAS-23, 1976, pp. 800-806.

Barnett, S.
[1] *Introduction to Mathematical Control Theory*. Oxford Univ. Press (Clarendon), London and New York, 1975.

Barnett, S., and Storey, C.
[1] Some results on the sensitivity and synthesis of asymptotically stable linear and nonlinear systems, in D.J. Bell (ed), *Recent Math. Developments in Control*, Acad. Press, 1973.

Basar, T., and Olsder, G.J.
[1] *Dynamic Noncooperative Game Theory*. Academic Press, 1982.

Bednarz, S., and Giergiel, J.
[1] Possibility of self-oscillations of airwing, Nonlin. Vibr. Problems, Vol. 4, 1962, pp. 149-158.

Bejczy, A.K.
[1] Robot arm dynamics and control, Tech. Memo 33-669, Jet Propul. Lab., Pasadena, California, February 1974.

Belusov, A.P., and Furasov, V.D.
[1] Stabilization of nonlinear control systems with incomplete information on object's state, Automat. Remote Control, Vol. 36, 1975, pp. 1385-1392.

Belytschko, T., and Hughes, T.J.R.
[1] *Computational Methods for Transient Analysis*. Elsevier Sci. Publ., 1983.

Bendixon, J.
[1] Sur les courbes définies par des équations différentielles, Acta Math., Vol. 24, 1901.

Benedict, C.E., and Tesar, D.
[1] Model formulation of complex mechanisms with multiple inputs, Parts I and II, ASME J. Mech. Design, Vol. 100, 1978, pp. 747-750, 755-760.

Bertsekas, D.P., and Rhodes, I.B.
[1] Recursive state estimation for a set-membership description of uncertainty, IEEE Trans. Autom. Control, Vol. AC, April 1971.
[2] On the min-max reachability of target sets and target tubes, Automatica, Vol. 7, 1971, pp. 233-247.

Bhatia, N.P., and Szegö, G.P.
[1] *Stability Theory of Dynamical Systems*. Springer, 1970.

Birkhoff, G.D.
  [1] Quelques theorems sur les mouvements des systemes dynamiques, Bull. Soc. Math. de France, Vol. 40, 1912.
  [2] Dynamical systems with two degrees of freedom, Trans. Am. Math. Soc., Vol. 18, 1917.
  [3] Dynamical systems, Am. Math. Soc. Coll. Publ., Vol. 9, New York, 1927.
  [4] Collected works, Vols. 1,2,3, Am. Math. Soc. Publ., New York, 1950, reprint Dover, 1968.
Blaquière, A.,
  [1] *Nonlinear System Analysis*. Acad. Press, New York, 1966.
Blaquière, A., Gerard, F., and Leitmann, G.
  [1] *Quantitative and Qualitative Games*. Academic Press, New York, 1969.
Blinchevski, V.S.
  [1] Existence of periodic solution of some autonomous system, Matem. Sbornik, Vol. 50(91), 1960, #1.
Blinov, A.P.
  [1] A method of estimating the domain of controllability in nonlinear systems, PMM USSR, Vol. 48, 1984, pp. 419-424, Russian text, pp. 593-600.
Bodley, C.S., Devers, A.D., Park, A.C., and Frisch, H.P.
  [1] A digital computer program for dynamic interaction simulation of controls and structure (DISCOS), NASA TP 1219, May 1978.
Bogusz, W.
  [1] On limit states of self-excited systems, Nonlin. Vibr. Problems, Vol. 3, 1962, pp. 21-30.
  [2] Inverse problems of stability of a certain mechanical system, Abhandlungen Deutsche Acad. Wiss. zu Berlin, Klasse fuer Math-Physik-Techn., 1965.
Bogusz, W., and Skowronski, J.M.
  [1] Kinetic synthesis of lumped mechanical systems, Mech. Teoret. Stos., Pol. Soc. Appl. Mech., Vol. 1, 1965, pp. 13-27.
  [2] Kinetic synthesis of general mechanical systems, Bull. Polytechnic Inst. Iassi, Romania, Vol. 13, 1967, pp. 409-413.
Bojadziev, G.N.
  [1] Controlled chain of pendulums, Trans. ASME J. Dyn. Syst. Measur. Control, Vol. 111, 1989.
  [2] *Nonlinear Vibrations* (in Bulgarian). Technica, Sofia, 1970.
Book, J.W.
  [1] *Modelling, Design and Control of Flexible Manipulator Arms*. Ph.D. Thesis, Mech. Eng. MIT, 1974.
  [2] Analysis of massless elastic chains, Trans. ASME J. Dyn. Sys. Meas. Control, Vol. 101, 1979, pp. 187-192.
Book, J.W., Maizza-Neto, O., and Whitney, D.E.
  [1] Feedback control of two beam two joint system with distributed flexibility, Trans. ASME J. Dyn. Sys. Meas. Control, Vol. 97, 1975, pp. 424-431.
Book, J.W., and Majette, M.
  [1] Controller design for flexible distributed parameter mechanical arms, Trans. ASME J. Dyn. Sys. Meas. Control, Vol. 105, 1985, pp. 245-254.
Bradshaw, A., and Porter, B.
  [1] Design of linear multivariable discrete time tracking systems for plants with inaccessible states, Int. J. Control, Vol. 24, 1976, pp. 275-281.
Breakwell, J.V.
  [1] Pursuit of a faster evader, in J.D. Grote (ed), *Theory and Application of Differential Games*, Reidel, Dordrecht, 1975, pp. 243-256.
  [2] A differential game with two pursuers and one evader, J. Optim. Theory Appl., Vol. 18, 1976, pp. 15-29.

Brogan, W.L.
  [1] *Modern Control Theory*. Quantum Publ., New York, 1974.
Brooks, R.A.
  [1] Solving the find-path problem by representing free space as
      generalized cones, AI Memo No. 674, MIT Artificial Intelligence
      Lab., May 1982.
  [2] Solving the find-path problem by good representation of free
      space, IEEE Trans. Systems Man Cybernet., Vol. SMC-13, 1983,
      pp. 190-197.
Budynas, R., and Poli, C.
  [1] Three dimensional motion of a large flexible satellite, Automatica,
      Vol. 8, 1972, pp. 275-286.
Bulgakov, B.V.
  [1] *Vibrations* (in Russian). Gos. Ind. Techn. Liter., Moscow, 1954.
Buligand, G.
  [1] Sur la stabilité des propositions mathématiques, Acad. Roy. Belg.
      Bull., Cl. Sci. (5), Vol. 21, 1935, pp. 277-282, 776-786.
Butchard, R.L., and Shackcloth, B.
  [1] Synthesis of model reference adaptive control systems by
      Liapunov's Second Method, *Proc. 2nd IFAC Sympos. Theory of Self-
      Adaptive Control Systems*, Teddington, 1966, pp. 145-152.
Byrne, R.M., and Wall, E.T.
  [1] Formulation of a definite class of Liapunov functions, IEEE Trans.
      Autom. Control, Vol. AC, 1975, pp. 174-176.
Calm, D.F., and Phillips, S.K.
  [1] Robnav-range based robot navigation and obstacle avoidance
      algorithm, IEEE Trans. Systems Man Cybernet., Vol. SMC-5, 1975,
      pp. 544-551.
Caroll, R.L.
  [1] A reduced adaptive observer for multivariable systems, *Proc. 15th
      Joint Aut. Control Conf.*, 1974.
  [2] New adaptive algorithms in Liapunov synthesis, IEEE Trans.
      Automat. Control, AC-21, 1976, pp. 246-249.
Caroll, R.L., and Lindorff, D.P.
  [1] An adaptive observer for single output single input linear system,
      IEEE Trans. Autom. Control, Vol. AC-18, 1973, pp. 428-435.
Cartwright, M.L.
  [1] Stability of solutions of certain differential equations of fourth
      order, Quart. J. Mech. Appl. Math., Vol. 9, 1956, pp. 185-193.
Cartwright, M.L., and Littlewood, J.E.
  [1] Nonlinear differential equations of second order, J. London Math.
      Soc., Vol. 20, 1945, pp. 180-189.
  [2] On nonlinear differential equations of second order, Ann. of
      Math., Vol. 48, 1947, pp. 472-494, and Vol. 50, 1949, pp. 504-505.
Cesari, L.
  [1] *Asymptotic Behavior and Stability Problems in Ordinary Differen-
      tial Equations*. Springer, 1959.
Cheprasov, V.A.
  [1] On controllability of nonlinear systems, SIAM J. Control, Vol. 8,
      1970, pp. 113-123.
Cherry, M.
  [1] Topological properties of solutions of ordinary differential
      equations, Amer. J. Math., Vol. 59, 1937.
Chetaiev, N.G.
  [1] *Stability of Motion*. Pergamon, London, 1961.
Chiang, H.D., Hirsch, M., and Wu, F.F.
  [1] Stability regions of nonlinear autonomous dynamical systems,
      IEEE Trans. Autom. Control, Vol. AC-33, 1988, pp. 16-26.
Chikriy, A.A.
  [1] Sufficient conditions for avoidance in nonlinear differential
      games, Dokl. Akad. Nauk, Vol. 241, 1978, pp. 547-550.

[2] Nonlinear differential evasion games, Soviet Math. Dokl, Vol. 20, 1979, pp. 591-595.

Chin, P.S.M.
[1] A general method to derive Liapunov functions for nonlinear systems, Int. J. Control, Vol. 44, 1986, pp. 381-393.

Choe, H.H., and Nikiforuk, P.N.
[1] Inherently stable feedback control of a class of unknown plants, Automatica, Vol. 7, 1971, pp. 607-625.

Chow, S.N., and Hale, J.K.
[1] *Methods of Bifurcation Theory.* Springer, New York, 1982.

Christensen, R.M.
[1] *Theory of Visco-Elasticity.* Academic Press, New York, 1971.

Chua, L.O., and Green, D.N.
[1] Synthesis of nonlinear periodic systems, IEEE Trans. Circuit & Systems, Vol. CAS-21, 1974, pp. 286-294.

Chukwu, E.N.
[1] Finite time controllability of nonlinear control processes, SIAM J. Control, Vol. 13, 1975, pp. 807-876.

Coiffet, P.
[1] *Modelling and Control.* Kogan Page, London, 1983.

Conley, C., and Easton, R.
[1] Isolated invariant sets and isolating blocks, Trans. Amer. Math. Soc., Vol. 157, 1971, pp. 1-27.

Corless, M.
[1] *Tracking Controllers for Uncertain Systems: Applicable to Manutec Robot,* J. Dyn. Syst. Meas. Control, Special Issue on Control Mechanics, 1989, to appear.
[2] Control of uncertain mechanical systems with robustness in the presence of unmodelled flexibilities, *Proc. 3rd Bellman Continuum,* Sophia-Antipolis, France, 1988.

Corless, M., Goodall, D.P., Leitmann, G., and Ryan, E.P.
[1] Model following controls for a class of uncertain dynamical systems, *7th IFAC Sympos. Identif. & System Parameter Estimation,* London, 1985, pp. 38-50.

Corless, M., and Leitmann, G.
[1] Continuous state feedback guaranteed ultimate boundedness for uncertain dynamical systems, IEEE Trans. Automat. Control, Vol. AC-26, 1981, pp. 1139-1142.
[2] Adaptive control of systems containing uncertain functions and unknown functions with uncertain bounds, J. Optim. Theory Appl., Vol. 41, 1983, pp. 155-168.
[3] Adaptive control for uncertain dynamical systems, in A. Blaquière and G. Leitmann (eds), *Dynamical Systems and Microphysics: Control Theory and Mechanics.* Academic Press, New York, 1983.
[4] Adaptive controllers for a class of uncertain systems, Ann. Found. Louis de Broglie, Vol. 9, 1984, pp. 65-95.
[5] Memoryless controllers for uncertain systems, in F. Gagliardi (ed), *Mathematical Methods for Optimization in Engineering.* Univ. of Cassino Press, Italy, 1985.
[6] Adaptive controllers for advoidance or evasion in uncertain environment, Computers Math. Applic., Vol. 18, 1989, pp. 161-170.

Corless, M., Leitmann, G., and Ryan, E.P.
[1] Tracking in the presence of bounded uncertainties, *Proc. IV Int. Conf. Control Theory,* Cambridge Univ., September 1984, pp. 87-92.
[2] Control of uncertain systems with neglected dynamics, in A. Zinober (ed), *Variable Structure Control Systems,* IEE Publ., London, 1987.

Corless, M., Leitmann, G., and Skowronski, J.M.
[1] Adaptive control for avoidance or evasion in an uncertain environment, Int. J. Computers Math. Appl., Vol. 13, 1987, pp. 1-30.

Cvetkovic, V., and Vukobratovic, M.
  [1] One robust dynamic control algorithm for manipulation systems,
       Internat. J. Robotic Res., Vol. 1, 1982, pp. 15-28.
Dailly, C.
  [1] Optimal workpiece positioning in a robot workspace, Tech. Notes,
       Engng. Dept., Univ. of Cambridge, 1985.
Dauer, J.P.
  [1] A controllability technique for nonlinear systems, J. Optim.
       Theory Appl., Vol. 37, 1972, pp. 442-451.
  [2] A note on bounded perturbations of controllable systems, J.
       Optim. Theory Appl., Vol. 42, 1973, pp. 221-225.
  [3] Controllability of nonlinear systems with restrained control,
       J. Optim. Theory Appl., Vol. 44, 1974, pp. 251-261.
  [4] Bounded perturbations of controllable systems, J. Math. Anal.
       Appl., Vol. 48, 1974, pp. 61-69.
Davidovitz, A., and Shinar, J.
  [1] Eccentric two-target model for qualitative air combat game
       analysis, J. Guidance, Vol. 8, 1985, #3.
Davison, E.J., and Kurak, E.M.
  [1] Computational method for determining quadratic Liapunov functions
       for nonlinear systems, Automatica, Vol. 7, 1971, pp. 627-636.
Davison, E.J., Silverman, L.M., and Varayia, P.P.
  [1] Controllability of a class of nonlinear time variable systems,
       IEEE Trans. Autom. Control, Vol. AC-12, 1967, pp. 791-792.
Davy, I.L.
  [1] Properties of solutions set of generalized differential equations,
       Bull. Austral. Math. Soc., Vol. 6, 1972, pp. 379-398.
Delfour, M.C., and Mitter, S.K.
  [1] Reachability of perturbed systems and min-sup problems, SIAM J.
       Control, Vol. 7, 1969, pp. 521-533.
Den Hartog, J.P.
  [1] *Mechanical Vibrations*. Wiley, 1957.
Desa, S., and Roth, B.
  [1] Mechanics: Kinematics & dynamics, in G. Beni and S. Hackwood (eds),
       Recent Advances in Robotics, Wiley, New York, 1985, pp. 71-130.
Devaney, R.L.
  [1] *Introduction to Chaotic Dynamical Systems*. Benjamin-Cummings
       Publ., 1986.
Diligenskii, S.N.
  [1] Estimation of attraction region of multi-dimensional nonlinear
       system by aggregation-decomposition method, Automation & Remote
       Control, 1975, pp. 645-658.
Donaldson, D.D., and Leondes, C.T.
  [1] Model referenced parameter tracking technique for adaptive control
       systems, Trans. IEEE Appl. and Industry, Vol. 82, 1963, pp. 241-
       262.
Drazenovic, B.
  [1] Invariance conditions in variable structure systems, Automatica,
       Vol. 5, 1965, pp. 287-295.
Drenick, R.F.
  [1] Optimization under uncertainty, in J. Stoer (ed), *Optimization
       Techniques*, Lecture Notes on Control and Information Sciences,
       Vol. 6, Springer-Verlag, Berlin and New York, 1978, pp. 40-58.
  [2] Feedback control of partly unknown systems, SIAM J. Control
       Optim., Vol. 15, 1977, pp. 506-509.
Dubovsky, S., and DesForges, D.T.
  [1] Application of model reference adaptive control to robotic mani-
       pulators, Trans. ASME Ser. G.J. Dynamic Systems Measurement
       Control, Vol. 101, 1979, pp. 193-200.

Dubovsky, S., and Gardner, T.N.
  [1] Dynamic interactions of link elasticity and clearance connections in planar mechanical systems, Trans. ASME J. Engng. Industry, Vol. 97, 1975, pp. 652-661.

Duffing, G.
  [1] *Erzwungene Schwingungen bei veraenderlicher Eigenfrequenz und ihre technische Bedeutung.* Braunschweig, 1918.

Duinker, S.
  [1] Traditors, a new class of nonenergetic network elements, Philips Res. Rep., Vol. 14, 1959, pp. 29-51.

Emelianov, S.V.
  [1] *Automatic Control Systems with Variable Structure* (in Russian). Nauka, Moscow, 1967.

Erdman, A.G., and Sandor, G.N.
  [1] Kineto-elasto-dynamics: A review of the state-of-the-art and trends, Mechanism and Machine Th., Vol. 7, 1972, pp. 19-33.

Erugin, N.P.
  [1] Design of a set of all systems of differential equations having given integral curve (in Russian), P.M.M., Vol. 16, 1952, pp. 659-670.

Erzberger, H.
  [1] Analysis and design of model following control systems by state space techniques, *Proc. JACC*, Ann Arbor, 1968, pp. 572-581.

Eykhoff, P.
  [1] *System Identification: Parameter and State Estimation.* Wiley, New York, 1974.

Filippov, A.F.
  [1] Existence of solutions of generalized differential equations, Math. Notes, Vol. 10, 1971, pp. 307-313.

Filippov, S.D.
  [1] Avoidance problem with incomplete information, Differential Equations, Vol. 13, 1977, pp. 1267-1272.

Fillod, R., Piranda, J., and Bonnecase, D.
  [1] Taking nonlinearities into account in model analysis by curve fitting of transfer functions, Trans. ASME, J. Eng. Ind., February 1976.

Flashner, H.
  [1] An orthogonal decomposition approach to modal synthesis, Int. J. Num. Math. in Eng., Vol. 23, 1986, pp. 471-493.

Flashner, H., and Guttalu, R.S.
  [1] Computational method for studying domains of attraction for nonlinear dynamical systems, Int. J. Nonlinear Mechanics, Vol. 23, 1988, pp. 279-295.

Flashner, H., and Hsu, C.S.
  [1] A study of nonlinear periodic systems via the point mapping method, Internat. J. Numerical Methods in Engng., Vol. 19, 1983, pp. 185-215.

Flashner, H., and Skowronski, J.M.
  [1] Control of Hamiltonian systems, Trans. ASME J. Dyn. Sys. Meas. Control, Vol. 111, 1989.
  [2] Adaptive Control of Hamiltonian Systems, *Control and Dynamic Systems*, Vol 35, 1990, Academic Press.

Flüge-Lotz, I.
  [1] *Discontinuous Automatic Control.* Princeton Press, 1953.

Foley, M.A., and Schmitendorf, W.E.
  [1] A class of differential games with two pursuers versus one evader, IEEE Trans. Automat. Control, Vol. AC-19, 1974, #3.

Frackiewicz, H.
  [1] Deformation of a discrete set of points, Arch. Mech. Stas. (Arch. Appl. Mech.), Pol. Ac. Sci., Vol. 18, 1966, #6.

413

Frederick, C.
  [1] *Modelling and Analogies of Dynamical Systems.* Houghton, Boston, Massachusetts, 1978.
Freund, E.
  [1] On design of multirobot systems, *Int. Conf. Robotics*, Atlanta GA, 1984, pp. 477-490.
Freund, E., and Hover, H.
  [1] Collision avoidance in multi-robot systems, *II Int. Symp. Rob. Research*, Kyoto, 1984, pp. 135-146.
  [2] On line solution on find path in multirobot system, *III Int. Symp. Rob. Res.*, Gouviex, 1985.
Fudjii, H., and Ishijima, S.
  [1] Mission function control for deployment and retrieval of a sub-satellite, J. Guidance, Vol. 12, 1989, pp. 243-247.
Fudjii, S., and Kurono, S.
  [1] Coordinated computer control of a pair of manipulators, *Proc. World Congress Theory of Machines*, Newcastle-upon-Tyne, 1975, pp. 411-417.
Furasov, W.D.
  [1] *Stability of Motion.* Nauka, Moscow, 1977.
Gabrielyan, M.S.
  [1] On stabilization of unstable motions of mechanical systems, PMM, Vol. 28, 1964, #3.
Galiullin, A.S.
  [1] Problem of constructing systems for a programmed motion, Automatic & Remote Control, 1970, pp. 32-37.
Galperin, E.A.
  [1] Asymptotic observers for nonlinear control systems, *Proc. IEEE Conf. Decision and Control (CDC)*, New York, 1976, pp. 1929-1300.
Galperin, E.A., and Skowronski, J.M.
  [1] Playable asymptotic observers for differential games with incomplete information - the user's guide, *Proc. 23rd IEEE Conf. Decision and Control (CDC)*, Las Vegas, 1984, pp. 1201-1206.
  [2] Geometry of V-functions and Liapunov stability, J. Nonlin. Analysis: Th., Methods & Applic., Vol. 11, 1987, pp. 183-197.
  [3] Pursuit-evasion differential game with uncertainty in dynamics, Int. J. Comp. Math. Appl., Vol. 13, 1987, pp. 13-35.
  [4] V-functions in control of motion, Int. J. Control, Vol. 42, 1985, pp. 361-367.
Gayek, J., and Vincent, T.L.
  [1] Manoeuverable sets, J. Optim. Theory Appl., to appear.
Genesio, R., Tartaglia, M., and Vicino, A.
  [1] An estimation of asymptotic stability regions: state of the art and new proposals, IEEE Trans. Autom. Control, Vol. AC-30, 1985, pp. 747-755.
Gershwin, S.B., and Jacobson, D.H.
  [1] Controllability theory for nonlinear systems, IEEE Trans. Automat. Control, Vol. AC-16, 1971, pp. 37-50.
Getz, W.M., and Leitmann, G.
  [1] Qualitative differential games with two targets, J. Optim. Theory Appl., Vol. 68, 1979, pp. 421-430.
Getz, W.M., and Pachter, M.
  [1] Capturability in a two target game of two cars, J. Guidance, Vol. 4, 1981, pp. 15-21.
Gevarter, W.B.
  [1] Basic relations for control of flexible vehicles, AIAA J., Vol. 8, 1970, pp. 666-678.
Gilbert, E.G., and Johnson, D.W.
  [1] Distance functions and application to robot path planning, Robot System Division, Univ. of Michigan, T.R. pp. 7-84.

Gilmore, R.
   [1] *Catastrophe Theory for Scientists and Engineers.* Wiley, New York, 1981.

Goldstein, H.
   [1] *Classical Mechanics.* Addison Wesley, 1980 (second edition).

Gondhalekar, V., and Holmes, R.
   [1] Design of electromagnetic bearing for vibration control of flexible transmission shaft, Rotordynamics Instability Problems in High Performance Turbomachinery, T & M Univ. NASA Conf. Publ. 2338, 1984.

Goodall, D.P., and Ryan, E.P.
   [1] Feedback controlled differential inclusions and stabilization of uncertain dynamical systems, SIAM J. Control & Opt., 1989.

Grantham, W.J.
   [1] *A Controllability Minimum Principle.* Ph.D. Thesis, Aerospace Engng., Univ. of Arizona, Tucson, 1973.

Grantham, W.J., and Chingcuanco, A.O.
   [1] Lyapunov steepest descent control of constrained linear system, IEEE Trans. Autom. Control, Vol. AC-29, 1984, pp. 740-743.

Grantham, W.J., and Lee, B.
   [1] Aeroassisted orbital transfer in an uncertain atmosphere, AIAA J. Guidance, Control & Dynamics, Vol. 12, 1989, pp. 237-242.

Grantham, W.J., and Vincent, T.L.
   [1] Controllability Minimum Principle, J. Optim. Theory Appl., Vol. 17, 1975, pp. 93-114.

Grayson, L.P.
   [1] Two theorems on the second method, IEEE Trans. Automat. Control, Vol. AC-9, 1964, pp. 587-562.
   [2] The status of synthesis using Liapunov's method, Automatica, Vol. 3, 1965, pp. 91-121.

Griffith, E.W., and Kumar, E.S.P.
   [1] On observability of nonlinear systems, J. Math. Anal. Appl., Vol. 35, 1971, pp. 61-69.

Guckenheimer, J.
   [1] Bifurcations of dynamical systems, in *Dynamical Systems*, CIME Lectures, publ. as Progress in Math., Vol. 8, 1978, Birkhauser, Boston.

Guckenheimer, J., and Holmes, P.
   [1] *Nonlinear Oscillations, Dynamical Systems and Bifurcations of Vector Fields.* Springer, 1983.

Gusev, S.V., Timofeev, A.V., and Yakubovich, V.A.
   [1] Algorithms of adaptive control of robot movements, Mechanism & Machine Theory, Vol. 18, 1983, pp. 279-281.

Gusev, S.V., and Yakubovich, V.A.
   [1] Adaptive control algorithm for a manipulator, Automation and Remote Control, Vol. 41(9), Part 2, 1980, pp. 1268-1277.

Gutman, S.
   [1] Uncertain dynamical systems - a differential game approach, NASA TMX - 73, April 1976, p. 135.
   [2] On optimal guidance for homing missiles, *Proc. 20th Israel Conf. Aviation and Astronaut.*, 1978, pp. 75-81.
   [3] Uncertain dynamical systems - a Liapunov min-max approach, IEEE Trans. Automat. Control, Vol. AC-24, 1979, pp. 437-443.
   [4] Remarks on capture-avoidance games, J. Optim. Theory Appl., Vol. 32, 1980, pp. 365-377.

Gutman, S., and Leitmann, G.
   [1] Stabilizing feedback control for dynamical systems with bounded uncertainty, *Proc. IEEE Conf. Decision and Control (CDC)*, New York, 1976, pp. 1-9.

Gutman, S., and Palmor, Z.
   [1]   Properties of min-max controllers in uncertain dynamical systems,
         SIAM J. Control Optim., Vol. 20, 1982, pp. 850-861.
Guttalu, R.S., and Flashner, H.
   [1]   A numerical method for computing domains of attraction, Int. J.
         Num. Methods in Engng., Vol. 26, 1988, pp. 875-890.
Guttalu, R.S., and Skowronski, J.M.
   [1]   Algorithms for degree of controllability of uncertain mechanical
         systems, Trans. ASME J. Dyn. Sys. Meas. Control, Vol. 111, 1989,
         pp. 600-604.
Ha, I.J., and Gilbert, E.G.
   [1]   Robust tracking in nonlinear systems, IEEE Trans. Autom. Control,
         Vol. AC-32, 1987, pp. 763-771.
Hadamard, J.
   [1]   Sur certaines propriétés des trajectories en dynamique, J. Math.
         Pures et Appl., Ser. V, Vol. 3, 1897, pp. 331-387.
Hagedorn, P.
   [1]   The potential energy surface for nonlinear equations of motion,
         CERN-PS/RH-6, Geneva, November 1954.
   [2]   Nonlinear Oscillations.   Clarendon Press, Oxford, 1981.
Hagedorn, P., and Breakwell, J.V.
   [1]   A differential game with two pursuers and one evader, J. Optim.
         Theory Appl., Vol. 18, 1976, pp. 15-29.
   [2]   A differential game with two pursuers and one evader, in G.
         Leitmann (ed), Multicriteria Decision Making and Differential
         Games, Plenum, New York, 1976, pp. 80-88.
Hagedorn, P., Kelkel, K., and Weltin, U.
   [1]   On active control of rotors with uncertain parameters, Proc.
         IFTOMM Intern. Conf. on Rotor-dynamics, Tokyo, 1986.
Hajek, O.
   [1]   Dynamical Systems in the Plane.   Academic Press, 1968.
   [2]   Pursuit Games.   Academic Press, New York, 1975.
Hale, A.L., and Lisowski, R.J.
   [1]   Characteristic elastic systems of time limited optimal maneouvers,
         J. Guidance, Vol. 8, 1985, pp. 628-636.
Hall, J.E.
   [1]   Quantitative estimates for nonlinear differential equations by
         Liapunov functions, J. Diff. Eqns., Vol. 2, 1966, pp. 173-181.
Hanafi, A., Wright, F.W., and Hewitt, J.R.
   [1]   Optimal trajectory control of robotic manipulators, Mechanism &
         Machine Theory, Vol. 19, 1984, pp. 267-273.
Hang, C.C., and Parks, P.C.
   [1]   Comparative studies of model reference adaptive control systems,
         Proc. 1973 Joint Autom. Control Conf., Columbus, Ohio, 1973,
         pp. 12-22.
Haque, S.I., and Monopoli, R.V.
   [1]   Discrete adaptive control of a radio telescope, Int. Workshop on
         Appl. Adaptive Control, Yale, 1979.
Hayashi, Ch.
   [1]   Nonlinear Oscillations in Physical Systems.   McGraw Hill, 1964.
Hayati, S.A.
   [1]   Dynamics and control of coordinated multiple manipulators, JPL
         Publ. 87-13, Workshop on Space Telerobotics, Vol. 3, pp. 193-204.
   [2]   Hybrid position-force control of multi-arm cooperating robots,
         Proc. IEEE Intern. Conf. Robotic and Automation, San Francisco,
         1986, pp. 82-89.
Heinen, J.A., and Wu, S.H.
   [1]   Finite time stability and inverted Liapunov problems, Proc. 13th
         Midwest Symp. on Circuit Theory, Univ. of Minnesota, 1970, VIII
         pp. 1-9.

Hemami, H.

[1] Kinematics of two arm robots, IEEE J. Robotics & Autom., RA2, 1986, pp. 225-228.

Hemami, H., and Wyman, B.

[1] Indirect control of the forces of constraint in dynamic system, ASME Trans. J. Dyn. Sys. Control, Vol. 101, 1979, pp. 355-360.

Higgins, T.J.

[1] A resumé of the development of nonlinear control systems, J. Franklin Inst., Vol. 272, 1961, #4, pp. 253-274.

Ho, J.Y.L.

[1] Direct path method for flexible multibody spacecraft dynamics, J. Spacecraft and Rockets, Vol. 12, 1977, pp. 102-110.

Ho, J.Y.L., and Herbert, D.R.

[1] Development of dynamics and control simulation of large flexible space systems, J. Guidance, Vol. 8, 1985, pp. 374-383.

Hogan, N.

[1] Adaptive control of mechanical impedance by coactivation of antagonistic muscles, IEEE Trans. Autom. Control, Vol. AC-29, 1984, #7.

[2] Impedance control: an approach to manipulation, Trans. ASME J. Dyn. Sys. Meas. Control, Vol. 106, 1985.

Hollerbach, J.M.

[1] A recursive formation of Lagrangian manipulator dynamics, IEEE Trans. Sys. Man. Cybern., Vol. SMC-10, 1980, pp. 730-736.

Hooker, W.W.

[1] A set of r dynamical attitude equations for an arbitrary n-body satellite having r rotational degrees of freedom, AIAA J., Vol. 8, 1970, #7.

[2] Equations of motion for interconnected rigid and elastic bodies, Celestial Mechanics, Vol. 11, 1975, pp. 337-359.

Hooker, W.W., and Margulies, G.

[1] The dynamical attitude equations for a n-body satellite, J. Astronaut. Sci., Vol. 12, 1965, pp. 123-128.

Hopf, E.

[1] Abzweigung einer periodischen Loesung von einer stationaeren Loesung eines Differentialsystemes, Berl. Math-Phys. Klasse, Akad. Wiss. Leipzig, Vol. 94, 1942, pp. 1-22 (English transl. Marsden-McCracken 1976).

Horowitz, R., and Tomizuka, M.

[1] An adaptive control scheme for mechanical manipulators, ASME Paper 80-WA/DSC-6, 1980.

Hsu, C.S.

[1] *Cell to Cell Mapping*. Springer, 1987.

Hughes, P.C.

[1] Dynamics of a chain of flexible bodies, J. Astronaut. Sci., Vol. 27, 1979, #4.

[2] Space structure vibration modes. *NASA-JPL Workshop on Space Telerobotics*, Pasadena, 1987, pp. 31-47.

[3] *Spacecraft Attitude Dynamics*. Wiley, 1986.

Hughes, P.C., and Skelton, R.E.

[1] Model truncation for flexible spacecraft, J. Guid. Control, Vol. 4, 1981, pp. 291-297.

[2] Controllability and observability of linear matrix second order systems, ASME J. Appl. Mech., Vol. 47, 1980, pp. 415-420.

Huston, R.L.

[1] Multibody dynamics including the effects of flexibility and compliance, Computers and Structure, Vol. 14, 1981, pp. 443-451.

Hwang, M., and Seinfeld, J.H.

[1] Observability of nonlinear systems, J. Optim. Theory Appl., Vol. 10, 1972, pp. 66-67.

Ibrahim, A.E., and Misra, A.K.
  [1] Attitude dynamics of a satellite during deployment of large
       plate-type structures. J. Guidance, Control & Dynamics, Vol. 5,
       1982, #5, pp. 442-447.
Ibrahim, A.M., and Modi, V.J.
  [1] A formulation for studying steady state/transient dynamics of a
       large class of spacecraft and its applications, *Proc. II Intern.
       Symp. Spacecraft Flight Dynamics*, Dormstadt, Germany, 1986, pp.
       25-30.
Irvin, M.C.
  [1] *Smooth Dynamical Systems*. Academic Press, New York, 1980.
Isaacs, R.
  [1] *Differential Games*. Wiley, New York, 1964.
Ishida, T.
  [1] Force control in coordination of two arms, *Proc. 5th Int. Joint
       Conf. on AI*, August 1977, pp. 717-728.
James, D.J.G.
  [1] Stability analysis of a model reference adaptive control system
       with sinusoidal inputs, Int. J. Control, Vol. 9, 1969, pp. 311-
       321.
Jarmark, B.C., Merz, A.W., and Breakwell, J.V.
  [1] The variable speed tail chase aerial combat problem, J. Guidance,
       Vol. 4, 1981, pp. 323-328.
Jayasuria, S., Rabins, M.J., and Barnard, R.D.
  [1] Guaranteed tracking behavior in sense of input-output spheres for
       systems with uncertain parameters, ASME Trans. J. Dyn. Sys. Meas.
       Control, Vol. 106, 1984, pp. 273-279.
Jerkowsky, W.
  [1] The structure of multibody dynamic equations, J. Guidance and
       Control, Vol. 1, 1978, #3.
Jewusiak, H., and Bigley, W.J.
  [1] Mechanical network analysis, Machine Design, Vol. 25, 1964.
Johnson, G.W.
  [1] Synthesis of control systems with stability constraints via the
       Direct Method of Liapunov, IEEE Trans. Autom. Control, Vol. AC-9,
       1964, #3.
Johnson, H.
  [1] *Model Reference Adaptive Control*. Univ. of Qld., Eng. Math.
       Thesis, 1978.
Jumarie, G.
  [1] Structural differential games, C.R. Hebd. Sciences Acad. Sci.
       Ser. A, Vol. 280, 1975, pp. 969-972.
Kalman, R.E.
  [1] On general theory of control systems, in *Automatic and Remote
       Control*, pp. 481-491, Butterworths, London, 1961.
Kalman, R.E., and Bertram, J.E.
  [1] Control system analysis and design via the second method of
       Liapunov, Trans. ASME J. Basic Eng., Vol. 1, 1960, pp. 371-393.
Kanai, K., and Uchikado, S.
  [1] Application of a new multivariable model following method to
       decoupled flight control, J. Guidance, Vol. 8, 1985, pp. 637-643.
Karnopp, D., and Rosenberg, R.
  [1] *System Dynamics: A United Approach*. Wiley, 1975.
Kaskosz, B.
  [1] A sufficient condition for evasion in a nonlinear game, Control
       Cybernet., Vol. 7, 1978, pp. 515-521.
  [2] On a nonlinear evasion problem, SIAM J. Control Optim., Vol. 15,
       1977, pp. 661-679.
Kauderer, H.
  [1] *Nichtlineare Mechanik*. Springer, 1958.

Kaufmann, H., and Berry P.
[1] Adaptive flight control using optimal linear regulator techniques, Automatica, Vol. 12, 1976, pp. 565-576.

Khatib, O.
[1] Real time obstacle avoidance for manipulators and mobile robots, Int. J. Robotic Research, Vol. 5(1), 1986.

Khatib, O., and Burdick, J.
[1] Motion and force control of robot manipulators, *Proc. 1986 IEEE Conf. Robotics*, San Francisco, April 1986.

Kinnen, E., and Chen, C.S.
[1] Liapunov functions derived from auxiliary exact differential equations, Automatica, Vol. 4, 1968, pp. 195-204.

Klamka, J.
[1] On global controllability of perturbed nonlinear systems, IEEE Trans. Autom. Control, Vol. AC-20, 1975, pp. 170-172.
[2] On local controllability of perturbed nonlinear systems, IEEE Trans. Autom. Control, Vol. AC-20, 1975, pp. 289-291.

Klein, C.A., and Briggs, R.
[1] Use of active compliance in control of legged vehicles, IEEE Trans. Sys. Man. Cybern., SMC-10, 1980, #7.

Klein, C.A., and Patterson, M.R.
[1] Computer coordination of link motion of a multiple-armed robot for space assembly, IEEE Trans. on Systems, Man., Cybern., SMC-12, 1982, pp. 913-919.

Klimentov, S.I., and Prokopov, B.I.
[1] Synthesis of asymptotically stable algorithm of adaptive system with reference model by Liapunov direct method, Automatic & Remote Control, Vol. 35, 1974, pp. 1625-1632.

Klir, G.I.
[1] *On Approach to General System Theory.* Von Nostrand, 1970.

Klotter, K.
[1] *Einfuerung in die Technische Schwingungslehre.* Berlin, 1951.
[2] Steady state vibrations in systems having arbitrary restoring and damping forces, *Proc. Symp. Nonlinear Circuit Anal.*, Brooklyn Polytechnic, 1953.

Koditschek, D.E.
[1] Exact robot navigation by means of potential functions, *IEEE Int. Conf. Robotics and Automation*, Raleigh, NC, 1987.
[2] High gain feedback and telerobotic tracking, JPL Publ. 87-13, *Workshop on Space Telerobotics*, vol. 3, pp. 355-364.

Koenig, H.E., Tokad, Y., and Kesavan, H.K.
[1] *Analysis of Discrete Physical Systems.* McGraw Hill, 1967.

Koivo, A.J.
[1] Adaptive position-velocity-force control of two manipulators, *Proc. 24th IEEE CDC*, Ft Lauderdale, 1985, pp. 1529-1532.

Koivo, A.J., and Paul, R.P.
[1] Manipulator with self-tuning controller, *IEEE Conf. Cybernet. and Soc. 1980.*

Koivo, A.J., and Repperger, D.W.
[1] Optimization of terminal rendezvous as a cooperative game, *Proc. 12th JACC*, St Louis, 1971, pp. 508-516.

Kon, S.R., Elliot, D.L., and Tarn, T.J.
[1] Exponential observers for nonlinear dynamic systems, Inform. and Control, Vol. 29, 1975, pp. 204-216.

Kononenko, V.O.
[1] Some autonomous problems of theory of nonlinear vibrations, *Proc. IUTAM Symp.*, Kiev, 1961(1962), pp. 151-179.
[2] *Vibrating Systems with Limited Power Supply.* Iliffe Books, London, 1969.

Konstantinov, M.S.
  [1] Inertia forces of robots and manipulators, Mechanism and Machine
       Theory, Vol. 12, 1977, pp. 387-401.
Kooleshov, V.S., and Lakota, N.A.
  [1] *Manipulators Control System Dynamics*. Energia, Moscow, 1971.
Kopysov, O.Y., and Prokopov, B.I.
  [1] Identification of parameters of nonlinear systems by adaptive
       model method, Automation & Remote Control, Vol. 39, 1978, pp.
       1803-1808.
Kostiukovski, Y.M.L.
  [1] Observability of nonlinear control systems, Automat. Remote
       Control, 1968, pp. 1384-1396.
Kosut, R.L., Salzwedel, H., and Emami-Naeini, A.
  [1] Robust control of flexible spacecraft, J. Guidance & Control,
       Vol. 6, 1983, pp. 104-111.
Krassovski, A.A.
  [1] Integral equations and the choice of parameters in regulating
       systems, *Teoria Automaticheskogo Regulirovanie*, Mashgiz, Moscow,
       1951.
Krassovski, N.N.
  [1] *Theory of Motion Control* (in Russian). Nauka, Moscow, 1968.
  [2] *Control Theory*. Nauka, Moscow, 1968.
  [3] Differential game of approach-evasion, Izv. Akad. Nauk SSSR,
       Tech. Cybernet., 1973, #2 and 3.
  [4] Game theoretic control and problems of stability, Problems
       Control Inform. Theory, Vol. 3, 1974, #3, pp. 171-182.
  [5] Game theoretic control under incomplete phase-state information,
       Problems Control Inform. Theory, Vol. 5, 1976, pp. 291-302.
Krassovski, N.N., and Subbotin, A.I.
  [1] *Positional Differential Games*. Nauka, Moscow, 1974.
Kreisselmeier, G.
  [1] Adaptive observers with exponential rate of convergence, IEEE
       Trans. Autom. Control, Vol. AC-22, 1977, pp. 3-8.
Kroc, L.
  [1] On synthesis of adaptive multiparameter control systems by
       Liapunov method, Kibernetica, Vol. 2, 1975, pp. 277-286.
Krogh, B.H.
  [1] Feedback obstacle avoidance control, *Proc. XXI Allerton Conf.*,
       Univ. of Illinois, 1983, pp. 325-334.
  [2] Generalized potential field approach to obstacle avoidance, *SME
       Conf. Rob. Res.*, Bethlehem PA, 1984.
Krotkiewski, J.M.
  [1] Private communication.
Krutova, I.N., and Rutkovskii, V.Y.
  [1] Dynamics of first order adaptive system with a model, Automatic
       & Remote Control, Vol. 25, 1964, #1-6, pp. 175-180.
  [2] Effects of integrals in laws governing variations of modified
       coefficients of the dynamics of a self-adjusting system with a
       model, Automatic & Remote Control, Vol. 25, 1964, pp. 441-449.
  [3] Investigation of the dynamics of a model reference adaptive
       system for a plant with nonlinear characteristics and variable
       parameters, Automatic & Remote Control, Vol. 25, 1964, pp. 795-
       802.
  [4] Analysis of a second order model reference self-adaptive system,
       Automatic & Remote Control, Vol. 26, 1965, pp. 73-86.
  [5] Parameter selection for self-organizing systems with models,
       Automatic & Remote Control, Vol. 26, 1965, pp. 222-232.
Krylov, N.M., and Bogolubov, N.N.
  [1] *Introduction into Nonlinear Mechanics* (in Russian). Acad. Sci.
       USSR Publ., Kiev, 1937.

Kudva, P., and Narendra, K.S.
  [1] Synthesis of adaptive observer using Liapunov direct method,
      Int. J. Control, Vol. 18, 1973, pp. 1201-1210.
Kulinich, A.S., and Panev, G.P.
  [1] Parametric optimization of the equations of motion of multi-link
      system and adaptive control algorithms, Automation & Remote
      Control, Vol. 40, 1979, pp. 1793-1803.
Kumar, A., and Waldron, K.J.
  [1] The workspace of a mechanical manipulator, J. Mech. Design,
      Vol. 103, 1981, pp. 665-672.
Kuntze, H.B., and Schill, W.
  [1] Methods for collision avoidance in computer controlled industrial
      robots, *Proc. Internat. Symp. Indust. Robots*, Paris, 1982, pp.
      519-530.
Lal, M., and Mehrorta, R.
  [1] Design of model reference adaptive control systems for nonlinear
      plants, Int. J. Control, Vol. 16, 1972, pp. 993-996.
Lanczos, C.
  [1] *Variational Principles of Mechanics*. Univ. of Toronto Press,
      1970.
Landau, I.D.
  [1] A survey of model reference adaptive techniques, Automatica, Vol.
      10, 1974, pp. 313-379.
  [2] *Adaptive Control: The Model Reference Approach*. Dekker, New
      York, 1979.
Landau, I.D., and Curtiol, B.
  [1] Adaptive model following systems for flight control and simula-
      tion, J. Aircraft, Vol. 9, 1972, pp. 668-674.
La Salle, J.P., and Lefschetz, S.
  [1] *Stability by Liapunov's Direct Method*. Academic Press, 1961.
Laskin, R.A., Likins, P.W., and Longman, R.W.
  [1] Dynamical equations of a free-free beam subject to large overall
      motions, J. Astronautical Sci., Vol. 31, 1983, #4, pp. 507-528.
Laskin, R.A., and Sirlin, S.W.
  [1] Future payload isolation and pointing system technology, J.
      Guidance, Vol. 9, 1986, pp. 469-477.
Leborgue, M., Ibarra, J.M., and Espiau, B.
  [1] Adaptive control of high velocity manipulators, *Proc. XI Internat.
      Symp. Indust. Robots*, Tokyo, 1981.
Lee, C.S.G., and Chung, M.J.
  [1] Adaptive perturbation control with feed forward compensation for
      robot manipulators, Simulation, Vol. 44, 1985, pp. 127-136.
Lee, E.B., and Markus, L.
  [1] *Foundations of Optimal Control Theory*. Wiley, 1967.
Lefschetz, S.
  [1] *Differential Equations: Geometric Theory*. Interscience Publ.,
      New York, 1957.
  [2] *Stability of Nonlinear Control Systems*. Academic Press, New
      York, 1965.
Leipholz, H.
  [1] *Stability of Elastic Systems*. Noordhoff, 1980.
  [2] An alternative to Liapunov stability method, Comp. Methods in
      Appl. Mech. Eng., 1984, #43, pp. 293-313, and #47, pp. 299-314.
Leitmann, G.
  [1] Sufficiency theorems for optimal control, J. Optim. Theory Appl.,
      Vol. 2, 1968, pp. 285-289.
  [2] Stabilization of dynamical systems under bounded input disturb-
      ances and parameter uncertainties, in E.O. Roxin (P.T. Liu and
      R.L. Sternbe, eds) "Differential Games and Control Theory", *Proc.
      II Kingston Conf. Differential Games and Control*, 1976, pp. 47-
      52.

  [3] Guaranteed asymptotic stability for some linear systems without
      bounded uncertainties, Trans. ASME, J. Dyn. Syst. Meas. Control,
      Vol. 101, 1979, pp. 212-216.
  [4] Guaranteed asymptotic stability for a class of uncertain linear
      dynamical Systems, J. Optim. Theory Appl., Vol. 27, 1979, pp.
      99-104.
  [5] Guaranteed avoidance strategies, J. Optim. Theory Appl., Vol. 32,
      1980, #4, pp. 569-576.
  [6] Deterministic control of uncertain systems, Acta Astronaut.,
      Vol. 7, 1980, pp. 1457-1461.
  [7] On efficacy of nonlinear control in uncertain linear systems,
      Trans. ASME, J. Dyn. Syst. Meas. Control, Vol. 103, 1981, pp.
      95-102.
  [8] *An Introduction to Optimal Control.* McGraw Hill, New York, 1966.
Leitmann, G., and Liu, H.S.
  [1] Evasion in the plane, *8th IFIP Sympos. Optim. Tech.*, Warzburg,
      1977.
Leitmann, G., and Skowronski, J.M.
  [1] Avoidance Control, J. Optim. Theory Appl, Vol. 23, 1977, pp.
      581-591.
  [2] A note on avoidance control, Optimal Control Appl. Methods,
      Vol. 4, 1983, pp. 335-342.
Lenarcic, J.
  [1] A new method for calculating the Jacobian for a robot manipulator,
      Robotica, Vol. 1, 1983, pp. 205-209.
Leondes, C.T. (ed)
  [1] *Control and Dynamical Systems,* Vol. 17, Academic Press, 1981.
Letov, A.M.
  [1] Die stabilitaet von Regelsystemen mit nachgebener Fuerung, *Proc.
      Heidelberg Conf. of Automatic Control*, pp. 201-211, Instruments
      Publ. Co., Pittsburgh PA, 1957.
  [2] Analytic design of regulators (in Russian), Avtom. Telemech.,
      Vol. 21, 1961, pp. 436-441, 501-568, 661-665, and Vol. 22, 1961,
      pp. 425-435.
  [3] *Stability of Nonlinear Control Systems* (in Russian). Fiz. Mat.
      Giz., Moscow, 1962.
Levi-Civita, T., and Amaldi, U.
  [1] *Lezioni di Meccanica Rationale.* Zanichelli, Bologna, 1922-1927.
Lewandowska, A.,
  [1] Controller synthesis for nonlinear plants with conditions on
      stability region, Control Cybernet., Vol. 5, 1976, pp. 5-19.
Lewis, R.A.
  [1] Adaptive control of robotic manipulators, *Proc. IEEE-CDC*, 1977,
      pp. 743-748.
Liapunov, A.M.
  [1] Problème générale de la stabilité du mouvement (reprinted), *Annals
      of Math. Studies*, Vol. 17, Princeton, 1947, first Russian edition,
      1892.
Lichtenberg, A.J., and Lieberman, M.A.
  [1] *Regular and Stochastic Motion.* Springer, 1983.
Liegeois, A.
  [1] Automatic supervisory control of configuration and behavior of
      multi-body mechanisms, IEEE Trans. Systems Man Cybern., Vol. SMC-7,
      1977, #12.
Liegeois, A., Fournier, A., and Aldon, M.J.
  [1] Model reference control of high velocity industrial robots, *Proc.
      JACC*, San Francisco, 1980, TP10-D.
Liegeois, A., Khalil, W., Dumas, J.M., and Renaud, M.
  [1] Mathematical and computer models of interconnected mechanical
      systems, in A. Morecki and K. Kedzior (eds), *Theory and Practice
      of Robot Manipulators*, Elsevier, Amsterdam, 1977.

Liénard, A.
  [1] Etude des oscillations entretenues, Revue Générale d'Electricité,
      Vol. 23, 1928, pp. 901-912, 946-954.
Likins, P.W.
  [1] Dynamic analysis of a system of hinge connected bodies with non-
      rigid appendages, NASA Tech. Report 32-1576, 1974.
  [2] Point-connected rigid bodies in a topological tree, Celestial
      Mechanics, Vol. 11, 1975, pp. 301-317.
Lim, J., and Chyung, D.H.
  [1] On control scheme for two cooperating robot arms, *Proc. 24th CDC*,
      Fort Lauderdale, FA, 1985, pp. 334-337.
Lindorf, D.P.
  [1] Control of nonlinear multivariable systems, IEEE Trans. Autom.
      Control, Vol. AC-12, 1967, pp. 506-515.
Lindorf, D.P., and Carrol, R.L.
  [1] Survey of adaptive control using Liapunov design, Int. J. Control,
      Vol. 18, 1973, pp. 897-914.
Lirov, Y.
  [1] On synchronised proportional motion of multirobot systems,
      Robotica, Vol. 4, 1986, pp. 151-154.
Litmanov, C.V.
  [1] Sufficient conditions for the existence of simultaneous choice of
      strategies in differential game with several players, Differen-
      tialnyie Vravnienia, Vol. 16, 1980, pp. 1760-1765.
Liu, R.W., and Leake, R.J.
  [1] Inverse Liapunov problems, *Proc. Internat. Sympos. Differential
      Equations Dynamic Systems*, Puerto Rico, 1965, pp. 75-80.
  [2] Exhaustive equivalence classes of optimal systems with separable
      controls, SIAM J. Control, Vol. 4, 1966, pp. 678-685.
Livesley, R.K.
  [1] Equivalence of continuous and discrete mass distributions in
      certain vibration problems, Quart. J. Mech. and Appl. Math.,
      Vol. 8, 1935, pp. 353-560.
Lobry, C.
  [1] Controlabilité des systèmes nonlinéaires, SIAM J. Control, Vol. 8,
      1970, pp. 573-605.
Lorenz, E.N.
  [1] Deterministic nonperiodic flow, J. Atmos. Sci., Vol. 20, 1963,
      pp. 130-141.
Lowe, E.H., and Rowland, J.R.
  [1] Improved signal synthesis techniques for model reference adaptive
      control systems, IEEE Trans. Autom. Control, Vol. AC-18, 1973,
      pp. 119-121.
Lowen, G.G., and Jandrasits, W.G.
  [1] Survey of investigations into the dynamic behavior of mechanism
      containing links with distributed mass and elasticity, Mechanism
      and Machine Th., Vol. 7, 1972, pp. 3-17.
Lozano-Perez, T.
  [1] Automatic planning of manipulator transfer movements, AI Memo No.
      606, MIT Artificial Ietelligence Lab., December 1980.
  [2] Spatial planning: a configuration space approach, AI Memo No. 605,
      MIT Artificial Intelligence Lab., December 1980.
Lozano-Perez, T., and Wesley, M.S.
  [1] An algorithm for planning collision free paths among polyhedral
      obstacles, Comm. ACM, Vol. 44, 1979, pp. 560-570.
Luders, G., and Narendra, K.S.
  [1] A new canonical form of an adaptive observer, IEEE Trans. Autom.
      Control, Vol. AC-19, 1974, pp. 117-119.
  [2] Stable adaptive schemes for state estimation and identification
      of linear systems, IEEE Trans. Autom. Control, Vol. AC-19, 1974,
      pp. 841-847.

Luenberger, D.G.
    [1] Observing the state of a linear system, IEEE Trans. Mil. Electron.,
        Vol. MIL-8, 1964, pp. 74-80.
    [2] An introduction to observers, IEEE Trans. Autom. Control, Vol.
        AC-16, 1971, pp. 596-602.
    [3] Dynamic systems in descriptor form, IEEE Trans. Autom. Control,
        Vol. AC-22, 1977, pp. 312-331.

Luh, J.Y.S.
    [1] Anatomy of industrial robots and their controls, IEEE Trans.
        Autom. Control, Vol. AC-28, 1983, pp. 133-153.
    [2] Conventional controller design for industrial robots - a
        tutorial, IEEE Trans Systems Man Cybernet., Vol. SMC-13, 1983,
        pp. 298-316.

Luh, J.Y.S., and Lin, C.S.
    [1] Optimum path planning for mechanical manipulators, Trans. ASME
        J. Dynamic Systems Meas. Control, Vol. 102, 1981, pp. 142-151.

Luh, J.Y.S., and Zheng, Y.F.
    [1] Motion coordination and programmable teleoperation between two
        industrial robots, JPL-Publ. 87-13, Workshop on Space Telerobotics,
        Vol. 2, pp. 325-334.
    [2] Computation of input generalized forces for robots with closed
        kinematic chain, IEEE J. Robotics & Automation, RA1, 1985,
        pp. 95-103.

Lukas, D.L.
    [1] Global controllability of nonlinear systems, SIAM J. Control,
        Vol. 10, 1972, pp. 112-126.
    [2] Global controllability of disturbed nonlinear equations, SIAM J.
        Control, Vol. 12, 1974, pp. 695-704.

Magnus, K.
    [1] *Ueber ein Verfahren zur Untersuchung nichtlinearer Schwingungs-*
        und-Regelung-Systeme. VDI-Forschungs Heft 451, Dusseldorf,
        1955.
    [2] *Dynamics of Multibody Systems.* Springer, 1978 (ed).
    [3] *The Multi-body Approach to Mechanical Systems.* Solid Mechanics
        Archives - Univ. of Waterloo Press, 1979.

Mahil, S.S.
    [1] On application of Lagrange's Method to description of mechanical
        systems, IEEE Trans. Systems Man Cybernet., Vol. SMC-12, 1982,
        #6.

Maizza-Neto, O.
    [1] *Model Analysis and Control of Flexible Manipulator Arms.* Ph.D.
        Thesis, Mech. Eng. MIT, 1974.

Maltz, M.
    [1] *Analysis of Chatter in Contactor Control Systems.* Ph.D. Diss.,
        Stanford Univ., Stanford, 1963.

Mansour, M.
    [1] Generalized Liapunov function for power systems, IEEE Trans.
        Autom. Control, Vol. AC-19, 1974, pp. 247-248.

Margolis, S.G., and Vogt, W.G.
    [1] Control engineering applications of Zubov's construction for
        Liapunov functions, IEEE Trans., Vol. AC-8, 1963, pp. 104-108.

Markov, A.A.
    [1] Stabilitaet in Liapunovschen Sinne und Fastperiodicitaet, Math.
        Zeitschrift, Vol. 36, 1933.

Markov, M.D., Zamanov, V.B., and Nenchev, D.N.
    [1] Trajectory modelling and teaching of robots under cartesian path
        control, Internat. J. Prod. Res., Vol. 21, 1983, pp. 173-182.

Markus, L.
    [1] Controllability of nonlinear processes, SIAM J. Control, Vol. 3,
        1965, pp. 78-90.

[2] *Lectures in Differentiable Dynamics*. Am. Math. Soc., Providence, 1971.

Mason, M.T.
[1] Compliance and force control for computer controlled manipulators, IEEE Trans. Systems Man Cybern., Vol. SMC-11, 1981, pp. 418-432.

McGhee, R.B.
[1] Control of legged locomotion system, *Proc. JACC*, San Francisco, 1977.

McGhee, R.B., Klein, C.A., and Chas, C.S.
[1] Interactive computer control of adaptive walking machine, *Proc. MIDCON*, Chicago, 1979.

McGhee, R.B., and Iswandhi, G.I.
[1] Adaptive locomotion of a multilegged robot over rough terrain, IEEE Trans. Systems Man Cybern., Vol. SMC-9, 1979, #4.

Mease, K.D., and Vinh, N.X.
[1] Minimum fuel aeroassisted coplanar orbit transfer using lift modulation, J. Guidance Control and Dynamics, Vol. 8, 1985, pp. 134-141.

Medvedev, V.S., Leskov, A.G., and Yushchenko, A.S.
[1] *Control System for Manipulator Robots*. Nauka, Moscow, 1978.

Meirovitch, L.M.
[1] *Methods of Analytic Dynamics*. McGraw Hill, New York, 1970.
[2] *Analytical Methods in Vibrations*. MacMillan, 1967.
[3] Stability of a spinning body containing elastic parts via Liapunov direct method, AIAA J., Vol. 8, 1970, pp. 1193-1200.

Meirovitch, L.M., and Baruch, H.
[1] The implementation of model filters for control structures, J. Guidance, Vol. 8, 1985, pp. 707-716.
[2] On the problem of observation spillover in distributed parameter systems, J. Opt. Th. Appl., Vol. 39, 1981, pp. 611-620.

Meirovitch, L.M., and Nelson, H.D.
[1] On the high spin motion of a satellite containing elastic parts, J. Spacecraft & Rockets, Vol. 3, 1966, pp. 1597-1602.

Meirovitch, L.M., and Quinn, R.D.
[1] Equations of motion for maneuvering flexible spacecraft, J. Guidance, Vol. 10, 1987, pp. 453-464.

Merz, A.W.
[1] Optimal evasive manoeuvers in maritime collision avoidance, Navigation, Vol. 20, 1973, #2.
[2] The game of two identical cars, in G. Leitmann (ed), *Multicriteria Decision Making*, pp. 421-442, Plenum, New York, 1976.
[3] The homicidal chauffeur, AIAA J., Vol. 12, 1974, pp. 259-260.
[4] To pursue or evade: that is the question, J. Guid., Vol. 8, 1985, pp. 161-166.

Meyer, K.R.
[1] Energy functions for Morse-Smale system, Amer. J. Math., Vol. 90, 1968, pp. 1031-1040.

Minc, R.M.
[1] Limit cycle in three dimensional space, Doklady A.N. SSR, Vol. 125, 1959, pp. 38-40.

Minorsky, N.
[1] *Introduction to Nonlinear Mechanics*. Edwards Bros. Inc., Ann Arbor, 1947.
[2] *Nonlinear Oscillations*. Van Nostrand, Princeton, 1962.

Mirza, K.B., and Womack, B.F.
[1] On controllability of a class of nonlinear systems, IEEE Trans. Autom. Control, Vol. AC-16, 1971, pp. 497-498.

Misra, A.K., and Modi, V.J.
[1] Dynamic and control of tether connected two body systems - a brief review, *Space 2000*, AIAA, New York, 1983, pp. 473-510.

Monopoli, R.V.
   [1] Model reference adaptive control with augmented error signal,
        Trans. IEEE Autom. Control, Vol. AC-19, 1974, pp. 474-484.
   [2] Model reference adaptive control using only input and output
        signals, *Proc. 4th CDC*, 1973, paper #WP3-6.
Moon, F.C.
   [1] *Chaotic Vibrations.* Wiley, 1987.
Moravec, H.P.
   [1] Obstacle avoidance and navigation in the real world by a seeing
        robot rover, Tech. Report AIM 340, Stanford Univ., September
        1980.
Morse, M.
   [1] Relations between critical points of a real function of indepen-
        dent variables, Trans. Am. Math. Soc., Vol. 27, 1925, pp. 345-396.
Movchan, A.A.
   [1] The direct method of Liapunov in stability problems of elastic
        systems, PMM (Moscow), Vol. 23, 1959, pp. 483-493.
Moser, J.
   [1] *Lectures on Hamiltonian Systems.* MIR, Moscow, 1973.
Muszynska, A.
   [1] On some problems connected with modelling mechanical systems (in
        Polish), Nonlin. Vibr. Problems, Pol. Ac. Sci., Vol. 9, 1968.
   [2] Whirl and whip rotor/bearing stability problem, J. Sound &
        Vibration, Vol. 110(3), 1986, pp. 443-462.
Muszynska, A., and Bently, D.E.
   [1] Fluid generated instabilities of rotors, Bently Rotor Dyanmics
        Research Corporation Corner, Tech. Note 01971, 10/13/87.
Muszynska, A., Franklin, W.D., and Bently, D.E.
   [1] Rotor active "anti-swirl" control, Trans. ASME, J. Vibr. Acoustics
        Stress & Reliability in Design, Vol. 110, 1988, pp. 143-150.
Myklestad, O.
   [1] *Vibration Analysis.* McGraw Hill, New York, 1956.
Nakai, B., Okada, Y., Matsuda, K., and Kibune, K.
   [1] Digital control of electromagnetic damper for rotating machinery,
        Proc. IFTOMM Intern. Conf. on Rotor-dynamics, Tokyo, 1986.
Nakano, E., Ozaki, S., Ishida, T., and Kato, I.
   [1] Cooperational control of the anthropomorphous manipulator MELARM,
        Proc. 4th Intern. Conf. Industrial Robots, 1974, pp. 251-260.
Narendra, K.S.
   [1] Stable identification schemes, in R.K. Mehra and D.G. Lainiotis
        (eds), *System Identification*, pp. 165-209, Academic Press, New
        York, 1976.
Narendra, K.S., and Kudva, P.
   [1] Identification and adaptation using Liapunov Direct Method, *Proc.
        4th CDC*, 1973, paper #TP6-1.
Narendra, K.S., and Tripathi, S.S.
   [1] Identification and optimization of aircraft dynamics, J. Aircraft,
        Vol. 10, 1973, pp. 193-199.
Narendra, K.S., and Valavani, L.S.
   [1] Stable adaptive observers and controllers, Trans. IEEE, Vol.
        AC-64, 1976, pp. 1198-1208.
Naumov, B.N.
   [1] *Theory of Nonlinear Automatic Systems* (in Russian). Nauka,
        Moscow, 1972.
Neimark, Yu. I.
   [1] A method of point-transformations in theory of nonlinear oscilla-
        tions, *Proc. IUTAM Symp. on Nonlin. Oscill.*, Kiev, 1961, publ. A.
        Sci. USSR, Kiev, 1963, Vol. 2, pp. 268-307.
Nemitzky, V.V.
   [1] Dynamical systems on limiting integral manifold, Doklady AN SSR,
        Vol. 27, 1945, pp. 555-558.

[2] On steady state regimes, *Proc. I IFAC Congress, Automatic & Remote Control*, Butterworths, London, 1961.

[3] Oscillatory regimes in multi-dimensional dynamical systems, *Proc. IUTAM Symp. Nonlin. Oscill.*, Kiev, 1961, publ. Ac. Sci. USSR, Kiev, 1963, Vol. 1, p. 308-314.

[4] Some modern problems in the qualitative theory of ordinary differential equations, Russian Math. Surveys, Vol. 20, 1965, pp. 1-34.

[5] Topological methods in theory of dynamical systems (in Russian), Uspiehy Math. Nauk, Vol. 20, 1965, #4, pp. 1-36.

Nemitzky, V.V., and Stepanov, V.V.

[1] *Qualitative Theory of Differential Equations*. Academic Press, New York, 1960.

Newman, W.S., and Hogan, N.

[1] High speed control and obstacle avoidance using dynamic potential functions, *IEEE Conf. on Robotics & Automation*, Raleigh, NC, March 1987.

Nguen, T.L.

[1] Synthesis of adaptive control systems using standard models, Engng. Cybernet., Vol. 9, 1971, pp. 386-394.

Nikolsky, M.S.

[1] On some differential games with fixed time, Soviet Math. Dokl., Vol. 19, 1978, pp. 591-597.

Nikonov, O.I.

[1] On combination of control and observation processes in game problems of evading motion, Differential Equations, Vol. 13, 1977, pp. 1053-1060.

Nitecki, Z.

[1] *Differentiable Dynamics*. MIT Press, Cambridge MA, 1971.

Noldus, E.

[1] A frequency domain approach to the problem of existence of periodic motion in autonomous nonlinear feedback systems, Z. Angew. Math. Mech., Vol. 3, 1969, pp. 167-177.

Noldus, E., Galle, A., and Josson, L.

[1] Computation of stability regions for systems with many singular points, Int. J. Control, Vol. 17, 1973, #3.

Nurre, G.S., Ryan, R.S., Scofield, H.N., and Sims, J.L.

[1] Dynamics and control of large space structures, J. Guidance, Vol. 7, 1984, pp. 514-526.

Okhocimski, D.E., and Platonov, A.K.

[1] Control algorithms for a stepping machine capable of avoiding obstacles, Tech. Kibernet., 1973, pp. 3-10.

Olas, A.

[1] Criterion of stability of trivial solution of a set of ordinary differential equations, Bull. l'Acad. Polon. Sci., Ser. Techn., Vol. 29, 1981, pp. 67-73.

[2] *Recursive Liapunov Functions*, ASME Trans. J. Dyn. Syst. Meas. Control Special Issue: Workshop on Control Mechanics, Vol. 111, October 1989.

Olas, A., Ryan, P.W., and Skowronski, J.M.

[1] Linear damping of free motion of strongly nonlinear double mechanical system, Bull. Acad. Polon. Sci. Ser. Sci. Tech., Vol. 15, 1967, pp. 97-100.

Olsder, G.J., and Breakwell, J.V.

[1] Role determination in an aerial dogfight, Internat. J. Game Theory, Vol. 3, 1974, #1, pp. 47-66.

Olsder, G.J., and Walter, J.L.

[1] A differential game approach to collision avoidance of ships, *Lecture Notes in Control and Information Sci.*, Vol. 6, 1978, pp. 264-271.

Orin, D.E., McGhee, R.B., and Jaswa, V.C.
 [1] Interactive computer control of six-legged robot, *Proc. IEEE Conf. Dec. Control*, Florida, 1976.
Orin, D.E., and Oh, S.Y.
 [1] Control of force distribution in robotic mechanisms containing closed chains, J. Dyn. Sup. Meas. Control, Vol. 102, 1981, pp. 134-141.
Ostapenko, V.V.
 [1] On linear avoidance problem, Kibernetika, 1978, pp. 106-112.
Ozguner, U.
 [1] Dual arm robotic system with sensory input, JPL Publ. 87-13, *Workshop on Telerobotics*, Vol. 3, pp. 289-298.
Pachter, M., and Getz, W.M.
 [1] Two target pursuit evasion differential games, J. Opt. Th. Appl., Vol. 34, 1981, pp. 383-403.
Parks, P.C.
 [1] Lyapunov redesign of model reference adaptive control systems, IEEE Trans. Autom. Control, Vol. AC-11, 1966, #3.
Parszewski, Z.A.
 [1] System configuration and its vibration response, A. Muszynska, J.C. Simonis (eds), *Rotating Machinery Dynamics*, ASME, 1988, pp. 71-78.
Parszewski, Z.A., and Chalko, T.J.
 [1] Chatter coupled vibrations of a large vertical lathe, rotating table and its tool slide, I. Mech. E., Paper #C275-84, 1984.
Parszewski, Z.A., and Krotkiewski, J.M.
 [1] Machine dynamics in terms of the system configuration parameters, Int. Conf. Rotordynamics, JSME-IFToMM, Tokyo, September 1986, pp. 239-244.
Paskov, A.G.
 [1] On sufficient conditions for nonlinear positional games of encounter, J. Appl. Math. Mech., Vol. 40, 1976, pp. 148-151.
Paul, R.P.
 [1] *Robot Manipulators: Mathematics, Programming and Control*. MIT Press, Cambridge Massachusetts, 1981.
 [2] Modelling trajectory calculation and surveying of a computer controlled arm. Memo No. 177, Stanford Artificial Intelligence Lab., September 1972.
Pearson, J.O.
 [1] Worst case design subject to linear parameter uncertainties, IEEE Trans. Autom. Control, Vol. AC-20, 1975, pp. 167-169.
Peczkowski, J.L., and Liu, R.W.
 [1] A format method for generating Liapunov functions, ASME Trans., J. Basic Engng., Vol. 89, 1967, pp. 433-439.
Peng, W.Y., and Vincent, T.L.
 [1] Some aspects of aerial combat, AIAA J., Vol. 13, 1975, pp. 7-11.
Petrosian, L.A., and Danilov, N.N.
 [1] *Cooperative differential games* (in Russian). Izd. Tomskogo Universiteta (Tomsk Univ. Press), 1985.
Petrov, A.A., and Sirota, I.A.
 [1] Obstacle avoidance by a robot manipulator under limited information about environment, Automat. i Telemeh., 1983, pp. 29-40.
Piedboef, J.C., and Hurteau, R.
 [1] A nonlinear model of a two degrees of freedom one-link flexible arm, ASME Trans., J. Dyn. Syst. Meas. Control, to appear.
Pliss, V.A.
 [1] *Nonlocal problems of the oscillation theory*. Nanka, Moscow, 1964.
Poincaré, H.
 [1] Mémoire sur les courbes définies par une équation différentielle, Oeuvres (Paris), Vol. 1, 1892, pp. 3-84, 90-161, 167-221.

[2] *Les Méthodes Nouvelles de la Mécanique Céleste*. Ganthier-Villars, Paris, 1892.

Popov, E.P., Vereshchagin, A.F., and Zinkevich, S.L.
[1] *Manipulational Robots: The Dynamics and Algorithms*. Nauka, Moscow, 1978.

Porter, B.
[1] *Synthesis of Dynamical Systems*. Nelson, New York, 1969.

Pozaritskii, G.K.
[1] Game problem of impulse hard contact in a position attraction field with an opponent who realizes a bounded thrust, J. Appl. Math. Mech., Vol. 39, 1975, pp. 185-195.
[2] Game problem of impulse encounter with an opponent limited in energy, J. Appl. Math. Mech., Vol. 39, 1975, pp. 555-565.

Prasad, U.R., Rajan, N., and Rao, N.J.
[1] Planar pursuit-evasion with variable speeds, J. Opt. Th. Appl., Vol. 33, 1981, pp. 401-432.

Pringle, R.
[1] On stability of a body with connected moving parts. AIAA J., Vol. 4, 1966, pp. 1395-1404.
[2] Stability of the free-free motions of a dual spin spacecraft, AIAA J., Vol. 7, 1969, pp. 1054-1063.

Prusty, S.
[1] A variable parameter Liapunov function for a class of nonlinear systems, Int. J. Control, Vol. 13, 1971, pp. 1117-1120.

Pshenichnyi, B.N.
[1] On the problem of avoidance, Kibernetika, 1975, #4.

Raibert, M.H., and Horn, B.K.P.
[1] Manipulator control using configuration space method, Industrial Robot, Vol. 5, 1978, pp. 69-73.

Rajbman, N.S.
[1] An application of identification methods in USSR - a survey, Automatica, Vol. 12, 1976, pp. 73-95.

Rakhmanov, Y.V., Strelkov, A.N., and Shvedov, J.N.
[1] Development of a mathematical model of a flexible manipulator mounted on a moving platform, Engng. Cybernet., Vol. 19, 1981, pp. 81-86.

Rang, E.R.
[1] Adaptive controllers derived by stability considerations, Memo No. MR 7905, Minneapolis-Honeywell Regulator Co., 1962.

Red, W.E.
[1] Minimum distances for robot task simulation, Robotica, Vol. 1, 1983, pp. 231-238.

Red, W.E., and Truong-Cao, H.V.
[1] Configuration space approach to robot path planning, *Proc. ACC*, San Diego, 1984, pp. 817-821.
[2] Vibrations due to motion discontinuities in hydraulically actuated robots, Robotica, Vol. 1, 1983, pp. 211-215.

Reiner, M.
[1] Phenomenological macro-theology, in *Rheology*, Acad. Press, New York, 1956.

Reissig, R., Sansone, G., and Conti, R.
[1] *Nichtlineare Differentialgleichungen hoeherer Ordnung*. Edizioni Cremonese, Roma, 1969.

Ridgely, D.B., and Banda, S.S.
[1] Decoupling of high-gain multivariable tracking systems, J. Guidance, Vol. 8, 1985, pp. 44-49.

Roberson, R.E., and Schwertassek, R.
[1] *Dynamics of Multibody Systems*. Springer, 1988.

Roitenberg, Y.Y.
[1] Observability of nonlinear systems, SIAM J. Control, Vol. 8, 1970, #3.

Rouche, N., Habets, P., and Laloy, M.
[1] *Stability Theory via Liapunov's Direct Method*. Springer, 1977.

Roxin, E.O.
[1] Stability in general control systems, J. Differential Equations, Vol. 1, 1965, pp. 115-150.
[2] On generalized dynamical systems defined by contingent equations, J. Differential Equations, Vol. 1, 1965, pp. 188-205.
[3] *Ordinary Differential Equations*. Wadsworth, Calif., 1972.

Roxin, E.O., and Spinadel, V.
[1] Reachable zones in autonomous differential systems, Contrib. Diff. Equations, Vol. 1, 1963, pp. 275-315.

Ryan, E.P., and Corless, M.
[1] Ultimate boundedness and symmetric stability of a class of uncertain dynamical systems via continuous and discontinuous feedback control, IMA J. Methods Control Inform., Vol. 1, 1984, pp. 223-242.

Ryan, E.P., Leitmann, G., and Corless, M.
[1] Practical stabilizability of uncertain dynamical systems - application to robotic tracking, J. Optim. Theory Appl., Vol. 47, 1985, pp. 235-252.

Sadegh, N., and Horowitz, R.
[1] Stability and robustness analysis of a class of adaptive controllers for robotic manipulators, to appear.

Sadler, J.P.
[1] On the analytical lumped mass model of an elastic four-bar mechanism, ASME J. Engng. for Industry, Vol. 97, 1975, pp. 561-565.

Sadler, J.P., and Sandor, G.N.
[1] A lumped parameter approach to vibration and stress analysis of elastic linkages, ASME J. Engng. for Industry, Vol. 95, 1973, pp. 549-557.

Salm, Y., and Schweitzer, G.
[1] Modelling and control of a flexible rotor with magnetic bearings, C274-84, *Vibrations in Rotating Machinery*, I. Mech. E., York, UK, 1984.

Sansone, G., and Conti, R.
[1] *Nonlinear Differential Equations*. Pergamon Press, New York, 1964.

Saridis, G.N., and Lobbia, R.N.
[1] Parameter identification and control of linear discrete time systems, IEEE Trans. Automat. Control, Vol. AC-17, 1972, pp. 52-60.

Schmitendorf, W.E.
[1] Min-max control of systems with uncertainty in the initial state and in state equations, IEEE Trans. Automat. Control, Vol. AC-22, 1977, pp. 439-443.
[2] Exact expression for computing the degree of controllability, AIAA J. Guidance, Vol. 7, 1983, pp. 502-504.

Schmitendorf, W.E., Barmish, B.R., and Elenbogen, B.S.
[1] Guaranteed avoidance control, Trans. ASME, J. Dyn. Sys. Meas. Control, Vol. 104, 1982, pp. 156-172.

Schmitz, D., and Kanade, T.
[1] Design of reconfigurable modular manipulator system, JPL-Publ. 87-13, *Workshop on Space Telerobotics*, Vol. 3, 1987, pp. 171-178.

Schroeder, P.S., and Schmitendorf, W.E.
[1] Avoidance control for systems with disturbances, Int. J. Control, Vol. 36, 1982, pp. 1033-1044.

Schultz, D.G., and Gibson, J.E.
[1] The variable gradient method of generating the Liapunov function, Trans. AIEE, Vol. 81(II), 1962, pp. 203-210.

Seibert, P.
[1] Ultimate boundedness and stability under perturbations, RIAS Tech. Report 60-7, Baltimore, 1960.

Seibert, P., and Auslander, J.
[1] Prolongation and stability in dynamical systems, Ann. de l'Institut Fourier, Grenoble, Vol. 14, 1964, #2.

Seraji, H.
[1] Adaptive control of dual arm robots, JPL-Publ. 87-13, *Workshop on Space Telerobotics*, Vol. 3, 1987, pp. 159-170.
[2] Adaptive hybrid control of manipulators, JPL-Publ. 87-13, *Workshop on Space Telerobotics*, Vol. 3, 1987, pp. 261-272.
[3] An exact expression for computing the degree of controllability, J. Guidance, Vol. 7, 1987, pp. 502-504.

Serebriakova, V.S.
[1] Conditions for existence for circular motions of connected Froude pendula, Diff. Uravnienia, Vol. 3, 1967, #3, Russian ed. pp. 2047-2052, English ed. pp. 1063-1066.

Seregin, V.N.
[1] Synthesis of an asymptotically stable algorithm for identification of a nonlinear nonstationary system by the Liapunov Direct Method, Automation & Remote Control, Vol. 39, 1978, pp. 486-490.

Sestini, G.
[1] Criteria di stabilita in un problema di mechanica non-lineare, Rivista Mat. Univ. Parma, Vol. 2, 1951, pp. 303-313.

Seydel, R.
[1] *From Equilibrium to Chaos*. Elsevier Publ., 1988.

Shaked, U.
[1] Design of a general model following control system, Int. J. Control, Vol. 25, 1977, pp. 57-79.

Shestakov, A.A.
[1] About distribution of critical points of a system of differential equations, Trudy Kazanskoho Aviationnogo Instituta, Vol. 27, 1953, pp. 41-50.

Shilds, O.
[1] The behavior of optimal Liapunov functions, Int. J. Control, Vol. 21, 1975, pp. 561-573.

Shinar, J., and Gutman, S.
[1] Recent advances in optimal pursuit and evasion, *Proc. IEEE CDC*, San Diego, 1979, pp. 960-965.

Shirokorad, B.V.
[1] On existence of a cycle in an absolutely stable set for three dimensional system, Avtomatika & Telemekhanika, Vol. 19, 1958, #9.

Simo, J.C., and Vu-Quoc, L.
[1] On dynamics of flexible beams under large overall motions, Trans. ASME, J. Appl. Mech., Vol. 53, 1986, pp. 849-863.

Skelton, R.E.
[1] Model truncation using controllability, observability measures, in K. Magnus (ed), *Dynamics of Multibody Systems*, Springer, 1978, pp. 331-344.

Skowronski, E.C., and Shannon, G.F.
[1] Conditions for nonoscillatory responses of multiple nonlinear systems, Int. J. Control, Vol. 15, 1972, pp. 957-960.

Skowronski, J.M.
[1] Problem of soft nonlinear spring characteristic in damped oscillations of an elastic panel, Czasopismo Techniczne (Cracow Polytechnic), Vol. 63, 1958, #3.

[2]   A method of qualitative analysis of vibrating mechanical systems
      with strong nonlinearity, Arch. Mech. Stos. (Arch. Appl. Mech.)
      Pol. Ac. Sci., Vol. 10, 1958, #5, pp. 715-726.
[3]   Oscillatory property of motion of strongly nonlinear mechanical
      systems, Arch. Mech. Stos. (Arch. Appl. Mech.) Pol. Ac. Sci.,
      Vol. 13, 1961, pp. 23-34.
[4]   Character of motion of strongly nonlinear mechanical systems,
      Nonlinear Vibration Problems, Pol. Ac. Sci., Vol. 4, 1962,
      pp. 5-52.
[5]   Local limit steady state for general mechanical lumped systems,
      Bull. Acad. Polon. Sci., Ser. Tech. Sci., Vol. 12, 1964, pp.
      571-578.
[6]   Periodical local steady states for general mechanical lumped
      systems, in *Les Vibrations Forcées dans les Systèmes Nonlinéaires*,
      CNRS, Paris, 1965, pp. 531-538.
[7]   Periodic limit motion for general structures, in *Dynamika Strojov*,
      Czechosl. Acad. Sci. Publishing House, Praha, 1965, pp. 303-307.
[8]   Optimally synthesable stability of limit domain for motion of
      general mechanical systems, Bull. Acad. Polon. Sci., Ser. Tech.,
      Vol. 13, 1965, pp. 107-110.
[9]   Some remarks about the character of motion of mechanical systems,
      Proc. IV Nat. US Congr. Appl. Mech., Berkeley, 1962, pp. 356-364.
[10] Structural investigation of vibrating nonlinear mechanical
      systems, Nonlinear Vibr. Problems, Vol. 6, 1964, pp. 253-310.
[11] Synthesizable stability of limit domain for response of general
      mechanical systems, Abh. Deutsch. Akad. Wiss. Berlin Kl. Math.
      Phys. Tech., 1965, pp. 344-346.
[12] Nonlinear mechanical lumped systems, Nonlinear Vibration Problems
      Pol. Ac. Sci., Vol. 7, 1965, pp. 7-224.
[13] Sufficient criterion for synthesizable stability of general
      physical lumped systems, Bull. Acad. Polon. Sci. Ser. Sci. Tech.,
      Vol. 14, 1966, pp. 425-428.
[14] *Elements of Geometric Dynamics*. Sci. & Tech. Publishers (PWN),
      Warsaw, 1967.
[15] Geometric aspects of kinetic synthesis of multiple lumped systems,
      Proc. IV Internat. Conf. Nonlinear Oscillations, Prague, 1968,
      pp. 257-268.
[16] Sufficient conditions for optimal synthesizability, *Proc.
      Internat. Cong. Math.*, Moscow, Vol. 12, 1968, pp. 385-386.
[17] Asymptotic equivalence method for kinetic synthesis of shape,
      Bull. Acad. Polon. Sci. Ser. Sci. Tech., Vol. 16, 1968, pp.
      623-628.
[18] System asymptotic equivalence as a method for kinetic synthesis
      of a set, Nonlinear Vibr. Problems, Vol. 10, 1969, pp. 31-35.
[19] Inverse delta method for synthesis of nonlinear lumped systems,
      Nonlinear Vibr. Problems, Vol. 10, 1969, pp. 27-30.
[20] *Multiple Nonlinear Lumped Systems*. Polish Sci. Publishers (PWN),
      Warsaw, 1970.
[21] Space delta method for analysis and syntehsis of nonlinear lumped
      systems, Int. J. Control, Vol. 12, 1970, pp. 109-120.
[22] An attempt to design qualitative differential games, Int. J.
      Control, Vol. 12, 1970, pp. 121-127.
[23] Conjectures on qualitative Liapunov games, Int. J. Control, Vol.
      16, 1970, pp. 501-507.
[24] On qualitative competitive Liapunov game, *Proc. V Hawaii Internat.
      Conf. Systems Sci.*, 1972, pp. 152-154.
[25] Game modelled lumped physical system, *Proc. IX Hawaii Internat.
      Conf. Systems Sci.*, 1976.
[26] Liapunov type playability for adaptive physical systems, *Proc.
      Nat. Systems Conf.*, Combatore, PSG College of Tech., India, 1977,
      pp. Q11, 1-5.

[27] Note on Liapunov design of systems in conflict with environment, *Proc. IFAC Sympos. Environmental Systems Planning Design & Control*, Kyoto, 1977, pp. 28-35.

[28] Adaptive identification of models stabilizing under uncertainty, *Lecture Notes in Biomath.*, Springer, Vol. 40, 1981, pp. 64-78.

[29] Collision with capture and escape, Israel J. Tech., Vol. 18, 1981, pp. 70-75.

[30] Deterministic identification of turbulent loads for general mechanical structures with uncertain state measurements, Mech. Res. Comm., Vol. 10, 1983, #6, pp. 345-350.

[31] Parameter and state identification in nonlinearizable uncertain systems, Internat. J. Nonlinear Mech., Vol. 19, 1984, pp. 345-353.

[32] *Nonlinear Liapunov Dynamics.* World Scientific, N. Jersey-London-Singapore, 1990.

[33] Adaptive identification of opposition in nonlinear differential game, Internat. J. Nonlinear Mech., Vol. 21, 1986, pp. 83-94.

[34] Nonlinear model reference adaptive control, J. Austral. Math. Soc. Ser. B, Vol. 28, 1986, pp. 23-35.

[35] Model reference control and identification of robot manipulators under certainty, X. Avula *et al.* (eds), *Proc. Internat. Conf. Math. Modelling*, Berkeley, 1985.

[36] Adaptive control of robotic manipulators under uncertain payload, in M. Hamza (ed), *Advances in Robotics*, pp. 40-48, Acta Press, Anaheim, California, 1985.

[37] Model reference adaptive control under uncertainty of nonlinear flexible manipulators, *Proc. 1986 AIAA Guidance Navigation and Control Conf.*, Williamsburg, Virginia, pp. 11-18.

[38] *Control Dynamics of Robot Manipulators.* Academic Press, 1986.

[39] Algorithms for adaptive control of two-arm flexible manipulators, JPL Publ. 87-13, *Workshop on Space Telerobotics*, Vol. 3, pp. 245-254.

[40] Nonlinear model tracking by robot manipulators, *Proc. ASME Winter Annual Meeting*, Boston, 1987, #87-WA-DSC-29.

[41] Closed form adaptive algorithms for nonlinear flexible double arm manipulator tracking a model under uncertainty, *Proc. AIAA Conf. Guidance Navig. Control*, Monterey, 1987, pp. 227-234.

[42] Adaptive tracking of dynamical model by uncertain nonlinearizable spacecraft, *Proc. AIAA Dynamics Specialists Conf.*, Monterey, 1987, Part 2B, pp. 861-867.

[43] Algorithms for adaptive control of two arm flexible manipulators under uncertainty, IEEE Trans. Aerospace & Electronic Systems, Vol. AES24, 1988, #5.

[44] *Control Theory of Robotic Systems.* World Scientific Publ. New Jersey-Singapore, 1989.

[45] *Control Mechanics.* Kluwer, Boston, 1991.

[46] Adaptive nonlinear model following and avoidance under uncertainty, IEEE Trans. Circuits and Systems, Vol. CAS-35, September 1988, pp. 1082-1088.

[47] Coordination control of independent two robot arms on moving platform attained via differential game, AIAA Conf. Aerospace Sci., Reno, Nevada, 1989.

[48] Nonlinear model tracking by robot manipulators, Trans. ASME J. Dyn. Syst. Meas. Control, Vol. 111, 1989, pp. 437-443.

[49] Adaptive identification and model tracking by flexible spacecraft, AIAA Conf. Aerospace Sci., Reno, Nevada, 1989, AIAA Paper #89-0541.

[50] Winning controllers for nonlinear air combat game with reduced dynamics, *Proc. AIAA Conf. Guidance Navig. Control*, Minneapolis, 1988, Part II, pp. 866-873.

Skowronski, J.M., and Guttalu, R.S.

[1] Real time attractors, J. Dyn. Stab. of Systems, to appear.

Skowronski, J.M., and Singh, H.

[1] Coordination control of Hamiltonian systems, *Proc. 4th Workshop on Control Mechanics*, Los Angeles, 1991.

Skowronski, J.M., and Stonier, R.J.

[1] The barrier in a pursuit-evasion game with two targets, Int. J. Computers & Math. Appl., Vol. 13, 1987, pp. 37-45.

[2] A map of two person qualitative differential game, *Proc. AIAA Guidance Navig. Control Conf.*, Monterey, 1987, pp. 56-64.

Skowronski, J.M., and Szadkowski, J.

[1] Limit domain of dynamical systems, in *Dynamika Strojov*, Slovak Ac. Sci., 1963.

Skowronski, J.M., and Vincent, T.L.

[1] Playability with and without capture, J. Optim. Theory Appl., Vol. 36, 1982, pp. 111-128.

Skowronski, J.M., and Ziemba, S.

[1] Applying Delta Method to investigation of strongly nonlinear mechanical vibrating system, Bull. WAT (Army Acad. of Tech.), Warsaw, 1958, pp. 13-105.

[2] Some complementary remarks on Delta Method for determining phase trajectories of systems with strong nonlinearity, Arch. Mech. Stos. (Arch. Appl. Mech.) Pol. Ac. Sci., Vol. 10, 1958, pp. 699-706.

[3] Certain properties of mechanical models of structures, Arch. Appl. Mech. Pol. Ac. Sci. (Arch. Mech. Stos.), Vol. 11, 1959, pp. 193-209.

[4] Boundedness of motions and the existence and stability of limit regimes in strongly nonlinear nonautonomous mechanical systems, in *Automatic and Remote Control*, Butterworths, London, 1961, pp. 906-912.

[5] Domain of boundedness of motions of strongly nonlinear nonautonomous mechanical systems with partially negative damping, *Proc. IUTAM Sympos. Nonlinear Oscillations*, Kiev, 1961, Vol. 2, 1963, pp. 356-364.

[6] Abstract machine applied to physical systems, Bull. Acad. Polon. Sci. Ser. Sci. Tech., Vol. 15, 1967, pp. 1-7.

Slotine, J.J., and Sastry, S.S.

[1] Tracking control of nonlinear systems using sliding surfaces with application to robotic manipulators, Int. J. Control, Vol. 38, 1983, pp. 465-492.

Smale, S.

[1] Diffeomorphisms with many periodic points, in S.S. Chairns (ed), *Differential and Combinatorial Topology*, Princeton Univ. Press, 1963, pp. 63-80.

[2] Differentiable dynamical systems, Bull. Am. Math. Soc., Vol. 73, 1967, pp. 747-817.

Snyder, W.E.

[1] *Industrial Robots.* Prentice-Hall, Englewood Cliffs, New Jersey, 1985.

Sparrow, C.

[1] *The Lorenz Equations.* Springer, 1982.

Spong, M.W., Throp, J.S., and Kheradpir, S.

[1] Control of robot manipulators using an optimal decision strategy, *Proc. XXI Allerton Conf.*, Univ. of Illinois, 1983, pp. 303-311.

Stalford, H.

[1] Sufficient conditions for optimal control with state and control constraints, J. Optim. Theory Appl., Vol. 7, 1971, #2.

[2] Stability conditions for nonlinear control process using Liapunov functions with discontinuous derivatives, J. Math. Anal. Appl., Vol. 84, 1981, pp. 356-371.

Stalford, H., and Leitmann, G.
  [1] Sufficient conditions for optimality in two person zero-sum differential game, J. Math. Anal. Appl., Vol. 33, 1971, pp. 650-654.
Starzhinskii, V.M.
  [1] *Applied Methods in the Theory of Nonlinear Oscillations*. Mir Publ., Moscow, 1980.
Stein, G., Hartmann, G.L., and Hedrick, R.
  [1] Adaptive control laws for F-8 flight test, IEEE Trans. Autom. Control, Vol. AC-22, 1977, pp. 758-767.
Sticht, D.J., Vincent, T.L., and Schultz, D.G.
  [1] Sufficiency theorems for target capture, J. Optim. Theory Appl., Vol. 17, 1975, pp. 523-543.
Stiepanov, V.V., and Tichonova, A.N.
  [1] Ueber die Raume der fastperiodischen functionen, Mat. Sbornik, Vol. 41, 1934, #1.
Stocker, J.J.
  [1] *Nonlinear Vibrations*. Interscience Publ., New York, 1950.
Stonier, R.J.
  [1] *Liapunov Theory applied to Control, Games and Boundedness of Generalized Dynamical Systems*. Ph.D. Thesis, Univ. of Queensland, 1979.
  [2] Liaponov reachability and optimization in control, J. Optim. Theory Appl., Vol. 39, 1983, pp. 403-416.
  [3] On qualitative differential games with two targets, J. Optim. Theory Appl., Vol. 41, 1983, pp. 587-598.
Storey, C.J.
  [1] Stability and controllability, in D.J. Bell (ed), *Recent Math. Developments in Control*, Acad. Press, 1973.
Stoten, D.P.
  [1] Adaptive control of manipulator arms, Mechanism and Machine Theory, Vol. 18, 1983, pp. 283-288.
Streit, D.A., Kronsgill, C.M., and Bajaj, A.K.
  [1] A preliminary investigation of the dynamic stability of flexible manipulators performing repetitive tasks, Trans. ASME J. Dyn. Sys. Meas. Control, Vol. 108, 1986, pp. 208-214.
Struble, R.A.
  [1] *Nonlinear Differential Equations*. McGraw Hill, New York, 1962.
Sunada, W., and Dubovsky, S.
  [1] The application of finite element methods to dynamic analysis of flexible spatial and co-planar linkage system, J. Mech. Design, Vol. 103, 1981, pp. 643-651.
  [2] On the dynamic analysis and behavior of industrial robotic manipulators with elastic members, J. Mechanisms Trans. Automation in Design, Vol. 105, 1983, pp. 42-51.
Synge, J.L.
  [1] On geometric mechanics, in *Encyclopaedia of Physics*, Vol. 111/1, Springer, Berlin, 1960.
Szegö, G.P.
  [1] On application of Zubov's method of constructing Liapunov functions for nonlinear control systems, Trans. ASME J. Basic Engng., Vol. 8, 1963, pp. 137-142.
Takegaki, M., and Arimoto, S.
  [1] Adaptive trajectory control of manipulators, Int. J. Control, Vol. 34, 1981, pp. 219-230.
  [2] A new feedback method for control of manipulators, Trans. ASME J. Dyn. Syst. Meas. Control, Vol. 102, 1981, #19, pp. 119-125.
Talay, T.A., While, N.H., and Naftel, J.C.
  [1] Impact of atmospheric uncertainties and real gas effects on performance of aeroassisted orbital transfer vehicle, AIAA Paper #83-0408, *Proc. AIAA Aerospace Sciences Meeting*, Reno, 1984.

Taran, V.A.
  [1] Improving dynamic properties of automatic control by means of
       nonlinear corrections and variable structure, Autom. and Remote
       Control, Vol. 25, 1964, pp. 128-137.
Tarn, T.J., Bejczy, A.K., and Yun, X.
  [1] Nonlinear feedback control of multiple robot arms, JPL Publ.
       87-13, *Proc. Workshop on Space Telerobotics*, Vol. 3, pp. 179-192.
  [2] Design of dynamic control of two cooperating robot arms, *Proc.
       IEEE Int. Conf. Robotics & Automation*, Raleigh, NC, 1987.
  [3] Dynamic coordination for two robot arms, *Proc. 25th IEEE CDC*,
       Athens, 1986.
Tarnave, I.
  [1] A controllability problem for nonlinear systems, in A.V.
       Balakrishnan, L.W. Neustadt (eds), *Math. Theory of Control*,
       Acad. Press, 1967, pp. 170-179.
Thom, R.
  [1] *Structural Stability and Morphogenese*. W.A. Benjamin, Reading,
       MA, 1975.
Thompson, J.M.T., and Stewart, H.B.
  [1] *Nonlinear Dynamics and Chaos*. Wiley, Chichester, 1986.
Timofeev, A.V.
  [1] *Design of Adaptive Systems of Control for Programmed Motion*.
       Energya, Leningrad, 1980.
Timofeev, A.V., and Ekalo, Y.V.
  [1] Stability and stabilization of programmed motion of a manipulation
       robot, Automat. i Telemeh., 1976, pp. 143-156.
Timoshenko, S.P.
  [1] *Vibration Problems in Engineering*. Van Nostrand, 1928 (further
       editions 1937, 1955).
Tkachenko, A.N., Brovinskaya, N.M., and Kondratenko, Y.P.
  [1] Evolutionary adaptation control processes in robots operating in
       nonstationary environments, Mechanism and Machine Theory, Vol.
       18, 1983, pp. 275-278.
Tomizuka, M., and Horowitz, R.
  [1] Model reference adaptive control of mechanical manipulators,
       IFAC, Adaptive Systems in Control and Signal Processing, San
       Francisco, 1983, pp. 23-32.
Tongue, B.H., and Flowers, G.
  [1] Nonlinear rotorcraft analysis, Int. J. Nonlinear Mechanics, Vol.
       23, 1988, pp. 189-203.
Tosunoglu, S., and Tesar, D.
  [1] A survey of adaptive control technology in robotics, *Proc. Space
       Telerobotic Workshop*, NASA-JPL-Pasadena, 1987.
Totani, T., and Miyakawa, S.
  [1] Mathematical model of hand transfer motion for application to
       manipulator control, Trans. ASME Dyn. Syst. Meas. Control, Vol.
       102, 1981, pp. 152-157.
Tribieleva, T.N.
  [1] On stability of periodic motion, Soviet Math. Doklady, Vol. 1,
       1961, pp. 856-895.
Truckenbrodt, A.
  [1] Effects of elasticity on performance of industrial robots, *Proc.
       2nd IASTED Danos Int. Sympos. Robotics*, Switzerland, 1982,
       pp. 52-56.
Turcic, D.A., and Midha, A.
  [1] Generalized equations of motion for the dynamic analysis of
       elastic mechanism systems, Trans. ASME J. Dyn. Sys. Meas. Control,
       Vol. 106, 1984, pp. 243-254.
Tsypkin, Y.Z.
  [1] *Adaptation and Learning in Automatic Systems* (in Russian).
       Nauka, 1968.

[2] *Fundamentals of Theory of Learning Systems* (in Russian). Nauka, 1970.

Udupa, S.M.
[1] *Collision Detection and Avoidance in Computer Controlled Manipulators.* Ph.D. Thesis, California Inst. of Tech., Pasadena, 1977.

Ueda, Y.
[1] Explosion of strange attractors exhibited by Duffing equation, in R.H. Helleman (ed), *Nonlinear Dynamics*, New York Ac. Sci., 1980, pp. 422-434.
[2] Randomly transitional phenomena in the system governed by Duffing's equation, in P.J. Holmes (ed), *New Approaches to Nonlinear Problems in Dynamics*, SIAM, Philadelphia, 1980, pp. 311-322.

Ueda, Y., Akamatsu, N., and Hayashi, C.
[1] Computer simulation of nonlinear ordinary differential equations and non-periodic oscillations, Electronics & Communications in Japan, Vol. 56A, 1973, pp. 27-34.

Uicker, J.T.
[1] Dynamic behavior of spatial linkages, Trans. ASME J. Engng. Industry, Vol. 91, 1969, pp. 251-258.

Ulbrick, H., and Anton, E.
[1] Theory and applications of magnetic bearings with integrated displacement and velocity sensors, C299-84, *Vibrations in Rotating Machinery*, I. Mech. E., York, UK, 1982.

Usoto, P.B., Nadira, R., and Mahil, S.S.
[1] A finite element-Lagrange approach to modelling flexible manipulators, Trans. ASME J. Dyn. Sys. Meas. Control, Vol. 108, 1986, pp. 198-205.

Utkin, V.I.
[1] Variable structure systems with sliding modes, IEEE Trans. Autom. Control, Vol. AC-22, 1977, pp. 212-222.

Vaiman, M.J.
[1] *Investigation of Systems Stable in-the-large* (in Russian). Nauka, 1981.

Van Amerongen, J., Nieuwenhuis, H.C., and Udink-ten-Cote, A.J.
[1] Gradient based model reference adaptive autopilots for ships, *Proc. 6th IFAC Congress*, Boston, 1975, pp. 58-1, 1-8.

Van der Pol, B.
[1] A theory of the amplitude of free and forced triode vibrations, Radio Review, Vol. 1, 1920, pp. 701-751.
[2] Ueber Relaxation schwingungen, Z. Hochfrequenzentechnik, Vol. 28, 1926, p. 178, and Vol. 29, 1927, p. 114.

Vannelli, A., and Vidyasagar, M.
[1] Maximal Liapunov function and domains of attraction, Automatica, Vol. 21, 1985, pp. 69-80.

Vassar, R.H., and Sherwood, R.B.
[1] Formation keeping for a pair of satellites in a circular orbit, J. Guidance, Vol. 8, 1985, pp. 235-242.

Vincent, T.L.
[1] Avoidance of guided projectiles, in J.D. Grote (ed), *Theory and Application of Differential Games*, pp. 267-279, Reidel, Dordrecht, 1975.
[2] Collision Avoidance at Sea. Lecture Notes in Control and Information Sci., Vol. 3, 1977.
[3] Control design of magnetic suspension, Opt. Control Appl. Methods, Vol. 1, 1980, pp. 41-53.

Vincent, T.L., and Skowronski, J.M.
[1] Controllability with capture, J. Optim. Theory Appl., Vol. 29, 1979, pp. 77-86.

Voitenkov, I.N.
    [1] A parametric plant identification method, Automation & Remote
        Control, Vol. 39, 1978, pp. 1637-1641.

Volpe, R., and Khosla, P.
    [1] Artificial potentials with eliptical isopotential contours for
        obstacle avoidance, *Proc. 26th IEEE CDC*, Los Angeles, 1987,
        pp. 180-185.

Voroneckaya, D.K., and Fomin, V.H.
    [1] To the problem of path tracking by a manipulator, Vestnik
        Leningrad Univ., Vol. 13, 1977, pp. 132-136.

Vu-Quoc, L., and Simo, J.C.
    [1] Dynamics of earth-orbiting flexible satellites with multibody
        components, J. Guidance, Vol. 10, 1987, pp. 549-558.

Walker, M.W., and Orin, D.E.
    [1] Efficient dynamic computer simulation of robotic mechanisms,
        J. Dyn. Sys. Meas. Control, Vol. 104, 1982, pp. 205-211.

Wang, P.K.C.
    [1] Stability of simplified flexible vehicle via Liapunov direct
        method, AIAA J., 1965, pp. 1764-1766.
    [2] Stability analysis of elastic and aeroelastic systems via
        Liapunov second method, J. Franklin Inst., Vol. 281, 1966,
        pp. 51-72.
    [3] Control strategy for a dual arm maneuverable space robot, JPL
        Publ. 87-13, *Workshop on Space Telerobotics*, Vol. 2, pp. 257-266.

Wazewski, T.
    [1] Systèmes de commande et équations au contingent, Bull. Acad.
        Polon. Sci., Sér. Sci. Math-Phys., Vol. 9, 1961, pp. 151-155.
    [2] Sur une condition équivalente à l'équation au contingent, Bull.
        Acad. Polon. Sci., Sér. Sci. Math-Phys., Vol. 9, 1961, pp. 865-
        867.
    [3] Sur un système de commande dont les trajectoires coincident avec
        les quasitrajectoires du système de commande donné, Bull. Acad.
        Polon. Sci., Sér. Math-Phys., Vol. 11, 1963, pp. 101-104.
    [4] On an optimal control problem, *Proc. Conf. EQUADIFF I (Diff.*
        Eqns. and Appl.), Prague, 1962, pp. 229-242, Academia Publ.
        House, Czecholslovak Acad. Sci. Prague & Academic Press NY,
        1963.
    [5] Sur une méthode topologique de l'examen de l'allure asymptotique
        des intégrales des équations differentielles, *Proc. Int. Congress*
        Math., Amsterdam, 1954, Vol. III, Noordhoff, 1956, p. 132.

Whittaker, E.T.
    [1] *A Treatise on the Analytical Dynamics of Particles and Rigid*
        Bodies. Cambridge Univ. Press, London and New York, 1964.

Wie, B., and Barba, P.M.
    [1] Quaternion feedback for spacecraft large angle maneuvers, AIAA
        J. Guidance, Control, Dyn. Syst., Vol. 8, 1985, pp. 360-365.

Willems, J.C.
    [1] Generation of Liapunov functions for input-output stable systems,
        SIAM J. Control, Vol. 9, 1971, pp. 105-133.
    [2] Dissipative dynamical systems, Arch. Rational Mech. Anal, Vol.
        45, 1972, pp. 321-351.
    [3] Consequences of a dissipation inequality in the theory of
        dynamical systems, in Dixhorn-Evans (eds), *Physical Structure in*
        Systems Theory, Academic Press, New York, 1974, pp. 193-218.

Willems, J.L.
    [1] Computation of finite stability regions by means of open Liapunov
        surfaces, Int. J. Control, Vol. 10, 1969, pp. 537-544.

Wilson, D.J., and Leitmann, G.
    [1] Min-max control of systems with uncertain state measurements,
        Appl. Math. Optim., Vol. 2, 1976, pp. 315-336.

Wilson Jr, F.W.
[1] The structure of the level surfaces of a Liapunov function, J.
     Diff. Eqns., Vol. 3, 1967, pp. 323-329.
Wilson Jr, F.W., and Yorke, J.A.
[1] Liapunov functions and isolating blocks, J. Diff. Eqns., Vol. 13,
     1973, pp. 106-123.
Winsor, C.A., and Roy, R.J.
[1] Design of model reference adaptive control systems by Liapunov
     second method, IEEE Trans. Automat. Control, Vol. AC-13, 1968,
     pp. 204-210.
Wittenberg, J.
[1] *Dynamics of Systems of Rigid Bodies.* Teubner, Stuttgart, 1977.
Wong, E.C., and Breckenridge, W.G.
[1] Inertial coordinates determination for a dual-spin planetary
     spacecraft, AIAA J. Guidance, Control, Dyn. Syst., Vol. 6, 1983,
     #6.
Xu, D.M., Misra, A.K., and Modi, V.I.
[1] Thruster-augmented active control of a tethered subsatellite
     system during its retrieval, J. Guidance, Vol. 9, 1986, pp. 663-
     670.
Yeh, S.C.
[1] *Locomotion of three-legged robot over structural beams.* MS
     Thesis, EE Ohio State Univ., 1981.
Yeshmatatov, B.
[1] Escape from collision problem in differential game with many
     players (in Russian), Vopr. Vycisl. i Prikl. Mat. Tashkent, Vol.
     53, 1978, pp. 55-69.
Yoshizawa, T.
[1] *Stability theory by Liapunov second method.* Publ. Math. Soc.
     Japan, Tokyo, 1966.
Yurkovich, S., Ozguner, U., Tzes, A., and Kotnik, P.T.
[1] Flexible manipulator control experiments and analysis, JPL Publ.
     87-13, *Workshop on Space Telerobotics*, Vol. 3, pp. 279-288.
Zeeman, E.C.
[1] *Catastrophe Theory: Selected Papers 1972-1977.* Addison-Wesley,
     Reading, MA, 1977.
[2] *1981 Bibliography on Catastrophe Theory.* Math. Inst. Publ. Univ.
     of Warwick, Coventry, UK, 1981.
Zemlyakov, S.D., and Rutkovskii, V.Y.
[1] Synthesis of an algorithm for the variation of adaptable coeffic-
     ients in an adaptive system with standard model, Dokl. AN SSSR,
     Vol. 174, 1967, pp. 47-55.
Zheng, V.F., and Luh, J.Y.S.
[1] Constrained relations·between two coordinated industrial robots,
     Proc. 1985 Conf. Intell. Systems and Machines, Rochester, Ml,
     April.
[2] Control of two coordinated robots in motion, *Proc. 24th CDC*,
     Fort Lauderdale, FA, 1985, pp. 1761-1765.
Zheng, Y.F., Luh, J.Y.S., and Jia, P.F.
[1] A real time distributed computer system for coordinated motion
     control of two industrial robots, *Proc. IEEE Internat. Conf. on
     Automation 1987*, Raleigh, NC.
Zheng, Y.F., Luh, J.Y.S., and Pei, F.J.
[1] Design of real time distributed computer system for coordinated
     motion of two industrial robots, *Proc. 1987 IEEE Internat. Conf.
     on Robotics and Automation*, Raleigh, NC, pp. 1236-1241.
Ziemba, S.
[1] Some problems considered by the Warsaw nonlinear vibration theory
     group, *Proc. Int. Sympos. IUTAM*, Kiev, 1961, Vol. 1, Publ. Acad.
     Sci. USSR, Kiev, 1963.

Zinober, A.S., El-Ghezawi, O.M.E., and Billings, S.A.
    [1] Multivariable structure adaptive model following control systems, *Proc. IEE-D*, Vol. 129, 1982, pp. 6-12.
Zubov, V.I.
    [1] *Math. Methods for the Study of Automatic Control Systems* (translation). Macmillan, New York, 1963.
    [2] *Methods of Liapunov and Applications.* P. Noordhoft, Gronungen, 1964.